Proceedings
of the 5th International Yellow River Forum on Ensuring Water Right of the River's Demand and Healthy River Basin Maintenance

Volume Ⅳ

Yellow River Conservancy Press

图书在版编目(CIP)数据

第五届黄河国际论坛论文集/尚宏琦,骆向新主编. —郑州:
黄河水利出版社,2015.9
ISBN 978 – 7 – 5509 – 0399 – 9

Ⅰ. ①第⋯ Ⅱ. ①尚⋯ ②骆⋯ Ⅲ. ①黄河 – 河道整治 –
国际学术会议 – 文集 Ⅳ. ①TV882. 1 – 53

中国版本图书馆 CIP 数据核字(2012)第 314288 号

出 版 社:黄河水利出版社
　　　　地址:河南省郑州市顺河路黄委会综合楼 14 层　　　　邮政编码:450003
发行单位:黄河水利出版社
　　　　发行部电话:0371 – 66026940、66020550、66028024、66022620(传真)
　　　　E-mail:hhslcbs@ 126. com
承印单位:河南省瑞光印务股份有限公司
开本:787 mm × 1 092 mm　 1/16
印张:149.75
印数:1—1 000
版次:2015 年 9 月第 1 版　　　　　　　　　印次:2015 年 9 月第 1 次印刷

定价(全五册):960. 00 元(US $155. 00)

Under the Auspices of

Ministry of Water Resources, People's Republic of China

Sponsored & Hosted by

Yellow River Conservancy Commission(YRCC), Ministry of Water Resources, P. R. China
China Yellow River Foundation(CYRF)

Welcome

(preface)

The 5th International Yellow River Forum (IYRF) is sponsored by Yellow River Conservancy Commission (YRCC) and China Yellow River Foundation (CYRF). On behalf of the Organizing Committee of the conference, I warmly welcome you from over the world to Zhengzhou to attend the 5th IYRF. I sincerely appreciate the valuable contributions of all the delegates.

As an international academic conference, IYRF aims to set up a platform of wide exchange and cooperation for global experts, scholars, managers and stakeholders in water and related fields. Since the initiation in 2003, IYRF has been hosted for four times successfully, which shows new concepts and achievements of the Yellow River management and water management in China, demonstrates the new scientific results in nowadays world water and related fields, and promotes water knowledge sharing and cooperation in the world.

The central theme of the 5th IYRF is "Ensuring water right of the river's demand and healthy river basin maintenance". The Organizing Committee of the 5th IYRF has received near one thousand paper abstracts. Reviewed by the Technical Committee, part of the abstracts are finally collected into the Technical Paper Abstracts of the 5th IYRF.

An ambience of collaboration, respect, and innovation will once again define the forum environment, as experts researchers, representatives from national and local governments, international organizations, universities, research institutions and civil communities gather to discuss, express and listen to the opportunities, challenges and solutions to ensure the sustainable water resources management.

We appreciate the generous supports from the co – sponsors, including domestic and abroad governments and organizations. We also would like to thank the members of the Organizing Committee and the Technical Committee for their great supports and the hard work of the secretariat, as well as all the experts and authors for their outstanding contributions to the 5th IYRF.

Finally, I would like to present my best wishes to the success of the 5th IYRF, and hope every participant to have a good memory about the forum!

Chen Xiaojiang
Chairman of the Organizing Committee, IYRF
Commissioner of YRCC, MWR, China
Zhengzhou, September 2012

Contents

H. Culture and Civilization Development in River Harness History

I. Sediment Management of High Silt – laden Rivers and Reservoirs

2

J. Application of Experiences and New Technologies of Water Resources Management(Ⅰ)

H. Culture and Civilization Development in River Harness History

The Connotation, Characteristics and Spiritual Core Research of the Yangtze River Three Gorges Water Culture

Wang Deguang

Three Gorges University, Yichang, 443002, China

Abstract: The Yangtze River Three Gorges Water Culture came into being in humanity background of the unique geographical environment and water conservancy construction in the local. In a broad sense, the Yangtze River Three Gorges Water Culture is created by the local people in the event of water environment construction and reform and it's the sum of material culture and the spiritual culture with water as the vector. In a narrow sense, the Yangtze River Three Gorges Water Culture refers to the spiritual culture created and contained in construction and the reform of Three Gorges water conservancy facilities by people. It is characterized by super nature and hyper – individuality, a long history and inheritance, a wide range of sociality and extensiveness, abstruseness and advanced nature of implication and so on. The Yangtze River Three Gorges Water Culture has conceived and formed the scientific spirit of explorations and striving for perfection, unyielding spirit of hard struggle, patriotic spirit of taking the whole situation into account, struggle spirit of hard working, innovative spirit of vigorousness, open struggle of tolerance and harmony, immigration spirit of striving for the top and so on. And forever it will encourage the outstanding Yangtze River Three Gorges sons and daughters to spare no effort towards a more splendid ideal goal.

Key words: The Yangtze River Three Gorges, Water Culture, characteristics, spirit, research

The Three Gorges area has not only the beautiful natural scenery of singular and elegant but also a long history and water culture of unique style. Besides the connotation, characteristics and the spiritual core of the Yangtze River Three Gorges Water Culture has important practical significance to make the construction smoothly and place Three Gorges immigrants reasonably. It is also helpful for the development of the Yangtze River Three Gorges region's cultural resources and the tourism resources. Moreover, it contributes to the promoting the economic construction of the Three Gorges region and social development in a healthy and order way.

1 The basic connotation of Three Gorges Water Culture

Any kind of culture is formed in a particular geographical environment and there is no exception for social context and the water culture, especially the Three Gorges Water Culture, which has been born in the Qutang Gorge, Wu Gorge and Xiling Gorge's natural environment (or the geographical environment) and the cultural background of the Three Gorges water conservancy construction. Three Gorges region is located in the upper reaches of the Yangtze River. According to the division about economic area of The Three Gorges area set by the State Council, the Three Gorges regions which have 21 counties from the east of Yichang City, Yichang County, Xingshan County and Zigui County of Hubei Province to the west of Chongqing Changshou, Jiangbei County, Baxian Jiangjin City, rely on the Yangtze River and shape long and narrow distribution. The Three Gorges regions are the important parts of the economic belt along the Yangtze River region with the area of about 59,900 km^2. These regions have the unique natural environment, extremely rich hydraulic, biological and tourism resources, superior water transportation condition and huge development potential. All these resources provide the advantageous guarantee for these regions' economic development. But these regions also have many prominent restrictions, such as rugged terrain, lack of cultivated land, frequent natural calamities, and weak ecological environment and so on. The district is located in the joint of three tectonic units, Dabashan mountain broken plait, Chuandong fold belt and ChuanEXianQian uplift fold belt, where mountain is high and valley is deep and a total area of more than 70% are mountains which are bad for agricultural reclamation and also increase the difficulties of industry, transportation and urban construction. With a large population but relative deficiency of the cultivated land resources, per capita in the area less than a mu, and most of the land are fragmentary lands, slops, thin soil and barren lands with a poor quality, and all these hinder the district's planting industry development. Due to complex geological features and uneven distribution of the precipitation in space and time, geological disasters such as landslide, collapse, debris flow and disastrous weather such as drought, heavy rain often happens, and the excessive plants that lead to serious water and soil erosion, increasing area of bare rock and bare earth, increasing river sediment concentration, especially the tendency of worsening ecological environment has not been changed fundamentally.

In summary, in a broad sense, the Three Gorges Water Culture is all kinds of cultural phenomena created in Water conservancy construction and renovation activities with the water as the carrier by the people in the Yangtze River Three Gorges area in geographical environment and the social activities. It is the combination of the material life phenomenon (That is, Three Gorges Water Conservancy Hub Project, Gezhouba Water Conservancy Hub Project, GeHeyan Water Conservancy Hub Project, Qingjiang Gaobazhou Water Conservancy Hub Project, Qingjiang shuibuya Water Conservancy Hub Project and The Three Gorges cultural relics and historic site) and the spiritual life phenomenon (That is, the stories and legends spread during the activity of controlling water, including culture, art, folk customs and beliefs, sacrifice, festivals and cultural classics such as novels, poetry, drama, prose and academic research such as academic works, folk art, inscriptions, paintings). And it is the culture collection taking water as the axis in national culture. In a narrow sense, the Yangtze River Three Gorges Water Culture refers to the spiritual culture created by the people in construction and reform of Three Gorges water conservancy facilities and it is the important part of Chinese Water Culture. The positive classic stories including Cattle Separates Gorges, Dayu Break through River for Three Gorges Reservoir, Nandi Helped Yu Separate River and Yu Jun Manages Water with a unique style have witnessed the hard inoculation and formation and maturity of the Yangtze River Three Gorges Water Culture.

2 The characteristics of the Yangtze River Three Gorges Water Culture

Water Culture is one subculture of China's advanced culture. That is no other than the Yellow River Culture that is on behalf of the beautiful mark or title of the subculture first, and the Yangtze River Three Gorges Water Culture is a bright pearl set in this beautiful mark or the crown. It is derived from the water activity, but also reflects the social consciousness of water activity. It is people's rational thinking crystallization of water activities and the culture collection with water as the axis in national culture. Just like the national culture, the Yangtze River Three Gorges Water Culture has a distinctive brand of the times and can promote the rapid development of modern social productive forces as historical accumulation. As an important part of the Chinese Water Culture, China's Yangtze River Three Gorges Water Culture plays an essential role in promoting the healthy development of the water resources industry in China through spiritual and intellectual support. Remarkable characteristics are as follows.

2.1 The supernature and the super individuality

Apparently, the stories and legends which are related to water governance in the Yangtze River Three Gorges had been handed down to present, take Cattle Separates Gorges, Dayu Break Through River for Three Gorges Reservoir, Nandi Helped Yu Separate River and Yu Jun Manages Water for example. Firstly, the most remarkable characteristic represents ancestors' great imagination and fighting capacity which is generated in the process of conquering and surpassing the nature. One of the stories of having the supernatural is Dayu Break Through River for Three Gorges Reservoir, and there is records like this, "The old legends recorded that cattle helped Yu to break through river for water governance and they finished it after nine years. All is true". We can conclude that cattle helped Yu separate gorges has been spread almost more than a thousand years, and it displays that ancestors' great imagination of conquering the nature as the same as other stories about Three

Gorges Reservoir. Moreover, the old cattle works very hard in Cattle Separates Gorges. It not only symbolizes ancestors' loyal and honest characteristic and the great strength of reconstructing the land of country, but also presentes the case that the ancestors engaged in water works and the specific situation that ancestors struggled against nature for surviving at that time. We also could see the true face of the social conflicts which had the very strong reality. People established cattle temple under the cattle block in honor of the cattle who succeeded in helping Yu to separate gorges. Secondly, it pays more attention to the power and sociality of people that is to say super individuality. Putting people into God not only represents people's tireless pursue of truth and the ideal of the vision, but also represents fighting spirit based on realistic life and optimistic characteristic of Chinese nation. Overall, the super nature and super individuality represented in the stories about water control in the Three Gorges area were the powerful evidence of the basic characteristic of the Yangtze Three Gorges Water Culture.

2.2 A long history and inheritance

Every culture generates and develops based on former generation of culture. Because of accumulating and inheriting from generation to generation, human culture is becoming richer and more progressing increasingly, besides Water Culture. For instance, Quyuan who was born in The three Gorges area was the first great patriotic poet in our country. His first political lyric and romantic work Li Sao was a brilliant pearl which was conceived in the environment of the Three Gorges in which the mountains are odd and the water current was strong and there are lots of related content describing the Three Gorges Water Culture. In addition, Yang Shoujing who was a famous scholar and was born in the Three Gorges had spend nearly sixty years on finishing shui jing zhu shu. He not only created the geographic water conservancy culture firstly, but also deduced and defined the Kui Gorge, Wu Gorge and Xi Ling Gorge which formed the Three Gorges. He used lots of parts for describing the content of the Three Gorges Water Culture, shui jing zhu shu became the oriental geographic spring according to it. Yang Shoujing became the first person who scientifically defined the inheriting of the Three Gorges Water Culture.

2.3 A wide range of sociality and extensiveness

Water is closely related to economic construction and social developing of the Three Gorges. The Three Gorges regional culture had been influenced or is being affected by the Three Gorges Water Culture. It not only contains physical water culture, water culture of scene and behavioral water culture, but also contains mental water culture. It includes rich ethnic water culture and folk water culture which is represented by TuJia culture, and it also includes environmental ecological water culture which is represented by the magnificent Three Gorges. It includes the culture of notable person represented by Qu Yuan and Zhao Jun, and it also includes military water culture represented by culture of three kingdoms. It includes the traditional water culture represented as Bachu culture. and the modern water culture represented as Three Gorges Dam. Three Gorges itself is a world famous water culture. The modern water culture project represented as Three Gorges Dam is showed and carried forward. It contains many different connotation of culture, such as high technology, environmental projection, and architectural aesthetics and so on. It has been the most beautiful water culture landscape of the Chinese nationality. The prestige of height Tan Ziling has been a rising scenic spots where tourist travel more and more. The developmental prospect will be fine. The content is wide and full of variety.

2.4 The profound and advanced meanings

The Yangtze River Water Culture had the historical legends and stories such as The Cattle Seperates Gorges, Yu Regulates the Water and Cattle Helps Yu Separate the Gorges in the ancient period. These had not only long historical figures, but also the profound connotation, the deep meanings and is affecting and promoting Three Gorges Water Culture's or regional couture's development in an order circle. For instance, the ancestors of the Three Gorges made the heroically and aspiring pattern with all of their feelings, hometown people were proud of Qu Yuan who is the son of Three Gorges. Meanwhile, there are lots of stories about him, and the number of the stories was second to none between all of the figures of the Three Gorges. In the view of so many the Three Gorges folk legends, its emphasis is different from official written and literature record obviously. It omitted his changes in the official circle, his poems creation and literature status, it focuses on narrating his growth progress in circle of the Three Gorges mother, the caring for The Three Gorges elders and hometown people's memory to him. For example, the San xing Zhao Ming Yue is that he was hard to study when he was a child and read widely. Zhao Mian jing is about the thing that he was strict with himself and cautious to be alone from a child. Mi Cang Kou is that he sent his own rice to folks secretly. Zhen Zhu Yan is that he gave up the green grass he found to folks in the depths of winter and so on. None of these talked about the story regard to QuYuan and the Three Gorges. That's the reason why he grew up to a strong people ("Yi yu xin zhi suo shan xi, sui jiu si qi you wei hui") and to a patriotic in a big wave and big ripples era because he sucked Three Gorges mother's milk and trained personality integrity, hone spirit and will and edify soul and conduct in the Three Gorges mother's embrace. He is not only the Chinese outstanding figure but also a son of Three Gorges. His tragedy of not have a way to serve their county and death to the bitlevel make the Three Gorges people for generations sigh and grief. The story of Quan Yuan Tuo used the romantic art to render this emotion fully. The sky was sorrow and the ground was miserable. Clouds stained both moon and sun. Mountains and rivers were crying for his dead of jumpping into the river. The big fish living in Dongting Lake sent his corpse to the Three Gorges, and the other fish came here to escort him. Phoenix kept screaming. The birds were chirping. The wave was blasting. The wind was shouting. Hundreds of bird aspersed all kinds of flowers to the river. The hometown people were crying bitterly to welcome his soul. The scene that was strange, solemn and stirring and contains deep implication and advancement was a scared funeral for Qu Yuan held by universe. It was expressed so strong that the people living in the Three Gorges have a complexity about Qu Yuan, adored the person of ideals and integrity and their sentiment likes and dislikes.

3 The spirit of the Yangtze River Three Gorges Water Culture

The spirit of culture is relative to the embody of culture, and it is the crystallization of the concise culture, as is the water culture. In ancient documents, "jing" means "essence, fine"; "shen" means "mysterious, subtle, and marvelous". "jing shen" means vital essence and active of the universe and the essence internal dynamics of the thing motion developing. The unique spirit is not only the core that grew from it, but also the culture spirit that created in the experience of managing the Three Gorges' landscape including the Three Gorges migrant. This spirit consist of scientific spirit of exploration, innovative spirit of striving constantly for self – improvement, patriotic spirit of paying attention to the interests of the whole and revitalizing the Chinese nation, struggle spirit of hard work, initiative spirit of vigorousness, open spirit of tolerance and harmony, immigration spirit of striving for the top. It is the value orientation of the Yangtze River Three Gorges Culture, and the deep meaning of the Yangtze River Three Gorges spirit and also the distinctive features that the Yangtze River Three Gorges Water Culture in the Yangtze River Three Gorges personality reflects on. It has both the enlightening impact and penetrating power and the influence of ameliorate the heart of evil; it can inflect most of these region's people and accepted by them, even become their basic life belief and conscious value pursuit, and also has the active impact on maintaining these region's national existence and promoting social progress. Otherwise, it cannot be called the spirit of the Yangtze River Three Gorges Water Culture.

3.1 Scientific spirit of "exploration and keep improving"

The Yangtze River Three Gorges Water Culture has contained this spirit from the early time. QuYuan, the first great patriotic poet in our coutry, created "lu man man qi xiu yuan xi , wu jiang shang xia er qiu suo". This is the desire and shouting to know and master the law of nature and society, and also the most primitive footsteps of Chinese nation to pursue scientific spirit. The modern big scholars, YangShouJing, spent nearly 60 years of time carefully writting Shui Jing Zhu Shu. In this book, he created a culture of its kind of water conservancy geography and was known as the "three JueXue" in qing dynasty. He became the first man who had scientific definite the Three Gorges Water Culture inheritance. It made the world attention with it's great scientific value, high levels of the excelsior and high levels of Science and technology. The Yangtze River Three Gorges Project is the most massive scale in human history of super engineering so far. The engineering and technical personnel with "as if near abyss and tread on thin ice" seeking spirit and serious cautious attitude, always adhere to the principle of scientific excellence to carefully argument, design and construction. All this reflected a kind of tough, rigorous and realistic scientific spirit. Only a nation advocates the scientific spirit, can it become a real ethnic group with hope. Toward modernization, Chinese nation must cultivate and carry forward the spirit of scientific and probe, and probe in the Yangtze River Three Gorges Water Culture.

3.2 The pioneering spirit of self – improvement

The spirit of Yangtze River Three Gorges Water Culture was reflected earliest in the Chu Culture, which was famous for the entrepreneurial spirit of single road to blue thread, leading to mountain forest. Chu's area was less than 100 m at that time. Because of the Chu people's self – improvement, courageous warriors, and constantly expanding soil extension, finally made the Chu Culture public to acme for the pioneering spirit of self – improvement. To the development pattern was "five miles area, with a car by millions, thousands, riding ten thousand horses, millet a decade, drinking at the Yellow River, winning the central plains". However, nowadays in every rainy season it often causes river flood in the water channel of the Yangtze River Three Gorges, leading to the mountain collapse or rivers burst. Home field can be engulfed instantly by flood beast, the earth to ocean. When conquering the Yangtze River flood disasters, people in Three Gorges s foster a characteristic of fine tradition with constant self – improvement and the pioneering spirit. In addition, it encouraged the generations of local people to lead to prosperity, reform and become the first. People in Three Gorges deeply understand the water culture essence, which has a clinging pursuit of innovation spirit with self – improvement and pioneering since the ancient times. This kind of spirit should and must be an important part of the water culture in Yangtze River Three Gorges. As long as they are determined to do, they will spare no effort and be implacable unless succeeding. For example, Three Gorges Dam has been standing and completed in YiChang City, millions of immigrants were smoothly finished and "flood fighting spirit in 1998" are the revealing of the powerful proof, which are the spirit portraiture of Chinese nation striving constantly for self – improvement, pioneering struggle, vigorous spirit of innovation in 21 century. Therefore, not only did it create a long history of Water Culture of Yangtze River Three Gorges, but also would always encourage the descendants in the Yangtze River Three Gorges towards a more splendid prosperity.

3.3 Patriotic spirit of paying attention to the interests of the whole and revitalizing of the Chinese nation

The Three Gorges Water Culture itself has strong patriotism traditional spirit. Qu yuan was the great patriotic poet always concerned about his country and his people acknowledged by the world, and The Three Gorges project construction also embodies the patriotism spirit of revitalizing the Chinese nation. "Developing the Yangtze river, constructing the Three Gorges" is a power dream of the Chinese nation near a century. Dr. Sun Zhongshan put forward to constructing Three Gorges and enforce strength to make the Chinese nation no longer not bullied by others. Chairman Mao Zedong outlined the grand blueprint of "Geng li xi jiang shi bi, heng duan wu shan yun yu, gao xia chu ping hu". The scientific and technical personnel took 70 years – to demonstrate, design, suffer sorrow, bear overworked, exert their utmost effort and even wear out iron shoes to give the last measure of devotion, and the spiritual pillar was buried in their heart in order to revitalize the

Chinese. The Three Gorges Project construction affects the heart of Chinese descent both at home and abroad. "Loving the Chinese, building the Chinese" has become the common aspiration of Chinese descent. The Three Gorges project condenses the patriotic emotion of the Chinese both at home and abroad. Culture belongs to the nation and the world. In fact, the author thinks that, the Yangtze River Three Gorges Water Culture spirit is the refining of the essence of the history and reality, material and spirit, individual and group in the Three Gorges area, and abstract the Chinese nation treasure from it. It is the true portraiture of history or coincides with the direction of the contemporary culture development. We must grasp the essence of the spirit and carry forward the spirit to produce the huge physical strength, so as to promote the economic and social development of the Three Gorges area along in a healthy and order way.

3.4　Struggle spirit of industrious and hard work

Faced the bad natural environment, people in the poverty countries of the Three Gorges region didnt feel that they were worthless or contemptible. Instead, they tried their best to change their fate. With struggle spirit of hard working, they broke over the reality that the living pattern and mode and the number of harvest and the quality of life were all decided nature and they were happy – go – lucky and not seeking better living in the past year. For example, the Tujia minority, originated from qingjiang basin, which can endure hardship endurance and have highly entrepreneurial spirit shows the heroic of brave, hardworking and the arduous struggle that "xiang tian wang zi yi zhi jiao, chui chu yi tiao qing jiang he". Chairman Jiang Zemin, captioned "Carry forward the arduous entrepreneurial spirit, build the grand Three Gorges Project" for the Three Gorges Project. The Three Gorges Project construction needed millions of immigrants of the Three Gorges Reservoir area to concede farmland and home, so they made a huge sacrifice and dedication for it. And this vast immigrant project is very hard to imagine for foreigners. With the completion of the Three Gorges Dam, millions of migration project has been successfully completed and has obtained many remarkable achievements. The struggle spirit of industrious, brave and hard work from them will be invaluable spiritual wealth for the Chinese nation.

3.5　Initiative spirit of Vigorousness

Vigorousness is not only the general principles of handling the relationship between nature and man and various interpersonal relations of the people in the Three Gorges area, but also the most concentrated theoretical generalization and value refinement of Three Gorges area people's positive attitude towards life. Daedal nature makes the magnificent and deep Three Gorges. And hardworking, brave, loyal and wise Three Gorges people created the Three Gorges of the Yangtze River Water Culture. The most precious treasures of the Yangtze River Three Gorges Water Cultural treasure – house are Gezhouba Dam Project and the Three Gorges Project. We should say that Gezhouba Dam Project and the Three Gorges Project is the foundation for initiative spirit of vigorousness, and the pioneering and innovative work by the brave and hardworking people with their hands. Grand Three Gorges Project, over millions of migration project, over 2,000 billion investment, the whole nation's concern and support and the world's eyes focus will undoubtedly enlarge the Yangtze River Water Culture in the pioneering and innovative spirit of hard work and bravery. With the magnification of its innovation spirit, Three Gorges Water Culture is constantly spreading, subliming, progressing and updating, and has become coagulator, lubricant, catalyst to accomplish emigration, stabilization, economic construction and social development, and will be an important part of socialism with Chinese characteristics in Chinese Culture. Crossing Chongqing and Hubei Province, the stabilization and development of Three Gorges Reservoir area requires reconstruction of its culture. The realization of a striding development and the scientific lofty goal of creating ecological civilization demonstration area needs inoculation and inspiration of this culture to keep pace with the times. We should view Three Gorges Water Culture, especially Three Gorges Reservoir area's Water Culture, as mainstream culture of this area, which is the requirement of its history and culture, especially its future development. What's more, the immigrant spirit led by such a huge immigrant project enriches and develops Three Gorges Water Culture and its spirit. With the completion of Three Gorges Project and Immigrant Project, and the desirability of Three Gorges Reservoir Construction, Three Gorges' Water Culture must update constantly and keep pace with the times. Besides, Water Culture must adapt the development of Three Gorges Reservoir's economic construction and social mainstream culture.

3. 6   The open and compatible spirit

Open, compatible and blending spirit is the core of Yangtze River Three Gorges Water Culture. The formation of unique Ba Chu Water Culture in the Three Gorges is due to compatibility and blend of Ba Culture and Chu Culture. Ba and Chu is inseparable, which presents a great compatibility. From the earliest times, Three Gorges is the foreign trade center. Located in the exit of Three Gorges, Yichang, once a foreign trade port, has become an open city along the Yangtze River. The Three Gorges Project as a first – class world super engineering, is open to the world at the very beginning of its construction. They purchased equipment all over the world and employ international experts as technical supervisors. All present openness and magnanimity of Chinese Culture.

3. 7   The natural spirit of being first – class

Water is substantial and natural. It will be full of spirituality and infectivity once integrated with humanistic culture. The Yangtze River Three Gorges have unique environment and picturesque scenery. After the completion of Three Gorges Dam, there are many big artificial lakes which give the nature clearness, lightness and beauty. Besides the beautiful scene, it has long history. It links Wuchu (Hunan Province) and Bashu (Sichuan Province) from east to west and also Xinagqian (Guizhou and Hunan Province) and Sanqin (Shaanxi Province) from south to north, and the advantaged location is a place of strategic importance. The five thousand years of united, first – class motivated history rolled on Three Gorges area, from the rise and decline of Bashu, the storm of Qin Dynasty and Han Dynasty, the fight of the Three Kingdoms, the coming of Yuan army, the uprising of the end of Ming Dynasty, the fight against UK in Wan City, eight years' war against aggression, the liberation army's enter to the southwest part to the completion of world first class dam. The Chinese culture and western culture are integrated here. The glorious traditional culture and modern noble civilization are met, mingled here. On the fertile land, Quyuan, the first great patriotisms poem, and Wang zhaojun, the ambassador of marriage for the union of Chinese nationalities, were born here, and so was Wang shoujing, the famous scholar. Furthermore, Libai, Dufu, Bai Juyi and Ou Yangxiu had left their precious footprints and famous works. The long history and glorious culture make the land more fertile. The temper is as gentle as water. People are beautiful and clever. The heart is tolerant and the people of Three Gorges and water are united. The characteristics of water fully display the beautiful, clever and tolerant nature of the people of the Three Gorges. The water culture is created during the contact between people of the Three Gorges and water; the most important content is the tolerant spirit of integration of people and water, the union and accordance, and the motivation of being first – class. It gives people positive enlightenment, inspiration and experience, and also people rewards water much spirituality.

In conclusion, to seek high and low, keep improvement, constantly strive to become stronger, pay attention to the interests of the whole, work hard, and to be strong and successful, to be allinclusive, to be the first class are the core of the water spirit of the Three Gorges. They are not only unique and relieve against each other, but also complement each other and making up a connected organic integrity. Also they reflect Three Gorges water's spirit from different aspects. The long history and glorious culture make the miraculous land more fertile. The offspring of the Three Gorges from the old time are facing many unique opportunities in the construction of modernization, without question, they will keep worrying, working, exerting the utmost effort, even do anything or give lives for the economic construction and social development of the area under the inspiration of the water spirit of the Three Gorges.

References

Li Zongxin. Brief Description of Definition of Water Culture [J]. Beijing Water Resources, 2002 (3),45.

Cao Shitu, Run Qinqin. On the Water of Three Gorges Culture [J]. People of the Yangtze River, 2007 (8), 37 – 38.

Li You. Carry Forward the Spirit of Three Gorges Culture Building World – class Cities [C]// Cultural Studies Series of Three Gorges. Wuhan: Wuhan Press, 2010.

Wang Shiyi. Water Culture Conceived of the Spirit of Jiangsu [J]. Jiangsu Culture, 2010 (5), 5.

On the Culture of Three Gorges Project

Luo Meijie[1,2] and *Huang Quansheng*[1]

1. Research Center for Reservoir Resettlement; The Yangtze River Three Gorges Development
Research Institute, Yichang, 443002, China
2. China Three Gorges University (CTGU) Water Culture
Research Institute, Yichang, 443002, China

Abstract: The Three Gorges Project is consisted of the flood – control engineering, resettlement engineering, transportation engineering, electricity generation engineering, culture engineering and so forth. As a comprehensive water control project, Three Gorges Project is a unity of tangible culture and human spirits that indicating a high degree integration of naturalism and humanism and rich culture. Through a theoretical summary and research which from the perspective of culturology, Three Gorges Project culture includes flood control culture, contending scientific verification culture, migration culture, construction culture, township (city) culture, transportation culture, ecological culture, corporation culture, film and television (art) culture, national image culture, institutional culture, tourism culture, education culture and so on. The complexity of Three Gorges Project determines the dual characteristics of it which contains both positive influence and negative impact. The cultural connotation of Three Gorges Project, as a symbolization of China's perseverance and China's rise, greatly influences national merits and makes a surge on the relevant culture research and philosophy mediation. One day Three Gorges Project will be added to the National Cultural Heritage list and to the World Cultural Heritage list just like Dujiangyan Irrigation System and the Grand Canal. Thus, the necessity and, as time goes on, the importance of conducting a research on Three Gorges Project will be found.
Key words: Three Gorges Project, Three Gorges Project culture, culture contents, cultural traits, cultural influence

Three Gorges Project as a unity of tangible culture and human spirits indicates a high degree integration of naturalism and humanism. And achievements of the project take great influence on the Chinese social system, people's behavior and mentality activities. When viewing Three Gorges Project from the perspective of culturology, people could find the profound meaning of it that relates to social behavior, public psychology and symbol of spirit which beyond the level of a merely concrete construction. Many indexes of Three Gorges Project breaks the world records of water conservancy project indicates the change of developing countries in dealing with water: from flood control in traditional agriculture to flood control in modern industry and then to industrial water conservancy construction of comprehensive water utilization in flood control, transportation, electricity generation, irrigation, industry and other prospects. The Three Gorges Project also has a vital strategic significance to the sustainable development of China and the whole world for that "energy supply improvement will bring benefits in economy growth, education and sanitary condition amelioration, vocational training and employment growth and creating opportunities in business efficiency, hence to reduce the poverty." Besides generating electricity, Three Gorges Project functions in various aspects such as irrigation, industrial water supply and domestic water supply, flood prevention, habitat maintenance and so on. And the influence of such an incredible construction on culture can not be neglected. Theoretically, Three Gorges Project takes on a natural formation process consisting of project proposing, demonstrating, arguing, building and maintenance. Now the operation of the dam tends to be in a mature stage, however, in consideration of the power of public opinions and media the academia does not discuss or research on the subject of Three Gorges Project. So as for Three Gorges Project, what the real image it presents? What's its position in China's history of water conservancy? What are the culture contents and their characteristics? All those questions will be solved with the following contents.

1 General introduction to the Three Gorges Water Conservancy Project culture

1.1 Three Gorges Water Conservancy Project

1.1.1 Gongcheng

There are three levels of defining "Gongcheng" in Ci Hai(the most comprehensive Chinese dictionary): the primary definition is engineering which is a generic term of various applied sciences combining principles of basic subjects in natural science and experience getting from scientific experiment with agriculture and industry practices like civil engineering, hydraulic engineering, and biological engineering; second level definition is project, in which bigger and more complicated devices are needed like Three Gorges Project; and the third level indicates a huge program in general that demands all aspects cooperation for its large scale and complexity.

1.1.2 Water conservancy

The activity of human beings in the adaptation, utilization, transformation, development and preservation of water environment of nature. And the engineering done during the activity is called hydraulic engineering, all sorts of efforts which are made during the procedure named water conservancy business. Conducting hydraulic engineering is the main way of realizing the target of water conservancy, which functions as promoting benefit and abolishing harm to people and society by means of controlling and allocating water underground or on surface. In addition, water conservancy also displays a predominant prop function in social civilization and economic prosperity.

1.1.3 Hydraulic engineering

"It is an engineering for eliminating flood damage,developing and utilizing water resources. In respect of functions of the constructions, those engineering could also be classified into flood control engineering, agricultural hydraulic engineering, channel and harbor engineering, water supply and drainage engineering, environmental hydro engineering and coastal reclamation engineering. Meanwhile it also could be called comprehensive hydraulic engineering for its functions of flood control, water supply, irrigation, and electricity generation and so on. Different types of hydraulic structures are indispensable in its function realization, involving dam, dike, spillway, sluice, inlet, canal, aqueduct, log passage, and fishway and so on. "Three Gorges Project is one with huge scale and complexity, and it requires all rounds cooperation on history, geography, transportation, cultural relic, resettlement and so on that cross departments, provinces and industries. Take an example of flood control work of Three Gorges Project, in order to achieve this target it should be taken into the consideration of the operations all around and coordination among Three Gorges Project, the Qingjiang Cascade Hydropower Station, the Danjiangkou Reservoir, Jingjiang River Flood Control and Flood Division Engineering. Therefore, the significance of Three Gorges Project is beyond what has been brought by the concrete hydraulic complex construction itself but particularly, Three Gorges Project plays as a whole that includes impacts brought by the project towards China, some other countries and even the whole world before, during and after carrying out the project and politics, economics and cultural activities in connection with Three Gorges Project.

1.2 Culture of Three Gorges Project

Currently, people attach great importance to the research on engineering. Engineering culture is a product of the engineering research, which can be defined in broad meaning and narrowed meaning. In a broad sense, engineering culture as well as scientific culture and human culture is regarded as a perspective of looking at the world and a way of thinking. As for the engineering culture discoursed by Douglas Lewin in his Engineering philosophy: The third culture, some scholars interpret it as a complex unit, which involves various elements, as a result of interactions among social environment, engineering process and the participants. Some people point out, in a narrow sense, the engineering culture means the code of behavior, code of ethics and values that formed in the process of the engineering activities, and values act as the core of the engineering culture.

Some scholars hold that the rich culture implicated in Three Gorges Project can be subdivided into three aspects, namely the Yangtze River water conservancy culture, engineering construction culture and Three Gorges Project resettlement culture. Liu Yutang indicates that Three Gorges Project culture is a kind of culture form, with the representation of hydraulic complex and electric power facility, that carrying, entailing and reflecting by modern engineering landscape full of modern science and technology and characteristics of culture and art. And the engineering landscape is mainly composed of landscapes of the dam, hydropower station, navigation channel, bridge and so on. As far as I am concerned, the definition of Three Gorges Project culture above is far from expressing the complete connotation and extension of its culture. Because narrowly speaking, Three Gorges Project culture belongs to the league of engineering culture, but due to specific causes and influences, especially, the connotation of the form of engineering culture it carries surpassing that of the engineering culture itself. Actually, Three Gorges Project is a great manmade engineering by making use of natural laws of the river within the Yangtze River basin that realizes irrigation, transportation, electricity generation and natural disaster reduction, the Yangtze River ecology development maintenance as well. And the concrete construction is the sign and symbol of culture of Three Gorges Project, which overlaps with the Three Gorges culture and plays as a part of culture of the Yangtze River. From the perspective of cultural morphology, Feng Tianyu regards culture as a formless unit including kernel and exterior. And he divided this unit into four layers. Start from its surface to core, as Feng points out: the outside level layer is the tangible culture made of various objects by man's creation on nature. And it is the sum of methods and products of people's material production activities and also it is the material foundation of culture creation; the second layer is institutional culture of social norms set up during the social practice of mankind; and then comes to the habitual culture of conventional habit mode formed by social activities, especially social interactions; and the core layer is mentality culture that is mainly composed of value orientation, aesthetic interest, thinking mode and so on, deriving from the social practice and conscious activity of human beings. From my point of view, culture of Three Gorges Project is an indivisible organic whole integrated by substance, system, behavior which are related to the Three Gorges Hydro Project and the four aspects of mentality culture, involving value orientation, aesthetic interest, thinking mode, which are related to Three Gorges Project.

2 Contents of Three Gorges Project

The all – embracing constructions that contain in Three Gorges Project overshadow others. As for Three Gorges Project, Gezhouba Water Conservancy Project plays as its ancillary work and the Three Gorges five – step ship lock and the Three Gorges reservoir extend to Jiangjin District, Chongqing City. Thus, Three Gorges Project is one that combined by all hydro – projects like constructions of dam, ship lock , reservoir, and the electricity transmission and transformation from Gezhouba in Yichang to Chongqing. And major works of Three Gorges Project are showed as follow.

2.1 Three Gorges Water Control Project

The Three Gorges Water Control Project is a project containing the hydraulic engineering in the upper basin of the Yangtze River (main and branch waterways of the Chuanjiang River) and the flood diversion project in the middle and lower reaches of the Yangtze River which are centre of Three Gorges Project. And the concrete construction of Three Gorges Project is the coordinate point, central point, initiation point and key point in the construction process of the entire water control project of the Yangtze River. From this point. Three Gorges Project can be regarded as the signal of the whole water control project of the Yangtze River. And the Three Gorges as an indispensable part, is not only involved in the water control project in the upper basin of the Yangtze River, but also belonged to the water control project in the lower reach of the Yangtze River.

2.2 The resettlement engineering of Three Gorges

Along with the construction of Three Gorges Project, the hugest engineering resettlement takes its present in the history of mankind. It could not define the resettlement as simple movement of population; it contains the relocation, arrangement, stability and better livelihood of people in Three Gorges Project area and other resettlements like cities and towns, paths, cultural relics, intangible culture heritage (e. g. folks), animals, plants, industries and so on as well. Thus, to this sense the resettlement is no longer confined to the population movement.

2.3 Three Gorges traffic engineering

Shipping transportation is a significant function that carried out by Three Gorges Project. And the transportation within the Three Gorges Reservoir connects not only with the actual water transportation but also with the fairway, navigation mark, shipping track, carriage extensions on the main stream and the branch of the Yangtze River, manufacturing of the water transportation, ship lock, creature channel (especially for fish), and the traffic roads in the reservoir area including roads, railways, mountain paths, ferries and so on.

2.4 Three Gorges electricity generation engineering

Electricity generation is the original purpose of constructing Three Gorges Project and just as Sun Yat – sen , the advocator behind Three Gorges Project, puts forward, "Grave hydro power can be produced by the Kuixia Gorge in the upstream of the Yangtze River. Some people have made a research on the power carried by the stream segment last from Yichang City to Wanxian County, and they found that the quantity of the power is more than 30,000,000 horsepower, which surpasses the sum of power generated in other countries. The power generated can be provided to trains, trolley buses, industries all over China, and can be used to the produce of the fertilizer. " Because the Gezhouba Project is a part of Three Gorges Project, so that the Three Gorges electricity generation engineering includes not only the electricity generation engineering which located in Sandouping County, but also the Gezhouba electricity generation engineering. From a broader perspective, those two electricity generating engineering are significant to the entire the Yangtze River basin, especially to the water conservancy project construction of electricity from the west to the east project in southwest China, and those two engineering play a vital role in improving the overall pattern, enhancing the technique and connecting the east and west.

2.5 Three Gorges ecological engineering

Three Gorges Project can bring positive influences, and negative influences as well, on ecology. And how to make good of this two – edged sword lies in how men weight pros and cons and whether men can make good use of the benefits and effects of hydro engineering around Three Gorges. Meanwhile, men should figure out the potential and probably eco – benefit brought by Three Gorges Project.

The negative influences brought by Three Gorges Project are manifested in the destruction of cultural relics and historic sites by water storage in the Three Gorges area, which has changed and has lowered the completeness of cultural value of the natural landscape in the Yangtze Three Gorges. And predictably, the negative influences would raise landslide problem, sediment problem, water quality problem and so on.

As for the ecology protection, it can be discoursed into the following four aspects: firstly, the greatest ecology engineering is flood control. Once come across the 100 – year flood, measures should be carried out to hold the maximum stream flow of Jingjiang River less than 56,700 m³/s and the water level no more than 44.5 m in Shashi District, Jingzhou City, thus can prevent devastation happening along the Jingjiang River to ensure the safety of 20 million people and 1.533×10^6 hm² farmland in Jianghan Plain and the edge river towns without turning to the Jingjiang River flood – diversion project. Once coming across the 1,000 – year flood, through the reservoir flood – control, it can be controlled within 80,000 m³/s of the maximum stream flow of Jingjiang River, and with the Jingjiang River flood – diversion project, the water level of Shashi District can be controlled less than 45 m to make sure the safety of flood passage in Jingjiang River Channel to avoid the burst of the stem dike. In addition, we should raise the flexibility and reliability of the flood – control operation in the middle and lower reaches of the Yangtze River and lower the loss caused by the flood which impacts on the plain area of the middle and lower reaches of the Yangtze River with the area of 126 thousand, with the population of 75 million and the large and medium cities and towns. Secondly, saving the water and the electricity, reducing the use of non – renewable resources and reducing the emission of the harmful assistances in the Three Gorges area is the second largest ecological benefit. Thirdly, measures like relocating, managing of the reservoir bank, landslide treatment and farm land returning to woodland are for the protection of the ecological environment of the Three Gorges area. Fourthly, enhancing the environmental awareness acts as an important invisible benefit in ecology protection. Fifthly, from the macro – perspective, adjusting the water level in flood season and flood period is a measure of hydrological ecological environment protection of the Yangtze River.

2.6 Three Gorges Cultural Project

Destructions or damages toward the cultural relics and cultural ecology are coming along with the construction of Three Gorges Project, which is an inevitable fact we're facing with. And our nation has already made it as a cultural project. Firstly, we've done an unprecedented work on cleaning, protection and research on the history and culture of the Three Gorges area. And the enhancement of the consciousness of protecting the culture and cultural relic within the reservoir area contribute to a deeper study on the Three Gorges culture. Secondly, we've taken various protection work of the cultural relic, such as setting up site relocation, overall cut, in – situ conservation, and building county museum and municipal museum by local civil services, various county personal museums or township personal museums especially the construction of the Chongqing Three Gorges Museum and these entire museums make a great contribution to the collection of Three Gorges cultural relic, and play an objective role in the promotion of conservation work of those cultural relic. Thirdly, the Chinese government takes actions to enforce the protection of the intangible cultural heritage, for instance, Yichang City as a national intangible cultural heritage protection center, has hold several significant academic conferences on Qu Yuan, Zhao Jun, Guan Gong and so on (they are some famous historic characters in China) and sacrifice rituals as well. And so many intangible cultural protection cultural achievements have been made in the Three Gorges. Fourthly, tremendous of cultural vice industries emerge along with the construction of Three Gorges Project, such as literature, history, archaeology, migration, economy, management, art, folk custom, collection, film and television and so on. And as a result, those vice industries bring a number of by – products like migration culture opera, Still Live, documentaries on the Three Gorges, photo albums, the Three Gorges rocks, the Three Gorges paintings and so forth. Fifthly, an upsurge of a wild academic boom has appeared accompanying Three Gorges Project, and a considerable amount of relevant scientific research achievements have taken their presence. In addition, there are abundant of research production on the Three Gorges area and the Three Gorges' nature, culture, science and technology and the other aspects: by 9 p. m. on Jan. 8, 2010, from CNKI one can precisely retrieve 27,018 articles (those that contain words of Three Gorges Project in titles) during the period from 1979 to 2010, which directly research the Yangtze Three Gorges, and according to the record there were 7,431 items of the articles' titles that contained words of Three Gorges Project. And by Oct. 20, 2009, there were

24,985 articles (containing words of the Three Gorges in the title), and within 80 d there had increased 2,033 articles, that is to say there had increased more than 25 articles relating to the Three Gorges per day on CNKI. All in all, there are thousands of books in connection with Three Gorges Project. Sixthly, so far, by making good use of the Gezhouba Dam, the auxiliary project of Three Gorges Project, Three Gorges Project has created many economy entities constructing numerous water conservancy constructions, such as China Gezhouba (Group) Corporation, China Three Gorges Project Corporation and so on. And many colleges and universities and scientific research entities are founded under the brand effect of the Three Gorges, like China Three Gorges University (CTGU), Chongqing Three Gorges University, and Yangtze Normal University. By now, as for the research base aimed at the Three Gorges culture, there are the two arts and social science bases in the CTGU: Social Development Research Center for Three Gorges Culture and Economy (it is about to declare the ministry of education humanities and social science base: Development Research Center for the Yangtze Three Gorges) and the Three Gorges Reservoir Resettlement Research Center; Chongqing Ministry of Education Arts and Social Science Base in Chongqing Technology and Business University: Development Research Center for the Upper Reach of Yangtze River; Provincial Arts and Social Science Base in Chong Qing Normal University: The Three Gorges Culture Research Institute; and Wujiang River Culture Research Center in Fuling. And as for the research institutes about the Three Gorges culture, there are Hubei the Three Gorges Culture Research Institute, CTGU Social Development Research Center for the Three Gorges Culture and Economy, Chongqing the Three Gorges Culture Research Institute, the Three Gorges Culture and Social Development Research Center in Chong Qing Normal University, the Three Gorges Culture Research Institute in Chongqing Three Gorges University and so forth. And for each year, plentiful research achievements come out of those research centers and institutes. Besides, those colleges and universities have not only cultivated qualified scientists and technicians but also sublimated the Three Gorges culture, especially Three Gorges Project culture, and seldom of us have ever thought over or noticed that all those above are tangible or intangible cultural by – products of Three Gorges Project. Whether it can get the recognition or not by the academia, the Three Gorges Study has already made its existence. And the Three Gorges Culture Research Boards are set up in Chongqing and Hubei respectively, the scholars that do research on the Three Gorges culture are members of them, moreover, for each place there will hold a relevant academic seminar targets on the Three Gorges culture at fixed period per year as agenda activities. Besides, the local magazines in Hubei and Chongqing such as China Three Gorges Construction, Three Gorges Tribune, Renmin Changjiang, Three Gorges Culture and so on, as well as university journals and internal publications in Yichang and Chongqing act as the cultural fronts for the members of the Three Gorges Culture Research Boards. While, we should admit or at least face a fact that Three Gorges Project also act as political project, image project, energy project , water transfer project, irrigation project and so forth. So we cannot define Three Gorges Project to a narrow sense, instead it is a huge and comprehensive project that demands all aspects cooperation in its construction, and it is a great achievement achieved by the joint efforts of the Chinese history. The construction of Three Gorges Project involves all fields' work of the nation and the local places. In a word, the extension of the connotation of Three Gorges Project is beyond our imagination.

3   Cultural types of Three Gorges Project

The Three Gorges Project has diversified culture types, which exert great impact on China's politics, economy and culture. The construction of project culture helps to improve China's project development and project capacity.

3.1   Flood control culture of Three Gorges Project

The Three Gorges Project is a national great cause, whose main function and means is the flood control, to get the purpose of water control. But Three Gorges Project is not independent, seizing the chance of project construction, Chinese are "constructing Three Gorges, developing the

Yangtze River", realizing the pattern of cascade development, rolling development, full river basin development, all round development in the Yangtze River. The constructing of Three Gorges Project drove a series of water conservancy projects in southwest region along the Qingjiang River, Wujiang River, Jinsha River, Dadu River, Jiangling River, Minjiang River, Nujiang River. The Three Gorges Project along with these projects are playing a united role in regulating of water in river basin, maintaining of water and land in southwest, diverting estuary deposit, "West – to – East Hydroelectric Power Project" and "South – to – North Water Diversion Project". A great water conservancy project has far – reaching influence on the local region even the whole nation. just as the recording from River Water in Commentary on the Waterways Classic: "after the completion of the Dujiang River Irrigation System, Zhuge Liang launched march to the north, irrigated the farmland by the system, the whole state was funded by the system, and recruited two hundreds of peoples to protect the system , and appointed an official to take charge of it. " so Zhuge Liang said: "Yizhou has narrow pass, inside is vast expanse of fertile land, which is self – subsistence and strategically located region. " *New Book of Tang*. Chen Ziang records that: "Shu is an important state in southwest and the national treasury, there are rich people, plants and millets there, the food can feed the whole nation people if transported along the river. " Dujiang River Irrigation System makes Sichuan "vast expanse of fertile land, self – subsistence and strategically located region", "a national treasury", which possess important position in the nation economy. Now there are three major water conservancy projects in west Hubei concerning the overall national economy safety and the national future development. They are Three Gorges Project, Jingjiang River Flood Control (Flood Diversion) Water Conservancy Project and Dan jiangkou River South – to – North Water Diversion Project. These projects are not only related to the national destiny, but also become the tourism and culture resorts to demonstrate, inherit and publicize Dayu Taming the Water and Li Bing's water conservancy culture of Dujiangyan River Irrigation System in ancient China. Three Gorges Project (Gezhouba Project included) Qingjiang River Cascade Hydropower Project (Geheyan water – control Project, Qingjiang Gaobazhou water – control project, Qingjiang Shuibuya water – control Project) Danjiangkou River water – control project, Jingjiang River Flood Control (Flood Diversion) Water Conservancy Project are all water – taming projects, and the important role of water – control is flood – control. The Three Gorges Project has outstanding role in flood – control. Dayu regulated water to stabilize the nation, so experience get from the Chinese history is that, the nation is stable when the water conservancy projects prospers, and vice versa. The hardness, far – reaching impact and outstanding achievements of water – control project construction in Chinese history are rare in the world, so from the point of view, Three Gorges Project is the succession of Chinese five – thousand – year water – control history, and the culture heritage of Dayu regulated water to stabilize the nation.

3.2 Contending scientific verification culture of Three Gorges Project

It's Sun Yat – sen who first came up with the idea of "developing Three Gorges reservoir energy". In the year of 1919, he came up with the idea in Industry Plan of National Construction Plan that to develop Three Gorges water conservancy, especially the shipping value. In the year of 1924, when explaining his Three People's Principles in Guangzhou, Sun Yat – sen said: "the water power of Kuixia in the upstream Yangtze River is greater, some people has investigated the water power from Yichang to Wanxiang County, and found that it can generate more than three thousand horsepower electricity, which is far more than the electricity in various nations. " Later the Kuomintang government and previous governments of People's Republic of China had debates on Three Gorges Project. Scholars said that: now the subject project of Three Gorges Project will be completed, with three primary realizations of water reservoir, water transportation, and water electricity. Looking back to the course of 70 years dream, 50 years survey and 40 years proof, we have a more deep understanding of the project. The proof process is to seek truth and facts, to rethink and make decision. And the problems are inevitable; the process of solving problems is the process to break the old balance and seek new balance . this is the reason of heating debates around the project and that the people who against the project made greatest contribution to the project.

Only in this way, Three Gorges Project can absorb all parties' opinion, show its nature, exert its functions and reach the water – control state of making best of the circumstances and benefiting the people.

3.3　Migrant culture of Three Gorges Project

The key to Three Gorges Project's success is the migrants and the touching part of Three Gorges Project is also the migratory. From the point of view of the migrants, Three Gorges Project culture is a mobile, convective and open, the creator is the migrants in all generations. The construction of the project broadened the research field of migration culture, and the investigation to the material culture, spiritual culture and psychological mechanism such as moving, settling down, management, new settlement culture construction and new living environment is helpful to the completion of Three Gorges Project. Migrant culture includes old settlement and new settlement, is the hybrid of traditional culture and modern culture, native land culture and migration land culture. The migrants come from the underdeveloped Gorge and river belt, they can survive and develop only by overcome the challenges in new living environment. The migrants bring their local culture from different areas and communicate and blend their culture, thus the migrants community has an open culture attitude.

In the period of planning economy, the government valued the project more than the migrant's benefit, so the migrants cannot make a living in a normal production for a long time, thus the migrant's problems occurred constantly and remained unsettled. In the new period though the compensation standards of newly – built hydropower station improved, the difficulty of migrants' resettlement increased due to the government's less value on the migrants and the migrants' expected value increase to the placement. In fact, the migrants' resettlement of Three Gorges Project is the most valued and successful among reservoir region immigration resettlement. In the year of 1993, the state council issued rules on Three Gorges Project migrants resettlement, providing law basis and support to the migrant's resettlement. Three Gorges Project adheres to the win – win principle of hydroelectric development and migrants benefit protection, constantly pushes forward the reform of migrants management system, taking developing the migrants economy and stabilizing the community as the main line, practices the migrants policy of "people – oriented" development concept, aiming at let the migrants "move out, stabilize, gradually become rich". The government calls for that" build a power station, drive the local economy, improve the local environment, benefit the migrants". Migrants' culture of Three Gorges Project at least includes: the moving out culture, the resettlement culture, the management culture, environment culture, material culture, spiritual culture, new settlement construction culture, psychological action culture, and all of them will become the spiritual culture legacy after project construction.

3.4　Construction culture of Three Gorges Project

The Three Gorges Project has many constructions, which are both the tangible forms and works of art, so the project has formed a unique construction culture. The water conversancy project construction and the new settlement with modern public facilities enabled the Three Gorges become a new combination of modern civilization and historical civilization.

3.5　Township (city) culture of Three Gorges Project

Liu Benrong pointed out that Three Gorges reservoir region has started forming a network of big, medium, and small cities, with a good city culture foundation. counties along the Changjiang three gorges will be relocated and reconstructed with the project construction and the speeding up of the migrant relocation , which provides a chance to city culture construction. It's a important theme for city image – building to maintain, give play to the unique advantage of environment culture, deepen, cultivate, develop and sublime the culture connotation, cast the culture spirit by strong culture sense, to express both the modern culture and historical culture in harmony in the city constructions.

3.6 Transportation culture of Three Gorges Project

Three Gorges has been a major transportation center since the ancient times, as the records of Fengjie from Volume 34 of Annals of Sichuan by in Yongzheng (the 5th Empire of Qing Dynasty) period: "the goods from Sichuan to Hubei and Guanngdong, and the goods from other provinces to Sichuan must enter and exit through the three gorges." Second Volume of Local Annals of Baxian by Guangxu records about Chongqing that: "the center of water and land transportation to upstream and downstream ports, and the prosperous business center", Chonqing's transportation position is raised as 175 m water level sail upstream to it after the completion of Three Gorges Project. the project changes not only the water transporting, but also the land transporting, such as, the Gezhouba Dam Bridge become the No. 1 Yangtze dam bridge, Yichang is the center of Hankou and Chongqing, also the center of Chuanhan Railway, so it is a good place to build the electricity plant. After the foundation of the new China, Gezhouba Sanjiang River Bridge was built to connect Xiba and Yichang urban area to speed up the project construction and ensure the project quality in Gezhouba project process. The bridge "is one of the major supporting projects of Gezhouba water − control ship lock project, located in Gezhouba Yichang, Hubei Province, across the cross − strait of artificial watercourse." The bridge is completed with the Sanjiang River ship lock project, the main pass to the main lock power station and dam tourism site from the urban areas. Sanjiang River ship lock: Solved the contradiction between ship transportation and train transportation, lifting movable bridges were built on the upstream bridge piers of No. 2 and No. 3 ship locks to with a net height increase of 18 m, the bus and train can pass through the dam when no ship pass through the lock. The tunnels triggered by the project make up a museum of tunnels in Three Gorges, which add content to the Yangtze River Transportation culture. Pass through Shu road is harder than go up to the heaven, now a deep chasm turned into a thoroughfare, which is unexpected.

3.7 Ecological culture of Three Gorges Project

Scholars pointed out that: the dam and environment are not opposite; we can seek a balance in the development. The coordinating principle of society, economy and environment can be used as a rule to judge and handle the relation between the water conservancy project and the ecological environment. The construction and protection of ecological environment are an important guarantee of social and economic development in the reservoir region. To construct a harmonious Three Gorges, we need to strengthen the ecological culture construction, prompting the balance between people and nature, providing people in Three Gorges a good living and working environment. It remains a concerned issue that the project and impoundment will destroy the beauty of Three Gorges, and affecting the environment along the Yangtze River. As a key project to develop and govern the Yangtze River, the project has far − reaching impact on the ecological environment . The three functions of flood − control; power − generation and shipping are all well known. The project itself has huge ecological and environmental benefits, which will exert its power in the process of impoundment and power − generation. In fact, the construction of Three Gorges Project strengthened the national sense of environment protection; for example, Chinese people's awareness of environmental protection increased since the starting of project, we stick to the win − win principle of developing hydropower and protecting the environment, the policy of returning farmland to forests in the west, adhere to the principle of developing in protection and promoting the protection by developing, highlighting the ecological and environmental protection , to reduce or avoid the side effect to the environment by scientific planning, optimal design and effective project measures. From the culture psychology, we can say that the project has created the national psychology of emphasis on the rivers' protection. But the problems of pollution treatment, desertification combating, geological disaster, soil erosion, reservoir and bank management, water quality and converting the land to forestry still remain to be addressed, concerning and solving the problem are part of ecological culture.

3.8　Corporation culture of Three Gorges Project

The culture of Three Gorges Corporation, China Gezhouba Corporation and other corporations (ship – manufacturers) and institutions set up in the preparation and construction process of Three Gorges Project is the component of Three Gorges Project culture. Three Gorges project entity, Three Gorges Corporation use the socialist market economy rule to organize the project construction, according to the characteristic, rules and construction procedure of the project, relying on technological improvement, focusing on the control objectives of quality, schedule and cost, developing the corporation in the direction of modern organization system. The mode of management and operation has instructive significance to China's economic reform and the industrial restructuring and upgrading.

3.9　Film and television (art) culture of Three Gorges Project

The Three Gorges Project gave birth to related screen culture works including films, operas videos and documentaries. Three Gorges Project closure garden displays Three Gorges Project closure documentary, reappearing the soul – stirring scene. Hubei Relics Protection Departments use 3D technology to show the relics' discovery and relocation and the history of Three Gorges Reservoir Region in videos. The documentary Three Gorges in thousand years debuted in 2004 won china's highest film award "Huabiao award". China Three Gorges University made teaching documentaries included the VCD of Three Gorges culture which was reproduced in 2004, introducing the history, geography, humanity and hydropower in Three Gorges. Other included Miraculous Three Gorges Summary, Sanxia Zongheng and so on. The situational drama Three Gorges in Heyday with large – scale casting and high input demonstrated the Yangtze River water – control culture, project construction culture and migrant's culture, musical play Grand Three Gorges, drama Migrant JinDahua, symphony Traversing Three Gorges. the picturing drama ChushuiBashan won the Splendor Award. Besides' Three Gorges Project hasten the national development of photography, arts, sculpture, painting, dancing, forming a new art creation focus.

3.10　National image culture of Three Gorges Project

Yichang is the world's No. 1 hydropower city and famous tourism city. But Three Gorges Project belongs no to Yichang and Hubei Province. Scholars pointed out that: Three Gorges Project is the representative and symbol of human modern civilization. The space explorers will be more surprised to discover a more attractive human spectacle—Three Gorges Dam and the thousand miles of Three Gorges Reservoir region soon after they saw the Great Wall and Pyramids. Three Gorges Project is a human heroic undertaking to adjust the nature , from which formed the significantly symbolic culture spirit in the process of human civilization and modernization . the construction of Three Gorges Project inherits the national "Great Wall" spirit and add the new spirit core of "innovation, truth – seeking, braveness to be number one" to the Chinese national spirit. The Three Gorges Project is certain to become the new "Great Wall", towering in the east of world. In recent years, a series of activities held by Three Gorges Corporation to promote Three Gorges Project image achieved good response, the grand events includes: Mid – Autumn Festival Party sponsored by CCTV International Channel and "Heart – to – heart" Performance in 2003 which used the dam as the stage – setting. In 2008, Beijing Olympic Games torch relay in Three Gorges section was co – completed by famous Hongkong singer Karen Mok and vice – general manager of Three Gorges Corporation Cao Guangjing. We still need to work hard to make Three Gorges Project become the world intangible and tangible heritage as Dujiangyan Irrigation System.

3.11　Institutional culture of Three Gorges Project

The level of China's hydropower industry in survey, design, construction, working, management and renovation will reach the international advanced level, and make great contributions to the national economic development and renaissance, which is the irresistible historical trend. The culture of modern science and technology is generic and creative, and is the most powerful force driving the material and spiritual civilization of the world. As the relatively independent subculture system, it includes four parts that is the physical level, institutional level, code of conduct and values.

In fact, the institutional culture innovation is the key to the successful construction of Three Gorges Project and to its operating and preservation, and even the solving of remaining problems. The Three Gorges Project abandoned the "Battle Type" construction model for a market operating model of corporation, and constructing headquarters for contract management in technological innovation. A marketing operation management model is applied in Three Gorges Project to get rid of the traditional planned economy model. Given consideration to its characteristics, law and constructing procedures, CTGPC (China Three Gorges Project Corporation), under the guidance of the basic rules of socialist market economy, is marching forward to a modern corporate system by technological progress and control on quality, schedule and price. Its immigration system has become the benchmark of immigration work in other regions and exerted an influence on China's immigration policies.

The Three Gorges Project insists on mutually beneficial relations between hydropower development and environmental protection. The government launches Grain for Green Project in environmentally sensitive area, and follows the principle of "develop in protecting and protect in developing". Much attention was paid to reduce or avoid the adverse impacts on environment by scientific planning, design and effective measures. The Three Gorges Project promotes the implementation of the policies to return farmlands to forests, lakes and to grasslands. Pan Jiazheng points out that the hydropower construction in China has reached the international advanced level, setting an example for other countries, especially those developing ones. The institutional culture of Three Gorges Project is one of the most valuable experiences for the world.

3.12 The tourism culture of Three Gorges Project

As a famous Hydraulic Engineering in the world become scenic spots, we can see the great role hydraulic engineering plays to its location and surrounding areas. The Three Gorges Project "not only gives people knowledge of water and its relation with human beings, but also shows the technological progress". Ouyang Yunsen suggested the development of its tourism culture, mass culture and various collections at the beginning of the construction of the Three Gorges. However, Li Jin put forward four brand cultures, that is the modern hydropower project, science and education, immigration and tourism in reservoir areas. In recent years, with the construction of the Three Gorges, local tourism is blooming. With its annual tourists hitting over one million for the third consecutive year, it is listed at the top of the national industrial tourism programs and is one of China's first Class 5A scenic spots. It is now a landmark of China's renaissance and a window for the world to learn more about China. As a Chongqing scholar put it, the construction of the Three Gorges is and will be attracting more domestic and overseas tourists. Yu Qiuyu says, "The Three Gorges will be one of the world's greatest scenic spots and it deserves more compliments in poems and paintings. " The hot tourism in Three Gorges adds to the tourism culture connotations in China, as well as evokes relating theory discussions. Meanwhile, Chongqing, Yichang and the local travel agencies and institutes of culture research are all busy in new tourism lines and brand cultures of the Three Gorges.

3.13 The culture of education of Three Gorges Project

Technology is to solve practical, cognitive and technical problems toward the world. The better the cognitive system conforms to the objective laws, the closer it is to the truth. Humanity, however, solves the value rationality aspect of science and technology and gives it direction. On the one hand, only under the direction of humanity can researches of basic sciences boost the development of technological sciences and engineering technologies. On the other hand, only under the direction of humanity can the new problems in technologies and research methods they provide put forward the development of basic science researches. Zhang Bo points out in Thoughts on setting up Engineering Culture Course: It is the need of general education in high vocational college in engineering to set up the engineering culture courseand course content, compile new textbooks and do teaching practices. He hopes to make certain contributions to the setting up of general education courses of high vocational colleges in engineering. He believes a engineering culture course should include the following chapters: Overview of engineering culture, environmental engineering culture, energy engineering culture, engineering design culture, mechanical engineering culture, electronic engineering culture, automotive engineering culture, information engineering culture, management engineering culture, construction engineering culture, ideological and political engineering culture, corporate culture, tourism engineering culture, convention engineering culture, business engineering culture and public welfare engineering culture. The engineering culture course aforementioned by Zhang Bo is beneficial to the education, publicity and cultural position of Three Gorges. The exquisite course the Three Gorges culture opened in CTGU takes Three Gorges as an important teaching section and launches an research on Three Gorges engineering culture (website: http://210. 42. 35. 1/sxwh/poll/indexgn_ck. php). The main lecturer analyses the website questionnaire. A "water culture research center" was set up in Three Gorges University to open courses on water culture, which has much to do with the history of its anticipation in Three Gorges Project of CTGU.

Founded in 1978, Gezhouba Institute of Hydroelectric Engineering was the formal name of CTGU. It was founded under the Ministry of Water Resources and Power to support the Gezhouba Project, the auxiliary project of the Three Gorges. In 2009, the Ministry of Water Resources signed an agreement with People's Government of Hubei Province to found CTGU. It has now developed into a well-known regional comprehensive university with distinct characteristics and advantage of hydroelectric power, and also solid foundation of education. The CTGU was founded to support Three Gorges, and became famous because of it, and will make contributions as it always did.

Of course, the Three Gorges engineering culture has covered too wide range of areas to list. For example, much attention is paid to the ecology of the reservoir area, excavation and protection, rescue and rebuilding of relics. Some experts and scholars excavated and collected 12 items of intangible cultural heritage, over 120 sub items, 62 kinds of folk art and 128 representatives of culture heritage in Ming Dynasty (1368 ~ 1644), and 2,470 folk artisans in Yichang City, Hubei Province. The People's Government of Yichang City raised 3×10^9 yuan to construct "China Three Gorges World Intangible Culture Heritage Exposition" in Yidu City, Hubei Province, and "China Three Gorges Intangible Culture Heritage Industry Park" in Zigui County, Yidu City. They are parts of the culture protection under the influence of the Three Gorges. We list the cultural characteristics of Three Gorges Project.

4 The influences and characteristics of engineering culture of Three Gorges Project

The Three Gorges Project has far-reaching impacts. The building of Three Gorges Project is the end of an old ear and a way of thinking as well as the beginning of a new era and way of thinking. However, we are not supposed to be over optimistic. The Three Gorges Project is the practice of the Chinese nation of its wisdom and speculative philosophy on water. *Xunzi · Wangzhi* puts it this way: "The water supporting a ship can also upset it." Lao Tzu says: "Behavior grows better before it grows worse; Misfortune might be a blessing in disguises." Only by realizing this, can Chinese nation fully play the role of the Three Gorges, reduce and avoid it negative impacts.

4.1 The influences of The Three Gorges

The Three Gorges has brought us great benefits. However, we should pay attention to the negative impacts coming along as well.

4.1.1 Negative Impacts

①Many culture relics were damaged; lands were flooded and the ecological environment of plants and animals changed; ②It is the voting result of the Committee of the People's Congress; ③It is the most controversial project since the founding of PRC in 1949, and will be disputed; ④The resettlement problem will exist for a very long time to come; ⑤It puts the protection of the ecological environment under great pressure; ⑥The military pressure of Three Gorges Project should not be ignored, etc.

4.1.2 Positive impacts

①The Three Gorges Project is the largest immigration ever organized by the government in China; ②It is the largest ecological protection project ever in China; ③It is the largest hydropower (and water resources) project in size ever since the founding of PRC; ④It is also the world's largest water resources and hydropower project, especially in developing countries; ⑤It is and will be a project that benefits people in China; ⑥It is the largest sand control project ever in China; ⑦It is the only water transportation project comparable to the Grand Canal; ⑧It is also China's largest culture relic's excavation and protection project in history; ⑨It trained many technical and management personnel; ⑩It triggered a series of academic studies about engineering philosophy, engineering culture and Three Gorges culture; ⑪It makes the Yangtze River Basin an integrated economic region and among others. Besides the above benefits, its cultural functions are tremendous.

4.2 Cultural functions

①The 100 – year dream of China finally came true; ②It reveals the improvement of China's technological levels; ③It witnesses a stronger China and its modernization drive; ④It mobilizes various social resources, especially those of water resources to work together; ⑤It triggers the Three Gorges culture craze and thoughts on engineering culture and philosophy; ⑥China's technologies of dam construction have reached the international advanced level; ⑦It is the symbol of China striving and becoming stronger just as the Great Wall for ancient China; ⑧Science, democracy and people oriented concept are fully presented; ⑨It sets up an example for the world in emission and reduction of the greenhouse gases; ⑩It pushed the third development of the world's hydropower; ⑪The spirit of the Three Gorges is a valuable spiritual wealth; ⑫The Three Gorges would be national and even the world's cultural heritage as Dujiangyan and the Grand Canal.

4.3 The cultural characteristics of Three Gorges Project

The four aspects of culture (that is the state of matter, institution, actions and mentality) of the Three Gorges closely rely on each other. Since the Three Gorges and its culture are impartibly and closely related, the characteristics of the Three Gorges, to some extent, are its cultural characteristics. ①Ecological; ②a national comprehensive project; ③exists in a certain time; ④scientific and technological; ⑤organized and planned; ⑥institutional; ⑦marketability; ⑧humanity; ⑨being a culture heritage in the future; ⑩controversial; ⑪educational; ⑫stimulating consumption; ⑬dynamic and developing; ⑭time – limited; ⑮culturally influential; ⑯national participation; ⑰part of the world's heritage; ⑱both a natural and man – made landscape; ⑲connects the history and creates the future; ⑳A combination of both traditional and modern elements. These characteristics of the Three Gorges and its culture help us learn about hydraulic engineering and its future.

5 Advices on the shaping, development, heritage and protection of Three Gorges Project culture

As a systematic project, the Three Gorges needs the participation of the whole country. And so does the Three Gorges culture. We can do in the following ways: ①The governments of Chongqing Municipality and Hubei Province should play leading roles and cooperate more tightly; ②Mass participation should be fully emphasized; ③The hydropower enterprises are also very important to the Three Gorges culture; ④Making full use of public opinions and advertisement; ⑤Scientific research institutions (colleges, nongovernmental and governmental institutions) conduct more investigative researches; ⑥Schools including elementary and secondary schools and colleges should open courses about the Three Gorges; ⑦Taking advantage of books, magazines and media; ⑧More academic meetings are necessary to promote culture development of the Three Gorges; ⑨International exchanges can be helpful; ⑩People should improve their consciousness of culture heritage and put into actions; ⑪ Take actions from hydrology, transportation, ecology, immigration and geology protection and sand control. Actions should go hand in hand with advertisements.

So far the Three Gorges Project is the greatest one of its kind ever in the world. Since its completion, the Three Gorges, with 175 – meter water storage, has benefited people in flood control, power generation and shipping. However, it also produces some natural disasters and ecological damages (such as sand, geological disasters in reservoir area and water quality, etc). The ability to solve these problems shows that Chinese descendants can overcome difficulties and make great achievements. Culture will lead and combine with science and technology as well as humanity to ensure that the Three Gorges will achieve the maximum profits. Hopefully, the paper is intended to attract the attention of the governments and study circles. The research on Three Gorges culture is a long lasting topic and will be more and more important. This paper is just one small step forward, and we sincerely looking forward to your comments.

References

Declaration of Hydropower and Sustainable Development in Beijing[J]. China Three Gorges Construction, 2005 (1).

Modern Chinese Ci Hai Editor Board. Ci Hai[M]. Beijing: Guangming Daily Press, 2002.

Encyclopedia of China Total Editor Board. Encyclopedia of China[M]. Beijing: Encyclopedia of China Press, 2002.

Lewin D. Engineering Philosophy: The Third Culture? [J]. Leonardo, 1983, 16(2):127 – 132.

Wang Xi, Wan Qing. Comparative Analysis on Engineering Culture and Enterprise Culture[J]. Future and Development, 2009 (9).

Hong Wei, Hong Feng. Study on Engineering Culture in Perspective of the Engineering Essence of the Construction Process[J]. Journal of Chongqing Institute of Technology, 2007(11).

Huang Fan. Prosperous Xiajiang County: A Cultural Work on Interpretation of the Three Gorges Project[J]. The Decision and Information, 2009(2).

Liu Yutang. Principal Contents of the Three Gorges Culture[J]. Journal of China Three Gorges Univesity, 2005(5).

Feng Tianyu, He Xiaoming, Zhou Jiming. The Chinese Culture History[M]. Shanghai: Shanghai People's Publishing House, 2005.

Sun Yatsen. People's Livelihood of the Three People's Principles, The Complete Works of Sun Yatsen, Vol. 9[M]. Beijing: Zhonghua Book Company Press, 1986.

Xiao Feng. Seeing the Project Culture Value from Culture Reason of Quebec Bridge Collapse[J]. Journal of Dialectics of Nature, 2006(5).

Zhang Haidong. Analysis on the Relationship between the Three Gorges Power – generating Efficiency and Rolling Development of Jinshajiang River [J]. China Three Gorges Construction, 2005(4).

Sun Yatsen. People's Livelihood of the Three People's Principles, The Complete Works of Sun Yatsen, Vol. 9[M]. Beijing: Zhonghua Book Company Press, 1986.

Yuan Guolin. Historical Review and Ecological Environment of Three Gorges Project[J]. Science and Technology Review, 2005(10).

Liu Yutang. The Connotations of the Three Gorges Culture[J]. Journal of China Three Gorges University (Humanities & Social Sciences), 2005(5).

Wang Jin. The Cultural Integration of Immigration of the Three Gorges Project[J]. Jianghan Tribune, 2006(3).

Zhang Guobao. Don't Demonize the Development of Hydropower [J]. China Three Gorges Constrution, 2008(9).

Xu Changyi. The Golden Age of Hydropower Development Is Coming – What Will China Do[J]. China Three Gorges Construction, 2008(11).

Liu Benrong. Constructing the Culturally Industrial System of Yangtzer Three Gorges[J]. Chinese Culture Daily, Oct. 18, 2000.

Song Xishang. Overview of Hydropower Generation in the Upstream of Yangzijiang River [J]. Yangtze River Quarterly, 1933 (1).

Wang Jie. Yangtzer River Dictionary[M]. Wuhan: Wuhan Press, 1997.

Yichang Committee on Geographical Names of Hubei Province, 1984.

Yu Xuezhong. Dam and Ecology: Opposites and Cooperation[J]. Encyclopedic Knowledge,2005 (7).

Jiang Zemin. Speech by Jiang Zemin at CPC 80th Anniversary Assembly[R], 2001.

Xu Changyi. Goldern Age of Hydropower Development is coming: What Will China Do[J]. China Three Gorges Construction, 2008(11).

He Gong, Chen Wenbin, Hu Shaohua. Overview of Eight Years' Experience on Construction and Management of the Three Gorges[J]. China Three Gorges Constrution, 2001(11).

Ma Shangyun. Thoughts on the Spirits of the Three Gorges Culture[J]. Yangtzer Tribune,1995 (5).

We Have Only One Chance to Stride from a Country of Tremendous Hydropower to a Country of Profound Hydropower[J]. China Three Gorges Construction, 2005(1).

He Yaping. Science and Technology Culture in Modern Society[J]. Studies in Science of Science, 1997(4).

Bai Jinfu. How Many Independent Innovations Has China Realized[J]. China Economic Weekly, 2006(31).

Gong He, Chen Wenbin, Hu Shaohua. Overview of Eight Years' Experience on Construction and Management of the Three Gorges[J]. China Three Gorges Constrution, 2001(11).

Xu Changyi. Golden Age of Hydropower Development is Coming , What Will China Do[J]. China Three Gorges Construction, 2008(11).

Han Lei, Pan Jiazheng. China Hydropower from the Perspective of Humanity[J]. China Three Gorges Construction, 2007(4).

Liu Shukun. Free Discussions[J]. China Three Gorges Construction, 2008(12).

Ouyang Yunsen. The Three Gorges Project and Its Cultural Development[J]. China Three Gorges Construction, 1997(2).

Jin Li. The Development and Construction Research of the Culture Brand of the Three Gorges[J]. Special Zone Economy, 2005(11).

Huang Fan. Great Yangtze Three Gorges: The Three Gorges Project[J]. Decision & Information, 2009(2).

Yu Qiuyu. Free Discussion on the Three Gorges[J]. China Three Gorges, 2008 (2).

Zheng Jingdong, Liu Fang. The Tourism Culture of Yangtze Three Gorges [M]. Chongqing: Chongqing Publishing House, 2002.

Huang Dayong. Strategy Research on the Development and Usage of the Tourism Culture Resources of Yangtze Three Gorges Reservoir Area [M]. Chongqing: Chongqing Publishing House, 2002.

Zhu Chuanyi. Education Innovation Exploration of Advanced Engineering Culture [J]. China Metallurgical Education, 2003(2).

Zhang Bo. Thoughts on the Opening of Engineering Culture Course[J]. Mechanical Vocational Education, 2008(7).

Luo Meijie, Huang Quansheng, Hu Junxiu, et al. A Cultural Analysis of Internet Questionnaire on Three Gorges Project Construction[J]. Three Gorges Tribune, (5).

Xu Xu. Refreshing the Culture Memory of Folk Culture[J]. China Three Gorges Construction, 2008(12).

Wang Zhihe. Postmodernism and China[J]. Seeking Truth, 2001(3).

Research on the Ancient Non – governmental Group Participation in Hydraulic Facilities[①]

Bi Xia, *Sun Weicheng*, *Zhao Ruping* and *Peng Hui*

School of Marxism, Public Administration School, Hohai University, Nanjing, 210098, China

Abstract: The construction and maintenance of irrigation facility has been emphasized by the government of every dynasty. Meanwhile, the participation of no – governmental group also played an important role in solving the Hydraulic disputes, and in the construction and maintenance of Hydraulic facility. It was determined by the ancient political structure and the financial ability of the ancient government. And the practical need of the non – governmental group is also an important factor.

Key words: ancient non – government group, hydraulic facility, participation

"Flood control society" theory holds that the political system in ancient China was closely related to flood control, in "these regions, agriculture production could be sustained successfully and effectively only when people adopted irrigation or flood control approach when necessary to overcome the water supply shortcoming and dissonance. Such projects always required large – scale collaboration which needed discipline, affiliation and strong leadership." Thus the "oriental despotism" came into being. The theory is worthy of falsification while the fact is that ancient China society was an agricultural society when courts valued the water conservancy which included the construction and maintenance of the water conservancy facilities as water was the lifeblood of agriculture and farmers were direct related to the water. And the non – governmental groups played an important role in the water conservancy facilities construction each dynasty, especially the Southern Song Dynasty. In the Song Dynasty, the non – government group had become increasingly prominent in the water conservancy facilities construction as the government lack of money. The civil organizations' participation provided a lot of manpower, material and financial resources to the ancient water conservancy. That solved the government's lacking of finance and acted as the rule hub between ordinary people and the government. The feudal official needed civil organizations to get varieties of social network and resources in the country. On the other hand, civil organizations also played an important role in addressing irrigation dispute.

1   Classification of the ancient non – governmental group involved in water conservancy

Modern water conservancy's connotation is more abundant. In the History, Present Situation and Outlook of Chinese Water Conservancy, Qian Zhengying indicated: "Water conservancy are the activity of human to adapt to, to use, to alert, to develop and to protect the water environment. Project carried out for this purpose is called water conservancy project and works undertaken for this purpose is irrigation competent." While in the ancient, the term of water conservancy was first seen in the Spring and Autumn Annals which came out in the late Warring States. But "to take the water benefit" here refered to fishing. Later long feudal society, the contents of water conservancy enriched continuously. The term of water conservancy contained the flood controlling, irrigation, navigation and so on.

In this article, we define the time before Qing Dynasty which as the time scale of the ancient water conservancy, which Qing Dynasty was also included. Water conservancy talked in this paper not only deal with water conservancy projection but also the ancient water organization and the resolving of the water dispute. Civil organizations participated in the ancient water conservancy in

---

① The funded project of annual national social science in 2011, Research on the network of water environment management modes (11CZZ039).

this paper was groups engaged in the activities of the ancient water conservancy in the history of our country. Their forms are complex and varied which contained the village elites, family clan, guilds, temples and so on.

1.1 Village elites

China had set up county since the Qin Dynasty which equaled to setting up offices and agents instead of branches. County reported to the central government directly. But "the administration management had not penetrated into the village level while the clan – specific forces maintain the stability and order of the villages under the rule of Qin Dynasty."

Village elites were made up of two groups. One group was the gentlemen who had passed through the civil service examinations. They mainly referred to these two types of people, one of them was the retiring officials and the rural relatives of the current officials. The other one was pupils of the prefecture or county government, Jiansheng of the Imperial Academy. They were all the moralists and representatives of the moral principles and obligations decided by the Confucian doctrine which set the provisions of China's social and interpersonal criteria.

The other group was the families based on kinship, the parents, group leaders in the clan as well as the leaders of the township. Ancient south China was more stable and there were fewer wars. Clan forces were strong and span many administration regions. This was helpful to the construction and management of the water conservancy. Local clan organizations participated in the water conservancy construction in the name of "Tang" or "Hui". When it came to the fees, clan organizations provided the most of them. Clan was the main source of funding of the construction project.

1.2 Water conservancy autonomous organizations

Water conservancy autonomous organizations were the farmer water conservancy self – management organizations. They might ranged from a few families to several families, which decided the water conservancy scale. Small farmland water government usually repaired and managed by the water autonomous organizations. Water autonomous organizations and village elite cooperate to maintain the water conservancy order.

Dujiang Weir had gradually formed an unique three – level management mechanism of "officials managing the weir and civil organizations managing the canals in the 2,200 years' practice. The Dujiang Weir, or the provincial government water conservancy segment managed the head works and set the weir official to manage the weir works. People who benefited from the water conservancy establish the non – governmental group to manage the branch canal irrigation works. Civil must funded by themselves to build the organization and select their leaders. And everyone had to comply with the weir or township regulations.

1.3 Some monks and tao of the Buddhist and Taoist temples were involved in the construction of the water conservancy

Bridges in Buddhism have the meaning of helping all living beings, so a lot of temples are willing to participate in the bridge construction. For example, Suizhou Dahongshan temple in Hubei Province was famous. Senator Liu Fengshi once bought the Furong shoal land and invite Dao Kai abbot. Dao Kai said that if the lake water was brought to the canal then thousands of lands would be fertile.

2 The way of the ancient non – governmental group participating in the water conservancy

The ancient non – governmental group played an important role in China's ancient water conservancy's development and maintenance. They participated in the water conservancy works in many ways.

2.1 Participate in handling the water dispute

Ancient China's legal system about water had not concerned the right of possession, the use and the dominance of water resources. Under the incomplete rules, a large number of courtiers, the soldiers and even ordinary people seized the lake. They used the water conservancy exclusively and didn't turn on the water when there were cases of floods. These acts violated the rules, which led to the dispute. People must solve these disputes firstly before constructing the water conservancy.

These water disputes could be divided into three types. The first was the dispute between individuals. This type of dispute was common and had a small scale. And the most prominent was about the competing of the water resources between people living in the upstream and people living in the downstream, such as what happened in the forty – second year of Qianlong time. At that time, Wang Jingui and some other people who lived in the Zhangye county and Wang Xixian and some people who lived nearby via over the water source. Occlusion hole scramble for water, resulting in litigation, by Gan state officials personally inquest, hearing, the parties dispute subsided. The second was the dispute between regions. For example, people lived in Wu Wei county want to build irrigation canals at the east bank of the Shi Yang River, which was against by the people lived in the He Jinyang dam. The last type is the dispute between individuals and collective. It was mainly about the individual damaging or encroaching the collective's interests. The concrete manifestation was the destruction of the water conservancy project and facilities. The most common was the non – benefit farmers encroaching on the beneficial people.

Something required special attention among all the three types of dispute were some strong classes in the village possessed the water source exclusively. They occupied the water upstream so that water couldn't vent and people did not dare to contest. Like choking the upstream, occupying the lakes and other water sources were also the trick adopted by the group. Besides, the religions fought for the water always. For instance, the Xue and the Lin in Chang Tai fought with arms during the Qianlong fifty – nine year. The arm confliction between people living in Yu Tian of Fuzhou province and people living in Dong Du caused 19 persons died. It had a negative effect on the livelihood of the whole clan of the two villages. In ancient China, the size of clan in the north was smaller than those in the south. The few large families were scattered in the spatial and temporal scales, and the external cohesion symbol (clan property, graves, genealogy and ancestral halls) was also not so obvious. This caused some civil organizations and religious organizations involved in the water conservancy dispute. Those enthusiastic in the water conservancy often played a dominant role in mediating in the civil water disputes. What they were in common was that they were all familiar with water affairs and solve disputes according to the township rules. Water autonomous organizations also played an important role in the water dispute resolution process. Water autonomous organizations often associated with the squire to handle the water dispute.

2.2 Participate the construction of the water conservancy facilities

Dynasties paid special attention of agricultural produce. And the water conservancy closely related to agriculture also got plenty of attention. The governments often funded the various water conservancy construction projets, such as the Zheng Drainage in Warring States period. The building of these lager – scale water conservancies always funded by governments, but the non – governmental groups would join in when the government was short of finance.

2.2.1 Irrigation facilities construction by cooperation of officials and civilians

Under the organization of office, non – governmental groups contributed funds and labors to build irrigation facility. Like the construction during the Song Dynasty. AD1137, county leader planed to do overall renovation. However, taking the poor situation into consideration, He appealed old people to offer their support. After a discussion, they decided to raise funs by old and reputable people, business in the field of bamboo, bits of woods, podsol under the organization of office. They used the considerable funds to build and maintain irrigation project. In the next year, they earned more than 3 millions silver dollar by not taking from civilization and government to build and maintain irrigation facility. Since the old and reputable people contributed many efforts, He celebrated with them by having a drink. In the year of 1,163, Zhao Yanyu did a renovation again with all the funds donated

2.2.2 Irrigation construction only by civilians

The non – governmental groups played an important role in the construction of irrigation facility when government did have enough financial budgets. Things could be divided into two sides. The first is built by sole . Individual or family funded the construction of water conservancy facilities. For example the East Pond Lake, recorded in the reign of Pujiang County, was built in Song Renzhong early days, by Qian Kan. In Huizong Grand two years, Qian Kan's great – grandson Minister Qian rebuilt the East Pond Lake and done inscription which was about his great – grandfather' merits and construction of the East Pond Lake. It is said, "In Tiansheng early days' the East Pond Lake had built as the embankment to block water. It was difficult, my great – grandfather was appointed as leader, to lead the corvees building the East Pond Lake for three years. " The second was jointly funded. Compared with individual or family funded, it is feasible for joint venture to build water facilities. Although many people did not fund fully, but acted advocate. Such as Fanwu who was from Linan in Song dynasty advocated to be volunteer worker and donated wealth to build pond. According to the Xianheng Linan records in volume 35, "since the pond had been built, there was no flooding. They was named Yonghe embankment by Prime Minister Fanguang".

2.2.3 Participation in the maintenance of water conservancy facilities

The water conservancy facilities needed maintenance repair work. However, the government often could not timely maintain due to various causes, such as financial problems. Some water conservancy facilities projects relatively small – scale the local government could not attend to the maintenance. On the other hand, the sustainability of these projects has an direct interest with the local area. In this regard, the government could often complete the maintenance work of water conservancy facilities by means of non – governmental forces through a variety of forms.

2.2.3.1 Entrust folk management

This participation form of construction was force to by the official organization of local non – governmental forces to participate. Government was often the funders of the water conservancy construction, but the water conservancy facilities are often faced with the sediment congestion situation. Because ordinary people can not effectively organize, the government hopes the help of non – governmental groups in local prestige which was appealed to come forward to organize a follow – up management. Records, such as the Song Dynasty the Baoqing "four Mingzhi" —East Qian Lakes' four banks and seven weirs were repaired many times, In Bao Qing two years , Prime Minister asked 15,000 Dan' rice from count to dredge river. After dredging river, the Prime Minister asked 3,000 Dan' rice from count for maintenance due to worrying about the river clogged.

2.2.3.2 Water groups involved in by the ties of religion – worship and Peisi this cultural mechanisms

This form sought to strengthen its characteristics as the Community Water organization from the side. For example, in Kang Xi 59 years, while Bei Ao Po hydraulic engineering was completed, Huang Housheng ancestral temple was built to place the memorial tablets of the country magistrate Huang Hongren and the predecessors who advocated to repair the project first and had made a contribution to the works. These memorial tablets mainly on officials, gentry and satrap who made an contribution to government affairs as well as the subscribers who donated above a certain number. The ritual act itself was a creation and re – creation process of community sense of identity. Held every spring, ritual and symposium activities created the identity of a water conservancy organization in the inner feelings of the water users , there was no doubt that the generation of identity made the order of the water conservancy groups better maintenance.

2. 2. 3. 3   The spontaneous participation of clan communities

Because a lot of water conservancy facilities had an direct interest relationship with the local families and clans, some even were constructed by themselves, so the families and clans were often participated in the repair work of water conservancy facilities spontaneously in the name of their own. For example, ancient water conservancy project that was known as "little Dujiangyan" in BaoTun ,Da Xi Qiao town of Xi Xiu area. It was made by a man named Bao Fubao who moved to Pu Ding Wei in Hongwu second year of Ming dynasty, residing Yong ' an Tun (now named BaoTun). In the same year' Bao Tun was created. At the same time, they started the key action of" reclaiming wasteland, guarding the territory, Maintaining an army" —the construction of water conservancy irrigation system. The management of the water conservancy project constructed independently was an difficult problem. Due to the superior condition and relatively closed, Bao Tun was said to be a small farming community. Bao Tun people carried out the project management and maintenance with their strength of family and clan, rural regulation or mores. When the old called for the families to pay channels cleaning and maintenance engineering, the people of BaoTun responded to participate in the maintenance of water conservancy project spontaneously.

3   The cause analysis of non – governmental groups involving in water conservancy

There were many various reasons that the non – government groups participated in the water conservation enterprise, they were inextricably linked with the ancient Chinese political system and social structure.

3. 1   The social form which was isomorphic of home and country was the institutional factors

Social patterns determine the social system, or affect each other. Social forms of ancient Chinese society were based on patriarchal family blood relationship. Both prehistoric and in the period of civilization, patriarchal system or family system had always maintained the most primitive human blood relationship, and had never been interrupted . All other relationships in ancient Chinese society were produced by this radiation. This isomorphism was produced on the basis of social systems of social systems. The home was the foundation of the country, the country was the kingdom of the house, and all other social organizations could be explained and replaced by home or country. All organizations and individuals were melting in the country. The family was the little family; the country was the big family. The family was the private; the country was a government, except the family was just the family. In this foundation it formed the autocratic power— Monarchical patriarchal supremacy. And the corresponding set of patriarchal family system was the basic embodiment and guarantee of ancient Chinese society's continuity.

Although some scholars believed that: "in the patriarchal family and the autocratic centralized political system of ancient China, a homogeneous political system was no borders, ran down to the bottom, devoured all the social levels. ", "The society after the Qin Dynasty, despite the rural social management system had experienced Township kiosk system, the village system , the neighborhood organization system, the Bao – Jia System and so on, the basic meaning of the rural management system had not fundamentally changed, the power structure of rural society still was with the color of clan Ethics , in other words, in village communities of rural clan – management model with a strong ethic of autonomy did not rely entirely on administrative agencies of the Government order. " That laid the foundation for community participation in water resources system, and there wais a trend of increasingly enhanced. " After the Qing dynasty, the government – controlled was more indirect, water conservancy organizations ware increasingly showing the folk trend in Hanzhong. "

3. 2　The limited power and financial resources of the government was practical factors

"Only the people who were leading the water social construction and the organization enterprise on the basis of a fixed income could work out the task. " Lacking of necessary budgetary authority and finance was the practical factors for folk organization to involve in water conservancy. The social form of Home – countries had not defined the rate between officer and people in public water conservancy investment. The public projects including water conservancy were closely related with the institution of taxes and corvee. Therefore, with the government's financial income decreased and financial embarrassment, the governments at all levels were forced to give up many inherent administrative powers gradually, which leaded to the exit of many local public affairs including water conservancy. As the middle class of country and rural society, the civilian power used their social network and various resources in the countryside to serve as the role of filling up.

3. 2. 1　The non – government groups gained profit themselves

Firstly, the non – government groups consolidated and rose in social rank by organizing public construction and social welfare undertakings, which was good for their participation in local affairs. The legitimacy of local authorities of the traditional Chinese society is neither granted by government and the control of private wealth, nor just by virtue of the glory of the degree. The establishment of the local authority must meet two basic preconditions, they must have the ability to promote a local community of interests and there must be a set of rules to maintain the cohesion of the community. In ancient society, the government emphasizes on the development of agriculture and despises the development of commerce, and civil society organizations precisely in the local agriculture had a direct interest in these water conservancy facilities, the construction of water conservancy facilities would not only promote the development of local agriculture but also to promote the stability of the local order of that it was also valued by the local officials as "political achievements", and then became the power officials could borrow. So that the local government was also ready to support their participation in these areas, which made their social status continue to consolidate and improve.

Secondly, there were special reasons for the rural elite groups, these people held the rights to launch proceedings against the official, act as the makers of village regulation, and master the right of organization and leadership to set up the local public welfare . These enabled them to participate actively in the construction of local water conservancy

Finally, the squire, clan, and farmers of these non – governmental groups had their own fields. Because water was the lifeblood of agriculture, they had the important interest relations directly with the water. They actively participated in the construction of water conservancy facilities becoming the important interests of the beneficiaries.

3. 2. 2　The influence of the Confucian culture

By the influence of Confucianism culture—regarding the whole society as one's own duty, the bereavement of parents and retired officials of the rural elites actively participated in the activities of the water conservancy, and striving to set an example for the local .

3. 2. 3　The power of religious philosophy

Buddhism preached "what goes around, comes around ", which echoes the thought of karma . The bridge in Buddhism had the significance of bringing salvation to all living beings in the reality, therefore many monks were willing to participate in the construction of bridges . In addition, as most people regarded the monks and priests as the selfless, it was easy for them to seek funding on the money and manpower compared to the officials and the general population . Therefore, the participation of monks and priests make the local public works be reached effectively and quickly.

In conclusion, the flood and drought disaster and the water conservation geography difference which formed the disaster with distributed non – uniformity and the formidable difference. In addition, in ancient China officials were appointed by the state directly, and this kind of appointment generally stopped at county level, which made the government have not the ability to govern every corner of the nation and provide a favorable opportunity for civil society organizations to take part in water management . After the Tang Dynasty , especially after the great historical transition in the Song Dynasty , non – governmental forces rose according to their local prestige , status, and their actively participation in the construction of water conservancy facilities , and to adjust the water disputes and the maintenance of local order in its unique way, which effectively promoted the development of China's ancient water conservancy.

References

Karl August Wittfogel,Shigu Xu, Ruisen Xi, et al. Oriental Despotism—Comparative Study on the Amount of Power[M]. Beijing: China Social Science Press,1989.

Xu Zhuoyun. Organization From a Historical Perspective [M]. Shanghai: Shanghai People's Publishing House,2006.

Goodall W. The Family[M]. Wei Zhangling. Beijing: Social Science Literature Press,1987.

The Monthly Report,Volume 1[M]. 1833(11):461.

Quote From the Officer Weir and Canal: Unique Chinese Ancient Water Resource Management System.

http://www. jmnews. com. cn/c/2004/08/10/11/c_341759. shtml

Huang Minzhi. Set of Social and Economic History of Buddhism in the Song Dynasty[J]. Taiwan Student Bookstore,1998:413 –442.

Wang Peihua. Three Discussions About Water Resources Between North and Northwest China in the Yuan, Ming and Qing Dynasty[M]. Beijing China Commercial Publishing House,2009.

Zheng Lisheng. The Clan Fight of Fujian Coast in the Ming and Qing Dynasty [J]. Journal of Fujian Normal University (Philosophy and Social Science Edition) , 2000(1):103 –108.

History of the Song Dynasty (15) One Hundred and Thirty – two Volunteers. 2939. Chuili Sun. Seeing Folk Strength in the Song Dynasty From the Function of Water Conservancy Undertakings[J]. Journal of Jinggangshan University,2006:91 –92.

Seeing Folk Strength in the Song Dynasty From the Function of Water Conservancy Undertakings [J]. Journal of Jinggangshan University,2006:91 –92.

Salty Chun Linan Volunteers Third Volume 18. http://tool. xdf. cn/guji/8450. html.

Hu Ju. Treasure a Four Bit – level Volume (16)[M]. Shanghai: Zhonghua Book Company, The Song Local Copy of the Periodicals, 1990.

Zeng XiaoHua. The Chinese Ancient Political System's Unique Types and Characteristics[J]. Journal of the Party School of CPC Zhejiang Provincial Party Committee,2005(6):23 –30.

Cao HaiLin. The Village for the Reconstruction of the Public Power Logic—the History of the Northern Village Experience Furnace Reflection and Realistic Choice[M]. Nanjing: Phoenix Publishing & Media Group, Jingsu People's Publishing House,2009.

Jia Hongwei. The Change of Water Resources Environment and Rural Social—in the Center of the Qing Dynasty Hanzhong County Drainage Water[J]. Historical Science,2005:14 –21.

Zhang Jing. The Grassroots Political Power—A Series of Problems of the Rural System [M]. Hangzhou: Zhejiang People's Publishing House,2000:24

Study on Historical Geography of Sanmenxia in the Yellow River Basin

Liu Youying, *Sun Liang*, *Chen Lizhi*, *Hou Junfeng* and *Duan Jingwang*

Management Bureau of Sanmenxia Multipurpose Project, YRCC, Sanmenxia, 472000, China

Abstract: The Yellow River starts from the eastern Tibetan Plateau, winding down, passing through Ningxia plain and the Loess Plateau, in the vicinity of Tongguan turns east into western Henan Canyon. Then the Yellow River continues to run east 130 km and arrives at the famous Sanmenxia. Since ancient times, Sanmenxia in the Yellow River Basin has been a section of the most precipitous gorge river and has been known as its unique natural landscape, human landscape, rich culture and history. Although in the end of the nineteen fifties, when the construction of the Sanmenxia Dam changed a part of natural landscape and a lot of human landscape, the profound Yellow River culture and the canal culture, which have formed several thousand years of precipitation, have been deeply infiltrated into the blood of the nation living on this land, and become the common spirit and the unique cultural gene of the whole nation. Therefore, the inspection, mining, sorting and refining to the historical geography and culture of Sanmenxia in the Yellow River Basin are great historically valuable and significant.

Key words: Sanmenxia in the Yellow River Basin, historical geography, the canal culture, the Yellow River culture

1 Original geography of Sanmenxia in the Yellow River Basin

Sanmenxia in the Yellow River Basin is located in the middle reaches of the Yellow River (in Fig. 1, Fig. 2), 130 km east of Tongguan. According to geologists' research, in ancient times, the area above Sanmenxia is a lake, east to Sanmenxia, West to Baoji, south to margin of North Qinling Mountains, north to Yumenkou – Longmen Mountain. Later as a result of tectonic movement and river erosion, Sanmenxia gradually formed its unique landform.

Fig. 1

Sanmenxia canyon with steep sides, the narrowest is only 250 m apart. Before there were three stone islands at the gap, from east to West in turn named People Door Island, God Door Island, Ghost Door Island. The river is divided into three strands by God Door Island and Ghost Door Island, which looks like opening three gates for the Yellow River, from east to west, called as

Fig. 2

People Door, God Door, Ghost Door. In northern and southern dynasty of China, famous geographer Li Daoyuan in his "Shuijingzhu" said: Before Yu ruled flood when hills stopped the water, so broke hills and let water go. The river divided, water run around hill, seen as a pillar. The river was divided into three parts, also called as three doors (as 'Sanmen' in Chinese pronunciation). Sanmenxia was hence named (in Fig. 3).

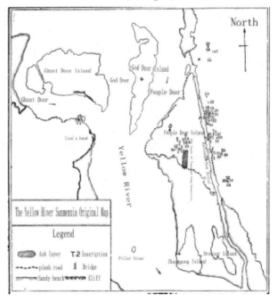

Fig. 3

Of the three doors, God Door the deepest, Ghost Door the most risky, People Door the shallowest, the only People Door can sail. People Door Island was to extend into the river before, and during the reign of Tang Kaiyuan, for ease of transport, a canal was dig in the southeast of island, namely Kaiyuan New River, full – length of 280 m, width of 6 ~ 8 m, the depth of 5 ~ 10 m. At the exit of three doors, there are three stone islands, from east to west, are the Dresser Island, the Zhanggong Island and the Pillar Stone, referred to as the "Three Door Six Peak". In the southern end of the Ghost Door, there is a fist shaped rock, looking from a distance, as a lion opening mouth, it is known as the Lion Head. Sanmenxia rock is mainly Diorite porphyrite, steel blue, hard texture. The Yellow River running from west become narrow suddenly at Sanmenxia, stone cliff on both sides, stones standing in the river, fast – flowing, swirling clouds, steep abnormally, therefore Sanmenxia has been the "three – door natural barrier" since ancient times. At south of Sanmenxia, there are Xiaoshan and Bear Ear Mountains etc, while at the north are zhongtiao, Wangwu mountains. On the high mountains and lofty hills, only the Yellow River connects East and West. So in the history of Sanmenxia it was of the military value, and it was said

"In the Yellow River between Shanxi and Shaanxi, Longmen is the most dangerous. While between Shanxi and Henan, Sanmenxia is the most dangerous".

In 1957, due to the Sanmenxia project construction, three islands of people, ghost and god, dresser, Kaiyuan New River are blown apart, leaving Zhanggong Island and Pillar Stone.

In local chronicles, "Shuijingzhu", poetry books written down in past dynasties, geographical landforms of Sanmenxia are recorded in the Yellow River Basin. Contrasting with "Sanmenxia Cannal Remains" written by Chinese Archaeology Institute of the Academy of Sciences in 1959, Sanmenxia original landform before the dam construction was same with local chronicles, "Shuijingzhu", poetry book written down in past dynasties. It shows the original Sanmenxia geography did not change much in two thousand years, only the buildings on it had changed.

2　The original human remains

In ancient times, the region above Sanmenxia was a Great Lake. It makes people believe, prehistoric Sanmenxia is a green water village. With "living beside water, fishing in water" habits of early human, in this land, it forms the original tribal clan as a unit combination. In 1957, on Ghost Island and ZhangGong Island Yangshao, Longshan and Yin Dynasty ruins were found, which indicated at that time human had been lived in this area. According to the terrain, the ancient human living here not only had inconvenient transportation, also were unable to survive. It can be supposed that the two islands may connect together with at least one shore land before Yin Dynasty, and later it separated with land.

3　Place name changes in history

About the name of "Three Door Six Peak", according to the relevant information on research, in the spring and Autumn Dynasty, it has been called Sanmen, and Sanmen is also called Pillar Mountain. Sanmenxia is referred to as Sanmen or pillar. Three "Doors" are in river centre, according to approximately North Song Dynasty cliff inscriptions, then three Doors named Ghost Door, YeCha Door and Jin Door. After Jin Dynasty, YeCha Door and Jin Door had gradually been replaced by People Door and God Door. In the Tang Dynasty, the Dresser Island was named Jin Door, ZhangGong island as Sandui, Pillar Stone as Tianzhu. To the Northern Song Dynasty, almost unknown the meaning of Pillar in ancient literature, the dresser is called the pillar for a period. It is transmitted wrongly, this may be because the pillar etched in the dresser, and pillar and Tianzhu are same only with one word different, very easy pseudo. Approximately in the Ming Dynasty, because of the destruction of pillar inscription, the dresser gradually wasn't called for the pillar and people moved it to Tianzhu. In addition, after the Jin Dynasty, the dresser was also called for LaoJunlu, maybe until late Qing Dynasty, the dresser returned the name. About the name of New River, the river was not named after digging, Tang Huiyao and old Tang book called "canal", TongDian as "Shiqu", Zizhi Tongjian as "Sanmen YunQu", a new book as "new gate", KaiTianChuanXinLu as "TianBao River", KaiYuan New River was named after North Song Dynasty, in the Jin Dynasty, was also named Princess River, and as Niangniang River at the late Qing Dynasty relative with folk tales.

4　Sanmenxia and canal transportation in the Yellow River Basin

Canal transportation is to transport the goods by boats in the river. The Yellow River canal transportation has a long history, from King Yu combating the flood. There is the legend "land line by car, water line by boat". In the Shang Dynasty oracle, there were records about this. Because of the big gradient and rapid flow of the middle reaches of the Yellow River, countercurrent navigation is difficult, especially the Sanmenxia natural barrier brings the obstacle to the Yellow River canal, causes huge transport cost. In order to ensure the normal operation of canal, every dynasty attached great importance to the Sanmenxia drainage treatment.

Historically, the real Yellow River canal and drainage treatment began in the Qin and Han dynasties. Qin, Han capital in Guanzhong, Guanzhong was called "Tianfu", but it produced little after all, so the imperial court still need to supply by Guandong. But in the ancient, because Xiao Han road was dangerous and difficult, land transport was very backward and the freight charge was also very heavy. The Han Emperor Liu Bang took the proposal "transport goods from everyplace to the west capital", food mainly rely on the water transport by the Yellow River to Weihe, west to Jingdu. The Yellow River transport must pass Sanmenxia. Emperor Wu of Han, in order to avoid the Sanmen natural barrier, canal transportation changed from Sujiang, Baoshui of Hanzhoung, through Qinling Mountains, land to Xieshui then into Weihe to arrive Jingdu. Tens of thousands of people cut the road more than 250,000 m, because there were turbulent stone in water. In the fourth year of Emperor Hongjia of West Han (17 BC), a large number of manpower excavated Sanmen, without success for "water rushing, the river is dangerous". From the late Eastern Han Dynasty to Wei – Jin Dynasty, based on summing up the experiences of cutting Sanmen in the Western Han Dynasty, they thought cuting Sanmenxia is impossible, from Emperor Huan of the Eastern Han, people began to cut the plank road. In AD 238 year, Kou Ci sent five thousand of craftsmen to repair plank road and remove river resistance from Sanmenxia to Wuhu. In AD 267 year, Emperor Wu of Jin Simayan sent Zhaogou, Leshi leading more than 5,000 people to repair and cutting the river, but the dangerous situation of the river had not improved. The plank road on the People Door's left side about more than 600 meters was built in this period.

The Sui and Tang Dynasties continued to capital in guanzhong. In June of AD 595 year, Emperor Wen of Sui commanded people on both sides of Sanmenxia cliff to carve plank road, so that the boatmen could pull ships countercurrent to enter beach. Later Sui Dynasty social was unrest, and some repaired plank road is destroyed. The establishment of Tan Dynasty, institutions to expand, personnel to increase, Jingdu need also to increases day by day, from the first year of Xianqing, it began to vigorously repair and cutting the Yellow River plank road. Tang Kaiyuan twenty – two years (AD 734 years), imperial court adopted Jing Zhaoyin and Pei Yaoqing's suggestion of in the east of Sanmen set Jijing warehouse, the west of Sanmen set Yan warehouse, between the two warehouses land road about eighteen miles was dug, from downstream of the river to transport grain to Jijing warehouse, after that through the eighteen land road to Sanmen West enter Yan warehouse, avoiding Sanmen risk, then by boat to Jindu. Tang Kaiyuan nineteen years (AD 741 years), in order to ship and avoid Sanmen, Li Qiwu led people adopting "Shaoshiwocu" method to cutting another canal in southeast rock of People Door, later referred to New River. Because of water level seasonal changes, the effect is not ideal, but played an important role in "pull ships by cable". Rebellion era, Sanmenxia plank road was destroyed, canal stopped. Later TangDaizong appointed Liu Yan to recovery the canal, Liu Yan vigorously undertook rectifying and reform, is mainly practiced to transport subsectionly. The Yellow River canal transportation had taken food shipped directly to Changan before, more time – consuming. Liu Yan set food warehouse at Qingkou, Heyin, Weikou, and according to the water level and the water flowing speed and so on, he took different ship types for transportation in different sections, segmented pick – up, transport time is greatly shortened, carriage ability into full play, accident greatly reduced, greatly improving the efficiency of transport. After Tang Xiantong, social unrest, canal was gradually cut off.

In Han and Tang Dynasties, the cannal transportation in Sanmenxia is the busiest section in our country inland rivers. After Tang Dynasty, economic center changed from the north to the southeast. After Five Dynasties, Song Dynasty located the capital in Kaifeng, and the direction of canal changed from Shanxi to Kaifeng for a period. Yuan, Ming, Qing Dynasty Capital in Beijing, economic main artery also subsequently moved to the North and south Canal. Sanmenxia lost its important role in transporting southeast grain to the capital, it evolved into the folk commercial activities or local economic activities, the boat through Sanmen is less and less. Afer Ming Wanli, it was given up regardless, disruption of development. To the Republic of China, as the railway was opened to traffic, the Yellow River canal transportation was only a small amount of transverse ferry and short range transport, longitudinal long – distance transportation interruption. It is worth mentioning that, Sanmen canal turned from upstream to downstream to transport Hedong Salt sold to Southeast around.

Known as above, the history of SanMen canal was lasted long, it was a grandiose project of delaying for a long time. The Yellow River canal transportation was economic lifeline in the Qin and Han Dynasties to the Tang Dynasty and Song Dynasty, and Sanmenxia was the throat on the lifeline, Sanmen canal in Chinese ancient transportation history had a very significant statue. SanMen canal took many times of renovations, also obtained certain result, but as the restrictions of historical conditions and technical level, couldn't fundamentally solve the problem.

From 1955 to1957, to coincide with the Sanmenxia dam building, Chinese Academy of Sciences and Archaeology Research Institute did cultural relics investigation near Sanmenxia, on the left of People Door found about 625 m, on the right of Ghost Door found the second ancient plank road, in the downstream they investigated for 10 km, on both sides of the Strait intermitten ancient plank road was found, and found the remains of Jijin warehouse, Yan warehouse and portion land road of eighteen miles, these canal heritage is precious relics to study China's east – west traffic history and the Yellow River canal history (in Fig. 4).

5 Changes of the Yu temple

The Legends say Sanmenxia was established when King Yu combating the flood, so in Sanmenxia to build a temple was one for worship, another for praying Da Yu God bless for hard transport.

About Yu temple's address, according to archaeological and left poems, in Tang and Song Dynasties, Yu temple was at People Door Island, the extant Yu temple is located at a platform on the north shore of the Yellow River, 1 km west to Sanmenxia. According to the local elderly people speaking, the highland was called as "Rich Mountain" and "Gold and Silver Mountain", presumably, approximately between Jin and Yuan or Yuan and Ming Dynasties , the temple on People Door Island destroyed then the temple migrated there. According to the record and textual research, on the People Door Island there was civilized temple, long gone. In SuiXingshi "Daheliushang", the highland on the north shore of Sanmenxia, had at least 4 temples, they were Laojun temple, Fire temples, Yu temple and Guanyu Temple (also called Three Door Temple), of which Yu temple existed the longest, the largest scale. Ming Wanli set "Sanmen Guanting tablets" recorded in inscriptions, "the mountain top, the Yu temple on north, before Chong Building, overlooking the river, the ancient scenery in eyes, and stretch as far as eye can see." According to the words, in the Qing Dynasty, Yu temple covered an equivalent to a village and had been preserved to the war of resistance against Japan, the Japanese destroyed. The site is now only the ruins. Residents of the neighborhood around the temple say, in Da Yu terrace, there were seven stone well dug. When Da Yu opened three doors, they were used for water. In history there is "Seven Well Three Door", around Yu Temple, the quantity of wells should be proportional to the daily demand of the local population. In those days, for begging lucky and eliminating disaster, numerous vessels anchored at the foot of the mountain before they went downstream through the SanMen canyon, boatmen, businessmen went to mountain, and showed their respect at the temple. After a long time, around Yu temple, the streets, markets formed. According to historical records, the Yu temple buildings, since the Tang Dynasty, had repeatedly to purchase and repair, costly. From the maintenance tablets, in Ming TainShun 3 years, Qing Qianlong Emperor 57 year there were the repair once, in 7 year, 17 year,18 year the Republic of China there were the repair once. The brilliant history of the Yu Temple proves the former peak and shipping scale of SanMen waterway from another side.

6 Poems, inscription and legend

From 1955 to1957, Chinese Academy of Sciences and Archaeology Research Institute found 107 sections of cliff inscription at People Door plank road, on both sides of New River, God Door Island, lion's head, Pillar Stone, Yu Temple, downstream plank road of Du Village. There are 14 tablets at Yu Temple, in which the well – known one is the Ming Dynasty Wanli inscription at lion's head "Steep cliff torrent stream, extraordinary as if done by the spirits". In addition, there are historical records, but have now lost his famous inscriptions. The book "Pillar Inscription" said by Wei Zheng of Tang Dynasty and written by XueChuntuo, the original engraved on the south of the dresser, "Words Achieve Ruler", article and calligraphy were superb. In the Northern Song Dynasty the rubbings were very rare, to YuanFu several tens of characters collapsed because of flood, probably in the early Ming Dynasty had lost dead. In 1957, in the archaeological investigation, archaeologists in the dresser had a careful search, has no writing. Liu Gongquan in Tang Dynasty had written "Pillar Inscription", suspected in the Yuan Dynasty it had been lost. However, the Northern Song Dynasty Huang Tingjian reproduced "Pillar Inscription", more than 600 words, spreaded so far. "Stand at pillar, see north Longmen; the old remains, a long source", it's the famous line in "Pillar Inscription".

Since ancient times, SanMen natural barrier is endowed with strong humanistic color, Sanmenxia left many beautiful legends. Legends say that the three doors were split by King Yu combating the flood, pillar stone is established for controlling the river by Da Yu, dresser is Wang Mu goddess' dressing place, in addition, also have the related legends about mother river, ZhangGong Island, alchemy furnace, rice soup ditch, live rock slope and other.

The SanMen natural barrier has attracted many people since the ancient times, including men of literature and writing, men and women with high ideals, emperor, national leaders etc. such as Wei Zheng, Liu Gongquan, Si Maguang, Yuan Haowen, Kang Youwei, Guo Moruo, He Jingzhi, Guo Xiaochuan, Dong Biwu and so on. They visited the scenery, expressed feelings, and left many eternal spiritual works at Sanmenxia in the Yellow River Basin.

7　Summaries

Sanmenxia in the Yellow River Basin has a long history and profound culture. In the end of 1950s, due to the Sanmenxia dam, a part of the natural landscape and a lot of human landscape were changed. However, the dam plays a great role currently in flood control, ice prevention, irrigation, power generation and so on. No doubt, if the dam was not built, Sanmenxia in the Yellow River Basin would have been become the representative of the Yellow River's iconic landscape, fully qualified for inclusion in the world heritage list. Therefore, the inspection, mining, sorting and refining to the culture and geography of Sanmenxia is great historically valuable and significant. The profound Yellow River culture and the canal transportation carried by Sanmenxia have been deeply infiltrated into the blood of the people on this land, and become the common spirit and the unique cultural gene of the whole nation.

References

Chinese Academy of Sciences and Archaeology Research Institute. Sanmenxia Canal Remains[M]. Beijing: Science Press, 1959.

Shu Shaochang, Ma Zili. The Anthology of Sanmenxia Scenic Spots[M]. Zhongzhou: Classics Publishing House, 1992.

Sui Xingshi. Daheliushang[M]. Shenzhen: China Book Publishing House, 2008.

Wei Jian, Shen Tongtai. Sanmenxia in the Yellow River Basin Folk Tales[M]. Beijing: China Yanhuang Culture Press, 2008.

A New – perspective Textual Research of Dayu's Flood Control

Huang Quansheng[1] and *Luo Meijie*[2]

1. Research Center for Reservoir Resettlement; The Yangtze River Three Gorges Development
Research Institute China Three Gorges University, Yichang, 443002, China
2. China Three Gorges University (CTGU) Water Culture Research
Institute, Yichang, 443002, China

Abstract: "Dayu (Yu, the Great) Controlling Water" is a famous legend relating to traditional Chinese water culture. On the basis of a thorough textual and logical research, this article makes a further study of Dayu's birthplace, methods of water control and contributions as well as the influence of the legend itself on today's construction of water conservancy project. Born in area of Shu (present – day Sichuan), Dayu married his wife and made his initial achievements in the area of Ba (the core area of the present – day Three Gorges areas), and succeeded in optimizing his methods of water control and controlling the floods. As his methods had became mature after gaining much experience in Ba, he then controlled floods in the whole country in a comprehensive way. The methods of water control of Dayu are featured with diversity and flexibility, among which the later generations have argued most about his utilization of "the five elements" and "Xirang". The five specific elements include water, fire, wood, metal and earth. In its narrow sense, water refers to the various forms of water in nature, wood is aquatic and terrestrial plants and animals, mental means all minerals on the earth except earth, fire refers to the sunshine, and earth the land and silt as well as the other materials on the surface of the earth, "Xirang" referring to the silt sediments in the rivers. Dayu, who has taken the advantages of the innate characters of water, such as softness, inclusiveness, mildness, powerfulness and kindness etc., owned the four virtues of water as "kindness, justice, bravery and intelligence", has exerted important influence on the following generations for his prominent contribution to water control and unification of the country and both his personality charm and the ideas and methods of water control as well are inspiring for water conservancy practice and projects in present era.

Key words: Dayu, water control, Ba and Shu areas, methods, cultural influence

    In the ancient times, the low productivity has led to a large number of myths and legends in the course of human's understanding and making use of the unpredictable world of nature, and because of the close relation between human and rivers, the myths and legends about floods or water control were more widely spread, which represent the determinativeness and perseverance of people in the ancient days in floods control. Among the myths and legends, the story of Dayu's Controlling Floods particularly demonstrates people's anxiety for water control and great will of people in the ancient times before the disastrous difficulties. According to the story, where there were great mountains and rivers where the floods occurred, there Dayu has made his efforts in water control. ① A passage quoted from *Guzi* in *Assorts of Art* states that when Guzi traveled to the East

---

① **Foundation Item**: Identified as the Ministry of Education's Arts and Social Sciences Youth Foundation Item "Water Conservancy History Research on Chuanjiang River Basin and Jingjiang River Basin (12YJC770041)"; Hubei Ministry of Education Item "Western Hubei Water Conservancy Culture Research and Water Resources Tourism Development Research (2012Q149)"; 2011 Yichang Science and Technology and Development Item "Yichang Communications Culture and Yichang Social Relationship Development Relationship Research (item number is A2011 – 302 – 33)"; China Three Gorges University (CTGU) Seeking Fund Three Gorges (Western Hubei) Water Conservancy Culture and CTGU Campus Culture Relationship Research.

Pool with Zihua, Zihua said to him: There are four virtues of water, take the water in the pool, for example, it is of kindness as to nourish all creatures, of justice as to wash away the dust and keep things in a clean state, of bravery and power in nature though it seems weak, of intelligence as to make the rivers into their forms with a self – control of its speed and flowage. Guzi replied: I have learnt much from you beside this pool. In this story, Zihua compares the water in the pool to a person with virtues. Actually, the four great virtues as the essence of the traditional Chinese values—kindness, justice, bravery and intelligence—has made a perfect manifestation on Dayu. His controlling the floods to save people out of danger stands for kindness; his justice was represented by the story that he once passed by his own home for three times but never got in during the years he was controlling the floods; his bravery is witnessed in facing the great floods fearlessly; finding new methods by comprehensive employment of dredging and blocking to tame the rivers and get rid of floods proved his intelligence. As a prominent figure of water culture in ancient China, Dayu deserves a further study Based on his legendary life, extraordinary qualities and virtues.

1　Place of Dayu's initial flood control

1.1　The area Shu—the birthplace of Dayu

Occurrence of the myths and legends relating to Dayu depends much on the social Background of the time of Dayu—as the old saying goes, "the times produce their heroes", i. e., the situation has made Dayu hero of flood control as the time demanded. According to the description in *Yao Anthority of The Book of History*, the flood disasters in the periods of the emperors Yao and Shun were extremely serious. It is stated in *Biography of Emperors in the Xia Dynasty* in *The Records of The Grand Historian that Gun*, the father of Dayu, was in charge of flood control; however, "it had been flooding for 9 a, and Gun failed to control the floods." Later Gun was killed for his failure when he was still taking charge of flood control. According to *Volume VI of The Works of Mencius*, having taken over the job of his father, Dayu "dug channels to lead the water into the sea and drove the dragons and snakes into the marsh, and had the water flow along the rivers⋯ Therefore, people could live on the field." Also according to *Biography of Emperors in the Xia Dynasty of The Grand Historian*, in the course of Dayu's control, he "traveled by chariot on earth and by boat in the river". *Ancient Chinese Traffic* recorded, according to the legend, Dayu once made a boat with a huge tree of one zhang (3.00 m) in diameter for flood control." There is also a textual record on this legend as *Volume 40 of Natural History and Sundries* quoted from *The Records of Shu*, which states that "Dayu wanted to make a dugout canoe, after he had learnt that in Mt. Nichen of Zitong County there was a catalpa tree of 1.2 zhang (4 m) in diameter, and ordered the craftsmen to cut down the tree. But the tree was afraid and turned into a kid, then Dayu became angry and had it cut down." According to such textual or legendary records, it is evidential that Dayu was born in the area of Shu and initially started his job of flood control in the area of Shu and then accomplished a nation – wide task.

In *The Annals of Wu and Yue*, it is recorded: "Dayu's home was in Xiqiang, exactly Shiniu, which is in Xichuan of the area of Shu State. In *The Records of Shu of The Chronicles of Huayang*, it is stated that "Shiniu was previously named Wenshan Prefecture. There Chongbo was in charge of flood control and married the woman called Youxinshi, who born Dayu in Kua'erping Terrace of Shiniu." In *Waterways* written by Li Daoyuan, the record goes as follows: "Dayu was born in Shiniu Village of Guangrou County in the state of Shu." According to *The Records of Geography of The Book of the Tang Dynasty*, "there was a mountain called Shiniu in Shiquan County, Maozhou Prefecture. Shiquan County belongs to present – day Long'an Prefecture, where there is a temple in memorial of Dayu. It is said that he was born here on the 6th of June in lunar calendar." *The Exterior Collection of Yang Shen'an* points out that "in the Sui Dynasty, Guangrou was renamed into Wenchuan, which is present – day Shiquan County", and "Shiquan of the Shu State, since it is the birthplace of Dayu, is also called 'Yuxue'⋯the two Chinese characters were inscribed on a stele by the famous poet Li Bai." Li XiShu's Little Discipline of Wenchuan also records that "⋯ there was a place called Kua'erping Terrace in Wenchuan County where there have been several families of the Qiang ethnic group. Dayu was said to have been born there. Hundreds steps away there is a temple in honor of Dayu by the Qiang people, and the temple is also called ' Qisheng Temple' ". I *Inscriptional Record of Dayu in Tushan* by Jia Yuan in the Yuan Dynasy pointed out that "Changyi, the second son of Yellow Emperor, married Shushanshi, and then had a son called Zhuanxu, Zhuanxu had a son named Gun, and Gun was Dayu's father. And Dayu got married in the area of Shu. " According to textual literature, the fact that Dayu was born in Shu is evidential. That is to say, Xia Culture may have originated in the state of Shu. Tong Enzheng holds that as early as in the Neolithic Period there started cultural association between Sichuan and the central area, on the Basis of which, Lan Yong puts with further inferences, there must have been some traffic route between the two areas. Even some experts suggest that the ancient Shu culture was developed earlier than the culture of the central area and the former is one of the sources of the culture the Xia Dynasty. Both the legends and textual records mostly illustrate that Dayu was born in the area of Shu, and it is there he had started his mission of flood control.

1. 2   Dayu and the area of Ba

Literature indicates that it is in the area of Ba that Dayu married his wife, exactly, in present – day Nan'an District of Chongqing City, where there's a temple called Tushan Temple. In Nan'an District, a book entitled *The Collection Dayu Culture Studies* which was specialized in research of Dayu Culture of Chongqing. In *The Chronicles of Huayang*, it is recorded, "when Dayu was in charge of flood control, the areas of Ba and Shu belong to Liangzhou Prefecture. Dayu married his wife in Tushan, and then was absent from home on mission for four years. Later his wife born a son named Qi who kept crying day and night; however, he had no time to see his child. During the four years, he had passed by his own home for three times, but due to emergency to deal with, he even didn't enter his home. The place where Dayu married his wife is just today's Tushan of Jiangzhou Prefecture, and there is a temple in memorial of the Emperor Yu". In *The Chronicles of Huayang*, it is stated that: "the Yu Prefecture (today's Chongqing), was the hometown of Dayu in Tushan. " According to *The Later Han Records*, it is cited in *Zuo Zhuan* (*The Zuo Documentary*) as "Ba was in Tushan, and it was the place Dayu married his wife". In *The Records of Counties of the Eastern Han Dynasty*, it states "Tushan was in Jiangzhou of the Ba Prefecture. " *Inscriptional Record of Dayu in Tushan* also pointed out that "Dayu was the people of Shu, as he was born and married there. " The historical facts that Dayu married his wife in Ba probably because the tribe under his leadership had received help and support from the native matriarchal society tribe of Tushanshi; hence they, the two tribes, united themselves by marriage. However, all these statements cannot exclude the possibility that he and his tribe fellows controlled flood in other places and united by marriage with other tribes. The textual proof in *Inscriptional Record of Dayu in Tushan* suggests that "*Tong Jian Wai Ji* records that ' Dayu got married with a Tushan woman and had a son named Qi, he went to meet the feudal chieftains in Tushan '; therefore, it can be inferred that Dayu married his wife and had a son, then he went to the south and met the other feudal chieftains. He married in one place, and met the chieftains elsewhere, the sequence is quite clear. Kuaiji is the spot where he stayed together with all his ministers or he died later, which was called ' Dayu Xue'; ···The areas of Dangtu and Kuaiji were not yet included in the territory of the country, therefore, it is impossible for Dayu to have married his wife there". It is inferable that Dayu gained more experience of flood control because of his marriage in the area of Ba, from which he traveled towards the lower – mid reaches of Yangtze River through the TG (Three Gorges) areas. Absolutely, it was a long – term process for Dayu to control the floods, and it covered more than one certain region. But it is of no doubt that it is in the area of Shu that Dayu started his career of flood control and in the area of Ba he gained more experience.

1. 3   From TG areas to the whole country

After Dayu married his wife Tushanshi and controlled the floods in Jiangzhou of the area of Ba, he went on with the mission of flood control in the Three Gorges areas where the disasters were even worse and the geographical conditions brought more challenges. In its introduction to the Yangtze River, Waterways states that the Three Gorges were excavated under the instruction of Dayu for drainage of water out of the Ba area; according the same literature, the Wu Gorge had already been under the governing of Dayu's subordinates, for his son, namely Qi later sent his subordinate Meng Tu to govern this area. When controlling the floods in the Three Gorges areas, Dayu received the help of the Goddess of Wushan. As stated in *Immortal Stories of Yongcheng*, "the Lady Yunhua was the 23rd daughter of the Queen Mother of the West, the younger sister of Taizhen Mrs Wang, named Yaoji … once during her return from her trip on the East Sea passing by the Mt. Wushan, which was tall and steep with exuberant woods on it, she was attracted by the beautiful scenery there…; then Dayu was staying at the foot of the mountain controlling the floods. He asked the lady for help, and was given a magic book, with which he could command several gods to help cut the huge rocks and drain the floods. Finally Dayu thanked the goodness as the problem had been solved. " This story has been widely known in the area of the Three Gorges. There is another story about Wangxia Peak in the WangxiaVillage of Wushan area: it is said that Yaoji killed 12 dragons in the river, and settled down there for she loved the beautiful scenery of the Mt. Wushan; there she helped Dayu excavated the Three Gorges, drove the wild beasts for the woodcutters, helped the farmers gain good harvest, planted herbs for the patients, kept the people in sailing safe. Day after day and year after year, her body turned into a rocky mountain. Everyday she was the first to welcome the morning glow, and the last to see off the sunset glow, therefore the mountain was named the Wangxia (morning or evening sunlight viewing) Mountain. " In Song Yu's *The Poem of Gaotang*, the preface introduces a story, "the late emperor dreamt a goodness to come to him and said that she lived in the Mt. Wushan. She had heard the emperor was in Gaotang, so she came here for an rendezvous with the emperor. " This poem taken from *The Selection* recited from *The Notes on the Book of Han*: "the Mt. Wushan is located in the south of Wuxian County of the South Prefecture … The clouds in the morning and rains make the beauty of the goddess" In *The Selection*, it is stated that the clouds in the morning and rains over the Mt. Wushan originally described the beauty of the goodness, but the later generations employed to refer to the matter of male and female. The phrase "Rains and Clouds over Mt. Wushan" became a synonym of love or matter between sexes. Some scholars believe that the phrase of "the clouds and evening rain" is a general description of the climate and scenery of the area. In fact, this phrase in connotation means generation and creation of things and creatures by water as well as giving birth to human civilization. In *Mountain and Sea Classics*, the text described as "over the remote and desolate place, there was a mountain called the Yunyu (Clouds and Rain) Mountain, on which there was a Koelreuteria tree rooted in a red rock, it is with yellow stem, red branches and green leaves. Dayu found that strange tree in Yunyu Mountain when he was cutting tress there. After that, all the emperors went to pick medicinal herbs there. " *The Poem of Gaotang* was composed after *Mountain and Sea Classics* in sequence and the idea of Dayu cutting trees in the Yunyu Mountain originally stands for flood control. In addtion, the stories that Dayu had received the help from the goddess in the course of flood control are of much similarity to the story on his marriage with Tushanshi. When controlling flood in the Three Gorges area, the magic book offered by the Goddess actually stands for the thinking of flood control, which basically includes Dayu's effective forms and methods of flood control.

Both the Tushan tribe in the area of Ba and the Goddess tribe in the Three Gorges area probably were tribes of later matriarchal society period that were good at flood control. When Dayu led his patriarchal tribes from the Mt. Minshan and Minjiang River area to the areas in control of the former tribes, they united together by inter – tribe marriages, and received help from the two tribes of matriarchal society for flood control skills. Dayu started his career of flood control in the Minjiang River area, exactly the upper reach of the Yangtze River, and via the areas of Ba and Shu, he had become very experienced in flood control ways and methods when he reached in the Three Gorges area, then he went to control water all over the country, and finally formed successfully a system of methods of flood control: drainage mainly with blockage as auxiliary means. Therefore, Dayu first went along the upper reaches of Yangtze River, the area of Ba in present – day Chongqing, and then to the Three Gorges area, finally to the lower – mid reaches of the Yangtze River and areas of the Yellow River and the Huai River. Later, Dayu united all the nations because of his contributions to flood control of the whole nation, and he finally established the state of the Xia Dynasty. Therefore, the future generations sincerely believed in that the one who could control water well could also bring peace to the country and the peace during Dayu's governing depended on the successful union among his tribe and the other ones, which actually indicates the historical tendency that tribes in the late period of the Patriarchal Society turned to unite with each other and finally made a state or states.

2 Methods of flood control by Dayu

2.1 The importance of leadership and main body in flood control

There was no doubt that Dayu was the hero of flood control, however, the flood was defeated by the wisdom and understanding united with his father Gun, assistant Yi and Houji, and various regions tribes like Tushan and Wushan goddess, and by the method of dredging mainly and stoppage secondarily. For instance, in the book *Guoyu*, it said: "Dayu controlled water successfully with the help of four great persons." In *Biography of Emperors in Xia Dynasty*, *the Grand Historian*, "So Yu, Yi and Houji took the order of Emperor Shun and lead the common people to begin the work of flood control." That is to say, Yu was the hero, so were the broad masses of the people. But the masses could only control water effectively under a good leadership. On the relics bronze of West Zhou Dynasty unearthed in 21*st* Century, there were 98 characters, phrases like "Dayu zhishui" (Yu controlled flood), "Yi zheng wei de" (govern the country by good virtue) were found on that. A great leader was not, nonetheless, appeared by chance. In *Rule Supreme*, *the Book of History*, it recorded the words of Qizi as "I heard that once upon a time, in order to deal with water, Gun used the five elements (fire, wood, water, gold, soil). The God got angry with that, and didn't give him the Nine Methods, so the country governing rules was destroyed. Gun was dead later, and Yu followed in his father's footsteps, then the God gave him the Nine Methods, and the rules for country government." Gun tried hard to control water, and tried to use the five elements materials to find an innovative approach which accumulated much experience for his son on flood control. Then he died, Yu followed his father and got the rules to govern country by the God. Qizi's words showed that in fact that before Gun's death, Yu had gotten the materials of flood control and its new methods from his father. He could get great achievement was because he followed his father's step. What's more, there couldn't be hero of flood control without the masses. As the leader of flood control, Yu was the inheritor of Gun, and the one who had the wisdom and contribution of the masses. All the statue, method and achievements were heredity and universal. The key point for him to be the leader of flood control and to get great achievement was that the methods of combining dredging and blocking together to dredge rivers, and his new innovations.

2.2 Flexibility of flood control by Dayu

In *one* chapter *of the Book of History*, it mentioned that Dayu had food and clothing poorly himself, while offered sacrifices to gods generously; Though lived in a tatty house, he invested a large sum of money to the irrigation works; He took carriage by land, boat by water, the muddy by sledges and a special shoe Wei by mountain road. The staff he always brought with was the water level tool and link marker used for opening up the Jiuzhou (nine provinces), opening roads, constructing dams, measuring the mountains. At the same time, he asked Yi to supply rice seeds and teach the masses to grow in dampness of a low – lying land, asked Houji to supply food the masses when had poor harvest, and asked the place had sufficient food to lend food to the poor places, and thus every place would get harvest equally. He looked over the products of various regions to define its contribution, and the situation of all the mountains and rivers as well. Yu also judged himself, "the flood was extremely large which had surrounded the mountains, flooded the hills, and devoured the masses. I've cut trees as the road sign along the mountain road by four medias of communication and sent birds and animals we caught to the masses with Boyi; I've also dredged the rivers of Jiuzhou to let the flood flow into the four seas, and the fassula in the field thus water enter the rivers; Planting with Houji was another thing I've done, we supplied grain and meat for the masses; What's more, I've developed trade for people to exchange their staffs. Therefore, the masses can settle down, and various regions begin to be governed." In *Biography of Emperors in Xia Dynasty, the Grand Historian*, it recorded as "Yu asked officers to conscript civilian worker for water controlling". In *the Works of Mencius*, "Dredging nine rivers to lead water flow into seas." In *Volume VI, the Works of Mencius*, "(He) dug channels to let the water pouring into the sea, droved the dragonsnakes to the marsh, and lead water flow along the rivers…thus the common people could live back on the flat field." Dayu lead Yi, Houji and other feudal princes finally defeated the great flood after years of arduous struggling by the ways of dredging, draining, digging, pulling and other methods. The method combining of dredging and blocking, was only one of the flood control methods, not the whole. In *Mencius Variorum*, it pointed as, "water, which can splash up even higher than a normal human as being hit, can flow back to the mountain as being blocked. Are these the nature of water? No, it's the situation." The nature of water is "flowing from higher place to lower". Thus in *Chapter 66, Taoist scriptures, Laotzu*: "the reason why the sea and ocean could be the king of water of hundreds of rivers from the valleys, is just because they are downstream the rivers." Dredging was the method which makes use of the nature of water—flowing down, thus lead it flow into river and sea, this was the general policy. And in *Master Sun's Art of War*, it pointed: "the tendency of water was avoiding the upwards (strong one), flowing to the lower." Ancient Chinese proverb said: "Man struggles upwards; water flows downwards."

Another nature of water is its inclusiveness, besides its nature of flowing to the lower. This kind of nature of water was also mentioned in the book *Guanzi*. In the ancient time, people usually took the advantage of its nature of inclusiveness to prevent flood, they first blocked the flood to lead the water flow into streams, rivers and lakes, the methods included dredging, leading, digging, blocking and etc., it seemed apparently dredging was the dominate method, in fact it was just a general principle, as from today's experience of Yellow River and Yangtze River water control, it is by using blocking to lead water flow into the sea. Even the water conservancy both home and abroad, none of them could make irrigation, transportation or power generation without building ditch and dam to block water to dredge it. That is to say, water control methods are neither good nor bad, it's up to the nature of water, we should control flood by blocking and dredging together or dredging to lead water or blocking to lead, they are all depend on its tendency, time, place and the flood. The methods should not run counter to the nature of water. In short, we should control water by its nature, but to get its potential as the precondition, dredging as the main idea. For instance, if we do not get the nature of water and run counter to it nature, the consequence would be quite terrible when water accumulated. In *LaoTse's Tao Teh Ching*, Chapter 78 mentioned: "there's nothing that can be as soft as water, but nothing could defeat it." In *Mencius Variorum*, someone used water to compare to personality, "personality, which is like the flowing water, it appears where you give it a way. There's no good or bad of personality like the water has no direction itself." People usually realize that when there's flood, but it's hard to practice it. Dayu, who realized and had the four virtues "kindheartedness, justice, bravery and intelligence" did realize that. And the virtues and nature of water have not only four virtues of human, but its nature of flowing to the lower, and its inclusiveness. We can dredge it by its nature of flowing to the lower, dredge and block together by its nature of inclusiveness, and dredge it depend on its tendency. Thus we could get great intelligence on controlling water.

2.3　On the materials of flood control: five elements (and "Xirang")

2.3.1　Flood control method: five elements method

It is pointed out in "Life of Hongfan" of *Book of History* that: "Five Elements represent water, fire, wood, metal and earth respectively. The water can moisten something downwards; the fire can burn something upwards; the wood can be bended or straightened; the metal can be made into different shapes; and the earth can be used to plant crops. Water downwards will be salt; fire upwards will be bitter; wood can be bended or straightened will be acid; the flexible metal will be hot; and the planted crops will be sweet. " This is the original five element theory which puts water first. In this theory, "People respect water and fire as priority, then earth. The wood and metal are the last. " Dong Zhongshu in Han Dynasty developed the five elements. He thought that: "There are five elements in the universe: wood, fire, earth, metal and water. The wood leads to fire, fire to earth, earth to metal, and metal to water. " (Details refer to *History of the Spring and Autumn*, *vol.* 10) Here we will not research how the Five Elements Theory developed into the theory of promotion and restriction of five elements. In this paper, we take water as the various forms water in nature, the fire as sunshine, wood as flora and fauna, metal as the minerals on the earth except earth, and earth as soil and sand exists on the earth and other floating materials.

According to the traditional five elements material concept, we find that, in terms of flood control, the head of the five elements, water, should be protected. This is needless to say. As for fire, too much sunshine will lead to global warming. If we want to protect water resource, we should prevent over – discharge of carbon dioxide and fade away of ozone. By doing these, we can reduce or stop the melting speed of glacier and snows, avoid the reduction of wet lands, marshes and lakes, and alleviate desertification. As for wood, we should take it as living beings. It is believe that wood is the center of the five elements. Because in *Book of History*, it is said that water leads to wood. Living beings must live in an environment with water. If there is no water, the living beings will die. However, the wood itself can also keep the ecological balance of the water resource. For example, the flora and fauna, especially the forest, can store water. If they can keep the balance of themselves, they can maintain the balance of water, stop the big changes of the metal and earth, and have positive effect on soil erosion, desertification and dust tornado. Different from water, metal is relatively stable. As an ancient saying in Three Gorges area goes that water is more dangerous than mountain. Once the flowing water is in power, it cannot be stopped. It seems that metal do little harms for human beings and water resource. It can turn into earth or water when affected by water, fire or wood. Most of the time, it turns into earth. When metal turns into water, it is easy to lead to some physics or chemical changes of water. Today's heavy water pollution has something with this matter. To exploit and use minerals in nature is the major suspect for water pollution. As for earth, it has close relationship with flood control, and the category is wide. The field, river load, sludge, desert, mud flat and dust tornado often are challenges and important materials for flood control. They will lead to soil erosion, desertification and petrifaction. Meanwhile, they are and always are important materials for flood control.

When regulating water, we should pay attention to the protection of forest flora, the dumping of industry and mineral waste and air pollution, and the climate warming and desertification. It is recorded in "Regulations of King" of *Book of Xunzi* that "To prevent plants and fishes from harm and keep them growing, a wise king should forbid people enter the forest with axe when it is the time for plants growing and prohibit people cast a net or poison into lakes when it is the time for fishes spawning. He should tell people not to miss the opportunity to cultivate in spring, weed in summer, harvest in autumn, and store food in winter so that the crops grow continuously and people can have enough food; he should forbid fishing in certain period so that the fishes grow continuously and people can have enough wealth; he should ask people not to miss the season to fell and cultivate trees so that the mountain will not be bare and people can have enough wood. " In "Life of Lianghui King" of *Book of Mencius*, it is also pointed out that "If military service and corvee get out of the road of agricultural production, fishing nets are not casted into lakes, and axes are taken into forest in specified season, there will be inexhaustible food, fishes and wood. " The ancient had already recognized that environmental protection is multifaceted. Flood control should be accordance with the laws of heaven and earth. For example, Wang Tingxiang pointed out in his "Book about Behavior" of *Cautious Words* that "The gas in heaven, earth and objects can produce entities by interacting. When the gas of sun in the universe combines with that of moon, they will be divided into two parts. One becomes sun, stars, thunder and lighting, the other becomes moon, cloud, rain and dew, which are seeds for water and fire. After being evaporated and tied, the water and fire will produce earth, and the metal and wood are the last ones which are produced by water, fire and earth. " In "Life of Hongfan" of *Book of History*, it recorded that "I heard that in the past, Gun blocked the flood with five elements. " This means that Gun, Yu's father, thought people should not just take water and earth into consideration in flood control. He considered five elements comprehensively, and dredged flood, which has laid foundation for Yu's flood control. That is to say, the flood control method that Gun left to Yu is not only to stop flood with earth, but also to stop flood with five elements. Taking climate, forest flora and mineral into consideration comprehensively, Yu completed the job of flood control in the end. Nowadays, the departments involve in flood control is similar to five elements. For example, the sectors relate to water are hydroelectric department, environmental protection department, weather department, geology department, city construction department, agricultural department, forestry department, shipping department, water supply department, and so on. However, the objects involve in today's flood control are not comprehensive. People often just solve the necessary problem, but not solve problems completely. To sum up, flood control must take each aspect of five elements into consideration, put the emphasis on cleaning sediment and sludge, and combine it with five elements method. If we regulate water by regulating the things relate to water, and do not emphasize one thing at the expense of another, there will no water cannot be regulated.

2.3.2 Flood control method: Xirang method

The old Chinese saying goes: "no matter what means the opposite side takes, the corresponding technique is always adapted". Therefore, in the *Mountain and Sea Classics* said: "Gun stole the soil of King Shun to bury the flood when deluge comes". However, Xunzi said: "Yu deposited the soil so that the world became flat". *Biography of Emperors in Xia Dynasty*, *the Grand Historian* recorded, "Yu lead the officials and common people to deposit the soil with Yi and Houji according to the command of Shun". The *Training of Huainanzi* recorded: "Yu buried the flood with the magical soil". All of them show that Gun and Yu ever controlled flood with the soil. Guo Pu explained *Shanhaiching* that the soil which grew limitlessly can bury flood. Lu Xun also considered that the soil was able to grow and never depleted in folklore. Obviously, the soil which Guo Pu and Lu Xun said did not exist. So what soil of earth is it?

It could be traced back to "Fu Tu" recorded by records of *Xunzi* and *Biography of Emperors in Xia Dynasty*, *the Grand Historian*, Fu means deposited. The verb "Fu" means "to distribute, adhere, or to make sticky, attached, spread, coated". "Fu Tu" is the verb – object structure, that is, Yu and his father controlled water with certain soil, which made the flood slowly subsidize and the whole world peaceful. What kind of magical soil was it? The book *Two of Forgotten Tales* recorded, "Yu tried to open ditch and lead water flowing into rivers, flatten the hillock and let the rivers go through with the help of yellow dragon which wagged its tail in the front and mysterious turtle which bore mud in the rear. " Here it is very easy to understand "Yu opens ditch and guides water flowing into rivers, flattens the hillock and lets the rivers go through". The sentence "Yellow dragon wags its tail in the front and mysterious turtle bears mud in the rear" seems to a little mythic. But you will not be confused if yellow dragon and mysterious turtle are taken as flood tools for bearing load or carrying luggage. The dragon and turtle are creatures in the water, and human beings cannot move on the water without means of boats. Thus the yellow dragon is likely to be the boat which used for digging silt in the water in the front while mysterious turtle to be the boat which loads silt. "Yu controlled flood by cart on the ground, by boat in the water, by sleighing in the mud and by Wei on the mountain" is recorded by the *Records of the Grand Historian*. That helps to prove that yellow dragon and mysterious turtle are the boats for controlling flood and digging mud. And "the soil which grows limitlessly" refers to the silt dug out from the river comes with flood the following year, that is why the soil grows limitlessly. The soil is the blue mud because of lot of forest vegetation and much vegetation sediments in the river at that time. That kind of soil which is rather fertile and very sticky builds levee together with stones and silt. But what is the relationship between objects and materials of controlling flood and this?

The "soil" is the key to disclose the success of flood control of Gun and Yu and development of human river civilization so far. It is also the key to solve the problem of the whole flood control with little effect, including constant flood control of the Yellow River and the the Yangtze River, the Yellow River continuing into the raised bed river, the Yangtze River as River Levee into Suspend River and Jing State having the same level as the Yangtze River. Today's many riverbeds rise up highly because of serious soil erosion. For instance, the ancient plank road hole has been under the water line or submerged by silt from Wuxi County to Jingzhu Dam in Ninghe of Three Gorges. Luckily, the mountains around Three Gorges are several thousand meters high so that the villages are not submerged. But the Yangtze River, the Yellow River, Huaihe River and the Haihe River are located in different stream segments in plains. People construct a dam with earth from other places after the arrival of silt in the middle and upper reaches and the riverbeds continue raised. Thus the Suspend River comes into being. If the same flood as several decades ago occurs the damages will be even larger because of the raised riverbeds. So how Gun and Yu and their assistants control flood?

The flood control heroes, Gun, Yu, Yu's son and their assistances used yellow dragons and turtles to excavate interest soil (the green mud of rivers) and salvage it on boats to control water. It was "Yu depositing soil" in *the Book of History*, which meant digging out the mud and put it on the bank. In this way, the dam level would be elevated and water flows deeper, which could shutoff water and solid dam, and then flood hazards, would disappear. Such kind of method was described in detail in *Volume VI. The Works of Mencius*, saying as "There was flood, Dayu was designed to control it. (He) dug channels to let the water pouring into the sea, droved the dragonsnakes to the marsh, and let water flow along the Yangtze River, Huaihe River, the Yellow River and Hanjiang River. Thus the common people could live back on the flat field without danger or drowned road. There were not any birds or beasts that could do harm on human. "Flood flew into the sea smoothly and the excess water was broken up to the sparsely populated lakes and swamp wetland after Yu emptied the sediment in the riverbed. Thus the villages, fields and cities would not be affected by flood for their levels are higher than the horizontal level. Yu took this measure to control flood of the Yangtze River, Huaihe River, the Yellow River, and the Hanjiang River and the scourge of flood disappeared eventually and people lived on the ground safely. Of course, Dayu built water conservancy. For instance, in particularly, people planted crops on the silt (interest soil and blue mud) dug from the Yangtze River, Huaihe River, Yellow River, and Hanjiang River, so later then common people had food to eat. It likes the description in *Biography of Emperors in Xia Dynasty. the Grand Historian* saying that Yu ordered his minister, Yi, to distribute all the rice and meat to people to eat; flood the field water into the river and flood the water of river into the sea; Ji and Yu gave people something which was hard to get to eat. Therefore, Sima Qian praised that "Yu was busy building ditches and conducting water···all the land in the country was well – collated and filled with fertility. " There were few fertilizers at that time, while the perennial interest soil was to be the fertilizers when people cultivated.

Take Yu for example, the flood control heroes had mastered the changing rules of the land and water. Firstly, salvaging the washed – away earth (blue soil) from the bottom of the river to maintain the depth and keep water flowing smoothly to decrease floods. Secondly, to use some silt build levee and the dam that would be enforced and heightened every year, which reduces floods. Thirdly, to return some silt flows into the farmlands nearby. As a result, it could prevent soil losing and soil thinning, maintain the near – farmlands' fertility, develop agricultural production and people could have enough food to eat. In this way, our national watering conservancy agriculture is improved creatively. Fourthly, to avoid water plants rotting and fermentating in the silt and to avoid water turbidity, salvage the mud from the bottom of river to maintain water's high quality. It not only provides high – qualified water for creatures especially for human beings in their agriculture, daily life and production, but also keeps the ecological balance fully. Earth moves in such kind of circle endlessly. Does not it be the legendary living soil? It is also one of the most important natures of human scientific use of water and embodiment of human's wisdom.

Huai Naizi says: "There is no object softer than water in the world. However, it is no marginal, unfathomable, edless and far to infinite···Living and disappearing with everything is the highest virtue of water. " *Laotze* said: "Water has the virtue of the nourishment to all the creatures. It profits everything but does not compete for interests with others. Thus, the noblest matter is no more than water in this world. " Water is impartial; the key is what human's attitude towards it is. Yu used the natural laws to service human being, which is reflected by the Du Jiang Yan Irrigation Project of Sichuan Province during the continuation of more than two thousand years. The situation in Nile River has been the same as Yu's living soil for thousands years. Nile floods from June to October every year regularly. In August, the river rises the highest, submerging the bank on both sides of the large fields. As a result, people have to move to higher places to have a temporary accommodation. When the flood subsides after October, it brings abundant soil of Nile, on which people plant the crops of cotton, wheat, rice, and dates, etc. It formed a "Green Corridor" in the arid desert regions. Egypt, the ancient civilized country for five thousand years, created the flourish Egyptian culture. That is the similar case in the basins of the Yellow River and the Yangtze River.

3 The influence of flood control by Dayu: controlling water and bringing peace to the country, creating Xia Dynasty

Dayu tamed the flood mainly by dredging combined with surroundings. Moreover, he honored ghosts and gods, constructing houses, digging channels, building boats, opening land and water transportation, draining off water into rivers, planting grains and regulating products from one place to another, etc. According to *The Book of History and Biography of Emperors in the Xia Dynasty* of *The Records of the Grand Historian* records on Dayu, The terrible waters surrounded mountains, and flooded hills and habitants. I, together with Boyi, took four kinds of vehicles, going along the mountain roads to send food to the people. I dredged nine rivers to pour into the sea and led the waters of ditches in the farmland to go into other rivers. Together with Houji, I planted crops to provide food for the people. In addition, I developed trade on mutual exchange of needed goods to settle down people. Therefore, each vassal state got well governance. Sima Qian gave Dayu the highest praise Dayu made his best to control floodwater and provided food to people. He went by his home for three times, but he didn't go into the door because of his work. He unified the country, so god gave him the Xuangui for his contribution to taming waters. By *Chuan Tzu*, Dayu regulated many rivers including 300 great ones, 3,000 branches and numerous small ones. Dayu, the hero of waters – taming in China, makes great achievements which are related with his love for people. By *Xunzi*, the ruler is the hope of people, whose behaviors influence the public concepts that decide the fate of a country. Therefore, to make a country strong, the ruler must pay much attention to talents, especially those able and virtuous ones who can govern a nation. *Xunzi* advanced the famous theory "the water that bears the boat is the same that swallows it up. " By *the Works of Mencius*, people submitting to one are as water falling down, and no one can hold back that turbulent momentum. The relationship between people and water explained in *Xunzi* is the same as those in *Xunzi*, which emphasis on the importance of people. Mencius warned rulers by water that one should carry out "Benevolent Governance", taking people's interests as primary. Finally, people are more than willing to submit to you because of your virtue. Also in *Guanzi*, it is indicated that water is the source of all things on earth including goodness, evil, virtue, etc. Dayu gained rich experience in taming water. His deeds laid a solid foundation for establishing Xia Dynasty and won the support of people. Dayu founded the first dynasty —the Xia Dynasty in China.

The book *Yu Gong* which records the deeds of Dayu has been listed as the Confucian Classic. *The Book of History* recorded Dayu unifying the country: Except dredging rivers, he arranged 6 officials to be in charge of tax system. He divided land into various grades and ranks, and people paid tax in accordance with their land ranks. He gave land to the people known for family name or virtue as a reward, and carried out political policies and advocated education. In addition, the state of Xia was vast in territory. So the God gave him Xuangui to praise his achievement. Both The *Records of the Grand Historian* and *Huainaizi* are seemingly the same as The Book of History that records Dayu's achievements on water – taming. Also in other chapter of *The Book of History and Record of Shu* as a volume of *The Chronicles of Huayang* records that thousands of leaders submit to Dayu with various treasures when he held the meeting in Kuaiji Mountain. This shows that Dayu has been the King of these feudal princes, and he conquered about 700 small countries. However, the small leaders passed the Three Gorges to Kuaiji with the fast waterway, which explains that the founding of Xia dynasty is closely related to Dayu's regulation on floodwater. Dayu unified the country by controlling flood to lay a foundation for a unified dynasty, which was a great success. An old saying goes, "All land on the earth belongs to the Monarch and all people for Monarch". From then on, Xia is the first dynasty in China. Yu became the first Emperor for his virtue. Moreover, he loved people and valued the role of people. Therefore, the rulers in all ages regarded water – control as a national policy. On the relationship between water and people, there is an old saying: "While water can carry a boat, it can also overturn it. " The one with virtues like those of water ruling people can be a wise man.

4 Myths of Dayu and the TG project

The amazing Three Gorges Dam stands in front of the Huangling Temple lying beside the Three Gorges canyon, was in the memory of Dayu of "kindness, justice, bravery and intelligence". Is this a coincidence of nature or an invisible arrangement by Dayu in the ancient time?

The Temple of Dayu (Huangling Temple) in the south of Xiba, a part of the Gezhouba Project, is a central island. As the place "Temple Mouth(Bazui) is "the part of Xiba that bulges into the river, it is named after its former Huangling Temple". Xiba Huangling Temple is recorded in *Donghu County Annals* and *Yichang County Annals*. These records and the Huangling Temple has become a cultural symbol of Three Gorges flood – control. This shows that, in people's opinion, especially those from Hubei and Sichuan provinence, Dayu was the god who drove the dragonsnakes, killed the evil dragon and controlled the flood. As described in *Volume VI. The Works of Mencius*, "There was flood, Dayu was designed to control it. (He) dug channels to led the water into the sea, droved the dragonsnakes to the marsh, and let water flow along the Yangtze River, Huaihe River, the Yellow River and the Hanjiang River. Thus the common people could live back on the flat field without danger or drowned road. There were not any birds or beasts that could do harm on human. "As the symbol of the water culture of the Yangtze River basin, Dayu was enshrined especially in Three Gorges area. Each county or town has a Temple of Dayu, people still consecrate Dayu in villages. For instance, in Xianfeng County Longping Town, it was rumored that there had been a lot of dragons···every time there was torrential flood, these dragonsnakes began to roar. In order to deter them, Li Qicheng and Gong Dexin with others raised money to build Temple of Dayu during Qing Emperor Daoguang Reign. The dragonsnakes kept roaring three days and left there after the temple being finish. Thus Longping got its name. " The Temple of Dayu in Jianlou Village, Wuxi County was "named after the consecration of Dayu. " The Temple of Dayu in Liangxiang Village, Wuxi County was "named after the consecration of Dayu of the early years. " The Temple of Dayu in Changyang County was "named after the Temple of Dayu in the early years. " In Jiulong Temple of Badong, "it was said that, Yu once passed there when he was controlling water, he found there are nine dragons half way up the mountain. Later, some people led by Gong Zhigang built the temple on that mountain and named it 'Jiulong (nine dargons) Temple'. " The Sanguan Temple in Yichang City, "was built in honor of the three emperors Yao, Shun and Yu in Qing Dynasty, thus it was named' Sanwang Temple' (three emperors). " Feudal rulers usually called themselves the Son of Heaven(the true dragon and Son of Heaven). Actually, the folk thought that Yu was the Dragon for he was good at controlling water. Moreover, Chinese nation takes the dragon as the god of water, thus only the rulers of "kindheartedness, justice, bravery and intelligence" could be the 'Dragon', or else the common people could kill him as the water could flood.

Of the Three Gorges building sites Sandouping and Gezhou Dam, the later one and Huangling Temple were selected as the building site of Three Gorges as early as in the Republican period, "a power station should have pump sump and outlet drain. Therefore after investigation and consideration, we chose Gezhouba and Huangling Temple that near Yichang as the suitable places of hydraulic electrogenerating, Since the Huangling Temple has a geologic structure mostly of granite, it also provides a possibility to build overflow dam there. From the early survey we select the place near Huangling Temple, which can be a complement to the Gezhouba plan. " In Zhuge

Liang's *Story of Huangling Temple*, it recorded: "Yu spent nine years to control flood. Later, people built a temple called Huangniu Temple in honor of his great achievement. "So, there must be some telepathy between Yu's great achievement thousands of years ago and today's Three Gorges Project.

It has been considered as the symbol of Chinese industrious wisdom, bravery in conquering nature. It is also the symbol of Chinese culture, especially Chinese water culture. Dayu left a lot places and legends in Three Gorges area of Kuimen, Huangniu Gorge, Bood – awarding Place, Dragon – cutting Place and the Twelve Peaks of Mount. Wushan etc. , which have been the hope of Chinese nation to control flood of the Yangtze River since the ancient time. Now, the Three Gorges Project, which turns the legends into fact, is a historical monument of the Chinese science and technology culture and human culture, reflects the spirit and wisdom of Chinese Yu culture in thousands of years, and replays the historical spirit of Yu who spent thirteen years to control flood and bring peace to the country. The Three Gorges Project is the highest embodiment of Chinese nation's subjective initiative and creativity in the process of nature remaking during preventing and controlling water, which also interprets the Chinese nation's spirit of "as heaven maintains vigor through movement, a gentleman should constantly strive for self – perfection. " Through the construction of the Three Gorges Project and its water culture, the Three Gorges Project is in its process of forming the symbol of the national culture and spirit of Dayu controlling flood. Thus, in Mao Zedong's poem Swimming, it mentioned: "Walls of stones will stand upstream to the west. To hold back Wushan's clouds and rain. Till a smooth lake rises in the narrow gorges. The mountain goddess if she is still there. Will marvel at a world so changed. "This great momentums verse expresses Chinese people's lofty sentiments and aspirations of construction of Three Gorges Project and remaking the nature.

Gezhouba (Xiba) and Zhongbaodao Island is described as "the gift of god to China" by foreign experts as they have profound historical culture and advantageous geographical position and geological conditions. This gift was perfectly explicated in 1924 Sun Yatsen's speech, "Kui gorge has extremely large hydraulic power. Some people once investigated the hydraulic power of the section from Yichang City to Wanxian County, which was found as large as more than three thousand horsepower of the electric power. This kind of electric power, which is much larger than today's other countries', could not only supply power to national trains, trolley bus and various factories, but manufacture major fertilizer. "As early as in his *General Plan to Found a State* Ⅱ, part four "improving the existing water and the canal", he put forward that "About 140 mi, from Yichang City to Sichuan Province, was called the Red Basin by geodetic scientists. "Today's "gift of god to China" actually was acquired by the spirit of Dayu of our Chinese nation, and the achievement we got today of water controlling will shock our ancestor Yu in thousands of years ago. Our water conservancy undertakings will get the blessing of our ancestor Dayu in the Huangling Temple (Temple of Dayu) near the Three Gorges Project, and what we need more is the spirit of Dayu who never give up, today's water controllers should be "kindheartedness, justice, bravery and intelligence".

5 Conclusions

Both water controlling and using by human are for its effective functions. Water has its beauty in softness, inclusiveness, unyielding characteristic, and kindness, but it also concludes nefarious sides with its fierce characteristic that can cause drought, water lagging etc. To take the advantageous sides of water and avoid its disadvantageous sides is what we call water conservancy— the subject of human water culture. When mentioning the flood control by Dayu, today's scholars usually only point "dredging" to criticize today's water conservancy project as building dam "blocking" water. But this garble understanding is unfair from the history of the flood control by Yu. We should, therefore, face the history objectively and fairly. The myths and legends of flood control by Dayu is a crystal of water culture in our nation, his great performance was the result of complete understanding of "kindness, justice, bravery and intelligence".

References

The Yangtze River Watershed Planning Office. A Brief History of Yangtze River Water Conservancy [M]. Beijing: Hydraulic And Electric Press,1979.

Wang Chonghuan. Ancient Chinese traffic[M]. Beijing: Commercial Press, 1996.

(Yuan Dynasty) Jia Yuan: Inscriptional Record of Dayu in Tushan, by Nan'an District, Chongqing City: DaDayu Culture Album.

Tong Enzheng. The Ancient Ba and Shu[M]. Chengtu: Sichuan People's Press, 1979.

Lan Yong. The History of Sichuan Ancient Traffic Routes [M]. Chongqing: Southwest Normal University Press, 1989.

Li Binghai. The Research of XiaChu Culture[J]. Tian Fu New Idea, 1990 (6).

(Yuan Dynasty) Jia Yuan: Inscriptional Record of Dayu in Tushan, by Nan'an District, Chongqing City: DaDayu Culture Album)

(Former Shu Dynasty) Du Gangting: Volume III Lady Yunhua, Immortal Stories of Yongcheng.

Leading group of Wushan County: Page27, Gazetteer of Wushan County, Sichuan Province, June 1983.

Editorial Board of Reprint of Records of Wushan County: Page 398, Kuanghsu Records of Wushan County, Volume 32, Poem Tang,1988 Edition.

Liang Xiaotong. Note – makers: Li Shan (Tang Dynasty), Page 876, Part 19th, Poem of Gaotang, Selected Work[M]. Shanghai: Shanghai Chinese Classics Press.

Lin Juan, Zhang Weiran. Goddess of Wushan Mountain: A Literary Imagery of the Geographical Origin[J]. Literary Heritage, 2004(2).

Huang Quansheng, Huang Baiquan. A View of Water Conservancy Culture from the Perspectives of Its Use and Properties[J]. China Three Gorges Tribune, 2011(2).

Fang Yizhi, Volume I, Physical Knowledge. Complete Works of Lu Xun, Volume II.

Place Annals of Yichang City, by Place Name Commission of Yichang City Hubei Province, Page 62, Feb, 1984.

Place Annals of Xianfeng County Hubei Province, by Place Name Commission of Xianfeng County, Page 152, July, 1984.

Place Annals of Wuxi County Sichuan Province, by Place Name Commission leading group of Wuxi County, Page 176,Oct, 1982.

Place Annals of Wuxi County Sichuan Province, by Place Name Commission leading group of Wuxi County, Page 335, Oct, 1982.

Place Annals of Changyang County Hubei Province, by Place Name Commission leading group of Changyang County, Page 48, Oct, 1982.

Place Annals of Badong County Hubei Province, by Place Name Commission leading group of Badong County, Page 410, Aug, 1983.

Place Annals of Yichang City, by Place Name Commission of Yichang City' Hubei Province, Page 73, Feb, 1984.

Song Xishang. A Survey of Hydroelectric Power of the Yangtze River[J]. Quarterly of Yangtze River,1933(1).

Sun Yatsen. The People's Livelihood of the Three People's Principles[J]. Complete Works of Sun Yatsen, 1986(9).

Sun Yatsen. Plans of National Reconstruction[M]. Beijing: Huaxia Press, 2002.

I. Sediment Management of High Silt – laden Rivers and Reservoirs

Resource Utilization of Reservoir Sediment and River Health

Jiang Enhui, *Cao Yongtao* and *Li Junhua*

Yellow River Institute of Hydraulic Research,
MWR Yellow River Sediment Lab, Zhengzhou, 450003, China

Abstract: Rivers are closely linked to social development. But sediment problems of reservoir always affect the healthy operation of rivers. In China, rivers, such as the Yellow River, the Yangtze River, the Pearl River and so on, all are faced with the problem that sedimentation leads to reducing reservoir storage which impacts the reservoir benefit and even makes reservoirs scrapped.

Now, sediment processing technologies of reservoir, including pipeline sediment technology, efficient desilting and dredging technology of deep reservoir, provide technical support for reservoir sediment transport. At the same time, such as silt back and solid barrier, silt filling embankment river, and making large rescue stone, steamed brick, sediment baked brick, also provide the effective way for sediment resources utilization.

Combined with the existing sediment transport and resource utilization technologies, considering the sediment – laden rivers of maintaining reservoir capacity and reducing sediment of the lower channel, the sediment treatment should be mainly resource utilization and the sediment in different reservoir area can be used in different purposes. The country should advance into a part of the fund as the start – up capital, and then use the profits to maintain long – run after the implementation of sediment resource utilization and reservoir power generation. For less sand river to maintain the stability of downstream channel and ensure the demand such as navigation, the sediment should be dealt with reasonable utilization of reservoir sediment release technology, keeping the reservoir sediment from reducing owing to the reservoir construction in recent years, and the investment mode of government investment for public welfare should be give priority to, reservoir power generation benefit compensation mode of an ounce of prevention is worth.

Key word: sediment of reservoir, sediment resource utilization, river health, operation mechanism

1 Preface

There are numerous rivers in China, which faces serious soil and water loss problem. Hydro – junctions are constructed on most of rivers. Besides playing comprehensive benefits like flood control, irrigation, water supply and power generation, all of the hydro – junctions will change the boundary conditions of sediment transportation, especially in the sediment – laden rivers. Therefore, what people value most is the benefits of sediment retaining and sedimentation reduction which played by the junctions.

Obviously, to sedimentation, there is a big difference of function, benefits and consequences of reservoirs between sediment – laden rivers and less sand rivers. At the same time, people's understandings to river health are different. Sediment – laden rivers have less runoff and too much sediment and there occurred the serious sedimentation in the lower reaches. So, reservoirs are hoped to retain more sediment in order to reduce sedimentation and flood protection pressure of the lower reaches. Also, reservoirs are hoped to chronically play their benefits of flood control and sedimentation reduction for river health considerations. But for less sand rivers, there relatively exist more water and less sediment. In order to ensure the downstream navigation safety and diversion safety and to maintain ecological health of estuary areas, reservoirs are hoped to release more sediment to avoid the beaches collapse lost, river channel incision and estuarine areas decreases because of sediment supply shortage. How to deal with reservoir sedimentation is an important problem in the process of reservoir operation and dispatch, which is related to reservoir service life and various functions playing limitations and has been concerned for a long time.

With the development of economic society and the deepening understanding of sediment resource properties, sediment, as a kind of silicate – based resources, is more and more valued by related research institutions, departments and communities. In recent years, because of advanced technology, utilization ways of sediment resources gradually increase and sediment resources are largely needed by the society. Besides traditional utilization as building materials, there are many new sediment utilization ways like making environmentally friendly building materials, creating land with fine sediment, extracting useful metals from coarse sediment etc.

What is more important is that, facing with continuous sediment, a new solution, changing from passive to active, must be formed. Due to the natural sorting of flow, reservoir becomes the best natural place for sediment sorting, which offers preconditions for sediment resources utilization. Development of sediment resources utilization technology and enhancement of economic society sediment need provide the possibility for large reservoir sediment resources utilization. If the sediment in reservoir can be used properly, it will contain the trend of reservoir sedimentation to varying degrees and will improve the position of sedimentation. It is the most direct and effective way to solve reservoir sedimentation problem and is the significant demand to give full play to reservoir and to maintain river health. As the end result of reservoir sedimentation processing, sediment resources utilization is its sublimation and final way. Sediment resources utilization is not only the significant demand to give full play to reservoir and to maintain river health, but also in line with our country's industrial policy. It has significant meanings of society, economy, environment, ecology and people's livelihood.

2 Overview of reservoir sedimentation in China

In China, reservoir sedimentation is a very serious problem. According to statistics, annual average silting loss of storage capacity is $8 \times 10^8 \text{m}^3$ and average annual siltation rate of reservoir is 2.3%, which is 3.2 times as reservoir sedimentation velocity in America.

2.1 Reservoir sedimentation in the Yellow River

The Yellow River is famous for its large amount of sediment. The quick sedimentation velocity of reservoirs in the Yellow River shocks the world. And it brings a series of reservoir sedimentation problems. Sanmenxia Reservoir, the first hydro – junctions constructed in the main stream of the Yellow River, is forced to rebuild for many times and to change its operation mode because of serious sedimentation. Sedimentation makes the reservoir still can not give full play to its functions.

During 1990 to 1992, a survey of reservoir sedimentation in the Yellow River was conducted. Until 1989, there were 601 reservoirs in the Yellow River. Total storage capacity of those reservoirs was $522.5 \times 10^8 \text{ m}^3$ and the silting loss of storage capacity was $109.0 \times 10^8 \text{ m}^3$, which was 21% of the total storage capacity. Silting loss of reservoirs in the main stream was $79.9 \times 10^8 \text{ m}^3$, which was 19% of those reservoirs total storage capacity. While in the tributaries, silting loss of reservoirs was $29.1 \times 10^8 \text{ m}^3$, which was 26% of those reservoirs's total storage capacity. Now, in most reservoirs of the Yellow River, sedimentation capacity is more than 50% of the total storage capacity, which greatly restricts the play of reservoir effectiveness and even makes reservoirs scrapped. Sanmenxia Reservoir, its total storage capacity is $96 \times 10^8 \text{m}^3$, but sedimentation was $71 \times 10^8 \text{ m}^3$ in 2005. Yanguoxia Reservoir, its total storage capacity is $2.20 \times 10^8 \text{ m}^3$, but sedimentation was $1.70 \times 10^8 \text{ m}^3$ in 2005. Bapanxia Reservoir, its total storage capacity is $0.50 \times 10^8 \text{ m}^3$, but sedimentation was $0.25 \times 10^8 \text{ m}^3$ in 2005. Tianqiao Reservoir, its total storage capacity is $0.7 \times 10^8 \text{ m}^3$, but sedimentation was $0.5 \times 10^8 \text{ m}^3$ in 2005. Other reservoirs existed less sedimentation. Longyangxia Reservoir, its total storage capacity is $247.0 \times 10^8 \text{ m}^3$, but sedimentation was $4.0 \times 10^8 \text{ m}^3$ in 2005. Liujiaxia Reservoir, its total storage capacity is $57 \times 10^8 \text{ m}^3$, but sedimentation was $16.52 \times 10^8 \text{ m}^3$ in 2005. Wanjiazhai Reservoir, its total storage capacity is 9.0×10^8 m^3, but sedimentation was 0.9×10^8 m^3 in 2005. Xiaolangdi Reservoir, which was operated in 2000, its total storage capacity is 127.5×10^8 m^3, but sedimentation was 26.3×10^8 m^3 in 2011.

2.2 Reservoir sedimentation in the Yangtze River

Although the sediment concentration of Yangtze River is 0.54 kg/m^3, annual sediment amount is about 5×10^8 t because of abundant water. So storage capacity reduction due to sedimentation also exists in the reservoirs of the Yangtze River. According to statistics in 1992, there were 11, 931 reservoirs in the upper Yangtze River, with the total storage capacity of 205×10^8 m^3. Total storage capacity of large reservoirs was 97.5×10^8 m^3. Annual sedimentation amount of reservoirs was 1.4×10^8 m^3 and annual sedimentation rate was 0.68%. Annual sedimentation rate of large reservoirs was 0.65%.

Similar to reservoir sedimentation, lakes in the lower Yangtze River like Poyang Lake, sedimentation is also a serious problem during flood seasons. Intersectional region of Poyang River and Yangtze River, of which length is 7,000 m and width is 1,700 m. Due to reverse irrigation of Yangtze River during flood seasons, sedimentation thickness is $50 \sim 100$ cm and sedimentation rate is $1.6 \sim 3.1$ cm/a. In addition, based on statistics during 1954 to 1985, annual average sediment amount brought into the Poyang Lake by five lakes around it was $2.419,8 \times 10^8$ t/a. About half of the sediment was silt in the lake. Total sedimentation amount was $1.209,8 \times 10^8$ t/a and average sedimentation height was 2.2 mm/a.

2.3 Reservoir sedimentation in the Pearl River

In the Pearl River, there are 32 large reservoirs, 279 medium – sized reservoirs and lots of small reservoirs. All the reservoirs face sedimentation problems to varying degrees, which affect the play of reservoir effectiveness and even makes reservoirs scrapped.

Sediment concentration of North and East River is small and storage capacity loss caused by sedimentation is not very serious. West River is the less sand river. But because of abundant water, its sediment discharge is large and the annual average amount is 7.18×10^7 t. So, storage capacity loss caused by sedimentation is still a serious problem. Baidonghe Reservoir was built in 1958 in Baise Guangdong Province, of which the sedimentation amount was 1.181×10^7 m^3 and was 32% of its effective storage capacity. Lanma Reservoir in Rongshui County became scrapped for the reason that effective storage capacity loss caused by sedimentation is 80% of the total amount. In the recent ten years, more than 1,400 hilly ponds were scrapped for sedimentation. Bianyang Reservoir was built in 1958 in Luodian County in Guizhou Province. Its designed effective storage capacity was 8×10^5 m^3. But the sedimentation was 6×10^5 m^3 in ten years and it made dam collapsed and rebuilt. Guanlu Reservoir in Zhenfeng County, of which the storage capacity was 5×10^5 m^3, became scrapped for sedimentation 12 years after it built. Xiaoshuiba Reservoir in Luliang County in Yunnan Province was built over 20 years. Its sedimentation amount was 7.12×10^6 m^3, which was 36% of its effective storage capacity. All the reservoirs above were built in West River. None of the reservoirs were set desilting facilities, so sedimentation could not be discharged, which made the serious effective storage capacity loss.

2.4 Reservoir sedimentation of other rivers

Rivers in other areas, especially in north areas, mostly originate or flow through the loess region where rainstorm always occurred in flood seasons and soil and water loss problem is very serious. Sediment concentrations of those rivers are quite high. So reservoirs built in those rivers all are faced with serious sedimentation problems.

Baishi Reservoir was built in Dalinghe River in Liaoning Province and its main projects were completed in 2000. Its total storage capacity was 16.45×10^8 m^3. But in 2004 its sedimentation was already 3. 293, 4 × 10^7 m^3 and 5 × 10^6 m^3 of sedimentation was in the tributaries. Tanghe Reservoir, built in the main stream of Tanghe, of which the total storage capacity was 7. 07 × 10^8 m^3 and its annual average sedimentation amount was 2. 828 × 10^5 m^3 since 1991.

Beneficial storage capacity of Fengman Reservoir in the Songhua River is 61. 7 × 10^8 m^3. Due to deforestation and steep wasteland, sedimentation of reservoir significantly increases. Average annual sedimentation amount in the early years was only 1. 45 × 10^6 t, but now is 6. 23 × 10^6 t, which is 4. 3 times as that in the early years.

Hongshan Reservoir in Inner Mongolia, its check flood level was 455. 10 m; correspondent storage capacity was 25. 6 × 10^8 m^3; normal water level was 433. 80 m and beneficial storage capacity was 8. 24 × 10^8 m^3. Its dead water level was 430. 30 m and dead storage capacity was 5. 1 × 10^8 m^3. During the 40 years from 1960 to 1999, total sedimentation amount of reservoir was 9. 41 × 10^8 m^3, which was 36. 8% of the total capacity. Sedimentation under the dead water level was 3. 46 × 10^8 m^3. The siltation loss rate was 67. 8%. Sedimentation amount of beneficial storage capacity was 1. 73 × 10^8 m^3. The siltation loss rate was 55. 1%.

Sedimentation amount of Guanting Reservoir in Yongding River in Hebei Province was 0. 842 × 10^6 t during 1950 to 2000. Because of the large sedimentation in reservoir, riverbeds of Yongding River, Yang River and Sanggan River continuously increased. The original underground river turned into "the aboveground river". Now, the No. 8 Bridge of Yongding River increased 13. 4 m. Riverbed of Jiahecun in Yang River increased 5 m. Riverbed of Jijiayingcun in Sanggan River increased 4 m. And Shuangshucun increased 3. 7 m where the riverbed is 1. 5 ~ 2 m higher than the external ground of the dike.

Daheiding Reservoir in Haihe River, its total storage capacity was 3. 37 × 10^8 m^3 and its effective storage capacity was 2. 24 × 10^8 m^3. Reservoir was impounded in 1979. And in June 2005, accumulative sedimentation amount was 0. 580, 6 × 10^8 m^3 and average annual sedimentation amount was 2. 23 × 10^6 m^3. Panjiakou Reservoir, its total storage capacity was 29. 3 × 10^8 m^3 and average annual sedimentation amount was 17. 2 × 10^6 m^3. Since it was operated in 1980 until 1994, total sedimentation amount was 1. 3 × 10^8 m^3, which was 4. 5% of its total capacity. Its average annual sedimentation amount was 8. 65 × 10^6 m^3.

Yiganqi Reservoir in Ye'erqiang River in Xinjiang Province, its designed storage capacity was 6. 2 × 10^7 m^3. During October 1956 to June 1966, sedimentation amount was 4. 44 × 10^6 m^3 and average annual capacity reduction rate was 0. 822%. Until June 1989, sedimentation amount increased 7. 66 × 10^6 m^3 and average annual capacity reduction rate was 0. 619%. Till August 2003, sedimentation amount again increased 2. 32 × 10^6 m^3 and average annual capacity reduction rate was 0. 340%. Storage capacity now is 4. 758 × 10^7 m^3, effective storage capacity is 4. 308 × 10^7 m^3 and dead storage capacity is 0. 45 × 10^7 m^3. Average annual capacity reduction rate is 0. 574% during the 47 operating years.

Geheyan Reservoir in Hubei Province has been operated for almost 10 years since 2005. Storage capacity of 200 m elevation correspondingly reduces 0. 971 × 10^8 m^3 and average annual sedimentation rate is 0. 35%.

3　The problem of the reservoir sedimentation

The sediment is deposited greatly, which will bring great effect to normal function of reservoir and river health is mainly observed in the following aspects.

3. 1　Influence on up – stream of reservoir area

The reservoir deposition is so excess that the end – position of reservoir deposition is extended to up – stream to submerge the end – blackwater area and expand flood losses. For example, the end – position of Kwanting Reservoir in Yong Ding River has extended 10 km, which makes underground water level rise 3 ~ 4 m and saline – alkali soil area of two banks expand 14 times.

Meanwhile, the sediment deposition of varying reservoir backwater region always results in the shortage of navigable depth and width, which influences navigation, especially to some large reservoirs. River situation of the varying region is in an unstable state and adverse for navigation due to the large variation of reservoir water level.

3.2 Influence on reservoir area and dam operation

Firstly, reduce reservoir capacity. The original beneficial goals can not be reached and some reservoirs have to change operational mode in order to slowdown deposition and to prolong the service life of reservoir.

Secondly, pollute water quality. Because sediment is the organic and inorganic pollutants' carrier, those deposited in the reservoir has influence on water quality.

Thirdly, influence normal operation of buildings in front of dam. Sediment, which is deposited in front of dam region, such as hydropower station water inlet, upper and lower approach channels, lock chamber etc. , may block sluice hole so as to threat engineering safety.

3.3 Influence on downstream channel and relationship of river and lake

Firstly, influence the downstream channel stability. After sediment retaining, discharged clear water can make downstream channel suffer long distance erosion, which results in beach land collapsing greatly and increase incidents. To reservoir desilting with high sediment concentration, if reservoir operation is not well, the disadvantage of "large flow with few sediment, small flow with numerous sediment" will happen, which will lead to deposition of downstream channel as to flood control extremely negative.

Secondly, influence the relationship of river and lake in downstream channel. Reservoir's operation has changed flow and sediment regime that is transported into downstream channel as well as change downstream channel boundary condition, which influence the health of a lake related to downstream channel. For Poyang Lake and Donting Lake of Yangtze River as an example, with Three Gorges Reservoir operation, downstream scoured extreamly and drawdown, then the lake area also reduced, leading to great influence of fisheries production, water supply and shipping.

Thirdly, influence the ecological environment of river and lake. With the decreasing of lake area, around environment is becoming more and more worse; for river mouth, especially less sand river, due to lack sediment supply, the area of river mouth will suffer serious coastal erosion and salt water now is backward, influencing ecological health of the area.

4　Comprehensive study on treatment and resource utilization technology of sediment of reservoir

4.1　Technology of sediment transportation in artificial pipelines

Sediment transportation technology that provides security for resource utilization has reached great achievement recently. Long distance sediment transportation is mainly by pipelines in Yellow River Basin. From the 1970s, it has been a relatively mature and practical technology by constantly studying and innovating and improving in more than 40 years. The power of transportation is dementedly high voltage power instead of diesel engine, and is multi – relay supply instead of single – stage transportation, the distance of which is up to 12,000 m from 1,000 m and daily maximum sediment transporting capacity of single ship is more than 5,000 m^3.

4.2　The efficient sediment flushing and desilting technology of deep – water reservoir

Jiang Enhui and her partners demonstrate, select and integrate the feasibility of desilting system of auto – suction pipeline and feasibility of the system of the jet – flushing type suction dredger in deep – water reservoir, based on sedimentation characteristics, transport law and performance characteristic of the Xiaolangdi Reservoir, and reference to home and abroad dredging results. Study points out that desilting system of auto – suction pipeline makes full use of natural power, structure of which is simple, and some of medium and small reservoirs have achievements and are easy to break through key technology. The option of pipelines has certain feasibility of desilting works of Xiaolangdi Reservoir area, and the comprehensive cost is preliminarily estimated about 1 yuan; the technology of the jet – flushing type suction dredger is the most economical and most practical in all mechanical dredging technology, characteristics of which are simple structure, lower cost of production, mature engineering experiences, independent domestic property rights of large – scale jet pump technology and equipment, when the maximum height is 80 m, the unit price of sediment flushing is 3 yuan per ton; disturb and pump sediment in front of reservoir, and flush fine sediment deposited in the reservoir based on siphon principle, by the potential absorbing and disturbing sand ships, under the existing and constant arrangement and buildings of Xiaolangdi Multipurpose Dam Project, the maximum height reaching more than 70 m, year desilting reaching 100,000,000 m^3, desilting unit price about 3.8 yuan, and initial investment about 200,000,000 yuan, the fixed number of year dead storage of Xiaolangdi can be prolonged to more than 10 years.

4.3 The technology of sediment resource utilization

For a long time, the researchers have been committed to the study of sediment resource utilization of Yellow River. Including silt arrester and embankment normally used, Institute of Yellow River Research has developed special environmental curing agent, according to the characteristics of fine grain and greater concentration of Yellow River sediment, in order to keep Yellow River sediment brick compact structure and high strength, put forward production process of steamed and sintered bricks, producing sample, testing performance, and develop large stone with emergency work, and has achieved some experience of comprehensive utilization of Yellow River sediment. Today, these product has been developed and most of these are the multiple and wide building materials in state's basic construction, as follows: sintered internal combustion bricks, ash solid bricks, sintered hollow bricks, sintered porous bricks(bearing hollow – bricks), building tiles, glazed tiles, wall and floor tiles, black pottery, painted pottery of Yellow River sediment,

In addition, Institute of Yellow River Researcher have developed the technology of topology interlocked bricks and exempt steamed aerated concrete bricks by theoretical analysis and laboratory test, and explored in some aspects, such as burn ceramisite, glass – ceramics, molding sand, and purification and processing chemical industry material with high additional value.

5 The different river reservoir sediment resources utilization modes and the dialectical relationship between them and river health

5.1 The sediment resource demand of river health

It has been pointed out that, for different rivers, according to the river health development demanding, the reservoir sediment treatment goals are different.

In order to maintain sandy river health, reservoirs need to intercept sediment, reduce the sedimentation of the lower river channel and corresponding flood pressure. On the other hand, for the long – term, excepting the sedimentation reduction, comprehensive benefits of flood control, power generation etc., reservoirs need to maintain the effective reservoir capacity as long as possible and reduce sediment silting in the reservoir area. Only the contradiction between them can be solved by utilizing and consuming the sediment resources.

To maintain less sand river health, on one hand, it needs to consider the stability of the river downstream of the reservoir, and the lead (take) water outlet and navigable water level, the ecological health of lakes and rivers and estuary areas and so on, which need to maintain sediment supply of the reservoir downstream. On the other hand, the reservoirs also need to minimize the capacity of silt loss in order to give full play to the comprehensive benefits. There is no contradiction between the two. What needs to be solved is the reservoir intercept sediment discharged technique, especially the efficient sediment diversion technique of deepwater reservoirs.

5.2   The sandy river sediment resources utilization modes

For sandy river reservoir sediment treatment, the previous researches and practices emphasized on discharging or dredging out the reservoir sediment but not utilizing them basically. This measure demanded the state to invest a lot of money but the benefit was only extended the reservoir service life. At the same time, the discharged sediment may deposit in the downstream river channel to aggravate flood control burden or accumulate near the reservoir area to pollute the environment, with a larger negative effects. In this mechanism, once the state invested funds run out, the variety of measures would immediately terminate, and not be benign long run. Therefore, in the face of a steady stream of sediment, we must put another thought to deal with it and change from passive to active, regard the reservoir sediment as a valuable resource treasure, through reservoir sediment reasonable resource utilization, extending the reservoir service life, mitigating the reservoir sediment to downstream flood control impaction, reducing environmental pollution, and creating substantial economic benefits and win – win results (especially increased hydroelectric benefits, which is the cleanest energy). The initial idea is as follows:

(1) The reservoir sediment resources investment mode.

The reservoir sediment resource utilization will firstly raise the reservoir flood control benefits and extend the service life of the reservoir. As one of the public welfare projects, the country should advance into the part of the funds as start – up fund. In the sediment resource utilization full swing, the resource utilization gains can be used to maintain the long – run funds, including the development and improvement of the sediment resource utilization technology. Meanwhile, because the sediment resource utilization extends the service life of the reservoir, enterprises should return a certain percentage of capital from the increase efficiency of power generation and maintain the reservoir sediment resource sustainable utilization .

(2) The reservoir sediment resources utilization.

For the reservoir sediment resource utilization, we can take the following ways: the coarse sand deposits in the reservoir tail, under the premise of the strict management and scientific planning, Because the water depth is shallow, the sand can be directly shipped out, as the application of building materials. For the coarse sand in the middle part of the reservoir area, according to the two sides terrain and the market demand conditions, using jet and suction sediment diversion or self – suction pipe sediment diversion technique, the sediment is transported t to the appropriate area to desilt and sort through the tube (drainage) road. The coarse sediment can be directly used as building materials and the fine sediment is used to silt field and improve soil, and the other sediment can be used to produce steam – cured bricks, topological interlocking structure bricks, flood control large stones and so on. And the sediment also can be stacked some suburban gully close to the river shore to provide construction land for urban development. The fine sediment deposited in front of the dam, can be directly transported to the sea or silt field to improve soil by creating artificial gravity flow method. Using sediment filling coal mine subsidence area is also a recent study made by one of the main sediment resources utilization presented by recent study.

In addition, in order to better sediment resource utilization, it can change the sediment separating effect by engineering technology. For example, building sand retaining weirs or rubber dams in the upstream and retaining sediment when the flood with high sand then utilizing it after the flood.

(3) The reservoir sediment resource utilization outfit.

The reservoir sediment resources utilization equipment, especially suitable for deep – water reservoir sediment resources utilization, is the premise of ensureing the reservoir sediment resource utilization smoothly carried out. On the basis of pipe desilting system and jet and suction dredging system in the present study proposed, in the next stage professional manufacturers should be chosen; machinery equipment should be manufactured. Particularly the silting up exhaust equipment, transport and control equipment should be designed, to ensure that the system finished products meet the requirements of the relevant design standards. For silt up exhaust system equipment such as water operation ship, power machinery, it needs to be optimized and selected; for the exhaust system of real – time monitoring and scheduling management of equipment, it needs to be selected, developed and researched. According to equipment design and running performance, practical construction technology and operation plan should be designed and a detailed and operable method and sediment diversion scheduling should be put forward.

5.3 The utilization modes of sediment resource of low sediment rivers

Maintaining the health of low sediment rivers, the sediment treatment of low sediment river reservoir is mainly to keep the reservoir sediment from reducing owing to the reservoir construction. For the technology of releasing sediment from the reservoir, in addition to regular sediment ejection by basic flow and emptying water, water and sediment relationship regulation and artificial shape density current technology developed by the Xiaolangdi Reservoir in recent is also a good method to eject sediment through the reservoir scheduling. For the using of machinery to eject sediment, self – suction sediment transport piping system and jet – flushing type suction system have good effect in ejecting sediment.

In addition, we can also consider engineering technology to deal with the sediment in reservoir, such as Japanese Tenryu River Miwa dam, to reduce the erosion on downstream river and collapse of beach . The Japanese built two river weirs at the end of dam reservoir, and excavated a tunnel through downstream of the dam. Through the regulation by the two river weirs transported sediment to downstream dam directly through the tunnel, avoiding the deposition in front of the dam and ensuring sediment supply in the upstream.

For the investment mode of dealing with sediment in low sediment river, public benefit input should be given priority to and the compensation by reservoir power generation benefit plays a subsidiary role. Under the premise of the health of the river, the redundant sediment should be used reasonably.

5.4 The dialectical relationship between river resource utilization and the health of river

The disaster and resource of river sediment is a pair of concomitant contradictions, and create mutual transformation in certain time and space conditions. In history, though the overflow of the Yellow River brought great disaster, created a lot of fertile land through irrigating sedimentation. Due to the increasing of people and the over utilization of nature continuously by people for years, the disaster of river sediment (especially high sediment river) becomes prominently relatively. At the same time, subjecting to the low technology of utilizing sediment resource, people have a favor of disaster in the cognition of river sediment. When dealing with sediment, it is just regarded as harmful material without utilization, and the health of river cannot be maintained owing to the lack of funds.

With the development of economic and people's high demanding for the health of river, on one hand, resources attribute of sediment is gradually known by people and equipment and products for utilizing sediment have been developed and applied to social production and practice. The utilization of sediment resource is gradually becoming an effective way to deal with sediment in high sediment river; On the other hand, the effect of sediment supply to maintain the health of reservoir downstream channel of low sediment river and estuary has gradually been known. From the two sides, sediment will no longer be "harm", but precious resources; therefore, we must change an idea to deal with sediment. For high sediment river, the utilization of sediment resource will be the main method to maintain the health of it; for low sediment river, sediment is also the precious resources to maintain the healthy development of reservoir downstream channel and estuary, and keeping the supply of sediment for these locations is also a kind of utilization of sediment resource.

In a word, sediment plays an important role in the health of river. No matter more or less sediment, it will have bad effect on maintaining the health of river. From the past practice, we can know that for the plan of river sediment, simple processing or utilization is unable to maintain the health of river. We must combine sediment treatment and sediment utilization organically on the basis of the existing technology of sediment transport and sediment resources utilization, and according to the needs of maintaining the health of the rivers, gradually establish a benign operation mechanism, which is an important method to solve the problem of river sediment fundamentally and the sublimation of river sediment treatment and the final way for dealing with sediment, having great significance on social, economic, environment ecology and people's livelihood.

6 Conclusions

The reservoir sediment treatment is in relation to the service life of reservoir and flood control and power generation and all kinds of the methods of reservoir operation. The sediment control for downstream channel, navigation, the safety of water supply, environment and all aspects of river regulation, should be given attention strategically. Formerly, people regarded sediment as a burden, taking it as the source of disaster; With the increasing needs of the development of the economic and social and the development of the sediment resource utilization technology, the reservoir deposited sediment being a kind of resources will gradually be more and more used of extensively, and we can achieve large - scale utilization of sediment resource and turn the harm into a benefit under the present national strength of our country and the technology. In consequence, to maintain the health of low sediment river, we must combine sediment treatment and sediment utilization organically and establish a benign operation mechanism, to solve the problem of river sediment fundamentally and give full play to the reservoir function, and to maintain the health of river.

References

Jiang Naisen, Fu Lingyan. [J]. Journal of Lake Science, 1997, 9(1): 1 -8.

Ma Yilin, Xiong Caiyun, Yi Wenping. Sedimentary Characteristics and Developing Trend of Sediments in Poyang Lake, Jiangxi Province[J]. Volcanology & Mineral Resources, 2003, 24 (1): 29 -37.

Yang Hongyun Reservoir Sediment Problems of the Water Power Projects of the Youngtie River[J]. Resources and Environment in the Yangtze Basin, 1993, 2(3): 267 -274.

Hu Chunhong, Wang Yangui. Study on Water - sediment Optimum Allocation in Upstream Basin and Comprehensive Measures of Sediment Control in Guanting reservoir II : Water - sediment Optimum Allocation and Dredging Schemes [J]. Journal of Sediment Research, 2004(2): 19 -26.

Chen Wenbiao, Zeng Zhicheng. Discussion on Reservoir Sediment Problem of Pearl River Valley [J]. Pearl River, 1997 (5): 26 -28.

Jiang Enhui, Cao Yongtao, Gao Guoming, et al. Strategy Operating Mechanism of Sediment Treatment and Utilization in the Yellow River [J]. China Water Resources, 2011 (11): 16 -21.

Leng Yuanbao, Song Wanzeng, Liu Hui. Dialectical Thinking on the Yellow River Sediment Resources Utilization [J]. Yellow River, 2012, 34(3): 1 -3,10.

Xiaolangdi Reservoir Sedimentation and Flushing Studies

Chih Ted Yang[1], *Jungkyu Ahn*[1], *Liu Jixiang*[2], *Zhang Houjun*[2] and *Fu Jian*[2]

1. Department of Civil and Environmental Engineering, Colorado State University,
Fort Collins, 80523, USA
2. Yellow River Engineering Consulting Co., Ltd., Zhengzhou, 450003, China

Abstract: The GSTARS3 computer model was modified for the simulation and prediction of the flushing and sedimentation processes of the Xiaolangdi Reservoir in China. The modifications included the estimations of recovery factor as a function of sediment particle size. The modifications also included the computation of tributary water and sediment discharges flow into and from the reservoir. Han' (1980) non − equilibrium sediment transport and the modified unit stream power equation for hyper − concentrated sediment flows by Yang et al. (1996) were used. The simulated and predicted results were in good agreement with measured bed profiles, channel geometry, bed material size distribution, and flushed sediment volume.

Key words: computer model, flushing, non − equilibrium sediment transport, recovery factor, reservoir sedimentation, unit stream power

1　Introduction

The Xiaolangdi Reservoir is located 40 km north of Loyang and 128. 42 km downstream of Sanmenxia dam on the main stem of the Yellow River. The drainage basin area is about 7.0×10^5 km^2, and average annual discharge at the dam site is 400×10^9 m^3. The average annual sediment load is 13.47×10^9 t. The top of the reservoir storage is at 275 m. There are more than 40 tributaries flowing into the Xiaolangdi Reservoir. Fig. 1 shows the Xiaolangdi Reservoir and 12 major tributaries and the approximate location of the "imaginary" tributary to account for the volume of all small tributaries. Fan and Morris (1992) classified hydraulic methods used in China to manage reservoir sediment deposition as:

Fig. 1　Plan view of the Xiaolangdi Reservoir and tributaries

(1) Sediment routing during floods.
(2) Drawdown flushing.
(3) Emptying and flushing.
(4) Venting density current.

Drawdown flushing has been applied to the Xiaolangdi Reservoir. Due to high concentration of fine materials in the Yellow River and its tributaries, Xiaolangdi Reservoir sedimentation and flushing mechanism is complex and special considerations must be taken in their modeling.

Modifications of the GSTARS3 computer model were required to simulate sedimentation and flushing processes in the Xiaolangdi Reservoir.

2　Theoretical background

2.1　Numerical models

Computer models can be classified as one – dimensional (1D), two – dimensional (2D), and three – dimensional (3D). A 1D model is suitable for long – term simulation of a long reach of river or reservoir with the least amount of data required for calibration and verification. 1D numerical solutions are relatively simple and stable. Yang and Simões (2008) concluded that the Generalized Sediment Transport model for Alluvial River Simulation version 3.0 (GSTARS3) computer model is suitable for long – term river and reservoir sedimentation studies. Therefore, GSTARS3 is used as the basis for the development of a numerical model for the Xiaolangdi Reservoir.

2.2　Sediment transport

Sediment routing computations in GSTARS models are based on the following Exner equation:

$$\eta \frac{\partial A_d}{\partial t} + \frac{\mathrm{d}Q_s}{\mathrm{d}x} = q_{lat} \tag{1}$$

where, η is volume fraction of sediment in a unit bed layer (one minus porosity); A_d is sediment volume in bed; t is time; x is distance; Q_s is volumetric sediment discharge; and q_{lat} is lateral sediment inflow.

A complete list of sediment transport equations used in GSTARS3 is given by Yang and Simões (2002). In this study, sediment transport capacity was calculated by using the modified unit stream power equation (Yang, et al., 1996) applicable to high sediment concentration laden flow in the Yellow River.

2.3　Non – equilibrium sediment transport

GSTARS3 applies the method developed by Han (1980) to account for non – equilibrium sediment transport. The non – equilibrium sediment transport rate is computed from

$$C_i = C_{t,i} + (C_{i-1} - C_{t,i-1})\exp\left[-\frac{\alpha\omega_s\Delta x}{q}\right] + (C_{t,i-1} - C_{t,i})\left[-\frac{q}{\alpha\omega_s\Delta x}\right]\left[1 - \exp\left(-\frac{\alpha\omega_s\Delta x}{q}\right)\right] \tag{2}$$

where, C_i is sediment concentration at cross section i; $C_{t,i}$ is sediment transport capacity at cross section i; q is discharge of flow per unit width; Δx is reach length; ω_s is sediment fall velocity; i is cross – section index (increase from upstream to downstream); and α is recovery factor. Han and He (1990) recommended a α value of 0.25 for deposition and 1.0 for entrainment, respectively.

3　Numerical model set up

GSTARS3 was modified for the Xiaolangdi Reservoir study. The modifications were based on field data collected between 2003 and 2006 by the Yellow River Engineering Consulting Company. Each data set was analyzed and some assumptions were made because some of the needed data were not available or not appropriate for the simulation.

3.1　Upstream and downstream boundary conditions

Fig. 2 shows the incoming discharge from upstream and water surface elevation at Xiaolangdi dam. Annual peak incoming flow usually occurs between May and September. Drawdown flushings occur between May and August each year. After a drawdown, the water surface rapidly rises back to fill the reservoir and then the water surface does not change significantly before another drawdown the following year. The cycle of reservoir operation can be divided into drawdown flushing, rapid rise, and stagnant stages. Fig. 3 shows incoming sediment load and water surface elevation variations. Incoming sediment loads from upstream were assumed to be the same as that released from the upstream Sanmenxia Reservoir. Water and sediment discharges from the upstream Sanmenxia Reservoir are high between May and September each year when the Xiaolangdi Reservoir drawdown flushing occurs.

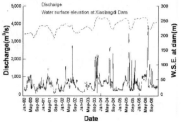

Fig. 2 Operation of Xiaolangdi Reservoir

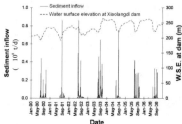

Fig. 3 Sediment load and water surface elevation

3. 2 Properties of bed material

For GSTARS3 modeling, information on bed material and incoming size distributions are needed. Bed material was surveyed several times between 2000 and 2006 for the main reservoir and some major tributaries. Comparisons among measured bed material data revealed that the bed material size changed significantly in the first three years during the initial reservoir filling operation. Fig. 4 shows an example of surveyed bed material size distributions.

Fig. 5 shows an example of the variation of bed material size measured in 2002 after flushing in May and water refill in October, respectively. After the May 2004 flushing, bed material became coarser because finer materials were eroded faster than coarser materials as shown in Fig. 5.

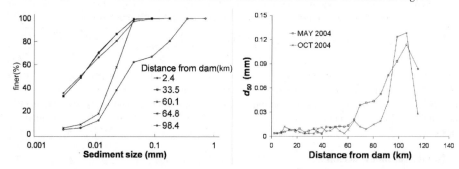

Fig. 4 Bed material size distribution surveyed between June, 10 ~ 13 2002

Fig. 5 Measured bed material sizes before and after flushing in 2004

Because dry density of bed materials in the Xiaolangdi Reservoir was not provided, recommended values by Yang (1996) are used in this study. There are four types of reservoir operation (U. S. Bureau of Reclamation, 1987), and the density of deposited sediment depends on reservoir operation type. Because sediments always submerged in the Xiaolangdi Reservoir, the reservoir reach can be classified as operation type 1. while that in the river reach is operation type 4. The dry density of four reservoir operation types are summarized in Tab. 1. It is assumed that consolidation effect is negligible in this study due to annual flushing operation.

Tab. 1   Initial dry density of four reservoir operation types

Operation	Initial dry density (kg/m^3)		
	Clay	Silt	Sand
1	416	1,120	1,550
2	561	1,140	1,550
3	641	1,150	1,550
4	961	1,170	1,550

3. 3   Incoming sediment quantity and size distribution

The GSTARS3 model requires not only the quantity of incoming sediment from the upstream but also its size distribution. To use Yang et al. (1996) method for the determination of sediment transport rate, information on the percentage of wash load is required. The percentage of wash load in a river depends on field conditions. Information on bed material and suspended material size distributions are needed to determine the percentage of the wash load fraction. Because information on the wash load fraction is not available, it is assumed that sediment size less than 0. 01 mm is wash load for the Yellow River (Yang, 1996 and 2003). Yang (1996 and 2003) found that even if the initially assumed wash load percentage is not accurate, his method can still give fairly good estimation of bed – material load in the Yellow River with high concentration of wash load. Therefore, bed – material load computation using Yang et al. (1996) method is valid for this study.

3. 4   Geometry data

The Xiaolangdi Reservoir and tributaries cross sections and longitudinal profiles were surveyed seven times, i. e. , January 2000, May and October 2003, May and October 2004, November 2005, and October 2006, respectively. Fig. 6 shows surveyed thalweg profiles of the Xiaolangdi Reservoir. There are 56 cross sectional data sets along the 120 km study reach in the main reservoir with about 2 km spacing between cross sections.

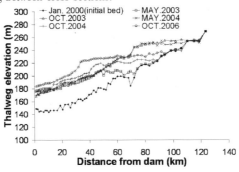

Fig. 6   Surveyed thalweg profile in the Xiaolangdi Reservoir

Drawdown flushing was conducted between May and September each year. The simulation started from May 2003 because field observations before and after flushing were incomplete between 2000 and 2003.

3.5 Water temperature data

No water temperature was measured for the Xiaolangdi Reservoir. Tab. 2 shows the assumed average monthly water temperature.

Tab. 2 Water temperature of Xiaolangdi Reservoir (assumed values)

Month	Jan.	Feb.	Mar.	Apr.	May	June	July	Aug.	Sep.	Oct.	Nov.	Dec.
Temp. (℃)	5	5	5	8	8	10	10	16	10	8	5	5

3.6 Roughness coefficient (Manning n)

Water surface elevation along the study reach was not measured. Consequently, roughness coefficient can not be calibrated in this study. Reasonable assumptions must be made for the determination of roughness coefficient.

Manning n values for all cross sections are computed using the following relationship.

$$n, \propto d_{50}^{1/6}, \quad n = \kappa \cdot d_{50}^{1/6} \, (d_{50} \text{ in mm}) \quad (3)$$

where, κ is a coefficient. Two different κ values are required for river and reservoir regimes, respectively. κ values of 0.063 and 0.022 for river and reservoir regimes, respectively, are assumed in this study. The Manning n values along the study reach with a transition between

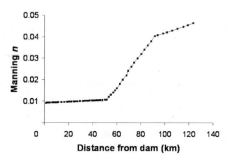

Fig. 7 Manning n values for the main reservoir routing

reservoir and river reaches shown in Fig. 7 are used for the GSTARS3 simulation.

4 Tributary influences

The Xiaolangdi Reservoir has complicated features with more than 40 tributaries. The inflows of water and sediment from tributaries to the main reservoir are very small, and can be ignored in this study. However, the total volume of all the tributaries is about 40% of the reservoir volume, and can not be ignored. The "level pool" concept as shown in Fig. 8 is used to determine the reservoir volume and discharge of tributaries.

During the reservoir water surface rising stage, water and sediment flow from the reservoir into the tributaries, i. e., lateral outflow. During the drawdown stage, water and sediment flow into the main reservoir from tributaries. The modified GSTARS3 can simulate water and sediment inflow and outflow between reservoir and its tributaries. The following assumptions were used for reservoir routing to simulate the influence of tributaries:

(1) Tributary mouth bed elevation is the same as that of the reservoir at the mouth of the tributary.

(2) Sediment concentration and size distribution of a tributary is the same as that in the reservoir at the mouth of the tributary.

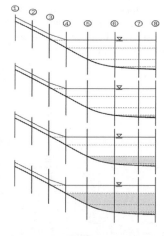

Fig. 8 Delineation of volumes to build the capacity table for tributaries

(3) During a sediment flushing period or reservoir water surface elevation falling stage, water

and sediment will be discharged into the reservoir.

(4) During a sedimentation or silting stage when the reservoir water surface elevation is rising, water and sediment will flow into tributaries.

(5) Reservoir water surface is horizontal and the discharge of water into or from a tributary is

$$\Delta Q = \Delta Vol/T \tag{4}$$

where, ΔQ is water discharge between a tributary and the reservoir; ΔVol is change of tributary volume due to the change of reservoir elevation in time T as illustrated in Fig. 8.

The volume of a tributary depends on water stage and bed elevation at the mouth. Therefore the tributary volume can be computed as

$$Vol = a(h - h_b)^b \tag{5}$$

where, Vol is volume of water in a tributary; h is water surface elevation; h_b is bed elevation at the mouth of a tributary; a and b is coefficients of a tributary.

The value of ΔQ in Eq. (4) is positive when the tributary discharges into the reservoir during the flushing or water surface elevation falling period. ΔQ is negative during a water surface raising stage with water flowing into a tributary. Sediment load, ΔQ_s, to and from a tributary is

$$\Delta Q_s = \Delta Q \cdot C_m \tag{6}$$

where, C_m is sediment concentration at the mouth of a tributary.

To compute ΔQ, the main reservoir routing should be carried out first to determine water and bed surface elevations and sediment concentration at the mouth of each tributary without considering the effects of tributaries. Using water surface elevations at the mouth of each tributary, ΔQ and ΔQ_s of each tributary are calculated, and then main channel or reservoir routing are repeated to calculate water and sediment discharges from or to the reservoir to update bed elevation in the main channel using ΔQ and ΔQ_s.

The total volume of small tributaries was combined into an "imaginary tributary" with location shown in Fig. 1 to simplify the simulation processes and save computational time. The relationship between water surface elevation and volume of tributaries can be computed by Eq. (5). These relationships are summarized in Tab. 3.

Tab. 3 Water stage and tributary volume relationships

Tributaries	Location of the mouth (distance from Xiaolangdi dam, m)	a	b	h_b(m)
Simengou	600	666.99	2.533,7	154.66
Dayuhe	4,225	4,404.6	2.435,7	149.90
Meiyaogou	6,350	102.89	2.978,6	159.80
Baimahe	10,355	187.71	2.709,9	169.70
Zhenshuihe	17,030	4,545.2	2.626,0	154.40
Shijinghe	21,680	3,177.5	2.464,3	160.40
Donyanghe	29,100	126.94	3.119,3	171.40
Xiyanghe	39,380	4,659.1	2.346,1	179.50
Ruicunhe	42,410	6,043.5	2.197,0	178.20
SZ	52,000	2,286.8	2.917,5	182.69
Yunxihe	54,570	39,380	2.047,2	196.10
Boqinghe	56,950	6,803.6	2.395,6	216.10
Banjianhe	61,590	4,158.2	2.215,4	208.40

5　Recovery factor

　　Different recovery factors have been suggested in the literature, either from an experimental or from a practical point of view. They include but are not limited those by Zhang (1980), Almanini and Silvio (1988), Zhou and Lin (1995), Han and He (1997), Wang (1998), Zhou and Lin (1998), Zhou, et al. (1997), Han (2006), and Dong et al. (2010). There is no consensus on the best value a modeler should use under different flow and sediment conditions. Among all the suggested parameters, sediment fall velocity or particle size is the most important one for the determination of recovery factor.

　　Wang (1998) conducted laboratory experiments by changing sediment size and flow characteristics. His results are shown in Fig. 9. It can be concluded that fall velocity, or sediment particle size, is the dominant factor for the determination of recovery factor. Because reservoir water temperature was not measured but assumed for the Xiaolangdi Reservoir, sediment particle size was used for the determination of recovery factor.

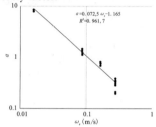

Fig. 9　Relationship between recovery factor and sediment fall velocity

　　A functional relationship between recovery factor and sediment size is assumed as

$$\alpha_k = \frac{\varepsilon}{d_k^\xi} \tag{7}$$

where, ε and ξ are site specific coefficients.

　　9 sediment size groups were used for sediment routing. Each group has a lower and an upper limit of sediment size and the representative size for each group is the geometric mean value of the lower and upper limit as shown in Tab. 4. In Eq. (7), two coefficients are required to determine α. More than 30 combinations of coefficients in Eq. (7) were tested. For the Xiaolangdi Reservoir, the calibrated value of ε and ξ were found to be 0.17 and 0.3, respectively. The recovery factor has the same value for scour and deposition. The need to consider the difference between scour and deposition suggested by Han and He (1990) was not found for the Xiaolangdi Reservoir. This conclusion seems reasonable because the difference between scour and deposition reflects the change of bed material size gradation. Tab. 4 summarizes the recovery factors used in this study.

Tab. 4　Recovery factor for both scour and deposition

Size group	Lower bound(mm)	Upper bound(mm)	Geometricmean (mm)	Recovery factor
1	0.002	0.004	0.002,8	0.989
2	0.004	0.008	0.005,6	0.803
3	0.008	0.016	0.011,3	0.652
4	0.016	0.031	0.022,3	0.532
5	0.031	0.062	0.043,8	0.434
6	0.062	0.125	0.088,0	0.352
7	0.125	0.250	0.176,8	0.286
8	0.250	0.500	0.353,6	0.232
9	0.500	1.000	0.707,1	0.189

6   Computational time interval (Δt)

GSTARS model uses a finite difference uncoupled scheme, which means that hydraulic properties are calculated first. Sediment routings and bed changes are computed after hydraulic computations, keeping all the hydraulic parameters frozen during the calculation (Yang and Simões, 2002). During a time step Δt hydraulic property and channel boundary changes should be small because the channel boundary is assumed to be fixed during the hydraulic property calculation.

GSTARS3 model calculates the change of cross section due to scour or deposition using finite difference method

$$\Delta A_{d,i,k} = \frac{\Delta t}{\eta_i} \cdot \frac{q_{lat}(\Delta x_i + \Delta x_{i-1}) + 2(Q_{s,i-1,k} - Q_{s,i,k})}{\Delta x_i + \Delta x_{i-1}} \qquad (8)$$

where, $\Delta A_{d,i,k}$ is change of volume of sediment size k at cross section i; and Δx_i is distance between cross section i and $i + 1$. $\Delta t = 3$ min is the optimum value for the Xiaolangdi Reservoir sedimentation simulation determined by trial – and – error method.

7   Distance between Cross Sections (Δx)

GSTARS3 uses an explicit method to solve the sediment routing equation and the stability criteria is (Yang and Simões, 2002)

$$\Delta x \geq c_s \cdot \Delta t \qquad (9)$$

where, c_s is the kinematic wave speed of the bed changes.

Both Δx and Δt should be determined to satisfy the stability criteria. Eq. (9) indicates that longer Δx may give a more stable solution with a given Δt.

Fig. 10 shows an example of comparison surveyed and simulated profiles after three years of flushing operations using 24 cross sections with $\Delta t = 3$ min. The surveyed results near the upstream Sanmenxia Reservoir is sensitive to the discharge of water and sediments from that reservoir.

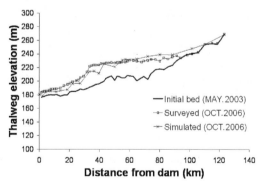

Fig. 10   Thalweg profile using 24 cross sections along the main reservoir after 3 years of flushing operation

8   Result of Xiaolangdi Reservoir sedimentation simulation

Fig. 10 shows good agreement between computed and surveyed bed profiles except near the upstream end affected by the operation of Sanmenxia Reservoir. Using more than one stream tube in the GSTARS3 application is not necessary due to the lack of lateral variation of channel geometry.

The bed material becomes coarser during and immediately after a flushing operation. During the silting period, the bed material becomes finer. The model predicts these trends reasonably well. Fig. 11 shows that after flushing in September and November, d_{50} of the bed material became coarser than that before flushing in April. During a rapid water surface rising stage, d_{50} decreased. Fig. 11 also confirmed that sediment routing by size fraction used in the modified GSTARS3 model can predict the decrease of bed material size in the downstream direction along the Xiaolangdi Reservoir very well.

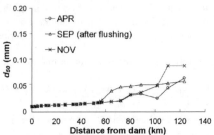

Fig. 11　Simulated d_{50} before and after flushing in 2005

Fig. 12 compares sediment volumes flushed out using three different methods. "Sediment IN – OUT (measured)" is based on the product of measured discharge and concentration at upstream and downstream boundaries, i. e., Sanmenxia dam and Xiaolangdi dam. "Sediment IN – OUT (simulated)" is based on simulated discharge and concentration. "Reservoir volume (measured)" is based on comparison between measured reservoir volume before and after flushing. Fig. 12 indicates that simulated sediment flushed volumes using the modified GSTARS3 agree reasonably well with measured results. Total volume of sediment flushed out from the Xiaolangdi Reservoir from May 2003 to October 2006 is about $11 \times 10^8 \sim 14 \times 10^8$ m^3.

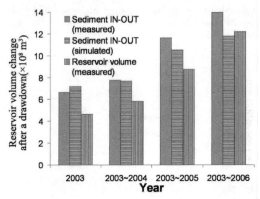

Fig. 12　Accumulated sediment flushed from year 2003 to year 2006

9　Summary and conclusions

The modified GSTARS3 was used to simulate and predict the variation of Xiaolangdi Reservoir longitudinal profile, cross section, bed material size, and the amount of sediment flushed out between May 2003 ~ October 2006. This study has reached the following conclusions:

(1) Empirical equations between roughness coefficient and sediment size as a function of distance from the dam were developed for the Xiaolangdi Reservoir.

(2) The recovery factor varies with sediment particle size of each sediment size fraction for the Xiaolangdi Reservoir. A functional relationship between sediment particle size and recovery factor was developed in this study with satisfactory simulation results. There is no need to use different

recovery factors for scour and deposition, respectively.

(3) The simulated yearly sediment flushing volumes obtained with the modified GSTARS3 are in close agreement with measured results.

(4) The simulated longitudinal bed profiles along the reservoir are in good agreement with surveyed profiles, especially in the lower reach of the reservoir. The release of water and sediments from the upstream Sanmenxia Reservoir can have a significant influence on the surveyed profile and bed material size distribution at the upper reach near the upstream Sanmenxia dam.

(5) Simulated and predicted bed material size using the modified GSTARS3 indicates that sediment size decreases in the downstream direction. This result is encouraging because the Xiaolangdi Reservoir can store coarser and flush finer sediments to the downstream Yellow River and eventually discharge the fine materials to sea.

Notation

The following symbols are used in this paper:

A_d = sediment volume in bed

a and b = coefficients of power function for volume of a tributary

C_i and $C_{t,i}$ = sediment concentration and transport capacity at cross section i, respectively

C_m = sediment concentration at the mouth of a tributary

c_s = the kinematic wave speed of the bed changes D = water depth

d = particle diameter

d_{50} = particle size at which 50% of the bed material by weight is finer

d_k = geometric mean diameter of sediment size group k

h = water surface elevation

h_b = bed elevation at the mouth of a tributary

i = cross-section index (increase from upstream to downstream)

j = time step index

k = size group index (increasing from small particle to large particle)

N = number of size fractions

Q_s = volumetric sediment discharge

q = discharge of flow per unit width

q_{lat} = lateral sediment inflow

T = time interval between water surface change in a tributary

t = time

U^* = shear velocity

V = flow velocity

Vol = volume of water in a tributary

x = distance

α = dimensionless recovery factor

α_d and α_s = recovery factors for deposition and scour, respectively

α_k = recovery factor of sediment size group k

ΔA = change of volume of sediment size k at cross section i

ΔQ = water discharge from a tributary to the reservoir

ΔQ_s = sediment load from or into a tributary

Δt = computational time interval

ΔVol = change of tributary volume due to the change of reservoir elevation

Δx = reach length

ε and ξ = coefficients of a power function for recovery factor

η = volume of sediment in a unit bed layer volume (one minus porosity)

κ = coefficient of d_{50} for Manning' n

ω_k = sediment fall velocity of sediment size group k

ω_s = sediment fall velocity

References

Armanin A, Sivio G D. A One – dimensional Model for the Transport of a Sediment Mixture in Non
– equilibrium Condition[J]. Journal of Hydraulic Research, 1998,26(3):275 –292.

Chen D, Acharya K, Stone M. Sensitivity Analysis of Nonequilibrium Adaptation Parameters for
Modeling Mining – Pit Migration[J]. Journal of Hydraulic Engineering, 2010,136(10):806
–811.

Fan J, Morris G L. Reservoir Sedimentation. I: Delta and Density Current Deposits[J]. Journal of
Hydraulic Engineering, ASCE, 1992,118(3):354 –369.

Han Q. A Study on the Nonequilibrium Transportation of Suspended Load[J]. Proceedings of the
International Symposium on River Sedimentation, Beijing, China, 793 –802.

Han Q. Reservoir sedimentation[M]. Beijing:Science Publication, 2003.

Han Q. Diffusion Equation Boundary Condition and Recovery Coefficient[J]. Journal of Changsha
University of Science and Technology, 2006,3(3):7 –19.

Han Q, He M. A Mathematical Model for Reservoir Sedimentation and Fluvial Processes[J].
International Journal of Sediment Research, 1990,5(2):43 –84.

Han Q, He M. A Study on Coefficient of Saturation Recovery[J]. Journal of Sediment Research,
1997(3):32 –40.

Julien P Y. River Mechanics[M]. New York: Cambridge University Press, New York, 2002.

U. S. Bureau of Reclamation. Design of Small Dams[M]. 3rd edition, Denver, Colorado,1987.

Wang Z. Experimental Study on Scour Rate and River Bed Inertia[J]. Journal of Hydraulic
Research, 1999,37(1):17 –37.

Yang C T. Sediment Transport: Theory and Practice, McGraw – Hill Companies, Inc. , New York,
1996 (reprint by Krieger Publishing Company, 2003).

Yang C T, Molinas A, Wu B. Sediment Transport in the Yellow River[J]. Journal of Hydraulic
Engineering, ASCE, 1996,122(5):237 –244.

Yang C T, Simões F J M. GSTARS Computer Models and Their Applications, Part I: Theoretical
Development[J]. International Journal of Sediment Research, 2008,23(3):197 –211.

Yang C T, Simões F J M. User's Manual for GSTARS3 (Generalized Sediment Transport model
for Alluvial River Simulation version 3. 0)[M]. Technical Service Center, U. S. Bureau of
Reclamation, Denver, Colorado, 202.

Zhang Q. Diffusion Process of Sediment in Open Channel and Its Application[J]. Journal of
Sediment Research, 1980:37 –52.

Zhou J, Lin B. 2 –D Mathematical Model for Suspended Sediment[J]. Journal of Basic Science
and Engineering, 1995,3(1):77 –97.

Zhou J, Lin B. One – Dimensional Mathmatical Model for Suspended Sediment by Lateral
Integration[J]. Journal of Hydraulic Engineering, 1998, ASCE, 12497: 712 –717.

Zhou J, Spork V, Koehgeter J, et al. Bed Conditions of Non – Equilibrium Transport of Suspended
Sediment[J]. International Journal of Sediment Research, 1997,12(3):241 –247.

Studies on the Key Problems of the Construction of Yellow River Water and Sediment Regulation and Control System[①]

Wang Yu, *An Cuihua*, *Li Hairong* and *Wan Zhanwei*

Yellow River Engineering Consulting Co., Ltd, Zhengzhou, 450003, China

Abstract: In this paper, the situation and existing problems of the Yellow River water and sediment regulation and control system were analyzed based on the root problem of Yellow River, which is summarize as insufficient water, excessive sediment and imbalance of the two. The overall arrangement of Yellow River Water and Sediment Regulation and Control System was presented here according to the demand of the storm flood and ice flood control, sediment deposition reduction and water resource configuration of the Yellow River. The regulated sediment – laden flood characters and operated ways of Yellow River water and sediment regulation and control according to the demand to maintain the capacity of release the flood and sediment, to assure the safety of the storm flood and ice flood control and to realize the effective management and configuration of Yellow River water resource. The built order of key projects was determined by analsing the development tasks, the status and role in Yellow River Water and Sediment Regulation and Control System and the urgent requirements of the regulation and development of Yellow River. Besides, the role of Guxian Reservoir in the Yellow River was analyzed, the result shows that it occupies a strategically important position in the regulation and development of Yellow River and plays an important role in the flood control, so it is vital necessary and urgency to accelerate the construction of Guxian Reservoir.

Key words: Yellow River, water and sediment regulation and control, characters, operated ways, built order

Generalized water – sediment regulation system means all the engineering systems and non – engineering system which are related to water increase, sediment reduction and water – sediment regulation. It includes water conservation and watershed basin water transfer measures, the soil and water conservation of Loess Plateau, reservoir sand block and sediment reduction measures and other non – engineering measures such as monitoring systems, forecasting systems and decision support systems. The general meaning of the Yellow River water and sediment regulation system includes engineering system and non – engineering systems which can effectively control the flood, runoff and sediment of the Yellow River. In this paper, water and sediment control system refers to the non – engineering system.

1 Less water and more sand, water and sediment discord is the crux of the Yellow River problem

The fundamental reason of the Yellow River to become the world's most complex and refractory rivers is the characteristics of less water, more sand and water and sediment discord. Mainly reflected in the following areas:

Firstly, there is less water, more sand and high sediment concentration. Yellow River's mean annual natural runoff is 53.5×10^9 m³, 1.6×10^9 t of sediment, the measured annual average sediment concentration up to 38 kg/m³ (1919 ~ 1960, Shanxian station). The runoff of the Yellow River is less than 1/20 of the Yangtze River, but the sediment amount if as three times of the Yangtze River. Compared with the sediment – laden river in the world, Bangladesh's Ganges River sediment load of 1.45×10^9 t, similar to the Yellow River, but the water up to 3.710×10^9 m³

① Fund Project: the Special Scientific Research of the Ministry of Water Resources for Public Welfare Industry (200901017,201001012)

which is seven times of the Yellow River, and the sediment concentration is only 3. 9 kg/m^3, far less than the Yellow River. Thus, the Yellow River is unique for its large amount of sediment.

Secondly, the water and sediment of the Yellow River are heterologous. Water comes from the upstream but sediment mainly comes from the middle reaches. The upper reaches' area of the Yellow River region accountes for 54% of the whole watershed. Water inflow accounts for 56%, but the sand was only 9% of the whole river; the middle reaches of the Yellow River Hekou Town to Longmen only accountes for 15% of the whole river and water inflow accountes for 14% of the river, but sediment accountes for 57% of the River. The middle reaches is the main sediment yield.

Thirdly, there are distribution imbalance of water and sediment in different year and different month of a year. In natural circumstances, the water yield during flood season accountes for 60% of the whole year's water yield and sand amount accountes for more than 85%. The concentration degree of sand is morera than water and of ten concentrated in a few torrential rains flood period. Looking from the water and sediment inter – annual distribution, the maximum annual water amount of the Yellow River main stream(Sanmenxia Station) is as four times as the minimum amount. The maximum annual sediment load is as 13 times as the minimum annual sediment load.

Fourthly, sediment coefficient (ratio of sediment concentration and flow) is very high. annual average sediment concentration in the region of Hekou Town to the Longmen Section is up to 123. 10 kg/m^3, sediment coefficient is 0. 67 kg · s/m^6, tributary of the Weihe River Huaxian station runoff and sediment concentration also reached 50. 2 kg/m^3, sediment coefficient is 0. 22 kg · s/m^6.

Due to water and sediment discord of the Yellow River, the great quantities of sand silt leads to the continuous raise of the riverway. In history, the Lower Yellow River dike breached, frequently river diversions, and serious flood sediment – related disasters. According to incomplete statistics, from the Qin Dynasty to the Republic (about 2540 years), the proliferation of downstream burst happen for 543 years, crevasse of 1,590 times and there were five major diversions and migration. In history, the Yellow River has brought grate disasters to Chinese peoples, which is known as the China Hardship. At present, the Lower Yellow River is 4 ~ 6 m higher than the levee back and some places even reaches more than 10m, forming a world – famous "hanging river". The two sides dike is a very important barrier to protect the 9×10^9 of population and $1. 1 \times 10^6$ mu of arable of the Huang – Huai – Hai Plain.

2　A comprehensive water and sediment control system is an effective way to solve the water and sand discordance

According to the water and sand characteristics of the Yellow River, we must take effective ways and coordinate the water and sediment relations in order to maintain the Yellow River's healthy life, to seek long – term stability, and to promote the economic and social sustainable development of the Yellow River's basin. Corresponding to the problems of the Yellow River, the ways to coordinate water and sediment relation are water increase, sediment reduction and water and sediment regulation. So – called water increase refers to measure which could increase the Yellow River's water resources, including outside basin water transfer and water – saving (relative increase in water); so – called the sediment reduction refers to a variety of measures to reduce sand amount, including soil and water conservation, reservoir storage and the main stream warping and other measures. In particular, we should try to reduce the coarse particles which seriously influence river siltation; so – called water and sediment regulation is based on the characteristic of the Yellow River water. We try hard to transport more sand into the ocean, reduce river siltation and expand and maintain the flow capacity of the river's main channel.

Among the three ways to solve the water and sediment uncoordinated problem of Yellow River, water increase mainly depends on the Western Route of the South to North Water Diversion Project, but water transfer needs adjustment of key projects in order to play a benefit.

Sediment reduction depends mainly on soil and water conservation, the key projects sediment trapping and large – scale warping of main stream. Soil and water conservation is the fundamental measure to reduce the sediment into the Yellow River. But the natural environment in Loess Plateau region is very harsh, so it is inpractical to reduce sand amount depending on water conservation measures in short term. Sand block in the reservoir backbone is the most effective and economic measure. What's more, large – scale warping also needs the adjustment of the reservoir backbone in order to achieve better results. Water diversion sand tone is a long – term implementation strategy, its status and role is very important. But the effect of a single reservoir is so limited that it needs some more reservoirs project and related non – engineering measures to play a role in the long term water and sediment regulation. Therefore, the construction of water and sediment regulation system is the most effective way of changing the Yellow River's Water and Sediment discordance.

3 Construction situation and limitations of the Yellow River water and sediment regulation system

There are 26 Cascade Engineering project under construction or have been build up, including four key projects—Longyangxia, Liujiaxia, Sanmenxia and Xiaolangdi, with a total capacity of $57.75 \times 10^9 \ \mathrm{m}^3$. These projects played huge comprehensive benefits in flood control, sedimentation reduction, water supply, irrigation and power generation. But because the Yellow River water and sediment regulation system has not been constructed perfectly, the current situationof key projects have a lot of limitations in coordinating economic and social development and maintenaning the needs of the Yellow River's healthy life. Mainly displayed as follows:

(1) Under the current conditions of engineering works, water flow could not provide adequate power conditions, the joint regulation of water flow and sediment is difficult so there is not a long time to maintain Xiaolangdi Reservoir storage period to restore the waterchannel.

Under current conditions, Wanjiazhai and Sanmenxia Reservoir's regulating storage is small so they couldn't provide plenty hydrodynamic conditions; the joint of reservoirs to regulate sediment is difficult. When Xiaolangdi use its single reservoirs to make the peak level, the water level is relatively high and sediment content is low so it can not fulfill its capactity. While the reservoir disilting, the using water level need to reduce and can not adjust enough water to meet the "flood with sand "requirements. So it is difficult for Xiaolangdi to regulate water and sediment with single reservoir. The mathematical model calculation results show that after the Xiaolangdi Reservoir storage capacity is full, the downstream channel in turn gradually atrophy, so it'll failed to maintain the needs of de – siltation water of downstream river channel.

(2) The operation of reservoir's status quo cannot solve the contradiction of water and sediment relation and water and electricity supply between Ningxia and Inner Mongolia Reach. There is river siltation and atrophy in Inner Mongolia Reach.

After the reservoirs' joint operation of Longyangxia and Liujiaxia, great benefits have come out, but it also changed the runoff distribution. With the rapid economic and social development, industrial and agricultural production and urban and rural domestic water are increasing, coupled with climatic factors, riverwater was significantly reduced. So the floods which are good to the transportation of river sediment are cut down for a large amount. Resulting in a lack of balance between Ningxia and Inner Mongolia river water, sediment aggravate siltation atrophy, small and medium – flow water level was elevated significantly. All are seriously threatening to the iceprevention and flood control.

(3) Tongguan elevation keeps high.

The changes of Tongguan elevation plays an important role to the siltation of lower Weihe River. Although many measures have been adopted in recent years such as river regulation, reservoir area cutoffs and Bait – casting dredging, it is still not effective to control Tongguan elevation.

4 The overall layout of the Yellow River water and sediment control system

4.1 Objectives of water and sediment control system

On the basic of in – depth analysis of the characteristics of the Yellow River incoming water and sand, combine with long – term practice of the Yellow River, considered to support economic and social development and the demand of maintaining the healthy life of the Yellow River, the goal of the joint control of the Yellow River water and sediment control system is: Firstly, coordinate relation of water and sediment, improve the river's sediment transport capacity, reduce the siltation of reservoir and river, long – term maintenance of flood sediment transport function of water channels in the river. The second is the effective management of floods, provide important protection for flood control and ice safety. The third is to optimize the configuration of the Yellow River water resources and the inflow of Western Route of South – North Water Diversion into the Yellow River, to protect the lives of urban and rural residents, agricultural, industrial, ecological and environmental water, to support the Yellow River Basin and related areas of sustainable economic and social development.

4.2　Overall layout of the water and sediment control system

According to the characteristics of the different reaches of the Yellow River and an important tributary of the water and sand, Consider the demand of the ice flood control, sedimentation reduction, water resources, configuration, combining the basin's economic and social development and water and sediment regulation goal.

The overall layout of the Yellow River water and sediment control system is: Longyangxia, Liujiaxia, Heishanxia, Qikou, Guxian, Sanmenxia, Xiaolangdi, etc as the main part, Haibowan, Wanjiazhai Reservoir as the supplement, Luhun, Guxian, Hekuocun, Dongzhuang, including another 4 controlling reservoirs, together constitute the engineering system of the Yellow River water and sediment regulation; constitute a non – engineering system of Yellow River water and sediment regulation by the monitoring system and forecasting system, and decision support systems, to provide technical support to the joint regulation of reservoir.

Heishanxia and Longyangxia, Liujiaxia Reservoir constitute the main body of the sub – system of Yellow River water regulation, heishan gorge's anti – regulation to the pouring water of longyang gorge and liujia gorge, to coordinate into the river water and sediment relations in Ningxia and Inner Mongolia, to improve the ice situation of Inner Mongolia River flood, the rational allocation of the Yellow River water resources and the Western Route Water Diversion water to protect the basin's economic and social development, while the middle and lower reaches of the Yellow River coordination of water and sediment relationship to water conditions. Haibowan Reservoir regulation capacity of the smaller, water and sediment limited ability to regulate, mainly with the reservoir ice scheduling, timely with the black gorge, water and sediment regulation replenishment use, improve the efficiency of water sediment.

Guxian and Qikou Reservoir and completed Xiaolangdi, the Sanmenxia Reservoir, etc. constitute the Yellow River flood sediment regulation and sub control system of the main control of the middle reaches of the Yellow River flood, sediment sources, significant abatement of the Yellow River downstream dilute case of flood protection of downstream flood control safety; reservoirs sediment retention and sediment regulation, can significantly reduce the downstream sediment, especially coarse sediment, slow down the Lower Yellow River siltation, optimize downstream water and sediment processes, and improve the transport capacity of flow and bed – making role of the river, long – term maintenance of the lower Yellow River flood of water channel sediment; can make a small North main stream waserosion, the water level at Tongguan significant decline; to reduce the chance of the Yellow River intrusion Weihe, reducing the Weihe River flood pressure in the middle and lower reaches; also thought that the large – scale warping of the small north main stream conditions to improve water supply conditions of the surrounding areas, to promote economic and social development.

The upstream sub system and midstream sub system are far from each other, Wanjiazhai Reservoir can play a connecting role, with the main stream of the reservoir water and sediment regulations in a timely manner supplement the hydrodynamic conditions. Tributary estuary village reservoir is an important part of the Yellow River downstream flood control system, the basic control of the flood of Qinhe, can also lower reaches of the Yellow River water and sediment regulation provide favorable conditions. Dongzhuang Reservoir can effectively control the Jing River sediment and siltation mitigation, reducing the flood pressure on the lower reaches of the Weihe River.

5　Water and sediment regulation index system

Water and sediment regulation index system includs the regulation index of maintenaning the flood discharge and sediment transportaion, regulaion of flood control and ice regulation and regulation of effectively control the management of water resources. Water and sediment regulation index system is the controlling conditions of watershed flood control, ice prevention, sedimentation reduction and water resources allocation control. It is also an important basis of the joint use of reservoirs.

5.1　Maintain the functional regulation of river flood discharge sediment indicators

The Sediment siltation problems of incongruous relation are mainly reflected in the lower Yellow River, Tongguan reach and Ningxia – Inner Mongolia Reach. The way to coordinate water and sediment relations is mainly to solve problems of these three rivers.

Study on the characteristics of different flow magnitude and sediment concentration level of lower Yellow River flow showed that low sediment concentration flow downstream's sediment concentration is less than 20 kg/m^3, scour development site with the the Huayuankou increased. When the the Huayuankou flow increased to 2,600 m^3/s, the full range of erosion of the river downstream, the flow rate increases to 3,500 m^3/s, the efficiency of full downstream erosion further improved. If sediment concentration is 20 kg/m^3 to 60 kg/m^3 floods, as the traffic level increases, the downstream gradually changed from siltation to erosion. When the flow reached 2,600m^3/s, or more basically showing erosion withflow level increases, increase the efficiency of the downstream channel erosion. If the flood's sediment concentration is 60 ~ 100 kg/m^3 or even more than 100 kg/m^3, with the traffic level increases, downstream siltation gradually reduced. Therefore, from the point of view of the downstream river sedimentation reduction, regulatory upper limit of flow should be selected more than 2,600 m^3/s. According tothe speed and flow relationship, the flow rate increases with the increase in traffic, and there is an inflection point whose size is similar with the bankfull discharge. Therefore, to improve the efficiency of water sediment transport, maintain the transport capacity and protect the beach area's safety, the maximum downstream regulation of traffic should be controlled between 2,600 m^3/s to 4,000 m^3/s, the capacity of the corresponding regulation approximately 0.8 × 10^9 ~ 15 × 10^9 m^3.

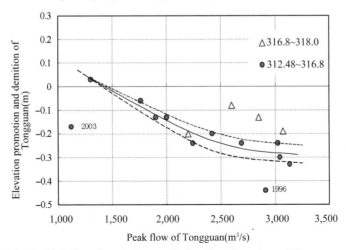

Fig. 1　Relation of peak flow and elevation demotion of Tongguan

Study on the relations of Tongguan elevation changes and Tongguan section flood magnitude nad flood volume showed that the Tongguan peak flow should be more than 2,500 m³/s when there is obviously erosion reduce in Tongguan elevation (shown in Fig. 1 and Fig. 2). There showed clear trend between Tongguan elevation change and Tongguan station magnanimity. The more value of magnanimity, the greater the water level dropped at Tongguan station. There is a goodrelationship between flood volume and peak flow on the 10th. The magnanimity shoud be abuot 1.3×10^9 m³ when peak flow is more than 2,500 m³/s.

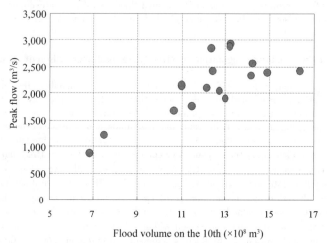

Fig. 2 **Relation of magnanimity and promotion and flood peak flow of Tongguan Station**

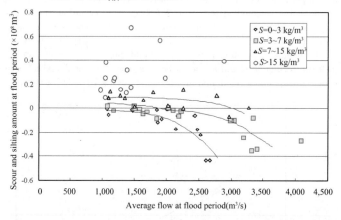

Fig. 3 **Relation of silting quantity and flow rate of Ningxia and Inner Mongolia river during flood period**

The study on the characteristics of Ningxia and Inner Mongolia river flood erosion and deposition showed that when flood's sediment concentration is less than 3 kg/m³, overall of Ningxia and Inner Mongolia Reach was in washed situation. When the flow is more than 2,000 m³/s, the erosion effect is obvious. If floods's sediment concentration is 3 ~ 7 kg/m³, flow is less than 2,000 m³/s, Ningxia and Inner Mongolia Reach of the overall was in slightly siltor red and silt situation. When flow is 2,000 m³/s to 2,500 m³/s, the river was in washed situation but erosion was less. If flow is greater than 2,500 m³/s, the erosion will increase with the flow a significant increase

(Fig. 3). Considering the objective condition of present stage that Ningxia and Inner Mongolia Reach of bankfull discharge of 1,500 m^3/s, it is better to take current regulation to 1,500 m^3/s or so, with the gradual recovery of the bankfull discharge, the regulation of flow can be increased to 2,500 m^3/s. After the west line of South to North Water Diversion Project was in effect, regulation flow could be increased to 3,000 m^3/s. For flow level from 2,500 ~ 3,000 m^3/s, Inner Mongolia Reach of the full range of scouring flood event lasted the minimum number of days to 14 days, and with the flood duration to increase its erosion amountincrease. Therefore, regulation of water should be not less than 3×10^9 m^3.

5.2 Flood and ice prevention regulation indicators

The target of Water and sediment regulation system is to control great flood and catastrophic flood. It is to say, when there is great flood or catastrophic flood in the Yellow River, we try hare to reduce losses according to scientific and rational flood – dealing plan and through the joint scheduling of reservoir engineering. We should do our best not let the flood exceed the flood protection standards of dike and detention basin.

5.2.1 Flood control regulation index

To the lower Yellow River, the minimum embankment fortification flow is 11,000 m^3/s (below Aishan section), and 10,000 m^3/s if Changqing—Pingyin mountain areas are not counted. When the flow rate is more than 11,000 m^3/s, we need to use Dongping Lake detention area to do flood devention. So flow rate at downstream Aishan should be controlled less than 10,000 m^3/s. The statistical analysis showed that when peak flow rate of Hua yuankou station is more than 8,000 m^3/s, most of the beach area has been flooded. Huayuankou station should try to control peak flow for small and medium – sized floods not more than 8,000 m^3/s.

With regard to Lanzhou river, the flow rate should be controlled under 6,500 m^3/s in order to ensure the safety of Lanzhou City. To Ningxia and Inner Mongolia River, the flow rate from Xiahe to Shizuishan should be 5,620 m^3/s and fortification flow of Sanshenggong to Putan is 5,900 m^3/s. Therefore, flood control indicators should not be larger than the the fortification flow of the river.

5.2.2 Ice prevention control index

To the Lower Yellow River, according to the ice prevention scheduling experiences of Sanmenxia and xiaolangdi Reservoirs, before the downstream river frozed, the flow of the Huayuankou should be controlled 500 ~ 600 m^3/s. Huayuankou's discharge volume controlled at 300 ~ 400 m^3/s after freeze – up. When doing ice prevention, first Xiaolangdi Reservoir, the end of December each year reserved for ice storage the capacity of 2×10^9 m^3. When the Xiaolangdi reservoir fills up, the Sanmenxia Reservoir began to control. Sanmenxia Reservoir's ice capacity is 1.5×10^9 m^3.

To the conditions of Ningxia and Inner Mongolia reach, combined with the Longyangxia and Liujiaxia Reservoir ice prevention and scheduling experience, control targets during ice flood season (December to March) is 460 m^3/s, 420 m^3/s, 360 m^3/s and 350 m^3/s. After Heishanxia Reservoir put into operation, considering Ningxia and Inner Mongolia Irrigation District winter irrigation to retire the water impact, the control indicators during ice flood season are 450 m^3/s, 420 m^3/s, 360 m^3/s and 350 m^3/s. After the commencement of a project of the Western Route of the South – to – North Water Diversion Project, ice controlvent flow of Black Gorge Reservoir is 550 m^3/s, 500 m^3/s, 450 m^3/s and 350 m^3/s.

5.3 Effective management of water resources regulation index

Gross control indicators of water river outside: in 2020, surface water consumption shall not be more than 40.18 $\times 10^9$ m^3, surface water consumption not more than 33.28 $\times 10^9$ m^3, groundwater extraction shall not be more than 12.37 $\times 10^9$ m^3. By the year of 2030 the level of water consumption of surface water shall not be more than 46. 85 × 10^9 m^3, surface water consumption not more than 40. 11 × 10^9 m^3, groundwater extraction not more than 12. 53 × 10^9 m^3.

Ecological and environmental water and the section within the river discharged water regulation indicators: the 2020 level, the average eco – environmental water Lijin section not less than 19. 36 × 10^9 m^3, Toudaoguai section an average of the ecological environment water of not less than 20 × 10^9 m^3; by 2030 level, the Lizin section average eco – environmental water not less than 18. 7 × 10^9 m^3, Toudaoguai section an average of the ecological environment water not less than 20 × 10^9 m^3; Lijin section the average eco – environmental water after the entry into force of the west line not less than 21. 1 × 10^9 m^3.

Index of Section flow control: According to the requirements of *Implementing regulations of the Yellow River Water Scheduling regulations* carried out in 2007 by the Ministry of Water Resources, section flow should be controlled to meet the requirement of main stream warning flow. The minimum flow rate of turn section is 250 m^3/s, Lijin section is 150 m^3/s.

6 Mechanism of joint use of the water and sediment regulation system

Mechanism of Joint use of Yellow River water regulation and control sub – system. Longyangxia, Liujiaxia and Heishanxia jointly use constitute the main part of Yellow River water regulation and control sub – system. Longyangxia and Liujiaxia Reservoir adjust the water yield into the Yellow River by the west line of South – to – North Water Diversion Project water of the Yellow River Water in order to add water sourse supply capacity during the Yellow River dry years, especially consecutive dry years, and improve the effectiveness of cascade power generation . Heishanxia Reservoir mainly doing anti – regulation to discharged water from upstream cascade power station. Combined with ice prevention and water storage, on the basis of meeting the need of the whole river's water allocation and Ningxia – Inner Mongolia river water amount, we adjust the surplus water in non – flood season to flood season, and to eliminate a large number of adverse effects bring by impoundment of Longyangxia, Liujiaxia Reservoir flood season which influence Ningxia and Inner Mongolia river, so to restore and maintain the capacity of the channel's sediment transportaion capacity. Haibo Bay hydro – junction mainly coordinates with the upstream backbone reservoir in ice prevention. During flood season and freeze – up period, the hydro – junction avoids flow fluctuations of Inner Mongolia River due to water – break in Ningxia irrigated area and river frozen up the Haibo Bay.

The joint use of regulation and control sub – system of the Yellow River flood sedimen mechanism, and the jiont use of middle reaches of Qikou, Guxian, Sanmenxia and Xiaolangdi water control, Constitute the main body of the Yellow River flood sediment control engineering system.

This plays an irreplaceable important role in flood management, coordination of water, sediment relations and supporting the sustainable development of regional economic and social.

The jiont use of Yellow River flood, sediment regulation sub – system, firstly, the joint management of the Yellow River floods can reduce the excess of the standard flood in the lower reaches of the Yellow River exceed the standard flood; when Yellow River floods, the joint regulate on the middle reaches of the flood process, try to shape the relations of coordination of water and sediment and give full play to the carrying capacity of the water in order to reduce the river main channel siltation and shape suitable for the middle and lower reaches of floodplain warping of water and sediment. When the Yellow River does not flood in a long period, the joint adjust to restore and maintain the water channel traffic; try to keep the sediment transport capacity of the water channel flood. Secondly, the use of reservoir storage capacity of the sand, try hard to retain coarse sediment which is most harmful to the lower Yellow River's sand deposition. Thirdly, joint runoff regulation protect the lower reaches of the Yellow River ice safety. Play the utilization efficiency of the industrial and agricultural water supply and power generation. The joint use of regulation and control sub – system of the Yellow River flood sedimen mechanism. The natural characteristics of the Yellow River Water and Sediment heterologous decided on wandering system of joint use must be organic in wandering system and form a complete system of water and sediment regulation. In water control, Years flow regulation of the upstream reservoir to compensate for the defects of the middle reaches of the reservoir regulation protect the middle and lower reaches of the dry season water supply security. In the coordination of the Yellow River water and sediment relations, upstream sub – body system configuration requirements to be based on the Yellow River water resources to arrange the discharged water in flood season and the process reasonably to create hydrodynamic conditions for the sediment of the water and sediment regulation of the wandering system regulation.

7 The order of the project to be built research

The Heishanxia, Qikou, Guxian water control key projects in the Yellow River water sediment regulation system to be built are social and public welfare which is given large static investment. In order to adapt to the requirements of the development of the national economy basin, considering the central investment and the finance ability of provinces along the Yellow River(area), we should arrange stage construction in priorities.

7.1 The Comparison of development order of Guxian, Heishanxia water control projects

Guxian Reservoir and Heishanxia water conservancy hub is the important backbone project in Water sediment regulation system. In the Yellow River management they play a very important role, each other also have close contact, but their tasks have different emphases. The function of Guxian water conservancy hub mainly reflects in harmonizing the relationship between runoff and sediment load in the lower Yellow River, reducing the sediment deposition of downstream channel, keeping flow capacity in medium flowing channel, and coordinating the relationship between runoff and sediment load in Xiaobei River, lowering Tongguan bed elevation and promoting economic and social development in the near area. The function of Heishanxia water conservancy hub mainly reflects in improving runoff and sediment conditions in Ningmeng reach, reducing the sediment deposition of Ningmeng reach, reducing the ice run disasters, allocating water resource rationally and so on. The role of them have different emphases. From the aspect that harmonizing the relationship between runoff and sediment load in the Yellow River and coordinating the overall requirement of the Yellow River basin and its related areas' economic and social development sustainable, it is very necessary that the construction of Guxian water control project and Heishanxia water control project should be as early as possible.

From the early work and local opinions, the exploitation plan of Heishanxia reach has not yet reached an agreement at present which restricted the preliminary work of the project. The construction of Guxian water control project get the active support of Shanxi and Shanxi provinces. The proposal for the project has been approved by the General Institute of Hydropower and Water Resource Planning and Design, and the project construction has good financing environment. Therefore, from the aspect of the objective need and possible of governance and development in the Yellow River, the construction of Guxian water control project is recommend recently.

7.2 The Comparison of development order of Guxian, Qikou water control project

Qikou, Guxian water control are the two key controlling projects which is located in the main stream of the Yellow River in Hebei Province and the same river upstream and downstream. They are all important parts in the system of harmonizing the relationship between runoff and sediment load. Coming out of the analysis of that, two key projects are necessary. Both of them have the function of flood control, sedimentation reduction, water supply, irrigation, power generation, but because of its differences on flood control and sediment conditions, two projects have some distinction on function.

Firstly building Guxian or qikou water conservancy hub could meet the requirement of flood prevention and sedimentation reduction in the lower Yellow River for 40 to 50 years. It basically can meet the recent regulating volume of irrigation and the requirements of the water supply on both sides of the energy base, and provide power grids for cross – straits load capacity, improve the power structure. Guxian Reservoir can better control of the Hekouzhen —Longmen reach flood and sediment and serve as Yutong reach of flood control, sedimentation reduction, reducing the Tongguan elevation greater than Qikou reservoir. And if make joint use with Sanmenxia, Xiaolangdi reservoirs, it could be better to maintain the medium flowing channel and long – term de – siltation reduction in the lower Yellow River. Meanwhile considering the opinions of the provinces on both sides, on the present basis of research achievements, we recommend Guxian water conservancy hub as the first development key project after Xiaolangdi project.

7.3　The comparison of development order of Heishanxia, Qikou water control projects

At present existent projects are difficult to coordinate sedimentation reduction, water supply, ice run protected and power generation of upstream Ningmeng reach. The conflict between ice run protected exists. The relationship between runoff and sediment load of Ningmeng reach is deteriorating, the main channel deposition is atrophying, ice run protection situation is very severe. The nearby energy base construction and ecological construction requirement source of water form Heishanxia reservoir. From the ansysis of the protection of ice run, improving the relationship between runoff and sediment load, economic and social development in the region nearby on Mongolia reach, it is in dire need of Heixhanxia reach project. Considering the urgency of governance on Ningmeng reach, we suggest that seize the day to do the early work. If the early work and national investment maturely, the construction should begin as early as possible. In order to express the effect of joint control subsystem of middle reach fully, Qikou reservoir can become effective before the storage period of Guxian reservoir.

In conclusion, the order of the adjustment system of water and sediment of the Yellow River is Guxian, Heishanxia and Qikou. As the relationship between runoff and sediment load gradually built to take effect, the function of governancing flood scientifically, coordinating relations between runoff and sediment load, optimizing the allocation of the water resources will be brought into fully play.

8　The strategic status and functions of Guxian hydraulic project construction in Yellow River water and sediment regulation and control system

Not only Guxian Reservoir has large capacity, but also can control of the Yellow River floods and the main source of sediment area, especially the coarse sediment source area. At the same time, it's close to Xiaolangdi project, which has the unique geographical advantage to block sand and regulate water – sediment joint with Xiaolangdi Reservoir . Once Guxian reservoir has built, it can not only effectively control the water sand of the upper reaches of the Yellow River, but also provided dynamic conditions of water flow for the lower reaches of Sanmenxia and Xiaolangdi Reservoir , which plays an key role in the general layout of Yellow River sediment regulation system, and has an very important strategic position.

Since Xiaolangdi Reservoir put into use in October 1999, it has plays an important role in the downstream channel of sedimentation reduction and restore ability of flood water de – siltation sediment by the reservoir block sand and the water – sediment regulation. At the beginning of the year 2010, sediment flushing of the downstream channel has accumulated to 1.815×10^9 t, de – siltation water flowing capacity is $4,000 \text{ m}^3/\text{s}$ or so, to some extent relief the catastrophe caused by the small flood situation of the downstream. But, the capacity of the Xiaolangdi Reservoir block the sand is limited, according to the normal coming water and sediment conditions estimate that around 2020 block reservoir sand will fill up, by then, although the reservoir can still play sedimentation reduction effect by water – sediment regulation, the deposition of the downstream channel is increased at the speed of 2.70×10^9 t/a, by 2026 or so, the flow of the downstream channel will drop to $4,000 \text{ m}^3/\text{s}$, and the grim situation of flood control will appear. For a long time, the elevation of Tongguan was too high to threaten the safety of flood control of the downstream of Weihe River , after 1986, the Tongguan elevation has a sharp rise to 328 m and keep up, for this reason, reduce and control the elevation of tongguan is one of the important targets of the Yellow River.

After the Xiaolangdi Reservoir, it's play an important role to construction Guxian hydraulic project such as manage the sediment of the Yellow River , slow down the downstream channel of sediment deposition, maintain de – siltation water, guarantee the safety of flood control and so on The results of the study show that Guxian Reservoir put into use in 2020, and apply with joint block sand and the water – sediment regulation of he Xiaolangdi Reservoir, could better coordinate water come into the sand of the downstream channel. The downstream channel with an average annual decrease to 0. 93 × 10^6 t occurred, could maintain the downstream with 4,000 m^3/s de – siltation water around for 50 years, the current condition of Xiaolangdi dam of which the capacity of block sand is limited and the channel has steadily declined has changed. The small north river mainstream can last flush, the maximum accumulation sluicingcapacity was 15. 13 × 10^6 t, The biggest drop of Tongguan elevation was 2. 57 m; Guxian Reservoir used in flood control, which can not only cut the flood flow of small north river, but also reduce the water level of storage of Sanmenxia Reservoir , and keep the flood not on the beach ; Guxian Reservoir still can significantly improve the conditions of water supply nearby , to promote local economic society development.

Mathematical model calculation results show that the earlier Guxian Reservoir put into use , the bigger function will play with the Xiaolangdi joint the water – sediment regulation to reduction sedimentation in the downstream channel. If Guxian Reservoir come into use prior to the Xiaolangdi Reservoir complete block sand , the joint function of two reservoir on water sediment regulation will be stronger, sedimentation reduction effect also even more significant. There are 1 × 10^9 m^3 the capacity of block sand of Guxian Reservoir in Xiaolangdi Reservoir, compared to the reservoir put into three years later after the sand capacity was filled , The sedimentation reduction effect will be reduced from 0. 8 × 10^9 t to 1 × 10^9 t of the Yellow River downstream , the number of the year when Maintain the flowing of downstream channel above 4,000 m^3/s will reduce to 11. Therefore, It is urgent to speed up the preliminary work of Guxian water control project, aiming that the project will have complete before the capacity of Xiaolangdi Reservoir was almost filled up.

9 Main conclusion

(1) Because short of water, too much sediment, the disharmonious relationship between water and sediment, the Yellow River become the most complex and hard govern river in the world. Because of the disharmonious relationship between water and sediment, it lead to channel deposition, embankment burst in downstream, channel diverted frequently, sediment disaster seriously, the people on both sides suffered severe disasters.

(2) According to water and sediment spacetime distribution characteristics of the Yellow River, in order to maintain healthy life of the Yellow River, seeking for the stability of the Yellow River, and promoting the basin and related areas of social and economic sustainable development, we must take effective ways to coordinate the relationship between water and sediment. As short of water, too much sediment and the disharmonious relationship between water and sediment, the way to harmonize the relationship between water and sediment is increasing water, reducing sediment, regulating the water and sediment. Building a consummate runoff sediment regulation system is the necessary condition of the effective runoff sediment relationship.

(3) Nowadays, it has built four backbone controlling projects as Longyangxia, Liujiaxia, Sanmenxia, Xiaolangdi dams. They play great comprehensive utilization efficiency on controlling flood, preventing ice runsedimentation reduction water – sediment regulation and water quantity optimization. But because the water sediment regulation system has not yet been completed constructed, the current use of the project is largely limited. Firstly, on present conditions, and the capacity of providing the flow of water dynamic conditions is insufficiency, water and sediment adjustment jointly currently is difficult, the recovered medium flowing channel of xiaolangdi dam can't maintain a relatively long period. Secondly, the existing reservoir can't solve contradiction between coordinating relationship between runoff and sediment load, water supply and power generation on Ningmeng reach. The Neimeng reach deposited atrophy seriously, flood control and ice run prevented are in the grim situation.

(4) The Yellow River water sediment regulation system is an overall layout, with the key projects as Heishanxia, Liujiaxia, Qikou, Guxian, Sanmenxia as structure, with HaiboBay, Wanjiazhai Reservoir as complement, and other four controlling projects which together constitute the water sediment regulation system. The Yellow River water sediment regulation non – engineering system which constitute by monitoring system, forecast system and decision support system, provides technical support for the reservoir united operation.

(5) Water sediment regulation index system includes river flood carrying and sediment transport regulation index, flood control and ice run prevent function regulation index and achieving the effective management of water resources of the control index. water sediment regulation system for flood control, ice run prevent and sedimentation reduction, and the benefit of the water resources allocation has important significances.

(6) According to development task, the status in water sediment regulation system and the urgent requirement of the Yellow River Governance and development, the order of the key projects to be built is Guxian Heishanxia and Qikou hydraulic projects.

(7) Guxian Reservoir in the general layout of the water sediment regulation system plays a key role. Its strategic position is very important. Guxian Reservoir comeing into service as early as possible, can give fully effective to reservoir sedimentation reduction on downstream channel. It can make the downstream channel maintain flood carrying and sediment transport abilities on medium flowing channel for a long period. It could lower Tongguan elevation, reduce pressure of flood control in the lower reaches of Weihe River. It also could promote the sustainable economic and social development in shanxi and shaanxi provinces. Therefore, we should speed up the process of the early work, build the project as soon as possible.

Mississippi River Management—A Perspective from Sediment Yield in Coastal Louisiana

Y. Jun Xu

School of Renewable Natural Resources, Louisiana State University Agricultural Center, Baton Rouge, LA 70803, USA

Abstract: The Mississippi River system discharges 610×10^9 m^3 of water annually into the Gulf of Mexico (GOM). In the past century river engineering and land use practices in the Upper Mississippi River Basin have changed dramatically. A large number of locks and dams were built along the major tributary rivers including Upper Mississippi River, Illinois River, Missouri River, Ohio River, Tennessee River, Arkansas River, and Red River, which has greatly contributed to the reduction in sediment yield to the continental shelf of GOM. Concurrently, the Louisiana Gulf coast has experienced the highest rate of relative sea – level rise of any region in the United States. In the past 50 years land loss rates in Louisiana's coast have exceeded over 60 km^2 per year, and in the 1990s the rate has been estimated to be between 40 km^2 and 56 km^2 each year. This change represents 80% of the coastal wetland loss annually in the entire continental United States. This paper synthesizes recent work and newly – derived estimates of sediment delivery from six major rivers along Louisiana's coast. The primary purpose is to review the current land loss situation, sediment availability, and discuss an alternative approach for sediment diversion to offset the sinking coast.

Key words: Sediment transport; Coastal land loss; Mississippi – Atchafalaya River; coastal Louisiana

1 Introduction

According to the Intergovernmental Panel on Climate Change (IPCC, 2007), the global average sea level has been rising from 1961 to the present time at an average rate of 1.8 mm/a. The Louisiana coast of GOM has experienced one of the highest sea – level rises over the past century (Dixon et al., 2006; Ivins et al., 2007). Within the past half century land loss rates have exceeded over 60 km^2/a, and in the 1990s the rate has been estimated to be between 40 km^2 and 56 km^2 each year (LDNR 1998). This loss represents 80% of the coastal wetland loss annually in the entire continental United States. The highest relative sea – level rise (RSLR) is 17.7 mm/a at Calumet station in St. Mary's Parish of Louisiana according to the U. S. Army Corps of Engineers tide gauge stations (Penland and Ramsey, 1990) compared to 10.4 mm/a at Grand Isle, Louisiana, 6.3 mm/a at Galveston, Texas, 2.3 mm/a at Pensacola, Florida, 2.2 mm/a at Key West, Florida, 1.7 mm/a at Cedar Key, Florida, 1.5 mm/a at Biloxi, Mississippi, and 3.1 mm/a eustatic sea – level rise.

Based on their assessment using long – term historical aerial photos and satellite images, Morton et al. (2010) found that much of the land loss in coastal Louisiana is caused by land subsidence, rather than erosion. While the land has been sinking and the sea level has been rising, sediment yields from the Mississippi River and other four major coastal rivers in southern Louisiana have been declining (Meade and Moody, 2010; Horowitz, 2010; Rosen and Xu, 2011). River engineering has confined sediment distribution to only the continental shelf of the northern Gulf of Mexico. Consequently, the Louisiana Gulf coast has been subject to the highest rate of relative sea – level rise of any region in the United States. Couvillion et al. (2011) found that approximately 4,900 km^2 of low – lying coastal land on Louisiana's delta plain have become submerged since 1932. Previous studies (Britsch and Dunbar, 1993; Barras et al., 2003) reported peak delta – plain land losses of 60 km^2/a to 75 km^2/a from the 1960s to 1980s.

Although much of Louisiana's coastline is eroding and sinking, the mouth of the Atchafalaya

River has been gaining land since the early 1970s. The river is the largest distributary of the Mississippi River and currently carries about 30% of the Mississippi's waters into the Gulf of Mexico. Several researchers (Shlemon, 1975; Rouse et al. , 1978; Roberts et al. , 1980) reported their observations of the early subaerial delta development in the shallow Atchafalaya Bay after the 1973 flood. Hupp et al. (2008) used clay pads at several transects within the basin and found a sedimentation rate ranging from about 2 mm/a to 42 mm/a. In a spatial sedimentation analysis with the river's sediment inflow – outflow, Rosen and Xu (in review) recently reported an average sediment retention of 9 MT annually and a higher sedimentation rate between 23 and 56 mm per year. The results of these studies imply that a large quantity of the riverine suspended sediment will not reach the continental shelf.

Riverine sediment is a precious resource to coastal Louisiana and its effective management is of long – term strategic importance. This paper synthesizes recent work and newly – derived estimates of sediment delivery from six major rivers along Louisiana's coast. The primary purpose is to review the current land loss situation and discuss an alternative approach for sediment diversion to offset the sinking coast.

2   Coastal Louisiana and the Mississippi – Atchafalaya River system

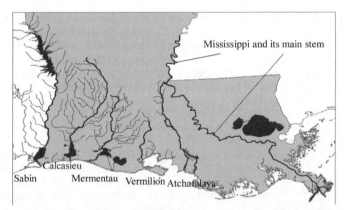

Fig. 1   Major river systems in southern Louisiana, USA, along the northern Gulf of Mexico—The Chenier Plain in the west (dashed line) and the Mississippi River Deltaic Plain in the east (solid line)

Coastal Louisiana is shaped in two distinctive physiographic regions, the Chenier Plain in the west and the Mississippi – Atchafalaya River Delta in the east (Fig. 1). The Chenier Plain comprises an area of approximately 5,000 km^2 and a west – east coastline of about 200 km. There are four rivers that flow north to south through the Chenier Plain into the Gulf of Mexico, including the Sabine River, Calcasieu River, Mermentau River, and Vermilion River. The river mouths are a combination of wave – dominated and tide – dominated, and the landscape is interspaced by chenier ridges composed of river mud – flat deposits and coarse marine and littoral sediments.

The Mississippi River Delta comprises a land area of about 12,000 km^2 and a coast line of approximately 250 km. At the present time, the delta boundary is commonly considered to include approximately the triangulated area east of the Atchafalaya and west of the Mississippi River (Fig. 1); however, the geological extension of the delta is beyond the boundary as the Mississippi River changed its course from the west to east several times during the past 8,000 years influencing the geomorphologic development of the entire southern Louisiana. One of the most engineered rivers in the world, the Mississippi River currently is regulated to flow mainly through two outlets into the Gulf of Mexico – the main stem channel and the Atchafalaya River. The main stem channel is 439 km long, confined by levees, and carries about 75% of the river's water through the currently active delta complex in the southeast Louisiana, also known as the bird's foot delta (Fig. 1). The Atchafalaya River is formed where the total flow from the Red River is combined with part of the Mississippi River's water that is diverted through the Old River Control Structure (ORCS) that was built by the United State Army Corps of Engineers during the 1960s. In the 1930s the Mississippi River discharged about 20% of its total flow into the Atchafalaya River. By the early 1950s it was observed that without intervention, the majority of the Mississippi River flow would be captured by the Atchafalaya River (Fisk, 1952). To prevent total capture of the Mississippi River by the Atchafalaya River, ORCS was built, to maintain 25 % of the Mississippi discharge flowing into the Atchafalaya River (Horowitz, 2010). Beginning from the control structure, the Atchafalaya River flows southward, first in a well – confined channel for about 87 river kilometers, and then in several braided channels through a low – land swamp area for another 100 river kilometers, discharging its water and sediment into the northern Gulf of Mexico through two outlets, the natural Atchafalaya River channel at Morgan City and the man – made Wax Lake Outlet. The river is confined by levees that surround it to the east and west at about 20 ~ 35 km in width. While the Mississippi – Atchafalaya is a typical river – dominated system, the flow of the Atchafalaya is slower and the bay water is much shallower than the main channel.

3 Sediment delivery to Louisiana's coast

On average, the Mississippi – Atchafalaya River system and the four Chenier Plain rivers transported a combined total of 181 mega tonnes of total suspended solids (TSS) each year to Louisiana's coast (Tab. 1). Much of the sediment yield was produced by the Mississippi River (70.1%) and the Atchafalaya River (29.6%). While all the six rivers showed a clear, similar seasonal trend with greatest TSS yield during the winter and spring months and lowest during the summer and fall months, corresponding to their seasonal flow conditions, interannual variability in TSS yield is very high with the four Chenier Plain rivers (up to 2,600%). Over the past two decades, the Chenier Plain rivers, except for Vermilion, showed a slightly declining trend of TSS yield, which is due mainly to the declining discharge corresponding to the decreasing trend in precipitation in the river basins. Vermilion received flow regularly from the Atchafalaya River and therefore, its long – term TSS yield was not affected by the regional precipitation.

Tab. 1 Annual average of total suspended sediment yield from major rivers to Louisiana's coast

River	TSS Yield (× 1000 metric tonnes)	
	Mean	Min ~ max
Sabine[1]	213	16 ~ 417
Calcasieu[1]	47	12 ~ 118
Mermentau[1]	40	14 ~ 93

Continued Tab. 1

River	TSS Yield (×1000 metric tonnes)	
	Mean	Min ~ max
Vermilion[1]	43	13 ~ 71
Atchafalaya[2]	53,590	22,540 ~ 76,960
Mississippi[3]	126,750	73,640 ~ 189,040

Note: 1. Estimated yields for Sabine near Bon Wier, TX, Calcasieu River near Kinder, LA, Mermentau River at Mermentau, LA, and Vermilion River at Perry, LA (Rosen and Xu, 2011).

2. Estimated yield combined from Morgan City and Wax Lake Outlet (Xu, 2010; Rosen and Xu, in review).

3. Estimated yield for Tarbert Landing (Rosen and Xu, in review).

A number of studies have reported a declining trend in the Mississippi River sediment discharge over the past half century following the 1950s ~ 1960s dam construction and river channelization (e. g. , Keown et al. , 1986; Kesel, 1988, 2003; Meade and Moody, 2010). Currently there are nearly 100 major dam/lock constructions on the main channels of seven major river tributaries (Tab. 2). It is, however, debatable, how the TSS yield of the Mississippi River should be interpreted. Our calculation shows that there was a downward trend through the early 1990s followed by a slight upward trend (Fig. 2). Horowitz (2010) suggested that the change started in 1993, while we believe the long – term decline of Mississippi River TSS yield may have ended in 1989. This trend was not caused by changes in discharge, but by changes in SSC. From 1980 to 1986 SSCs were elevated compared to the rest of the time period. Following 1986, the Mississippi River at TBL went through dramatic suspended sediment concentration reduction from an average of 323 mg/L in 1986 to a low of 162 mg/L in 1988. This reduction of SSC and discharge could have been related to a severe drought that affected the Midwestern United States that did not officially break until 1990 (Trenberth and Guillemot, 1995). Another possibility is that the opening of the Auxiliary Structure at ORCS in December 1986 could have affected sediment concentrations by diverting a greater portion of the sand load (Copeland and Thomas, 1992). Since 1989, there has been great fluctuation but an overall general increase in average suspended sediment concentration from 163 mg/L in 1989 to 289 mg/L in 2010. The years 2008, 2009, and 2010 are notable because SSCs are all in the top ten for the time period (1980 ~ 2010). The increase may reflect suspended sediment slowly reaching equilibrium following the severe drought of 1988 ~ 1990, opening of the Auxiliary Structure in 1986, and large flood of 1993. The trend suggests that without any further major alterations to Mississippi River engineering, current sediment load of the Mississippi River would likely remain stable for the foreseeable future.

Tab. 2 Dams and locks constructed in the main channels of seven major tributaries of the Mississippi River

Tributary	Length (km)	Drainage area (km^2)	Discharge (m^3/s)	# of dams
Upper Mississippi River	2,000	490,000	5,796	27
Illinois River	439	72,701	657	8
Missouri River	3,767	1,300,000	2,478	15
Ohio River	1,579	490,601	7,957	21
Tennessee River	1,049	105,868	1,998	9
Arkansas River	2,350	435,123	1,147	18
Red River	2,189	169,890	1,614	1

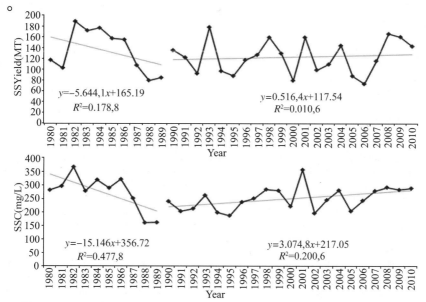

Fig. 2 **Trend of annual suspended sediment yield in million tonnes (top) and mean suspended sediment concentration in milligrams per liter (SSC, bottom) for the Mississippi River at Tarbert Landing, MS**

4　Spatial pattern of coastal land change

The Mississippi River main stem delivered nearly 1.27×10^8 t of suspended sediment each year to the southeast coast of Louisiana. Assuming an average fresh sediment bulk density of 1,200 kg/m³, this annual sediment yield would create a volume of 1.06×10^8 of sediment, or a sedimentation of about 4 cm over an area of 2,500 km², i. e., approximately a 10 – km – wide area across the 250 – km long Mississippi River Delta Plain (Fig. 1). Instead of gaining land, however, the region lost about 1,089 km² land from 1978 to 2000, according to a land change assessment with satellite imagery and historical aerial photography by Barras et al. (2003). The land loss occurred overwhelmingly along the coastal shoreline, while the vast quantity of sediment from the river was lost to the deep waters as the river discharge dominates the system.

The four Chenier Rivers delivered a total of 343,000 t of suspended sediment each year, which could create a sedimentation volume of 285,833 m³, assuming a bulk density of 1,200 kg/m³. According to Barras et al. (2003), however, the Chenier Plain lost 361 km² of land cumulatively from 1978 to 2000. Much of the land loss occurred in the interior area. There was no significant change in the coastal shoreline.

The Atchafalaya River delivered approximately 0.54×10^8 t of suspended sediment annually to the Atchafalaya Bay. The sediment yield could create a volume of about 0.45×10^8 m³ of sediment (assuming a bulk density of 1,200 kg/ m). From satellite imagery assessments Xu (2010) reported a total of 18.4 km² of newly created land in the bay in 1984; by 2004 the land area extended to 80.6 km², an increase rate of 3.1 km²/a. A recent study (Rosen and Xu, in review) found a reduced rate of land accretion (2.1 km²/a) from 2004 through 2010 due mainly to storm surge disturbance by Hurricanes Katrina and Rita in 2005 and Hurricane Gustav in 2008.

5　River diversion experience

Over the past decade much of the coastal restoration efforts have focused on river diversion to introduce freshwater and sediment. Up to now there have been three major river diversions constructed in southeast Louisiana: Caernarvon, Davis Pond, and West Bay at river kilometer 132, 190, and 5 of the Mississippi River, respectively. Although careful work has been done, none of the diversion projects have been successful in terms of creating new land or maintaining current wetlands. For instance,

(1)The Caernarvon Diversion was opened in 1991. Despite careful planning and many years of operation, a recent study (Kearney et al., 2011) found no significant changes in either relative vegetation or overall marsh area from 1984 to 2005 in zones closest to the diversion inlets. After Hurricanes Katrina and Rita in 2005, these areas sustained even larger losses in vegetation and overall marsh area, when compared to similar marshes of the adjacent reference sites (Howes et al., 2010; Kearney et al., 2011).

(2)The West Bay was opened in 2003 to capture sediment. But it has been reported that the diversion could be impacting navigational interests in the area while not producing the desired land growth (Brown et al., 2009; Barras et al., 2009; Heath et al., 2010). This has prompted USACE to close the diversion.

New sediment diversions further up river are under study (Myrtle Grove) or proposed for future investigation (CPRA, 2012), but uncertainty exists about how much and when the most sediment is actually available. It is also questionable whether channelized diversion is the best approach to effectively address various interests, such as capturing sediment while still maintaining navigation, and managing fisheries and wetlands. In general, despite our knowledge of long – term annual suspended sediment yield from the Mississippi River, there is a knowledge gap concerning the actual divertible quantity and variability of the riverine sediment. This was highlighted by Allison and Meselhe (2010) questioning the length and interannual variation of flow periods during which riverine coarse material can be obtained.

6　Future strategies for sediment management in coastal Louisiana

Based on the past river diversion experience discussed above, the channelized sediment diversion is ineffective to utilize riverine sediment. It is likely that the coastal area in Louisiana would be worse or the same in the future if the current situation continues. A completely new approach must be employed to offset the widespread land loss in the region. An alternative approach could be taken where levees are periodically dropped to a certain river stage allowing overbank flow to mimic the natural process of sediment replenishment over large areas. High sedimentation rates of coarse particles have been found during floods on floodplain areas bordering the river channel (Walling and He, 1998; Hupp, 2000). Natural overbank flow has been documented to benefit river – floodplain system, marshland, and fisheries (Baumann et al., 1984; Bayley, 1991; Nyman et al., 2006). This approach can be especially useful for the Chenier Plain, where interior land loss has been occurring and the area is sparsely populated. Most buildings in the coastal area of Louisiana are elevated above the ground (Fig. 3) to prevent damage from river flood and tropical storm surge, making the area suitable for sediment diversion via controlled overbank flow (COF), while incurring minimal damage. To implement COF, the following two aspects need to be considered.

6.1　Sediment availability under different flow conditions

The knowledge about annual total suspended sediment yield from the Mississippi – Atchafalaya River is well established. It is also clear that not all riverine sediment can be diverted for land creation. Few studies have looked at actual sediment availability under different river flow conditions, for example, the amount of sediment during high winter and spring flows. In general, there is a knowledge gap concerning the quantity and variability of suspended sediment from the Mississippi – Atchafalaya during its flood pulse, which is critical for developing management strategies to maximize sediment capture. We have recently quantified suspended sediment load of the lower Mississippi River at Tarbert Landing during three river flow conditions: Moderate Flood Stage (16. 8 m), Flood Stage (14. 6 m), and Action Stage (12. 1 m). The assessment provided two relevant findings: ① High sediment load occurred at Action Stage and Flood Stage, providing approximately 50% of the total annual suspended sediment yield over a period of only 120 d, This implies that sediment diversion outside of this period would be impractical highlighting the need to manage diversions to follow the natural flood regime; and ② The river showed significantly higher suspended sediment concentration and suspended sediment load during the rising than the receding limb, indicating that the most effective sediment diversions will have to rely on discharge during the rising limb of flood pulses.

(a)   (b)

Fig. 3   Elevated residential (left) and public (right) buildings in coastal Louisiana to prevent flooding from rivers and tropical storm surges (Photos taken in south Cameron, Louisiana, USA)

6. 2   Optimization of COF for sediment diversion

Controlled overbank flow can be implemented by lowering a stretch of river levee, as illustrated in Fig. 4. The flow will act similarly like a rectangular weir, whose flow rate can be approximated as a function of hydraulic head. Assuming that overbank flow spills over the land in a sectorial shape (Fig. 4), the flooded area and time needed for the flood depend on a number of factors including, among others, hydraulic head and width of the lowered levee, topography, soil physical characteristics, and vegetation of the land.

To illustrate the influence of hydraulic head and overbank flow width on the time that is needed for creating a certain depth of sediment, we computed flows for three different hydraulic heads and widths, whereby the assumptions were made: ① flooding area is a sector – shape flat plane with no shear stress, infiltration, and vegetation effects; ② hydraulic head and flow velocity remain constant. Based on the geometry, the formula below can be used to determine the sectorial flooding area (A):

$$A = \frac{\beta}{2} \{ b / [2\tan(\frac{\beta}{2})] + L \}^2 - b^2 / [4\tan(\frac{\beta}{2})]$$

where, $\beta = 180 - 2\alpha$ in radians; b is overbank flow width; and L is distance from the bank to the flood front edge (see Fig. 4).

To determine time for flooding the area, overbank flow rate is needed, which may be estimated as follows:

$$Q = k \frac{2}{3} \sqrt{2g} \, bH^{3/2}$$

where, k is discharge coefficient; g is gravitational acceleration; b and H are the width and

94

hydraulic head (Fig. 4).

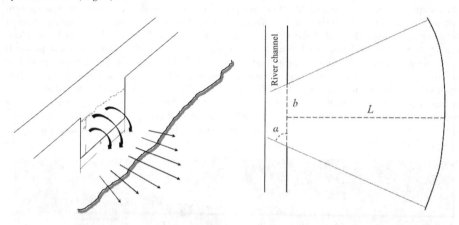

Fig. 4 Diagram of overbank flows via a lowered levee and the geometry of hypothetic sectorial flooding area

With the information on A and Q, flow volume and time can be estimated for a certain flooding depth. Tab. 3 summarizes computational results on the time that would be needed for creating 1 mm depth of sediment for the flooding area under different hydraulic heads and flow widths. In the calculation, k is assumed 0. 7, riverine sediment concentration 313 mg/L, and sediment bulk density 1,200 kg/ m.

Tab. 3 Effects of hydraulic head (H) and overbank flow width (b) on flow rate (Q), flooding area (A) and the time (T) that would be needed to create 1 mm depth of sediment over a sectorial area with a distance of 20 km (L) and an outflow angle of 120° (i. e. , $\alpha = 60°$, Fig. 4)

H & b(m)		$Q(\text{m}^3/\text{s})$	$A(\text{km}^2)$	$T(\text{d})$
when H = 0. 5	b = 1,000	730	228	13. 8
	b = 1,500	1,096	237	9. 6
	b = 2,000	1,461	246	7. 5
when b = 1,000	H = 0. 3	339	228	29. 8
	H = 0. 8	1,478	228	6. 8
	H = 1. 0	2,066	228	4. 9

The calculation for hypothetic overbank flow shown here is a simplification of highly complex hydraulic and fluvial processes. The assumptions for zero shear stress, zero infiltration, and a sectorial flooding progression are not exactly representative of the reality. Nonetheless, the results do indicate a plausible relation between and the effects of overbank flow width and hydraulic head on the time needed for flooding. When compared with flow width, hydraulic head shows a stronger effect on the time for flooding. On the other hand, flow width affects flooding area because of a geometric change. Hydraulic head can be regulated based on river stages, while flow with can either structurally or operationally regulated. Optimization of the two can help maximize sediment capture using COF.

7 Conclusions

Extending from the Chenier Plain at the Texas border to the Mississippi River Bird Foot Delta, coastal Louisiana is a maze of deltaic plains, bayous, fresh and salt marshes, and barrier islands. These coastal areas are home to over 2 million people, they support a quarter of the U. S. energy supply, and provide vital habitat for wildlife and fisheries. However, this region has experienced one of the highest rates of land loss globally, and so its restoration is a top priority in Louisiana.

Until recently, one of the largest underused resources for coastal restoration was sediment from rivers that drain into the northern Gulf of Mexico. Although there has been much research on the quantity and trend of sediment discharge from the Mississippi River and three major river diversion projects in the past 20 years, neither has been successful in creating new land or maintaining disappearing wetlands. Most of the vast riverine sediment were lost and continue to be lost to deep water through the channelized rivers. To offset the current land loss in coastal Louisiana, we believe that a completely new approach for utilizing the riverine sediment must be developed. The approach should mimic natural processes of sediment replenishment over the broad coastal areas rather than through the river channels to a few small bays.

As first step for developing the approach, we assessed the total amount of sediment input to Louisiana's coast and the quantity of potentially divertible sediment under different river stages. We found the Mississippi – Atchafalaya River system produces 99.8% of the total annual sediment (1.8×10^8 t), and that suspended sediment concentration and load were maximized during Flood and Action Stages, accounting for approximately 50% of the total annual sediment yield even though duration of these stages accounted for only one – third of a year. The annual sediment yield estimated by a hydrograph – based approach can produce a sediment volume of approximately 0.75×10^8 m^3 (assuming a density of $1,200$ kg/m^3). The current river diversion is ineffective leaving vast quantity of sediment lost to deep water. A new approach that can mimic the natural process of sedimentation is needed to maximize sediment capture. Here we presented COF as an alternative that can be periodically used to replenish sediment over large land areas. More studies are needed to examine technical details and socioeconomic feasibility of the approach.

Acknowledgements

I am thankful for the following agencies that have provided part of the long – term river discharge and sediment records used in this paper: U. S. Army Corps of Engineers, U. S. Geological Survey, and Louisiana Department of Environmental Quality. Thanks also go to my graduate student, Timothy Rosen, for his assistance with data analysis.

References

Allison M A, Meselhe E A. The Use of Large Water and Sediment Diversions in the Lower Mississippi River (Louisiana) for Coastal Restoration[J]. Journal of Hydrology, 2010, 387 (3 – 4): 346 – 360.

Barras J, Padgett W C, Sanders C B. Aerial and Bathymetric Spatial Change Analysis of the West Bay Diversion Receiving Area, Louisiana, for U. S. Army Engineer District, New Orleans (MVN), Report MR – 03, pp. 1 – 39, Mobile District Operations Division, Spatial Data Branch, Mobile, Alabama, 2009.

Barras J, Beville S, Britsch, D. , et al. Historical and projected coastal Louisiana land changes: 1978-2050. U. S. Geological Survey Open – File Report 03 – 334, 39 p. (Revised January 2004), 2003.

Baumann R H, Day J W, Miller C A. Mississippi Deltaic Wetland Survival: Sedimentation Versus Coastal Submergence[J]. Science, 1984, 224(4653): 1093 – 1095.

Bayley P B. The Flood Pulse Advantage and the Restoration of River – floodplain Systems[J]. Regulated Rivers: Research and Management, 1991(6): 75 – 86.

Britsch L D, Dunbar J B. Land – Loss Rates: Louisiana Coastal Plain [J]. Journal of Coastal Research, 1993(9):324 – 338.

Brown G, Callegan C, Heath R, et al. ERDC Workplan Report – Draft, West Bay Sediment Diversion Effects[R]. Coastal and Hydraulics Laboratory U. S Army Engineer Research and Development Center, Vicksburg, MS, 2009.

Coastal Protection and Restoration Authority of Louisiana (CPRA). Louisiana's Comprehensive Master Plan for a Sustainable Coast. Coastal Protection and Restoration Authority of Louisiana. Baton Rouge, LA.

Copeland, R R. , W. A. Thomas. Lower Mississippi River Tarbert Landing to East Jetty Sedimentation Study, Numerical Model Investigation, Department of the Army Waterways Experiment Station, Corps of Engineers, Technical Report HL – 92 – 6[R]. 1992.

Couvillion B. R. , Barras J. A. , Steyer G. D. , et al. Land area change in coastal Louisiana from 1932 to 2010. U. S. Geological Survey Scientific Investigations map 3164, scale 1:265,000, 12 p. pamphlet. 2011.

Dixon, T. H. , Amelung F. , Ferretti A. , et al. Subsidence and Flooding in New Orleans. Nature, 2006: 587 – 588.

Fisk H. N. Geological Investigation of the Atchafalaya Basin and the Problems of Mississippi River Diversion. USACOE, Miss. Riv. Comm. , Vicksburg, Ms. , 1952: 145.

Heath R. E. , J. A. Sharp, C. F. Pinkard Jr. 1 – Dimensional Modeling of Sedimentation Impacts for the Mississippi River at the West Bay Diversion[J], Paper Presented at 2nd Joint Federal Interagency Conference, Las Vegas, NV, 2010.

Horowitz A. J. A Quarter Century of Declining Suspended Sediment Fluxes in the Mississippi River and the Effect of the 1993 Flood[J]. Hydrological Processes, 2010(24): 13 – 34.

Howes N. C. , D. M. FitzGerald, Z. J. Hughes. Hurricane – induced Failure of Low Salinity Wetlands, PNAS, 2010, 107 (32): 14014 – 14019.

Hupp C. R. , Demas C. R. , Kroes D. E. , et al. Recent Sedimentation Patterns within the Central Atchafalaya Basin, Louisiana, 2010. Wetlands, 2008, 28 (1): 125 – 140.

Hupp C. R. Hydrology, Geomorphology and Vegetation of Coastal Plain Rivers in the South – eastern USA[J]. Hydrological Processes, 2000, 14(16 – 17), 2991 – 3010.

IPCC. Intergovernmental Panel on Climate Change. Working Group I. , Climate Change 2007 : the Physical Science Basis: Contribution of Working Group I to the Fourth Assessment Report of the Intergovernmental Panel on Climate Change. Cambridge University Press: Cambridge ; New York, 2007; http://www. ipcc. ch/ipccreports/ar4 – wg1. htm.

Ivins E. R. , Dokka R. K. , Blom R. G. Post – glacial Sediment Load and Subsidence in Coastal Louisiana. Geophysical Research Letters, 2007.

Kearney M. S. , J. C. Alexis Riter, R. E. Turner. Freshwater River Diversions for Marsh Restoration in Louisiana: Twenty - six years of Changing Vegetative Cover and Marsh Area [J]. Geophysical Research Letters, 2011(38): 1 – 6.

Keown M. P. , E. A. Dardeau Jr. , E. M. Causey. Historic Trends in the Sediment Flow Regime of the Mississippi River[J]. Water Resources Research, 1986, 20 (11): 1555 – 1564.

Kesel R. H. The Decline in the Suspended Load of the Lower Mississippi River and its Influence on Adjacent Wetlands [J]. Environmental Geology and Water Sciences, 1988, 11 (3): 271 – 281.

Kesel R. H. Human Modifications to the Sediment Regime of the Lower Mississippi River Flood Plain[J]. Geomorphology, 2003(56): 325 – 334.

LDNR. Coast 2050: Toward a Sustainable Coastal Louisiana: Report of the Louisiana Coastal Wetlands Conservation and Restoration Task Force and the Wetlands Conservation and Restoration Authority [R]. Louisiana Department of Natural Resources, Baton Rouge, La, USA, 1998.

Meade R. H. , Moody, J. A. Causes for the Decline of Suspended – sediment Discharge in the Mississippi River System, 1940 ~ 2007[J]. Hydrological Processes, 2010(24): 35 – 49.

Morton R. A. , Bernier J. C. , Kelso K. W. , et al. Quantifying Large – scale Historical Formation of Accommodation in the Mississippi Delta[J]. Earth Surface Processes and Landforms, 2010, 35 (14): 1625 – 1641.

Nichol S. L. , Boyd R. , Penland, S. Hydrology of a Wave – dominated Estuary: Lake Calcasieu, Southwest Louisiana [J]. Gulf Coast Association of Geological Socieites, 1992 (42): 835 – 844.

Nyman J. A. , R. J. Walters, R. D. Delaune, et al. Marsh Vertical Accretion via Vegetative

Growth[J]. Estuarine, Coastal and Shelf Science, 2006(69): 370 – 380.

Penland S. , Ramsey K. E. Relative Sea – Level Rise in Louisiana and the Gulf of Mexico – 1908 ~ 1988[J]. Journal of Coastal Research, 1990, 6(2): 323 – 342.

Phillips J. D. Toledo Bend Reservoir and Geomorphic Response in the Lower Sabine River[J]. River Research and Applications 2003(19): 137 – 159.

Roberts H. H. , Adams R. D. , Cunningham R. H. W. Evolution of Sand – dominant Subaerial Phase, Atchafalaya Delta, Louisiana [J]. American Association of Petroleum Geologists Bulletin, 1980, 64 (2): 264 – 279.

Rosen T. , Xu Y. J. Riverine Sediment Inflow to Louisiana Chenier Plain in the Northern Gulf of Mexico[J]. Estuarine, Coastal and Shelf Science, 2011, 95 (1 – 2): 279 – 288.

Rosen T. , Xu Y. J. Two – decadal Sedimentary Delta Growth of the Atchafalaya Bay: Implications for sediment management in the Mississippi River Deltaic Plain. (in Preparation).

Rosen T. , Xu Y. J. A Hydrograph Based Sediment Availability Assessment with Implications for Mississippi River Diversion Management[J]. Water Resources Research (in Review).

Rouse Jr. L. J. , Roberts H. H. Cunningham, R. H. W. Satellite Observation of the Subaerial Growth of the Atchafalaya Delta, Louisiana[J]. Geology 1978(6): 405 – 408.

Trenberth K. E. , C. J. Guillemot. Physical Processes Involved in the 1988 Drought and 1993 Floods in North America[J]. Journal of Climate, 1995, 9(6): 1288 – 1298.

Walling. D. H. , Q. He. The Spatial Variability of Overbank Sedimentation on River Floodplains [J]. Geomorphology, 1998, 24(2 – 3): 209 – 223.

Shlemon R. J. Subaqueous Delta Formation – Atchafalaya Bay, Louisiana[C]// Broussard, M. L. (Ed.), Deltas. Houston Geological Society, Huston, Texas, 1975: 209 – 221.

USACE. Vermilion River Ecosystem Restoration Report 2/6/2004[R]. United States Army Corps of Engineers, New Orleans, 2004: 1 – 16.

US EPA Region 6 (EPA). Total Maximum Daily Load (tmdl) for TSS, Turbidity, and Siltation for the 15 Subsegments in the Vermilion River Basin. http://www. epa. gov/waters/tm dldocs/ ACFB9BF. pdf

Xu Y. J. Long – term Sediment Transport and Delivery of the Largest Distributary of the Mississippi River, the Atchafalaya, USA[C]// K. Banasik, A. Horowitz, P. N. Owens, M. Stone, and D. E. Walling (eds.): Sediment Dynamics for a Changing Future, 2010: 282 – 290, IAHS Publication 337, Wallingford, UK.

"Sediment – water Separation and Resource Utilization" in the Yellow River Management

Yang Guolu[1,2] , *Lu Jing*[1,2] , *Zhu Senlin*[1,2] , *Liu Linshuang*[1,2] ,
Xia Runliang[2,3] and *Luo Xin*[1,2]

1. State Key Laboratory of Water Resources and Hydropower Engineering Science,
Wuhan University, Wuhan, 430072, China
2. Sewage Sludge and Silt Research Centre, Wuhan University, Wuhan, 430072, China
3. Yellow River Institute of Hydraulic Research, Zhengzhou, 450003, China

Abstract: It is a general thought that concept has been the guiding precedent thinking in river management. The confirm of management method has been the key problem in the Yellow River management. The reason that the Yellow River sediment control is so difficult is determined by the ideal scientific nature, limitation of practicality and the actual feasibility. Against the condition of "little water and much sediment" in the Yellow River, the paper recognized that management of the Yellow River from past to now is profound. All the previous and existing methods were limited by their prior recognition of "little water and much sediment" and proposing "sediment reduction" measures among closed river systems. Based on this, the paper proposed the idea of "Sediment – water separation" to solve the problem of "little water and much sediment" in the Yellow River, which reconstructed water and sediment proportion, changed the state of water and sediment transport and expanded sediment provenance through "resource utilization". Mixing this with "Soil and water conservation" and "Water and sediment regulation", a "point – line – surface" river management strategy can be built. The basic purpose of the paper is to provide an innovative idea which provides reference and needs further scientific research for solving sediment problems in the Yellow River.

Key words: sediment of the Yellow River, soil and water conservation, water and sediment regulation, sediment – water separation, resource utilization, concept of management

1 Introduction

From past to now, river management in the Yellow River attaches great importance to the fact that the Yellow River is the only big river which flows through arid and semi – arid area in northwest and north China and is famous for its huge sediment load. As our mother river, the Yellow River nurtures the land of China and also brings frequent disasters. The Yellow River is named after nature, and the "yellow color" is caused by overwhelmed frequent flood, huge sediment load, imbalance of resource supply and demand and damage of ecological environment. Owing to different source, uneven distribution and harmonious proportion of water and sediment, flooding, sediment deposition, bed elevation, environmental degradation, ecological atrophy and other disaster issues arise which makes it difficult to control flood and sediment in the Yellow River.

The existing river management methods in the Yellow River had one common characteristic, that is all the control measures were carried out in the limited river system. Da Yu improved the "enclosing and blocking" method in Zhang River by Gong Gong and Gun and proposed the diverting strategy. In Western Han Danasty, Jia Rang proposed three river management strategies, they are abandoning old river and diversing the north, diversing water by channel and guiding flood to Zhang River and building dyke. In Eastern Han Danasty, Wang Jing put forward to widening river to remove flood. In Ming Danasty, Pan Jixun advocated that dykes should be built on north and south river sides, and flood should be constrained among the dykes and governance priority should be placed on sediment. All the above strategies were based on the precondition that the flood has come and were carried out for the purpose of controlling flood, also all measures were carried out among the limited river system. For half a century, comprehensive methods of "blocking, discharging, placing, adjusting and digging" have successfully used in the Yellow River. In the upstream "blocking" was used and in the downstream "discharging" was used, they together can control the flood effectively. Zhang Hongwu advocated that soil – water conservation and water – sediment control in river management should be integrated, and water can be transferred from other river basins to solve the problem of water stress, this is unique. Modern river management has pushed research and governance in the Yellow River to a new stage, and great progress has been made. But frequent drying up, flood, sediment and environment disasters still exist which makes it a truth that concept of river management is inadequate. Methods of "blocking, discharging, placing, adjusting and digging" are focus on the inflow condition of "little water and much sediment", water and sediment control in the main stream and tributary, but they do not tend to change the inflow water and sediment structure, what is more sediment is not utilized as a resource.

The Yellow River is a rare river which has a huge sediment load in the world and its sediment problem is so complex and difficult. The main problems are flood, imbalance of water supply and demand, sediment deposition and deterioration of ecological environment. Among them the most crucial problem is sediment deposition caused by the inflow condition of "little water and much sediment". Truly, runoff and sediment erosion are determined by effectiveness of watershed control which identifies the inflow condition and plays important role in sediment transport and deposition. On the other hand, essence of sediment transport and deposition is sediment management, then it is crucial to explore a new method and technology of sediment management which is not making sediment deposit on the floodplain or discharge into the ocean. So, after learning the concept of river management proposed by previous researchers, based on scientific thinking of resource and environment, a new method that is "water – sediment separation and resource utilization" is proposed. Using the comprehensive engineering measures of "blocking, discharging, separation, adjusting, utilizing and cooperating", a "point – line – surface" river management strategy can be built which regards "soil and water conservation" as foundation, "water and sediment separation" as essence, "water and sediment regulation" as quantity and "resource utilization" as destination. Due to the complexity and difficulty and inadequate theoretical research of sediment problems, the basic purpose of this new method is to provide an innovative idea for solving sediment problems in the Yellow River.

2　The source of "sediment – water separation and resource utilization"

2.1　Expansion of modern river management

Modern river management in the Yellow River established a noble idea that is the maintenance of healthy life of the Yellow River and employed comprehensive methods of "blocking, discharging, placing, adjusting and digging". Among them, "blocking, discharging, placing and digging" strengthened blocking, discharging and placing sediment. "Blocking, discharging and adjusting" played key role in flood control. The combination of them made it possible to effectively treat sediment and control flood and great progress has been made. From the perspective of river dynamics, the above concept is complete and perfect, however, from the perspective of resource and environment, the above concept is inadequate and it placed too much emphasis on engineering measures in stream, weakened "blocking" method which played important role in sediment management and ignored sediment utilization. "Blocking" was carried out mainly in coarse sediment yield area which is about $100,000 \text{ km}^2$ in the middle reach of the Loess Plateau region using the method of integrating engineering measures with biological measures, blocking sediment with large and medium reservoirs, blocking part of the sediment in the gullies and dispersing sediment on the plain. Zhang Hongwu also advocated "blocking" by the means of developing soil and water conservation in the Loess Plateau region with sediment retention structures, check dams and retaining walls. Meanwhile, with blasting method, the hillside fields can be changed into relative plain and gullies can be filled up. It is feasible to carry out soil and water conservation project, and we support it, yet it is better to develop a method which combines sediment control from the source with sediment utilization, and sediment – water separation with machinery after which the isolated sediment being developed into renewable resources, like the planting soil, erosion resistance stabilized soil etc.

2.2 Origination of "sediment – water separation"

From past to now, river management in the Yellow River has not systematically touched the field of "sediment – water separation". River management in the Yellow River should inevitably consider the process of watershed runoff and sediment, channel conveyance of sediment and water and ocean acceptance of sediment and water which is a complete and complex river management system reflecting source, process and destination of water and sediment. From the perspective of connotation, application of water and sediment separation has many vivid examples. Rehabilitation Project in Loess Plateau Watershed conserves sediment by vegetations in the hillsides. Depositing sediment on the plain is the process of sediment settlement with gravity. Hydropower projects raise the water level before the dam, slow down the flow kinetic energy and focus on the cumulative deposition of sediment. Water – sediment separation of these projects is all carried out with gravity. Sediment in the Yellow River has fine particle size and flocculation phenomenon is weak, and then effect of sediment – water separation is not obvious. Despite this, history proves that various forms of sediment – water separation play a key role in river management of the Yellow River. In recent years, silt research in lakes and rivers indicate that with the technology of conditioning and modulation, researching and developing sediment coagulation agent which is able to dramatically transform the internal structural properties, reduce the internal resistance, increase floc density, accelerate sediment aggregation of sandy water, sediment – water separation with large – scale and high efficiency can be successfully carried out. Fig. 1 and Fig. 2 have clearly illustrated this.

Fig. 1   Sediment enrichment separation

Fig. 2   Mechanical separation of muddy water

2.3 Source of "Resource utilization"

Any idea about the Yellow River management system without a scientific study of sediment source or destination is far away from the actual problem. The solution of sediment will be incomplete. Sediment and engineering problems of the Yellow River can not depart from the subject of water conservancy project. Water has benefits and disadvantages. Water conservancy makes the benefit maximization and disadvantage minimization through engineering measures. Too much water and sediment cause disasters, which is not the "disaster" attribute of the Yellow River. The drainage area comes less water than sediment, excessive sediment deposit on floodplain, reservoirs and coastal ocean which would lead another disasters. Inadequate use of sediment contributes to disaster, so resource utilization through sediment control and disaster treatment is the best way to handle with sediment.

Sediment is as wealthy and harmful as water. Mud and water hazards resonance is the crux of the Yellow River. Unreasonable structure of industrial and agricultural water in the Yellow River

Basin makes water shortage worse. Less use and settlement incorrectness of the sediment makes the disasters heavier. Premier Zhou Enlai issued instruction in 1964 on that the Yellow River water and sediment should be conductive to the production whether in the upper, middle or lower reaches. Due to the understanding of the connotation of factors to production and material is not profound enough, though half a century has passed, resource utilization of sediment is ineffective. Such as: in coarse sediment yield area, implementation of the engineering measures and biological measures were supplemented by large and medium – sized sediment trapping reservoir to intercept sediment at ditches and valleys, which aim at blocking sand and reduce the sand into the Yellow River. Combined with main engineering of flood control and water utilization, using dead storage to block sediment, decrease damage caused by sediment deposition upstream and downstream is effective. Making silting beach and embankment is also benefit for reduction of sediment. Although these projects have something about sediment, they had not put the sediment as resource to utilize. Gully in Loess Plateau need backfill soil, serious erosion areas need to improve the environment vegetations, levee reinforcement need solidified soil, embankment toe erosion and surface carrying all need constructed road of sediment consolidation. Rural farming and grade highway pavement also need sediment consolidation bricks which have certain mechanics properties. The Yellow River sediment can be raw material and made into loose soil sediment, sediment gel soil, sediment erosion – resisting bricky system as well as all kinds of sediment ecological materials through consolidation by the quenched and curing of the cement matrix. Such can be not only used in river basin water and soil conservation project, but also used for the river flood control project. What's more, sediment will be kind of valuable and marketable resource to society, combined with social road engineering, foundation engineering, civil engineering and garden engineering, which makes profit and find a new solution for the Yellow River sediment at the same time.

2.4 Scientific development concept of the Yellow River management

The past Yellow River management philosophy has a distinctive characteristic of the times. From the legendary Da Yu "containing water" practice, Jia Rang in Western Han Dynasty proposed three strategies, and Pan Jixun in Ming Dynasty proposed "bunching water and dealing with sediment" strategy to contemporary Yellow River management thought, concepts are all closely related to the political, culture and military. Based on the previous river management concepts, "maintaining the Yellow River healthy life" comes into being after a few generations of experience and effort. Four main signs of the ultimate goal are "no dyke breaches, no river flow drying up, no water quality exceeding the standards and no rising of the river bed". Thoughts and contents are not only limited to river water and sediment control, river dynamics, fluvial process of learning areas, but included and extended to the subject of resources and environment. The Yellow River flood and sediment control takes scientifically and comprehensively consideration of the watershed runoff and sediment environment, which is undoubtedly correct. Based on the Yellow River Valley facing such worsening situations as water disaster severity, the contradiction between supply and demand and eco – environ degradation, countermeasures for Short – Term Management of the Yellow River plan set up flood prevention and sedimentation reduction system, which try to control sediment, unify management of water resources and scheduling system and the quantity of pollution. According to statistics, in the early 1990s, about 4.2×10^9 m^3 of pollution were put into the Yellow River, which is doubled than the early 1980s and beyond the pollutant carrying capacity. In 2000, water quality monitoring results show that in the 7,247 km of the main stem of the Yellow River, only 38.7% belong to the class III water quality standard, 20.1% belong to class IV water quality standard, 41.2% belong to class V and worse than class V water quality standard. It clearly indicates that river management is not just about the sediment problem, water pollution treatment also needs to be noted. We must transfer the "sediment" named in river dynamics into "resources" and "pollutants" named in resources and environment disciplines. Sediment in the Yellow River is very fine, and has strong adsorption effect, carry pollutants to second transfer in the process of their own transport, gather pollutants in the riverbed when purify the water, such would constitute a new source of pollution after a period of "catastrophic". This will inevitably lead to the comprehensive management of water environment of sediment. The substance of water – sediment separation concept is solid – liquid separation. Both corrupted sedimentary bodies and solid can be separated from water which could achieve the reduction of sediment, silt and pollution at the same time, reflecting the scientific development connotation of sediment control and pollution control.

3 Connotation of water – sediment separation and resourceful utilization

3.1 Connotation

Sediment – water separation and resource utilization is put forward as a new connotation combined with water conservancy, resource and environment subject, which belongs to water environment engineering and environment project.

Thoughts of water – sediment separation and resource utilization are based on the inflow condition of sediment and water. It can scientifically and mechanically separate sediment and water, reduce the amount of sediment, and maintain the flow. After multi – objective optimization, The Yellow River stream sediment and water quality conditions would be reorganized, a new water and sediment environment would be constructed, the pattern of sediment and water would be changed which would enhance purification capacity and improve the environment of ecological restoration. Combined with water conservancy and hydropower projects and water – sediment regulation, trying to transfer short – term sedimentation reduction, long – term micro – deposition, long – term siltation trend to scouring is possible. Simultaneous implement "water" and "sediment" can develop into resource utilization. Applying new materials technology, curing technique and biological technology, sediment can be used for the construction, soil conservation, water conservancy, embankment works, road engineering, agricultural engineering and land reclamation according to the specifications.

3.2 Technical measures

Technical measures including comprehensive management of "blocking, discharging, separation, adjusting, utilizing and cooperating". "Blocking" means "basin blocking" and "river blocking". "Basin blocking" means sediment reduction, use sediment separated from "river blocking" to backfill drainage gully and hill land, control soil and water erosion by biological engineering. "River blocking" construct environmentally friendly hydroelectric projects (hereinafter referred to as "EH Project") for two usage. Firstly, for flood detention, flood control and flood regulation; secondly, for intercepting sediment and creating the environment of sediment and water separation. "Discharging" means that excess sediment intercepted and separated from environment and electronic power engineering is discharged into river and transferred to the sea. "Separation" means that sediment and water are separated and the sediment intercepted from environment and electronic power engineering is taken out after conditioning by technology to reduce the sediment discharged into downstream as well as to reduce sediment, to reduce deposition, and to reduce pollution. "Adjusting" means that "environment and electronic power engineering" and "environment and electronic power engineering" are combined to adjust water and sediment at the same time, in order to raise the level of sediment transport, to reduce cumulative deposition of river sediment, to increase the quantity of sediment into the sea. "Utilizing" means that the separated water and sediment will be recycled effectively in the way that "clear water" backflows to river and "dry sediment" is used to backfill the plateau gully to strengthen the ability of soil and water conservation and sediment interception as well as to reinforce the bank. "Cooperating" means that technology and method are combined to realize comprehensive treatment. "Separation", "adjustment of water and silt" and "soil and water conservation" play fully technology advantage to control point, straighten line and manage surface respectively. In this way, a kind of flood control and sediment reduction system in the Yellow River combined with "point – line – surface" is built.

4 "Separation" is a great method to solve both superficial and fundamental problems

The crux of the Yellow River is sediment and the difficulty of the Yellow River management that is to find a method to solve this crux. Sediment in the Yellow River comes from Loess Plateau and end up into the sea. The difficulties of transporting sediment in the Yellow River without harmfulness is to find scientific, effective and realistic management ideas, technologies and engineering measures those can balance transportation of water and sediment. People are agreed with that scientific control of plateau soil and water loss is the fundamental method to manage the Yellow River sediment. To realize conservation of plateau soil and water and control of soil and water loss, we need to fulfill gullies and smooth slope surface to create ecological environment that can breed vegetation on one hand, and to select and cultivate plateau biotic population to make sure that living beings and vegetations can develop benign on the other hand. However, practice from the past half century shows that even the correct ideas of water and soil conservation can not produce effect as expected in longer time. The surface runoff in plateau carrying sediment still infuses into the Yellow River. To different assorted attribution of upstream water and sediment, the Yellow River is oppressed to transport water and sediment, no matter how its own condition is. Because of the long time overloaded transportation of water and sediment, the Yellow River falls sick and is in the condition of sub – health, which leads to frequent disaster of water and sediment. Therefore, with a fixed management level of conversation of water and sediment in plateau, the water and sediment going into the Yellow River is fixed. Altering the quantity of water and sediment infused into the Yellow River from river basin will certainly change the transport attributions of water and sediment in the Yellow River. Altering the quantity of sediment infused into the Yellow River and keeping the quantity of water unchanged will change water and sediment assorted attributions of upstream water and sediment in the Yellow River, which will certainly change the transport attributions of sediment in the Yellow River. Altering the quantity of coarse sand infused into the Yellow River from river basin and keeping the quantity of fine sand and water will certainly change the transport situation of water and sediment in the river. Dewatering uses new materials to condition, extracts sediment infused into the Yellow River from river basin in real time by mechanical methods to alter the condition of upstream water and sediment in the Yellow River, and put the sediment outside the river system, in which solving superficial problems is reflected in the reduction of sediment in basin and solving fundamental problems is reflected in the transportation of water and sediment in the river. Honestly, building a dewater system to optimize the separated quantity of solid and liquid or to separate coarse and medium sand can guarantee the sediment concentration of upstream water in an extremely low state and the Yellow River will certainly be in a sub – saturated transport state, which will be extremely good to the reduction of deposition and will keep the cumulative elevation of river bed in limits.

The highest state of modern new ideas of maintaining a healthy life of the Yellow River is to regard the Yellow River as an animated and healthy river. The final goal of new ideas which can be explained as no bank breaks, no river flow drying up, no pollution exceeds and no river bed lifts – will be realized by many ways and engineering methods. Actually, using "the Yellow River" to name this big river in the history has already branded yellow color of silt on blood of the Yellow River. And things are going worse. The problems of the Yellow River are complicated and hard to solve and it is in innate sub – health state. The basic symbols of above are shortage of water resource, innate damage of source that maintains life, shortage and cutoff of water and impurity of water quality. This problem can be an analogy of human body, in which shortage of water resource equals bad hematopoietic function, water equals blood and water quality equals blood quality. This kind of sub – healthy river has characteristics of leukemia and the management of this kind of river is harder than management of leukemia. It is worried that methods such as interception, drainage, release, adjustment, dig and comprehensive management can not be able to solve such problems. Zhang Hongwu regards the Yellow River as a whole to study manage counter plan and takes neighbor basin as a system into consideration, allocating water into the Yellow River from the upstream of Hanshui River and Huaihe River to sweep sediment, reduce deposition and improve ecological environment. Though this kind of water diversion and supplement – compared with blood transfusion – costs a lot of money, the manage effect is remarkable. Another effective method to manage sub – healthy river with characteristics of leukemia is dialysis. The dewater system keeps extracting water and sediment infused into the Yellow River from basin, separating both sediment and pollution attached to sediment outside the Yellow River system and clear water returns to the Yellow River system, which will not only reduce sediment and deposition, but also purify the water quality and maintain water quantity.

5 "Resource utilization" is the best policy to solve the crux

The crux of the Yellow River is sediment and the source and home of managing the crux are two major key problems. People are ordered in the source of sediment but are confused in the home of silt. In the ancient time, dredge and gather of water power are used to transport sediment into seas in the management of the Yellow River. In the 1950s, water storage and sediment interception in sections are used. Sediment interception leads to reduction of deposition, which will lead to increase of sediment drainage. Sediment still ends up to seas. In the past fifty years, engineering methods based on conservation of water and silt and combined interception, drainage, release, allocation and dig are carried out to manage the Yellow River and silt is intercepted in the dead capacity of reservoir. Arrangement of sediment is multifarious such as draining into seas, releasing to beach, allocating water and sediment, digging trough and so on. Though many advantages, these methods can not solve problem in a fundamental way and the obvious characteristics are as follows: ① no matter capacity of reservoir or beach, sediment is arranged inside the Yellow River system and the conservation of the silt quantity will certainly make the transportation happen inside the river system; ② just for reduction of sediment and deposition, sediment is arranged temporarily. Interception to reduce deposition, drainage to push sediment into seas, releasing to reduce sediment concentration in the river and allocating sediment to move sediment will eventually make sediment stay in coast, bed, beach and capacity of reservoir. Given the little water and much sediment, sediment certainly will lead to disaster everywhere in the river system in a long run; ③ the disasters of the Yellow River sediment are impressive. More attentions are paid to the disadvantage of sediment before, and fewer are paid to its advantage. Especially in digging innate potential of sediment, ideas of managing the Yellow River in the past dynasties have seldom involved.

Sediment is a kind of source and it can be in favor of production. Arranging and utilization of sediment as source in scientific ways can turn sediment in the Yellow River into useful, valuable and worthy material. Instead of draining into seas, depositing on the bed and beach, we will find expansive space as home of sediment by pushing and digesting them in rive, basin and society in the way of industry. Those sediment abstinently arranged at reservoir, bed, beach and cost along river nowadays is hard to collect, classify, dispose and utilize. "Dewatering and utilization as source" can dispose sediment in centralized ways by system separating solids and liquids and utilize sediment in dispersive ways by classifying and parting. For instance, sediment scattered materials with good water stability (Fig. 3) can be used as backfill soil for gully in plateau; sediment solidified materials with good mechanical properties can be used to reinforce river bank; sediment indurations with high strength (Fig. 4) can be used as retaining wall in plateau, bank steady protection, slope protection, road protection and guardrail to bear road. Apart from those talked above, the most important meaning is that those recycling sediment materials digest silt in the Yellow River and reduce the ability to lead to disasters by using in industries. After arranged as resource, sediment becomes valuable and useful resource. Not only sediment can be used in soil and water conservation project, hydraulic and hydro – power engineering, bank engineering and road foundation engineering, but also we find social homes for sediment. Thereby sediment will be no longer transported or deposited inside the river system as usual. At the same time, sediment becomes valuable and useful resource. Not only sediment can expressed its engineering application, but also the Yellow River management and social capital market are connected and social capital market of the Yellow River is built by treating sediment as resource.

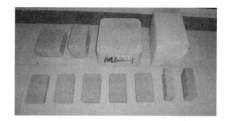

| **Fig. 3   Sediment resource soil** | **Fig. 4   Sediment consolidation brick** |

6   Reality of ideas expressed as "separation and resource utilization"

The theory and practice of the proposed ideas of the Yellow River management must be made aimed closely and practically at the crux of sediment. Sediment is the crux of the Yellow River; the source and home are the problems; sediment transportation in river is the key; the quantity and attribution of upstream water and sediment are the environmental condition of the crux. "Little water and much sediment" is the comprehensive expression and result of natural law of water and sediment in the Yellow River Basin and it is also the crux of the Yellow River disasters caused by sediment. The virtual expression of the damage of little water and much sediment to disasters in the Yellow River is deposition in long series and distance, which further causes the elevation of bed year after year, the frequent flood, the ill and collapse of bank and so on. Therefore, "systematic reduction of sediment" in the Yellow River turns into the core of technological methods of the Yellow River ideas.

"Systematic reduction of sediment" must consider problems of sediment reduction in basin, river and coast at the same time. The essence of sediment reduction in basin is to reduce the loss of water and sediment to realize sediment reduction by biological vegetation engineering and ecological maintenance. At present, to the Loess Plateau, modern technologies and crafts of conservation of water and sediment although has a few effects, yet sediment reduction in basin remains an expectation. We have fixed confidence that sediment reduction in basin is fundamental, but the crux is sediment reduction in river system before the expectation comes true. Sediment reduction of the whole river system is one of the most realistic and guileless idea which are taking out a kind of sediment with disastrous property from the river system. Instead of participating in river sediment transportation, these sediment will translate into resources placed in or outside the system perimeter. This is the real sense of the Yellow River Basin sediment reduction which is different from the normal concept that the amount of the downstream sediment reduced by sediment retaining of upstream; the amount of the sediment reduced by river dredging; the amount of sediment emission to the downstream reduced by placing them on the bottomland. Only by water – sediment transporting in the Yellow River and water – sediment regulation of water conservancy and hydropower projects can we solve the long series of systematic sediment deposition problem.

From past to now, the Yellow River which flows through north and northwest china is named for acceptance and transportation of sediment. By that we can have the concept and achievement of harnessing the Yellow River. Nowadays soil erosion is not the only problem of Loess plateau, effect of runoff, sediment, silt and sludge produced by rainfall become more and more serious. In addition the past tribal type group living in the Yellow River border nowadays becomes medium – sized and large cities, sewage wastewater increases dramatically and discharges into the Yellow River. Mixture of runoff, sediment, silt and sludge challenges the old name of "Yellow River". Instead of sediment problem, the key of defending the Yellow River is comprehensive treatment of turbid water in the Yellow River. As is known to all, pollutants in turbid water will be adsorbed by sediment, suspension transport and deposition of sediment will transfer pollutants inevitably. Sediment – water separation and resource utilization can separate sediment and extract water pollutants simultaneously which not only reducing the sediment discharging into the Yellow River, but also reducing the pollutants into the river and achieving the purpose of reducing sediment and pollutants. So sediment

– water separation has the huge prospective potential in harnessing the Yellow River.

7 Conclusions

On the basis of absorbing the essence of science and technology directed by soil and water conservation and water – sediment transporting, including management thoughts from river dynamics, resources and environment subjects, a new idea of harnessing the Yellow River is put forward according to the Yellow River sediment problems which are flooding, riverbed deposition, water quality deterioration and contradiction between supply and demand of water resources caused by a long series of "more sediment with less water". This management idea highly lays stress on considering blocking and reduction of sediment in the basin, reduction of sediment and silt in the river and seaside as a whole sediment reduction system which can reduce sediment into the Yellow River through sediment – water separation and reduce sediment deposition by scientifically restructuring the conditions of sediment and runoff flowing into the river. Meanwhile sediment can be utilized as a useful resource for plateau basin, river levee, beach protection and road base. A "harnessing sediment by sediment, harnessing pollutants by sediment and harnessing river by sediment" sediment resources circulation system and a new way of treating the sediment can be constructed simultaneously. Then, systematic sediment reduction can be achieved. This idea can be seen as a supplement, perfection and sublimation of "soil – water conservation" and "sediment – water regulation" which is worth further research and practice.

References

Li Guoying. Production and Practice of the Yellow River Control[M]. Beijing: China WaterPower Press,2003.

Yao Wenyi, Shi Mingli, Cui Changjiang. Review of the Yellow River Sediment Problem Research [J]. Journal of Yellow River Conservancy Technical Institute, 2004, 16(1): 1 – 3, 10.

Jing Ke, Li Fengxin. Sediment Disaster Type and Formation Mechanism Analysis [J]. Journal of Sediment Research, 1999,2(01):12 – 16.

Zhou Yuelu. Exploration and Practice on Innovation and Development of Soil and Water Conservation of the Yellow River in the New Era[J]. Soil and Water Conservation in China, 2006(10):5 – 9.

Li Rui, Yang Wenzhi, Li Bi. Research and the Prospect of the Chinese Loess Plateau [M]. Beijing: Science Press,2008.

Zhang Xingchang, Gao Zhaoliang, Peng Keshan. Success and Treatment Measures of Soil and Water Conservation with Chinese Character[J]. Chinese Journal of Nature, 2008,1.

Zhou Yuelu. Exploration and Practice on Innovation and Development of Soil and Water Conservation of the Yellow River in the New Era[J]. Soil and Water Conservation in China, 2006,10(295): 5 – 8.

Liu Wangshuan. The Scientific System of Soil and Water Conservation in the Yellow River Basin [J]. Yellow River, 1996,7(7): 5 – 9.

Jia Zhefeng, Sun Taimin, Lang Wenlin. Thought on Soil and Water Conservation and Ecological Construction and Sustainable Development of Yellow River[J]. Soil and Water Conservation in China, 2005,12(02).

Xu Jiongxin. Recent Tendency of Sediment Reduction in the Middle Yellow River and Some Countermeasures[J]. Journal of Sediment Research, 2004, 4(2): 5 – 10.

Tang Keli. Characteristics and Control Way of the Soil Erosion on the Loess Plateau in Regional [M]. Beijing: China Science and Technology Press , 1991.

Zhang Shengli. Effect of Soil Conservation on Decrease of Water and Sediment in the Middle of the Yellow River [R]. Research of the Yellow River Water Conservancy Scientific Research Institute(ZX – 9908 – 18), 1999.

Milliman J D, Meade R H. World – wide Delivery of River Sediments to the Oceans [J]. Journal of Geology, 1983,91:1 – 21.

Wang Houjie, Yang Zuosheng, Saito Y, et al. Interannualand Seasonal Variation of the Huanghe (Yellow River) Water Discharge over the Past 50 Years: Connections to Impacts from ENSO Events and Dams[J]. Global and Planetary Change,2006,50(3 – 4):212 – 225.

Wang Houjie, Yang Zuosheng, Saito Y, et al. Stepwise Decreases of the Huanghe (Yellow River) Sediment Load (1950 – 2005):Impacts of Climate Change and Human Activities[J]. Global and Planetary Change, 2007,57(3 – 4):331 – 354.

Li Guoying. Spatial Scale Based Yellow River Water and Sediment Regulation[J]. Yellow River, 2004,2(26):1 – 4.

Xu Guobing, Zhang Jinliang, Lian Jianjian. Effect of Water – sediment Regulation of the Yellow River on the Lower Reach[J], Advances in Water Science, 2005,4(16):518 – 523.

Liu Hongbing, Wang Hongyao, Feng Hongguan. Discussion on Sanmenxia Sediment Control of Water Conservancy Hub[J]. Pearl River, 2007,2:67 – 69.

Wang Ying. Sanmen Gorge Reseroir Sediment Deposition and Management Measures[J]. Shanxi Water Resources, 2008,2:69 – 71.

Han Qiwei. Study on Preliminary Operation of Xiaolangdi Reservoir and Flow – sediment Regulation of the Yellow River[J]. Journal of Sediment Research, 2008,6(3):1 – 17.

Yue Zhao,Zhang Anlu, Liu Yun. Discussion on Science and Technology Information of Water – sediment Regulation in the Yellow River[J]. 2009,02(a). 156 – 157.

Zhou Yinjun, Liu Chunfeng. Research Progress of Water – sediment Regulation in the Yellow River [J]. Haihe Water Resources, 2009,12(06): 54 – 57.

Zhang Jinliang. Practice of Water and Sediment Regulation of the Yellow River[J]. Journal of Tianjin University, 2008,9(41):1046 – 1051.

McGhee T J. Water Supply and Sewerage[M]. New York:McGraw – Hill, 1991.

Dallmann W. Mechanische Entwasserung von Klarschlamm (Mechanical Dewatering of Sewage Sludge)[J]. Chem Technik – 45Jg – Heft, 1993;1:42 – 49.

Xin Zhou, Gong Jun, Huang Jian – long. Application of Hydrocyclone Separation Technique in Silt Separation of Yellow River' wate[J]. Journal of Lanzhou University of Technology, 30(4):67 – 69.

Wang Yonghu. Based on Fuzzy Control of GA Yellow River Sediment Centrifugal Separation Process[D]. Master Dissertation 2003.

He Xiaojun. Present Situation of Water Resources Utilization and Water Pollution Prevention and Control Countermeasures[J]. Science & Technology Information, 2007,18: 247 – 248.

Li Mei, Zhu Li, Wang Hongbo. Present Situation of Water Resources Utilization and Water Pollution Prevention and Control Countermeasures[J]. Water Resources Protection, 2008,1 (24):31 – 34.

Shi Qingjun. Discussion of Water Resources Recycling[J]. Ground Water, 2008,3(30):117 – 123.

Dashi Qingyan. Separation Science and Technology [M]. Beijing: China Light Industry Press, 1999.

Liu Fangqing, Fan Deshun, Huang Zhong. Solid – liquid Separation and Industrial Water Treatment[M]. Beijing: China Petrochemical Press, 2001.

Qian Ning. Properties of Hyperconcentrated Flow[M]. Beijing: Tsinghua University Press, 1989.

Jing Ke, Chen Yongzong, Li Fengxin. The Yellow River Sediment and the Environment[M]. Beijing: Science Press, 1993.

Xie JiangHeng, Zhao Wenlin. The History and Current Situation of the Yellow River Sediment Problem[M]. Zhengzhou: Yellow River Water Conservancy Press,1996.

Zhang Junhua, Zhang Hongwu, Chen Shukui. Influence, Reasons and Countermeasures of Yellow River Cutoff[M]. Beijing: Ocean Press,1999.

Yao Wenyi, Zhao Ye'an. Research of the Yellow River Downstream Flow Channel[J]. Advances in Water Science, 1999(2).

Xie Jiaze. Downstream of Yellow River Governance Issues [M]. Beijing: China Science and Technology Press,1995.

Zhang Hongwu. Solutions to the Problems of the Yellow River[J]. Science Times, 1999, 3(24).

Zhao Ye'an, Fan Xianti. The Strategic Position of Sediment Research in Yellow River Harnessing and Development[M]. Zhengzhou: Yellow River Conservancy Press,1996.

Zhang Junhua, Zhang Hongwu. Opinion of Xiaolangdi Reservoir Sedimentation Reduction and Downstream Sediment Deposition[M]. Beijing: Ocean Press,1999.

Zhang Hongwu, Zhang Junhua, Yao Wenyi. The Exploration and Practice of Yellow River Harnessing Strategy[J]. Sediment Research,1999,8.

Analysis of Riverbed Deformation in Infrastructure Zones

P. H. Baljyan and **A. A. Sarukhanyan**

Yerevan State University of Architecture and Construction, 105/1a Teryan Street,
Yerevan, 0009, Republic of Armenia

Abstract: Riverbed formation deformations usually proceed slowly. However during spring and autumn floods the processes are developed quickly. Erection of structures regulating flow in riverbeds or providing engineering services of other nature (dam, bridge passage, etc.) in a given section of the bed initiates artificially caused riverbed forming. They, indeed, are quicker than the previous but die very quickly. However regardless of causes stimulating riverbed forming, sometimes their consequences reach such large sizes, which threaten ecological conditions not only areas of the bed and its surrounding but also nearby residential and industrial areas. Therefore predictions of dangerous riverbed forming process and its final result assume great scientific and practical importance.

This paper presents a physical model of riverbed deformation development caused at roads built over rivers and bridge passages erected for other engineering services which enabled to develop conceptions of mathematical description of phenomena related to the riverbed deformation.

When a bridge passage is designing and constructing over a river the natural width of the riverbed is narrowing. As a result in that section the flow rate is developed and the stream begins washout process of the riverbed. With time the riverbed non – stable formation causes initiation of largest deformations decreasing the flow rate. In the section of a bridge passage the disturbed stability of the riverbed is recovered but in newly formed riverbed conditions. Hence, it can be assumed that transported by the river silt per route remains constant $Q_T = $ const. This concept of the silt balance maintaining is an important condition of the stabilized state.

In evaluation of riverbed formation phenomena riverbed deformations final result is of primary importance.

Until now the stabilized state was presented by a non – stable flow which was described by a system of non – linear differential equations. The latter are solved for various assumption conditions, limiting the obtained results application frameworks.

A new approach for describing the problem is presented, according of which such a pattern of the newly formed stable riverbed is sought when the riverbed formation process runs as a stable one.

Therefore the system of equations for a stable state presents a set of differential equations describing stable state. On the basis of the obtained results of their integration parameters of the shaped riverbed are determined and consequences of riverbed forming are evaluated.

Key words: river, riverbed, deformation, river structure, riverbed forming

The problems related to riverbed formation at mountain zones bridge passages are not sufficiently studied, which explains bridge failure disturbing statistics according to which 2/3 of bridges have been failed by floodwater – triggered erosion and washout around bridge footings. Generally, bridge structures over small rivers are erected only on two riverside piers, often without protecting long dams. For such transitive sections deformation describing a few semi – empiric relationships enables to with a specific accuracy calculate riverbed's floodwater maximum dimension without predicting the place of its origin.

In this paper a mathematical model has been developed for determining river beds floodwater at mountain rivers bridge passes having only near river bank piers.

Let us suppose a bridge passage of b_0 opening and γ_0 length placed over a river with B width and i_0 gradient (Fig. 1). When Q flood flow passes through the opening, the bed washout phenomenon starts. Obviously, due to drastic narrowing of B width close – to – coast jets near the bridge piers come out of smoothly changing range.

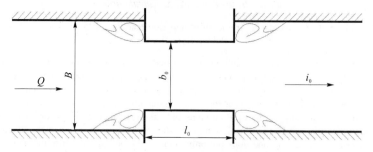

Fig. 1　The plan of the bridge passage

Therefore flow pattern shown in Fig. 1 should be replaced by adequate to reality one when at distant L from the bridge the flow width decreases from B_0 to b_0, at l_0 section b remains constant ($b = b_0$), and after the passage increases from b_0 to river B width (Fig. 2(a)).

Let us now present, for this physical model, mathematical description of bed formation final, steady state phase. On the basis of the carried out development a suggestion on the canal bed z coordinate determination will be presented.

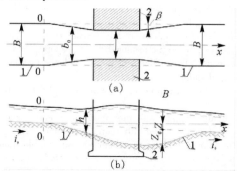

1—Natural bed; 2—Bridge piers

Fig. 2　Plan and profile of the canal and the stream running through it after ending of deformation

In case a selected initial point for calculation (Fig. 2) and designating the jets widening or narrowing angle by β in bridge passage section for stream b width, we have

$$0 \leqslant x < L' \text{ interval}$$
$$b = B - 2x\tan\beta \qquad (1)$$

$$L \leqslant x \leqslant L + L_0' \text{ interval}$$

$$b = b_0$$

$$L + l_0 < x \leqslant 2L + l_0' \text{ interval}$$

$$b = b_0 + 2x\tan\beta$$

where for L section we have

$$L = \frac{B - b_0}{2\tan\beta} \tag{2}$$

By reason of non – steady nature of flood wash phenomenon its behavior as well as final consequences usually until now were described by non – linear differential equations. On the other hand, at the end of non – stationary phenomenon it turns to steady one, which easily can be described by simple linear equations. Using the above principle for the case under discussion this phenomenon can be described by the following equations:

for non – uniform flow

$$\frac{dz}{dx} + \frac{dh}{dx} + \frac{d}{dx}\left(\frac{V^2}{2g}\right) = \frac{V^2}{C^2 R} \tag{3}$$

for continuity of flow

$$Q = AV = \text{const} \tag{4}$$

for silt balance

$$Q = \text{const or } S = \text{const} \tag{5}$$

According to Eq. (5) condition we get

$$\frac{x}{x_0} = \left(\frac{A}{A_0}\right)^2 \tag{6}$$

where, x is wetted perimeter of effective cross – section, the values with "0" subscription indicate natural sections of the river, the values without subscription indicate the bridge section.

Once the flood wash comes to an end and the canal becomes steady to obtain its z coordinate change regularity h, V, R, C values in Eq. (3) presenting by b width, of which relation according to x is presented by known B, b_0, β values Eq. (1). For Z coordinate according to Fig. 2(a) we have

$$Z = Z_r + Z_g \tag{7}$$

where, Z_g is the washout depth, Z_r is the coordinate of the river bed, for which we have

$$Z_r = i_0 x \tag{8}$$

Assuming that the cross – section of the bridge passage is of rectangular shape, we get

$$A = bh \tag{9}$$
$$x = b + 2h \tag{10}$$

Using Eq. (6) condition with Eq. (4), Eq. (9) and Eq. (10) relationships in consequence of particular modifications we have

$$h = \frac{\varphi_0}{b^{\frac{2}{3}}} \tag{11}$$

$$R = \frac{b\varphi_0}{b^{\frac{5}{3}} + 2\varphi_0} \tag{12}$$

$$V = \frac{Q}{b^{\frac{1}{3}}\varphi_0} \tag{13}$$

Employing Manning's formula to obtain Shezi C factor and substituting h, V, R values in motion Eq. (3), we have

$$\frac{dz}{dx} + \varphi_0 \frac{d}{dx}\left(\frac{1}{b^{\frac{2}{3}}}\right) + \frac{Q^2}{2g\varphi_0^2}\frac{d}{dx}\left(\frac{1}{b^{\frac{2}{3}}}\right) = \frac{Q^2 n^2}{\varphi_0^{\frac{5}{3}}}\left(\frac{b^{\frac{5}{3}} + 2\varphi_0}{\sqrt{b}}\right)^{\frac{4}{3}} \tag{14}$$

Taking into account Eq. (7) and Eq. (8), and replacing Eq. (14), we get

$$\frac{dz}{dx} + i_0 + \left(\varphi_0 + \frac{Q^2}{2g\varphi_0^2}\right)\frac{d}{dx}\left(\frac{1}{b^{\frac{2}{3}}}\right) = \left(\frac{Qn}{\varphi_0^{\frac{5}{3}}}\right)^2\left(\frac{b^{\frac{5}{3}} + 2\varphi_0}{\sqrt{b}}\right)^{\frac{4}{3}} \tag{15}$$

To obtain channel deformations z in the bridge passage section using Eq. (15), it is necessary to integrate this equation taking into account conditions of the bed width change expressed in Eq. (1).

References

Balshakov V A, Boukhin M N, et al. Hydrological and Hydraulic Calculation of Small Road Constructions. High School, 1983;280.

Andreev O V. Planning of Bridge Passages – M. : Транспорт, 1980;437.

Neill C R. Measurements of Bridge Scour and Bed Changes in a Flooding Sand – bed River[J]. Proceedings of the Institution of Civil Engineers, 1965, 30(2) : 415 –435.

Балджян П. О. Определение зависимости между гидравлическими параметрами потока при постоянстве их наносонесущей способности. – Известия НАН РА и ГИУА (серия ТН), т. 58, № 2, Ер. : 2005, стр. 380 –385.

Optimization and Regulation of Operational Mode in Xiaolangdi Reservoir during Later Sediment Retaining Period

Zhang Junhua, *Ma Huaibao*, *Dou Shentang*, *Wang Ting* and *Zhang Fangxiu*

Yellow River Institute of Hydraulic Research, YRCC, Zhengzhou, 450003, China

Abstract: The designed sediment storage capacity in Xiaolangdi Reservoir is 7.5×10^9 m^3. The reservoir deposition morphology during the whole sediment retaining period is as follows: in initial sediment retaining period, the deposition morphology is a delta. With the continuous deposition in reservoir, the delta vertex moves gradually downstream to dam. And at last, the deposition morphology will be a cone. After that, deposition surface will increase and reach designed elevation. During later sediment retaining period, in the process of water and sediment regulation, in most time the reservoir is in water storage condition. The delta deposition morphology can keep relatively large storage capacity. With the same amount of sediment and water storage, compared with cone deposition morphology, the water lever before dam of delta deposition morphology is obviously lower, the backwater length is obviously shorter. The analysis shows that delta deposition morphology is obviously superior to cone deposition morphology in the aspects of optimizing outlet water and sediment process, effectivly utilizing tributary storage capacity, trapping the coarse sand and discharging the fine sand, maintaining effective storage capacity for a long time and so on. The results of mathematical model and physical model preliminarily show that optimizing operational mode of reservoir can keep delta deposition morphology for a relatively long period, slow reservoir deposition velocity, extend sediment retaining life, and increase the amount of sand into the sea. At the same time, it does not obviously decrease sedimentation reduction effect of downstream Gao – cun reach.

Key words: reservoir deposition morphology, reservoir optimization operation, density current sediment delivery, free flow sediment delivery

The operational mode of multi – year sediment regulation and man – made precipitation washout at right occasion is recommended as reservoir operational mode in later sediment retaining period. The characteristic of the mode is, in general water and sediment condition, gradually increasing water level, trapping the coarse sand and discharging the fine sand and carrying out water and sediment regulation, especially in flood period, decreasing water level and carrying out precipitation washout. During later sediment retaining period, in the process of water and sediment regulation, in most time the reservoir is in water storage condition. The most water storage is 1.3×10^9 m^3. In reservoir backwater zone, the flow pattern of sediment transport is free flow or density current.

The designed sediment storage capacity of Xiaolangdi Reservoir (XLD) is 7.5×10^9 m^3. The reservoir deposition morphology during the whole sediment retaining period is as follows: in initial sediment retaining period, the deposition morphology is a delta; with the continuous deposition in reservoir, the delta vertex moves gradually downstream to the dam; finally, the deposition morphology will be a cone. The longitudinal slope always looks sediment transport equilibrium slope as a target value and dynamically adjusts. At last, deposition surface will increase and reach designed elevation. There is no doubt that reservoir deposition morphology and its change have important influences on water and sediment transport characteristics, and the former is closely related with reservoir regulation. Therefore, in later sediment retaining period, it is necessary to study influence and limitation between deposition morphology and sediment transport process, select deposition morphology in favour of water and sediment regulation, and study reservoir operational mode maintaining the morphology as long as possible. They are of important significance to make full use of the comprehensive utilization benefit of reservoir.

1 Influence mechanism and function of reservoir deposition morphology on water and sediment transport

If the deposition morphology in Xiaolangdi Reservoir is different, its storage capacity distribution will have great difference. For example, In October, 2008, the amount of sediment in Xiaolangdi Reservoir was 2.411×10^9 m^3; the deposition morphology was a delta; its vertex was 24.43 km apart from dam; the elevation was 220.25 m; the storage capacity below vertex was 1.0×10^9 m^3, and the deposition surface elevation before dam was at 185 m. With the same amount of sediment, if the deposition morphology is a cone, then the eigenvalues is different (Tab. 1). As can be seen, compared delta deposition morphology with cone deposition morphology, if water storage is same, then the water level of the former is lower; if the water level is same, the backwater length of the former is shorter. The researches based on sediment transport in Xiaolangdi Reservoir show that, with the same amount of sediment, in the process of water and sediment regulation, the delta deposition morphology which can keep more storage capacity near dam (downstream Balihutong reach, which is 24 km apart from dam) is superior to cone deposition morphology.

Tab. 1   Storage capacity eigenvalues of different deposition morphology

Elevation (m)	Storage capacity($\times 10^8$ m^3)		Backwater length(km)	
	Delta	Cone	Delta	Cone
210	5.845	2.618	19	19
215	8.228	5.142	23	34
220	10.95	8.497	24	51

1.1   Delta deposition morphology is beneficial to optimizing output water and sediment combination

During later sediment retaining period, the operational mode is multi – year sediment regulation and man – made precipitation washout at right occasion. In general water and sediment condition and in the process of water and sediment regulation, in most time the reservoir is always in water storage state, and the water storage changes between 0.2×10^9 and 1.3×10^9 m^3 regulating storage capacity. Obviously, the delta deposition morphology can make storage capacity of water and sediment regulation more near to dam and more beneficial to shaping density current. The density current is superior to backwater free flow in aspects of increasing sediment ejection effect and volume in water storage state and slowing output sediment polarization. Muddy water reservoir caused by density current can further optimize output water and sediment combination.

1.1.1   Delta deposition morphology is beneficial to plunging of density current

The water depth of density current plunging point can be expressed as follows:

$$h_0 = \left[\frac{1}{0.6\eta_g g} \frac{Q^2}{B^2} \right]^{1/3} \tag{1}$$

The water depth of density current uniform flow can be expressed as follows:

$$h_n' = \frac{Q}{V'B} = \left(\frac{\lambda'}{8\eta_g g} \frac{Q^2}{J_0 B^2} \right)^{1/3} \tag{2}$$

where, η_g is correction coefficient of gravity; Q is flow; B is average width; J_0 is bottom slope of reservoir; λ' is drag coefficient of density current and taken as 0.025.

If water depth h_n' of uniform flow formed by density current is less than h_0, namely $h_n' < h_0$, the plunging of density current is successful. Otherwise, the water depth of density current will exceed the water depth of clear water, and the density current will float upward and disappear.

When water depth h_n' is equal to h_0, correspondingly, the critical bottom slope of reservoir is 0.001,875, that is, $J_{0,c} = J_0 = 0.001,88$. In general, besides satisfying plunging condition Eq. (2), bottom slope J_0 also should be bigger than $J_{0,c}$, then density current will form. Therefore, once cone deposition morphology is formed, in most time bottom slope J_0 is little than $J_{0,c}$. So it is difficult to shape density current.

1.1.2 In the same water storage condition, sediment delivery rate of density current is bigger

In backwater zone, there are two kinds of flow patterns of sediment transport, which are free flow and density current, among which the formula of sediment ejection of backwater free flow is as follows:

$$\eta = a \lg Z + b \tag{3}$$

where, η is sediment delivery rate; $Z = \left(\dfrac{V}{Q_o} * \dfrac{Q_i}{Q_o} \right)$, is backwater index; V is water storage volume of calculation time; Q_i, Q_o are reservoir inflow and outflow; a, b are constants.

In order to compare sediment ejection effect of free flow with density current, we select several sediment ejection process of density current during water and sediment regulation period, which are of low control water level and flow state in delta top slope section. With the same incoming water and sediment process and water storage condition, we suppose that sediment ejection pattern is backwater free flow; sediment ejection amount is calculated by Eq. (3), and the calculated values are compared with the practically measured values of density current (Tab. 2). We can see, when density current formed, because of small muddy water flow area and large flow rate, the sediment ejection effect was obviously superior to backwater free flow. Especially after 2007, delta vertex was relatively near to dam, this meant running distance of density current was short, and the effect was more remarkable.

Tab. 2 Comparison of sediment ejection of backwater free flow and density current

Year	Period (month. day)	Inflowing sediment amount ($\times 10^8$ m^3)	Outflow sediment amount ($\times 10^8$ m^3)	
			Calculated values	Measured values
2006	6.25 ~ 6.28	0.230	0.052	0.071
2007	6.26 ~ 7.2	0.613	0.161	0.234
	7.29 ~ 8.8	0.834	0.153	0.426
2008	6.28 ~ 7.3	0.741	0.157	0.458

1.1.3 Turbid water reservoir can further optimize output water and sand combination

After density current moves to the front of dam, the sediment particles suspended in flow is very fine and the concentration is very high, and the turbid water reservoir settles very slowly. By using this characteristic, according to incoming water and sediment and the law of sediment transport of the lower Yellow River, the discharge tunnels at different elevation can be opened so as to optimize output water and sand combination.

1.2 Delta deposition morphology is beneficial to trapping the coarse sand

The longitudinal slope of delta foreslope reach is 10 times more than cone deposition morphology. During later sediment retaining period, in the process of water and sediment regulation, with the same water storage, the backwater length of delta deposition morphology is about half of cone deposition morphology (Tab. 1). If deposition morphology of reservoir is a cone, not only coarse sand deposits in the river reach of backwater zone, but also a large amount of fine sand will deposit along reservoir. If deposition morphology is a delta, in delta continent surface section, the flow is free flow or low backwater state. When meeting high sediment – laden small flow process coarse sediment sorts and deposits, after plunging of density current, the moving distance of density current is relatively short, and therefore the sediment delivery rate of fine sand is relatively high.

1.3　Delta deposition morphology is beneficial to effectively utilizating tributary storage capacity

When tributary is in the downstream river reach of main river delta vertex, the deposition patterns of main river and tributary are flowing backward of density current and tributary longitudinal deposition distribution are smooth, without obvious sand bar, and tributary storage capacity can be used to water and sediment regulation. If deposition morphology is a cone, tributary is equivalent to horizontal extension of main river riverbed. Especially during a long low water period, the sand bar is easy to form at tributary estuary, and storage capacity below the elevation of sand bar is always difficult to effectively use. Zhenshui is the biggest tributary of Xiaolangdi Reservoir, of which the original storage capacity is 1.75×10^9 m^3 and the distance from dam is 18 km. Tributary Shijinghe and Dayuhe are 20 km and 4 km from the dam respectively. The total original storage capacity of these three tributaries is 2.883×10^{10} m^3 and accounts for 52.7% of total tributary. Maintaining delta deposition morphology and distance of vertex to dam more than 20 km are beneficial to using storage capacity of several big tributaries near dam to carry out water and sediment regulation.

1.4　Delta deposition morphology is beneficial to maintaining effective storage for a long time

When sediment retaining period is over, the deposition morphology will be in high beach and deep channel. According to the design, 1.0×10^9 m^3 channel storage capacity below beach 254 m will be used to carry out water and sediment regulation, and 4.1×10^9 m^3 over beach will be flood control storage. In fact, with the change of flow process, channel often constantly adjusts. Long – term small flow process makes channel atrophy, and in backwater zone deposition is very obvious. When a larger flow process occurs, it is inevitable to appear overbank deposition and loss beach storage. If delta deposition morphology can be maintained in a long period, the most reach will be apart from backwater zone; atrophy will be slow, and sediment in the downstream reach of reservoir is easier to discharge.

2　Optimization principle and regulation mode of deposition morphology in Xiaolangdi Reservoir

2.1　Optimization regulation ideas

Because unharmonious incoming water and sediment and complex topography in Xiaolangdi Reservoir, deposition morphology of main river and tributary and its change are of diversity, randomness and complexity. In addition to natural factors, reservoir deposition morphology mainly depends on regulation mode.

Near vertex of deposition delta, the coarse sand in sediment – laden flow sorts and deposits, and the flow carrying fine sand forms density current and moves towards dam. Therefore, the bed material near dam, mostly being fine sand, is viscous deposit. When the deposits are still in the unconsolidated state, it can be seen as Bingham fluid and described by rheological equation $\tau = \tau_b + \eta \dfrac{du}{dy}$. When the shear stress of deposits along a certain sliding surface exceeds its yield shear stress τ_b, the slump will happen and storage capacity will recovery.

In the process of reservoir operation during sediment detaining period, when meeting proper flood process, by decreasing water level, retrogressive erosion will occur. Scouring of density current deposition reaches before dam and erosion recession of delta can effectively recovery storage capacity below delta and possibly maintain delta deposition morphology gradually increase (Fig. 1). At the same time, in the process of moving toward dam, the sediment scoured from delta will sort once again, so that the finer sand will be discharged from reservoir. Because sand discharged from reservoir comes from beach and downstream reach of reservoir, the sand is relatively fine. Therefore, when storage capacity is recovering, at the same time, it is of little effect on the lower Yellow River.

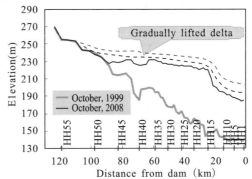

Fig. 1   Schematic diagram of deposition morphology in XLD

2.2   Optimization regulation principles

During the process of reservoir sediment retaining, delta deposition morphology should be maintained and gradual increase and large regulating storage capacity should be kept near dam. When scouring and recovering storage capacity occurs, at the same time, combination and process of outlet water and sediment should be controlled, and adverse effect should be reduced or eliminated on the lower Yellow River reach.

2.3   Regulation optimization measures

Discrimination index of reservoir deposition morphology should be established, mainly including delta vertex position, near dam reach slope, near dam storage capacity, delta area, average water depth, and other key parameters. They are related with reservoir control instructions so as to optimize regulation.

Based on operational mode of "multi-year sediment regulation and man-made precipitation washout at right occasion" during later sediment retaining period, two methods are used to associate reservoir deposition morphology with control instruction. Firstly, when reservoir morphology index do not satisfy the set requirements, increasing sediment ejection opportunity of reach near dam. For example, at the end of flood season, it is needed to carry out a precipitation washout at right occasion. The influence of adverse output water and sediment combination on the lower Yellow River reach can be eliminated or decreased by preflood water and sediment regulation in the next year. Secondly, it is not to increase new scour opportunity, but when executing instruction of "man-made precipitation washout at right occasion", appropriate extension or development are needed (e. g. increasing flood days or optimization control index, et al.).

Optimization measures include: in the period of full water storage and making flood peak, according to basic flow in Toudaoguai, distinguishing whether 0.2×10^9 m^3 water storage is still retained. If incoming water of the next six days are more than 0.4×10^6 m^3, this means that empty reservoir can be quickly supplied, then reservoir will no longer retain water storage. Emptying scour goal is to make regulating storage capacity near dam satisfy the requirements, and it also can be controlled by increasing empty scour time. By the end of August, if regulating storage capacity near dam is less than demand index, then according to reservoir water storage or incoming water, it is needed to carry out a precipitation erosion process between September 1 and September 10.

3　Regulation and effect of reservoir optimization scheme

In view of the research is a further optimization (in short, Optimization scheme) of operational mode in Xiaolangdi Reservoir during later sediment retaining period (in short, Foundation scheme). By means of mathematical model of one dimensional unsteady flow, calculations of the said two schemes include the same series of water and sediment regulation process in Xiaolangdi Reservoir and water and sediment transport process in the lower Yellow River respectively, and results are also compared. At the same time, according to water and sediment condition calculated by mathematics model, physical model test in Xiaolangdi Reservoir is carried out.

3.1　Calculation results of mathematical model

3.1.1　Xiaolangdi Reservoir

The initial boundary conditions of tests are all the terrains of October 2007. The incoming water and sediment condition is 20 year series of less water and sediment, that is the series of 1990 ~ 1999 + 1956 ~ 1965 based on design level year of 2020. The results are shown in Tab. 3. By the end of 20th years, the amount of sediment of optimization scheme is 6.686×10^9 m^3 and less than 7.5×10^9 m^3 which is the end index of sediment retaining period. The storage capacity near dam is 1.658×10^9 m^3, and the storage capacity of water and sediment regulation below 254 m elevation is 1.425×10^9 m^3. Obviously, after 20 years, the amount of sediment of optimization scheme is less than foundation scheme, and correspondingly, storage capacity index are relatively bigger. The accumulated amount of sediment and the terrain contrast diagram of two schemes are shown in Fig. 2 and Fig. 3 respectively.

Tab. 3　Amount of sedimemt and storage capacity of foundation scheme and optimization scheme　　　　　　　　　　　　　　　Unit: $\times 10^8$ m^3

Scheme	Accumulated amount of sediment			Total storage capacity of main river	Storage capacity of main river near dam	Storage capacity of water and sediment regulation
	Reservoir	Main river	Tributary			
Optimization scheme	66.86	48.95	17.91	29.66	16.58	14.25
Foundation Scheme	77.07	55.51	21.56	23.13	14.01	9.30

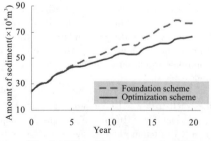

Fig. 2　Deposition process in XLD

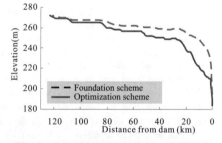

Fig. 3　Longitudinal thalweg profile of mainstream in XLD

3.1.2 Lower Yellow River

The sediment into the lower Yellow River of optimization scheme is more 1.327×10^9 t than foundation scheme. Accumulated full face volume of scour of two schemes in the lower Yellow River are 3.23×10^8 t and 9.74×10^8 t respectively, and the difference is 6.51×10^8 t. Accumulated main channel volume of scour are 4.35×10^8 t and 1.03×10^9 t respectively, and the difference is 5.95×10^8 t. Sediment into the sea of optimization scheme is 6.67×10^8 t more than foundation scheme.

For different reaches, in Fig. 4, the difference of two schemes is mainly in Xiaolangdi—Sunkou reach, especially in Xiaolangdi—Jiahetan reach. In Xiaolangdi—Huayuankou reach, accumulated volume of scour of foundation scheme is 8.4×10^7 t, while accumulated volume of sediment of optimization scheme is 1.97×10^8 t, and the difference is 2.81×10^8 t. In Huayuankou—Jiahetan reach, accumulated volume of scour of foundation scheme is 1.49×10^8 t, while accumulated volume of sediment of optimization scheme is 1.7×10^7 t, and the difference is 1.66×10^8 t. For reach downstream Gaocun, the accumulated volume of scour of foundation scheme and optimization scheme are 6.09×10^8 t and 4.80×10^8 t respectively, the latter is 1.29×10^8 t less than the former. The scour and silting amount in Xiaolangdi—HuaYuanKou reach are shown in Fig. 5. As we can see, the most difference occurs in 14th years, and the amount of sediment of optimization scheme is more than 3.7×10^7 t.

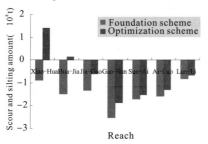

| Fig. 4 | Accumulated full face scour and silting amount in different reach | Fig. 5 | Full face scour and silting amount Between Xiaolangdi – Huayuankou |

3.2 Physical model test of Xiaolangdi Reservoir

According to water and sediment condition supplied by mathematical model, physical model test of Xiaolangdi Reservoir was carried out. Physical model and mathematical model can complement and confirm each other.

By the end of 20th years, the total amount of sediment of optimization scheme is 6.124×10^9 m^3. The storage capacity near dam is 1.655×10^9 m^3, and the storage capacity of water and sediment regulation below 254 m elevation is 2.227×10^9 m^3. The results of optimization scheme and foundation scheme are shown in Tab. 4 and 18th year longitudinal thalweg profiles of main river are shown in Fig. 6.

Tab. 4　Amount of sedimemt and storage capacity of foundation scheme and optimization scheme　　　　　　　　　　　　　　　　Unit: $\times 10^8$ m^3

Scheme	Accumulated amount of sediment			Total storage capacity of main river	Storage capacity of main river near dam	Storage capacity of water and sediment regulation
	Reservoir	Main river	Tributary			
Optimization scheme	61.24	44.52	16.73	30.39	16.55	22.27
Foundation Scheme	73.91	53.68	20.23	21.23	11.00	13.17

By contrasting results of physical model test with mathematical model calculation, we can see, conclusions are consistent, but each index is slightly different. By the end of 20th years, the amount of sediment in reservoir is relatively less; the corresponding storage index is large.

4 Epilogue

By analyzing two deposition morphologies, it can be seen that the delta deposition morphology is more beneficial

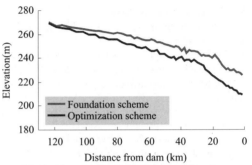

Fig. 6 Longitudinal thalweg profile of mainstream in XLD

than cone deposition morphology. By optimizing reservoir operational mode, it is possible to maintain delta deposition morphology for a long period.

The results of mathematical model show that optimization scheme of Xiaolangdi Reservoir can effectively slow deposition speed, and the amount of sediment of 20th years is less about 1.021×10^9 m^3. Optimization scheme can maintain delta deposition morphology for a long period The storage capacity near dam at the end of 20th year is 1.658×10^9 m^3, and the channel is obviously lower about 10 m than foundation scheme. Optimization scheme have great influence on slowing sand bar uplift at tributary estuary downstream delta vertex and is beneficial to effective use of tributary storage capacity. The qualitative conclusion of physical model of Xiaolangdi Reservoir and mathematic model are basically consistent. As for the amount of sediment of 20 years in reservoir, the optimization scheme is about 1.267×10^9 m^3 less than foundation scheme, and optimization scheme is obviously large than foundation scheme in the aspects of main river, near dam and water – sediment regulation storage capacity.

The sediment into the lower Yellow River of optimization scheme is 1.327×10^9 t more than foundation scheme. The amount of scour of whole the lower Yellow River is less 0.651×10^6 t, but sedimentation reduction effect of reach below Gaocun does not obviously decrease. However, sediment into the sea increases. The difference of two schemes is concentrated on several years with more water and sediment. Therefore, it is necessary to further optimize reservoir operational mode.

Acknowledgements

The work was supported by the National Natural Science Foundation of China(NSFC) (NO. 51179072) and public welfare industry research special fund of the Ministry of Water Resources (MWR) (No. 200901015).

References

Fan Jiahua. Research and Application of Density Current[M]. Beijing: Hydraulicand Electric Power Press, 1959.
Han Qiwei. Reservoir Sedimentation[M]. Beijing: Science Press, 2003.
Tu Qihua, et al. Analysis and Calculation of Sediment Scour and Silting and the Effective Capacity in Xiaolangdi Reservoir [R]. Zhengzhou: Institute of Survey, Planning and Design, YRCC, 1983.

Estuary Sedimentation Control Using "UTSURO" (Tidal Reservoir)

Yoshiya Ogawa[1], *Kenji Sawai*[2] and *Kazuaki Akai*[3]

1. Setsunan University, 17-8, Ikeda-nakamachi, Neyagawa, Osaka, Japan
2. Setsunan University, 17-8, Ikeda-nakamachi, Neyagawa, Osaka, Japan
3. NGO "UTSURO" Research group, 201, Ogura, Wakayama, Wakayama, Japan

Abstract: The Yellow River which is the second largest in China is worldwide famous for a river with much sediment discharge and the river bed aggradation is occurring rapidly and the risk of the flood overflow increases.

The authors study on river bed sedimentation control using "UTSURO" (a tidal reservoir surrounded artificially). In this method, a tidal reservoir is connected to the river in the tidal area, and river bed is scoured by the flow caused by the reservoir which saves water at the flood-tide and this saved water is discharged from the tidal reservoir at the falling tide. In this study, the river bed change in the Yellow River estuary part (lower than Lijin) for a long term is analyzed by numerical simulation. The obtained result of this study is summarized as follows:

(1) Without tidal reservoir, the Yellow River overflows at the big flood.

(2) Tidal reservoir is effective in the river bed decline or sedimentation restraint.

(3) The longer the distance between the connecting point and estuary, the bigger the effect of the tidal reservoir to the upper stream reaches of the river as long as the reservoir is affected by the tide.

(4) However, the longer the distance between the connecting point and estuary, the smaller the degree of the river bed decline in the downstream.

(5) The tidal reservoir is able to become one of the Yellow River improvement measures.

Key words: sedimentation control, estuary, tide, UTSURO, the Yellow River

1 Introduction

The Yellow River is a river with much sediment discharge worldwide, yearly average effluent discharge 5.80×10^8 m³ yearly average sediment discharge 1.6×10^9 t and average sediment density 35 kg/ m³.

According to the statistics (see Tab. 1) after 1950, yearly average outflow to the Yellow River and yearly average inflow to the sea decrease by the reduction of the precipitation and rapid development with the population growth increased water resources using quantity. In addition, the development availability rate of water resources increases rapidly after the 1970s, and can confirm high dependence to river water.

According to a certain survey, it is reported that the water level equivalent to the discharge 3,000 m³/s of the downstream section increase from 1.3 m to 1.7 m at flooding period between 1985 and 1996, and it is the serious problem that is the decline of the river improvement security degree at the time of the flooding with the river bed rise in the downstream section.

Therefore we examined to solve this problem by numerical analysis about possibility estuary sedimentation control using tidal reservoir.

2 Characteristic of river bed sedimentation control using a tidal reservoir

Let us suppose a tidal reservoir "UTSURO" connected by a channel to the river near an estuary as shown in Fig. 1.

Tab. 1　Yearly average outflow and inflow to the sea of the Yellow River

	Year	Yearly average outflow quantity(A) ($\times 10^6$ m^3)	Inflow to the yearly average sea(B) ($\times 10^6$ m^3)	A/B(%)
①	1950 ~ 1959	611. 6	480. 5	78. 6
②	1960 ~ 1969	679. 1	501. 1	73. 8
③	1970 ~ 1979	559. 0	311. 0	55. 6
④	1980 ~ 1989	598. 1	285. 8	47. 8
⑤	1990 ~ 1999	437. 1	140. 8	32. 2
⑥	2000 ~ 2009	288. 4	45. 7	15. 8
⑦	Average	528. 9	294. 2	55. 6
⑧	⑥/⑦(%)	54. 5	15. 5	—

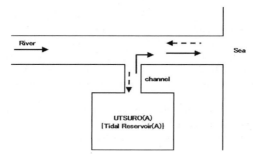

Fig. 1　Tidal current between a reservoir and an estuary

　　Here, water flows in and out if connected part does not have the structures such as gates because tidal reservoir is connected to the channel in tidal area. It is thought that the following influence extends to the waterway by this tidal reservoir.

　　(1) In falling tide, It is added discharge from the upper stream and tidal reservoir.

　　(2) Depending on a scale of tidal reservoir, the flow of the flood scale occurs.

　　(3) Because the tide occurs twice a day, the flow of the flood scale produces twice a day.

　　(4) Depending on a flood, a river bed shape changes.

　　(5) Because a flood occurs repeatedly, the quantity of river bed change grows big.

　　(6) The influence that it occurred in the downstream side from a joint extends to the side of upper reaches.

　　It is not affected by the quantity of water from the upper reaches because water flowing in tidal reservoir is mainly back water at the time of a high tide.

3　Numerical simulation

3. 1　Analyzing method

　　This study examine the most suitable connection position and a scale of tidal reservoir to the decline the river bed in the estuary part less than Lijin. In addition, a period assumed around 100 years.

　　Therefore the analysis method adopt one-dimensional river bed change analytical method. This analytical method is superior technique for a long term calculation, because there is less calculation number of times than two-dimensional river bed change analysis.

The flow demands from an expression Eq. (1) and Eq. (2).

$$Q = const \tag{1}$$

$$z_b + h + \frac{v^2}{2g} + h_l = const \tag{2}$$

where, Q is discharge; z_b is river bed level; h is the depth of the water; v is velocity; g is gravitational acceleration; h_l is head loss.

Next, the river bed change demands from an expression Eq. (3) to Eq. (8).

$$q_{b*} = 17\tau_*^{3/2}\left(1 - \frac{\tau_{*c}}{\tau_*}\right)\left(1 - \sqrt{\frac{\tau_{*c}}{\tau_*}}\right) \tag{3}$$

$$q_{b*} = \frac{q_b}{\sqrt{(\sigma/\rho - 1)gd^2}} \tag{4}$$

$$q_{su} = 0.008\left[0.14\frac{\rho}{\sigma}\left(14\sqrt{\tau_*} - \frac{0.9}{\sqrt{\tau_*}}\right) - \frac{wf}{\sqrt{(\sigma/\rho - 1)gd}}\right]\sqrt{(\sigma/\rho - 1)gd} \tag{5}$$

$$\frac{wf}{\sqrt{(\sigma/\rho - 1)gd}} = \sqrt{\frac{2}{3} + \frac{36v^2}{(\sigma/\rho - 1)gd^3}} - \sqrt{\frac{36v^2}{(\sigma/\rho - 1)gd^3}} \tag{6}$$

$$\frac{Q}{B}\frac{dc}{dx} = q_{su} - wf \cdot C \tag{7}$$

$$\frac{\partial z_b}{\partial t} + \frac{1}{(1-\lambda)B}\frac{\partial Bq_b}{\partial x} + \frac{1}{(1-\lambda)}(q_{su} - wf \cdot C) = 0 \tag{8}$$

where, τ_* is dimensionless tractive force, $\tau_* = u_*^2/\{(\sigma/\rho - 1)gd\}$; τ_{*c} is dimensionless critical tractive force; q_b is bed road sediment; σ is density of sand; ρ is density of water; d is grain size; λ is porosity; B is width; u_* is friction velocity, $u_* = \sqrt{gRI_e}$; R is hydraulic radius, $R = Bh/(B + 2h)$; I_e is energy gradient, $I_e = (n^2Q^2)/[R^{4/3}(Bh)^2]$; n is roughness coefficient; q_{su} is unit area suspended sediment surfacing quantity; wf is settling velocity; v is kinematic viscosity; C is suspended sediment concentration.

3. 2　Analyzing condition

A conditions of various kinds examined based on a document of the past and it assumed the analysis condition of this study. In this study, but the sediment which flowed into Bohai Bay from the Yellow River do sedimentation and drift sand by tides, the influence of the tide ignore because I expect that the influence of the river bed decline extends to the joint, the upstream and downstream side of tidal reservoir. Because the width of Bohai Bay is infinite wider than the Yellow River, it open a little to avoid destabilization of the analysis. In addition, but the calculation result is consider more than the height of the dike, it does not consider overflow and flood levee breach because the dike is high enough. The downstream edge of the analysis assume the offshore 10 km spot of Bohai Bay, the average tide level approximately 2 m and the average tide range give tide of approximately ± 1. 25 m. In this study, from the Lijin to Bohai Bay assume that, distance approximately 104 km, width approximately 1 km and river bed incline about 0. 7/10, 000. In addition, in this section, there is not the junction from other rivers. The particle size constituting a river bed is 0. 05 mm occupies most.

At the edge of upper reaches (Lijin) do known discharge. A yearly average outflow quantity of water is approximately 120×10^6 m^3 and approximately 60% are concentrated from July to October. In addition, about the discharge at the time of the flood was calculated discharge of the Lijin based on the record at the Huayenkou (Tab. 2) . In calculation, it is consider to the Sonkou is narrow segment between both sections. In late years, the discharge include it because adjustment of quantity of water and the quantity of sediment is carried out by Xiaolangdi dams. In addition, the flood starting date is $0:00$ on August or September 1 , and It can get closer to flood discharge for one day. In duration, because it is an average of ten days by the record at Huayuankou after 1950 , the duration adopts ten days. The flood end date and time can get closer to discharge in the normal afterwards for 24 hours at $24:00$ on August or September 11 .

Next, the connection position of the tidal reservoir is three places; the dried tide level neighborhood, the average tide level neighborhood, the middle of the average tide level and the high tide level. But the scale varies according to a connection position, it is establishes so that the Yellow River discharge of the neighborhood of joint in the tide height of becomes 2,000 m^3/s to 15,000 m^3/s. In addition, if a flood occurs, the outflow case from tidal reservoir is not performed. In addition, the sand does not go in and out of tidal reservoir because there is enough distance between tidal reservoir and the river.

Tab. 2   A scale and outbreak time of the flood discharge in the Lijin spot
(edge of upper reaches)

Year	August	September	Year	August	September	Year	August	September
1	7,850	2,000	35	6,550	2,000	69	3,500	2,000
2	6,550	2,000	36		2,000	70		2,000
3	6,185	2,000	37		2,000	71		2,000
4	3,500	2,000	38		2,000	72		2,000
5		2,000	39	3,500	2,000	73		2,000
6		2,000	40		2,000	74	3,500	2,000
7		2,000	41		2,000	75		2,000
8		2,000	42		2,000	76		2,000
9	3,500	2,000	43	6,185	2,000	77		2,000
10		2,000	44	3,500	2,000	78		2,000
11		2,000	45		2,000	79	3,500	2,000
12		2,000	46		2,000	80		2,000
13	3,500	2,000	47		2,000	81		2,000
14		2,000	48		2,000	82		2,000
15		2,000	49	3,500	2,000	83	6,185	2,000
16		2,000	50		2,000	84	3,500	2,000
17		2,000	51		2,000	85		2,000
18		2,000	52		2,000	86		2,000
19	3,500	2,000	53		2,000	87		2,000
20		2,000	54	3,500	2,000	88		2,000
21		2,000	55		2,000	89	3,500	2,000

Continued Tab. 2

Year	August	September	Year	August	September	Year	August	September
22		2,000	56		2,000	90		2,000
23	6,185	2,000	57		2,000	91		2,000
24	3,500	2,000	58		2,000	92		2,000
25		2,000	59	3,500	2,000	93		2,000
26		2,000	60		2,000	94	3,500	2,000
27		2,000	61		2,000	95		2,000
28		2,000	62		2,000	96		2,000
29	3,500	2,000	63	6,185	2,000	97		2,000
30		2,000	64	3,500	2,000	98		2,000
31		2,000	65		2,000	99	3,500	2,000
32		2,000	66		2,000	100		2,000
33		2,000	67		2,000			
34	3,500	2,000	68	6,550	2,000			

3.3 Analysis result

Fig. 2 is a result of calculation start one year in the tidal reservoir presence. Without the tidal reservoir, the water level continued rising from a calculation start and exceeded the dike height at the time of a flood of August, and the water level decreased once and re-increased slightly. In addition, but the river bed level continue rising until July, it is decrease at the time of the flood of August and September by scouring. With the tidal reservoir, a water level is not beyond the dike height because a water level decline of approximately 3 m was confirmed in comparison without the tidal reservoir at the time of a flood of August. But the river bed level decreased by scouring in the downstream from a joint until July, it was able to confirm that the river bed level decreased by scouring after depositing at the time of the flood of August and September. The upper stream side becomes the tendency that is reverse to the downstream side from a joint. In addition, without the tidal reservoir, but tractive force showed a tendency toward the decline to become bed slope is easy slope in comparison with an initial incline, With the tidal reservoir, tractive force was a tendency of the increase in a steep grade.

Next, Fig. 3 shows a result of the dike height and the high water level. Without the tidal reservoir, the high water level exceeded the height of the dike height in section of No. 28 ~ No. 70. With the tidal reservoir, but high water level approaches to number cm of dike height (Q_{max} = 15,000 m^3/s) in a side of upper reaches in No. 95, it followed that the dike height did not exceed No. 67 and No. 77. In addition, but the water level becomes higher because the river bed incline is easy slope in a side of upper reaches, the water level lowered because the river bed decreases in the downstream side greatly under the influence of days. In No. 67 and No. 77, the larger the scale of the tidal reservoir, the smaller the water level and the river bed level.

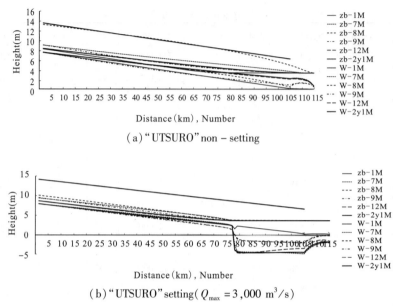

(a) "UTSURO" non – setting

(b) "UTSURO" setting($Q_{max} = 3,000$ m^3/s)

Fig. 2 A change of river bed level and water level during initial one year

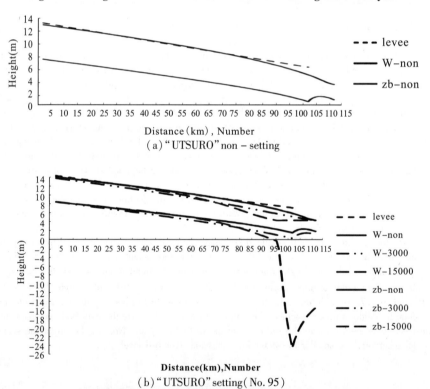

(a) "UTSURO" non – setting

(b) "UTSURO" setting(No. 95)

Fig. 3 Water level and bed level during a big flood

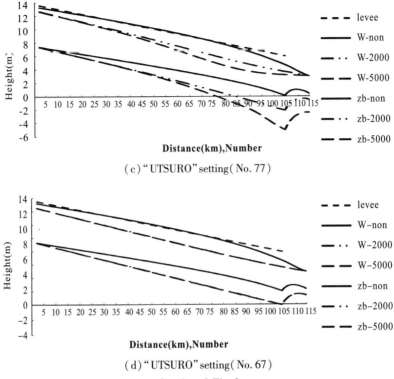

(c) "UTSURO" setting (No. 77)

(d) "UTSURO" setting (No. 67)

Continued Fig. 3

Fig. 4 is a result with the tidal reservoir. At an edge of upper reaches, a river bed level and water level rose approximately 70 cm, and at estuary (No. 104), a river bed level approximately 3 m and a water level approximately 1 m rose. In addition, the surface of the water is a backwater, and the river bed level and the water level increases in future.

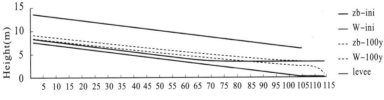

Distance (km), Number

Fig. 4   River bed change without tidal reservoir

Next, Fig. 5 is a result with the tidal reservoir in low water level neighborhood (No. 95). The main river discharge of the tidal reservoir joint in falling tide assumed five cases of 3,000 m³/s, 5,000 m³/s, 10,000 m³/s, 13,000 m³/s and 15,000 m³/s. The figure illustrated a result of 3,000 m³/s and 15,000 m³/s. But in the downstream side from a joint, the larger the scale of the tidal reservoir, the smaller the river bed level, in the upper reaches side, the river bed level and the water level rose tendency. In addition, the surface of the water is a backwater, and the river bed level and the water level increases in future.

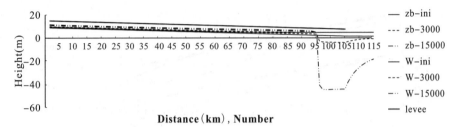

Fig. 5 River bed change with tidal reservoir at No. 95 (the smallest water level)

Fig. 6 is a result with the tidal reservoir in average tide level neighborhood (No. 77). The main river discharge of the tidal reservoir joint in falling tide assumed four cases of 2,000 m³/s, 3,000 m³/s, 4,000 m³/s and 5,000 m³/s. The figure illustrated a result of 2,000 m³/s and 5,000 m³/s. In result, but the river bed level and the water level decreased in the neighborhood of No. 0 ~ No. 50 of the side of upper reaches in comparison with the initial river bed level and initial water level, in the neighborhood of No. 50 ~ No. 70, the river bed level and the water level increased if a scale is big. In addition, the river bed level decrease from the edge of upper reaches to the estuary part in comparison with a initial river bed. But the surface of the water form shows a tendency toward the decline backwater in the neighborhood of joint, but it shows a tendency toward the backwater from it of upper reaches.

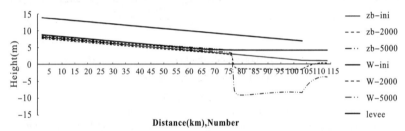

Fig. 6 River bed change with tidal reservoir at No. 77 (the average water level)

Fig. 7 is a result with the tidal reservoir in middle neighborhood (No. 68) of an average tide level and the high water level. The main river discharge of the tidal reservoir joint in falling tide assumed four cases of 2,000 m³/s, 3,000 m³/s, 4,000 m³/s and 5,000 m³/s. The figure illustrated a result of 2,000 m³/s and 5,000 m³/s. In result, the river bed level and the water level decrease from the edge of upper reaches to the estuary part in comparison with a primary river bed and a primary water level. In addition, if a scale was small, it is back water and if a scale was big, it was the decline backwater.

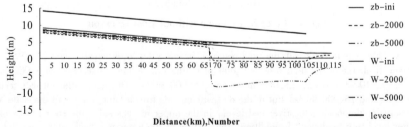

**Fig. 7 River bed change with tidal reservoir at No. 67
(the middle of the average and highest water level)**

Next, Fig. 8 shows about a variation of river bed level and the water level in the edge of upper reaches. In No. 95 which is near to the estuary part, the river bed level and the water level increase a tendency from a calculation start. In No. 67 and No. 77, the river bed level is smaller than scale of the tidal reservoir is big. In addition, in 5,000 m³/s of No. 67 and 2,000 m³/s of No. 77, the river bed level and the water level is same equal.

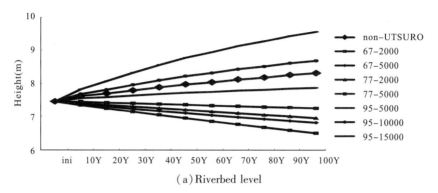

(a) Riverbed level

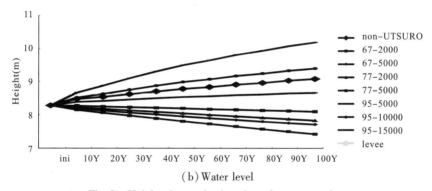

(b) Water level

Fig. 8   Height change in the edge of upper reaches

Next, Fig. 9 show about the change by the connection position of the tidal reservoir . But if a scale is smaller in the downstream side than a joint, river bed level is the same the level, the river bed decline become small when the tidal reservoir set the side of upper reaches even if a scale is big. In addition, but the river bed decline of the downstream side advanced even if a scale is big, the influence was small on side of upper reaches.

4   Conclusions

In this study, the effect of the tidal reservoir "UTSURO" in the Yellow River was analyzed by numerical simulation.

The obtained result is summarized as follows:

(1) Without tidal reservoir, the Yellow River overflows at the big flood.

(2) Tidal reservoir is effective in a river bed decline or sedimentation restraint.

(3) The longer the distance between the connecting point and estuary, the bigger the effect of the tidal reservoir to the upper stream reaches of the river as long as the tidal reservoir is affected by tide.

(4) However, the longer the distance between the connecting point and estuary, the smaller the degree of the river bed decline in the downstream.

(5) On the other hand, the size of the tidal reservoir is needed to be big when the connecting position is far from estuary. Therefore, it is preferable to connect the reservoir near No. 77 (28 km from the river mouth) where the bed level is about the mean tide level.

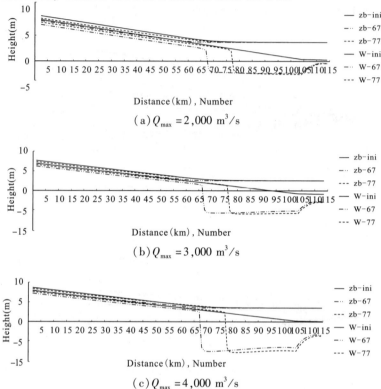

Fig. 9　Change of river bed and water level in the connection position.

(6) The effect of the flow discharge 5,000 m³/s at No. 67 (18 km from the river mouth) and the effect of 2,000 m³/s at No. 77 are almost same. From an economical viewpoint, it is preferable to reduce the size of the reservoir. Therefore, we recommend to connect the reservoir at the point No. 77.

As the annual mean flow rate of the Yellow River is almost 200 m³/s, the discharge supplied from the reservoir is needed $2,000 - 200 = 1,800 (\text{m}^3/\text{s})$.

(7) The needed area to generate this amount of discharge is estimated to be about 100 km².

(8) Moreover, in order to prevent from the sediment entering into the reservoir, the length of the connecting channel must be not shorter than 62 km.

Reference

Li Guoying, Kazuo Ashida, Kenji Sawai, et al. Restoration of Yellow River[M]. Kyoto University Press, 2011.

Experiment Study on Performance of Self – Excitation Pulse Jet Equipment in Submerged Condition

Gao Chuanchang and *Liu Xinyang*

North China University of Water Resources and Electric Power, Zhengzhou, 450011, China

Abstract: In order to study the influence of the self – excitation inspiratory and non – inspiratory pulsed jet equipment on the striking force in the submerged condition, the experiment has been carried out to study the performance of pulse jet equipment on the independently developing experimental platform. The striking force and inspiratory capacity of jet equipment were getting under different submerged water depth. The results of experiment show that the striking force of the inspiratory equipment is higher than the non – inspiratory equipment. As the increase of the working pressure, the raised margin of self – excitation pulse inspiratory equipment is higher than non – inspiratory equipment. The inspiratory capacity and relative inspiratory capacity increase along with the working pressure, while the working pressure is beyond the certain value and the change of relative inspiratory capacity is tend towards stability.

Key words: self – excitation, inspiratory pulsing jet, submerged condition

1 Introduction

Water jet technology has been widely used in ground – breaking, rock breaking, cleaning, cutting and other related engineering field (Li, et al. , 2009; Wang, et al. , 2009; Ma, et al. , 2005; Xiang, et al. , 2009). Now, the research for submerged jet mainly concentrates in the range of medium pressure (≥10 MPa), high pressure (≥ 35 MPa) and ultra – high pressure (≥ 140 MPa) at home and abroad. These water jet belongs to the scope of high – pressure and small flow. Self – excitation pulsed jet equipment is a low – pressure (≤ 2 MPa) and large flow jet equipment and has been studied and developed constantly(Tang, et al. ,2002; Li, et al. ,2004; Wang Leqin, 2004), but the study is mostly limited to non – submerged condition. Generally, under the non – submerged condition, generation mechanism of the self – excitation pulsed jet for low – pressure large flow is that when the pressure in both sides of the jet core is below the atmospheric pressure in the chamber, the cavitation is formed and affects the jet, and the pulsed striking force is produced. But cavitation is difficult to produce in the submerged condition, so the striking force is obviously less than non – submerged condition. At this time, working pressure must be increased to form the cavitation in the chamber, and only in this way the pulsed striking force can appear. On the basis of previous studies, the author developed a new self – excitation inspiratory pulse jet equipment that applies to the low pressure high flow in submerged conditions. The working pressure conditions of the jet equipment was the same as non – submerged one. The independently developing experimental platform was adopted to study the striking force and the inspiratory capacity of the new jet equipment, and the results were contrasted and analyzed with the non – inspiratory condition.

2 Test equipment and content

2.1 Test equipment

Fig. 1 shows the self – excitation inspiratory pulsed jet equipment. The equipment consists of water inlet pipe, upper nozzle, chamber, inspiratory pipe, collision body and the under nozzle, in which the number of inspiratory pipe is 4 and evenly arranged along the chamber body. The location of inspiratory pipe is close to the collision body, and the diameter is 6.5 mm. Submerged jet test system is shown in Fig. 2. Pressure vessel tank is the main carrier to simulate underwater environment.

The test process is as follows. The centrifugal pumping is used to intake water; electromagnetic flow meter is used to measure water flow; the gate valve is used to control the working water pressure; the water is injected into the pressure vessel tank; the automatic control cabinet, pressure transmitter and safety valve are used to control the confining pressure. When the water flow through the jet equipment, it began to inhale the gas. Water jet impacts the target plate and produces the pressure signal. And the signal is transmitted to the data acquisition system through the pressure transmitter. The inspiratory capacity is measured through the gas turbine flow meter and is stored and analyzed through the computer.

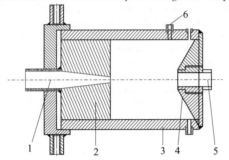

1—Water inlet pipe; 2—Upper nozzle; 3—Chamber;
4—Collision body; 5—Under nozzle; 6—Inspiratory hole

**Fig. 1   Self – excitation inspiratory pulsed
jet equipment**

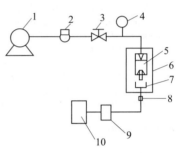

1—Centrifugal pump; 2—Electromagnetic
flow meter; 3—Gate valve; 4—Pressure
gauge; 5—Jet equipment; 6—Pressure
vessel tank; 7—Target plate; 8—Pressure
transmitter; 9—Test system; 10—Computer

Fig. 2   Submerged jet test system

2. 2   Test content

The test content includes inspiratory and non – inspiratory cases of self – excitation pulsed jet. The equipment structure parameters are as follows. The upper nozzle diameter is 8 mm; the under nozzle diameter is 14 mm; the chamber diameter is 85 mm; the chamber length is 40 mm; the target distance is 30 mm. In the non – inspiratory condition, the working pressure is 0. 8 MPa, 1. 0 MPa, 1. 2 MPa, 1. 4 MPa, 1. 6 MPa, 1. 8 MPa and 2. 0 MPa; the corresponding work flow is 6. 9 m³/h, 7. 7 m³/h, 8. 4 m³/h, 9. 1 m³/h, 9. 7 m³/h, 10. 3 m³/h and 10. 8 m³/h. The inspiratory work pressure is decided according to the jet equipment inspiratory process. The confining pressure used in the experiment is 0. 1 MPa, 0. 2 MPa, 0. 3 MPa, 0. 4 MPa, 0. 5 MPa and 0. 6 MPa. Experiments were done under the different working pressures and submergence water depth; the corresponding striking force was got, and inspiratory capacity was measured.

3   Results and analysis

3. 1   Performance comparison between the inspiratory and non – inspiratory equipment

Fig. 3 shows the comparison chart of the striking force between the inspiratory and non – inspiratory condition for self – excitation pulse jet equipment under different submergence water depth. When the water depth is 10 m and working pressure is in the range of 0. 5 ~ 2. 0 MPa, the internal pressure in the chamber is lower than atmospheric pressure, then the equipment begins to inhale the gas, and the striking force of the inspiratory condition is higher than that of the non – inspiratory. The reason is that the vortex ring air mass is formed on both sides of the jet core in inspiratory jet equipment as shown in Fig. 4, which is equivalent to the cavitation area in the non –

submerged condition. After inhaling the gas, the internal flow becomes from single – phase flow to gas – liquid two – phase flow. The internal structure of the flow field and pressure field are changed and the jet is ultimately affected. When the water depth is more than 20 m and the working pressure is greater than a certain value, the pressure in the chamber is below atmospheric pressure and the jet equipment begins inhale the gas. Form Fig. 3, when the water depth is 20 m and the working pressure is 0. 7 MPa, the jet equipment begins to inhale the gas; When the water depth is 30 m and the working pressure is 1MPa, the jet equipment begins to inhale the gas; When the water depth is 40 m and the working pressure is 1. 2 MPa, the jet equipment begins to inhale the gas; When the water depth is 50 m and the working pressure is 1. 4 MPa, the jet equipment begins to inhale the gas; When the water depth is 60 m and the working pressure is 1. 8 MPa, the jet equipment begins to inhale the gas. As long as the jet equipment inhales the gas, the striking force is higher than that of the non – inspiratory condition. In different water depth, when the working pressure is low, the increase value of striking force after inhaling the gas is not obvious compared with the striking force in non – inspiratory condition. Such as water depth is 40 m, when the working pressure is 1. 2 MPa, the striking force improves only 0. 004 MPa; While the working pressure is 2 MPa, the striking force improves 0. 155 MPa. But with the increase of working pressure, the raised margin of the inspiratory striking force is larger than the non – inspiratory striking force and the inspiratory striking force increases rapidly. The main reason is that as the work pressure improves, the internal low – pressure area further enlarges, accordingly the inspiratory volume increases (Fig. 5); the vortex rings air mass on both sides of the jet core change bigger, and the influence on the jet is greater and leads to the increment of striking force.

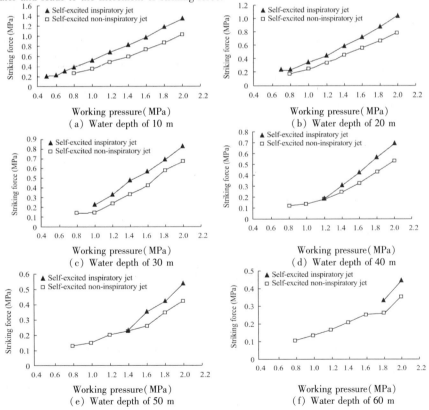

Fig. 3　Comparison of striking force between inspiratory and non – inspiratory jet in different submergence water depth

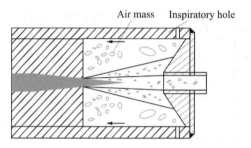

Fig. 4 Vortex ring air mass in chamber

3.2 Inspiratory capacity and the relative inspiratory capacity

Fig. 5 shows the relationship of inspiratory capacity, the relative inspiratory capacity and the working pressure under different water depth. The relative inspiratory capacity is the ratio of inspiratory capacity with the work flow. Under the same water depth condition, the inspiratory capacity increases with the increment of working pressure, The same reason is that as the work pressure improves, the internal low – pressure area increases. As the water depth increases, the inspiratory capacity and relative inspiratory capacity show a trend of decreasing. When the water depth is 10 m, the inspiratory capacity and relative inspiratory capacity reach maximum, and when the water depth is 60 m, the minimum value is got. Initial relative inspiratory capacity changes faster with the working pressure, but when the working pressure exceeds a certain value, the curve of relative inspiratory capacity changes gentle under the different water depth. When the water depth is 10 m and the working pressure exceeds 0.8 MPa, the relative inspiratory capacity is between 30% ~ 35%. When the water depth is 20 m and the working pressure exceeds 1.2 MPa, the relative inspiratory capacity is between 20% ~ 30%. When the water depth is 30 m and the working pressure exceeds 1.6 MPa, the relative inspiratory capacity is between 20% ~ 25%.

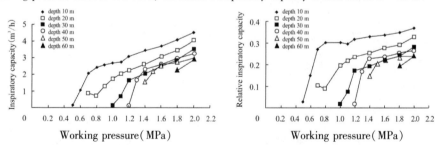

Fig. 5 Change of inspiratory capacity and relative inspiratory capacity in submergence conditions

4 Conclusion

This article studied the striking force and inspiratory capacity of self – excitation pulsed jet equipment under the inspiratory and non – inspiratory conditions, and the following conclusions was got.

(1) As long as the pulse jet equipment inhale the gas, the striking force is higher than non – that of the inspiratory condition in different water depth and working pressure, and with working pressure increasing, the raised margin of striking force is also increased.

(2) At different water depth, the inspiratory capacity and relative inspiratory capacity increase with the increment of working pressure and when the working pressure exceeds a certain value, the

change trend of relative inspiratory capacity gets gentle.

(3) The experiment proved the striking force of the inspiratory equipment is higher than that of the non – inspiratory equipment, and the size of vortex ring air mass in the chamber has big impact on the striking force. It is also need to be further researched and explored that how to increase inspiratory capacity and enlarge vortex ring air mass by changing the structural parameters, so as to improve the striking force.

References

Li Gensheng, Liao Hualin. Rock Damage Mechanisms under Ultra—High Pressure Water Jet Impact [J]. Journal of Mechanical Engineering,2009,45(10):284 – 292.

Wang Zonglong, Hu Shougen, Yao Wenlong. Experimental Research of Ultra – high Pressure Abrasive Water Jet Cutting Rock in Submerged Environment [J]. Journal of Hydrodynamics, 2009, 31(6):151 – 155.

Ma Fei, Zhang Wenming. Equation of Soil Bore Enlarging with Submerged Water Jet and Experimental Study [J]. Journal of University of Science and Technology Beijing, 2005, 27 (3):268 – 271.

Xiang Wenying, Li Xiaohong, Lu Yiyu. Cavitation Capacity of Submerged Abrasive Water Jets [J]. Journal of Chongqing University, 2009, 32(3):299 – 302.

Tang Chuanlin, Liao Zhenfang. Theory of the Self – excited Oscillation Pulsed Jet [J]. Journal of ChongQing University (Natural science Edition), 2002, 2(1):24 – 27.

Li Jiangyun, Xu Ruliang, Wang Leqin. Numerical Simulation of Mechanism of the Self – excited Pulse Nozzle [J]. Journal of Engineering Thermophysics, 2004, 25(2):241 – 243.

Wang Leqin, Wang Xunming, Xu Ruliang. Experimental Study on Structural Parameters Optimized Design of the Self – excited Oscillation Pulsed Jet nozzle [J]. Journal of Engineering Thermophysics, 2004, 25(6):956 – 958.

Comparative Roles of Roughness Reduction and Bed Erosion on Downstream Peak Discharge Increase

Wei Li[1] , *Zhengbing Wang*[1] , *Huib de Vriend*[1] and *Wu Baosheng*[2]

1. Department of Hydraulic Engineering, Faculty of Civil Engineering and Geosciences,
Delft University of Technology, P. O. Box 5048, 2600 GA Delft, the Netherlands
2. State Key Laboratory of Hydroscience and Engineering, Tsinghua University,
Beijing,100083, China

Abstract: Highly silt – laden flow in the Yellow River (China) tends to exhibit abnormal features during floods, among which the phenomenon of a downstream increasing discharge peak is observed in the Lower Yellow River. This has become an issue of significant interest to hydraulic engineers and scientists, also in relation to the so – called man – made floods. The increasing discharge entails an increasing load on the flood defence system and an increasing flood risk as the flood peak moves downstream. Yet, the basic mechanisms underlying this phenomenon remain to be unraveled, in order to be able to mitigate its potential damage. The many existing explanations and the complexity of this phenomenon call for a comprehensive understanding. Following the perspective of turbulence and effective bed roughness reduction in hyperconcentrated flow, we have conducted a numerical model study for the hyperconcentrated flood of August 2004 in the Xiaolangdi-Jiahetan reach. In that model, we used a simple power law relation between Manning's n and the volumetric sediment concentration. In order to focus on the essentials of the phenomenon, the river was schematised to a 1 – D straight channel with a rectangular cross – section of constant width and with the bed consisting of uniform sediment. The feedback between the bed deformation and the turbid flow, however, is fully accounted for, in the constituting equations as well as in the numerical solutions. The model successfully reproduced the phenomenon of downstream flood peak increase when considering the hyperconcentration – induced bed roughness reduction. As the hyperconcentration lags shortly behind the flood peak, later parts of the flood wave experience less friction and can overtake the wave front. The model results also suggest that the contribution of bed deformation is much less important to the magnitude of the flood peak and its downstream evolution, at least for 'normal' hyperconcentrated floods without extreme bed erosion and sediment concentrations (e. g. $800 \sim 900$ kg/m^3).

Key words: hyperconcentrated flow, mathematical model, flood peak increase, the Yellow River

1 Introduction

The Yellow River, the second longest river in China, is famous for the high sediment load in its middle and lower reaches. The annual average sediment discharge at Huayuankou hydrological station was 1.6×10^9 t before 1970 s, which however decreased dramatically over the past 40 years due to intensive human disturbances (e. g., dam construction, soil conservation etc.). In particular, the annual average sediment discharge decreased by 90% to 0.13×10^9 in 2000 s after the construction of the Xiaolangdi Reservoir (IRTCES 2000 ~ 2009). Nonetheless, highly silt – laden flow still occurs whenever the Xiaolangdi Reservoir flushes turbidity currents in flood seasons. This leads to hyperconcentrated floods in the lower Yellow River, with sediment concentrations ranging from $100 \sim 400$ kg/m^3. Unlike most hyperconcentrated floods in earlier years, those floods sometimes have a downstream increasing peak discharge without floodplain inundation in the Xiaolangdi—Huayuankou (X—H) reach. Five cases of flood peak increase have been observed in the years 2004 ~ 2010. Especially in 2010, the measured peak discharge at Huayuankou hydrological station was $6,680\ \text{m}^3/\text{s}$, which was nearly twice as large as that of $3,490\ \text{m}^3/\text{s}$ at the Xiaolangdi hydrological station, while the tributaries just accounted for $86\ \text{m}^3/\text{s}$. The phenomenon also happened in the period 1970 ~ 1990, combined with higher sediment concentration and floodplain inundation, but not as frequently as in the 2000 s. The discharge increase entailed an increasing load on the flood defense. Since 2000, the average discharge increase is about 50%. A comprehensive understanding of the mechanisms underlying the downstream peak discharge increase is therefore very important, and so is the development of strategies to mitigate the impacts.

Even when taking into account a measuring error of about 5% (Wang et al. , 2009), the tributary contribution is far from sufficient to explain the phenomenon, so there should be other reasons. In the period 1970 ~ 1990, this phenomenon was always observed with floodplain inundation and main channel deformation, whereas it occurred without overbank flow since 2000. In the former case, the floodplain inundation and channel deformation may have had a strong effect: it is suggested that the water returning from the floodplain into the main channel during the waning stage of the flood would cause the downstream discharge increase (Chien and Wan, 1983; Qi, 1992; Qi and Zhao, 1993; Qi and Li, 1996). Also, the channel deformation from a wide – shallow to a narrow – deep profile may have accelerated the flood propagation, due to which later parts of the flood wave would overtake earlier parts, thus causing flood peak increase (Wang et al, 2009). After 2000, the hyperconcentration effects on the effective hydraulic drag may have contributed to the phenomenon. During the operation of the water – sediment regulating scheme at Xiaolangdi, clear water flushing in the first stage coarsens the bed material and increases the bed roughness. When in a later stage more turbid water is flushed, the hyperconcentration – induced bed roughness reduction becomes important and accelerates the flood propagation. If the hyperconcentration part lags behind the flood peak, later parts of the flood will experience less friction and may overtake the front waves, thus yielding an increased peak discharge. Jiang et al. (2006) numerically reproduced the downstream peak discharge increase of the man – made flood of August 2004 by considering the effect of sediment concentration on bed roughness (Zhao and Zhang, 1997). The laboratory experiment by Zhu and Hao (2008) further demonstrated bed roughness reduction in hyperconcentrated flow. Based on turbulence perspectives, Winterwerp et al. (2009) also suggested that for fine suspended sediment turbulence damping in hyperconcentration reduce the effective hydraulic drag.

In this paper, we ignore the floodplain influence and focus on the essentials of the phenomenon by conducting numerical experiments for a 1 – D schematized channel of dimensions similar to the Xiaolangdi—Jiahetan reach. The mechanisms of the downstream peak discharge increase in the 2004 hyperconcentrated flood are investigated by analyzing the relative importance of bed roughness reduction and bed erosion in this mathematical model.

2 Mathematical model

2.1 Governing equations

The governing equations for mass, momentum and sediment conservation constitute a non – linear system. In conservative form it reads

$$\frac{\partial \overline{U}}{\partial t} + \frac{\partial \overline{F}}{\partial x} + \frac{\partial \overline{G}}{\partial y} = \overline{R} \tag{1}$$

$$\overline{U} = \begin{bmatrix} h \\ hu \\ hv \\ hc \\ \phi \end{bmatrix}, \quad \overline{F} = \begin{bmatrix} hu \\ hu^2 + 0.5gh^2 \\ huv \\ huc \\ huc \end{bmatrix}, \quad \overline{G} = \begin{bmatrix} hv \\ hvu \\ hv^2 + 0.5gh^2 \\ hvc \\ hvc \end{bmatrix} \tag{2}$$

$$
\bar{R} = \begin{bmatrix} R_1 \\ R_2 \\ R_3 \\ R_4 \\ R_5 \end{bmatrix} = \begin{bmatrix} \dfrac{E-D}{1-p} \\[2ex] gh(S_{0x}-S_{fx}) - \dfrac{(\rho_s-\rho_w)gh^2}{2\rho}\dfrac{\partial c}{\partial x} - \dfrac{(\rho_0-\rho)(E-D)}{\rho(1-\rho)}u \\[2ex] gh(S_{0y}-S_{fy}) - \dfrac{(\rho_s-\rho_w)gh^2}{2\rho}\dfrac{\partial c}{\partial y} - \dfrac{(\rho_0-\rho)(E-D)}{\rho(1-\rho)}v \\[2ex] E-D \\ 0 \end{bmatrix} \tag{3}
$$

where, t is time; x, y is horizontal coordinates; h is water depth; u, v is depth – averaged flow velocity in x and y direction respectively; c is depth averaged volumetqric sediment concentration; z_b is bed elevation; E, D is sediment entrainment and deposition fluxes respectively; S_{0x}, S_{0y} is bed slopes in x and y direction respectively; S_{fx}, S_{fy} is friction slopes in x and y direction respectively; ρ_s is sediment density, $\rho_s = 2,650$ kg/m^3; ρ_w is water density, $\rho_w = 1,000$ kg/m^3; ρ is density of sediment – laden flow, $\rho = \rho_w(1-c) + \rho_s c$; ρ_0 is density of saturated bed, $\rho_0 = \rho_w p + \rho_s(1-p)$; p = bed porosity; g is acceleration of gravity, $g = 9.8$ m/s^2; ϕ is conservative variable, $\phi = (1-p)z_b + hc$. The feedbacks of bed deformation on turbid flow are considered in the mass (R_1) and momentum (3^{r}d terms of R_2 and R_3) conservation equations respectively (Cao et al. , 2011).

2.2　Empirical relations

The friction slopes are estimated using Manning roughness n

$$
S_{fx} = \frac{n^2 u \sqrt{u^2+v^2}}{h^{4/3}}, \qquad S_{fy} = \frac{n^2 u \sqrt{u^2+v^2}}{h^{4/3}} \tag{4}
$$

The net sediment flux between the turbid flow and river bed is determined by

$$
E-D = \alpha(\omega_s c_* - \omega_s c) \tag{5}
$$

where, ω_s is effective sediment settling velocity; c_* is volumetric sediment transport capacity; α is non – equilibrium adaptation coefficient computed by (Wang and Xia, 2001)

$$
\alpha = \begin{cases} 0.001/\omega_s^{0.3}, & c > c_* \\ 0.001/\omega_s^{0.7}, & c \leqslant c_* \end{cases} \tag{6}
$$

The effective sediment settling velocity is computed by (Richardson and Zaki, 1954),

$$
\omega_s = \omega_0 \left(1 - \frac{c}{1-p}\right)^5 \tag{7}
$$

where, ω_0 is sediment settling velocity in tranquil and still water according to e. g. , Zhang's formula (Zhang and Xie, 1993); p is 0.45 (Winterwerp et al. , 2003).

The volumetric sediment transport capacity is determined by (Wu et al. , 2008)

$$
c_* = \frac{K}{\rho_s}\left(\frac{\rho}{\rho_s-\rho gh\omega_s}\frac{u^3}{}\right)^m \tag{8}
$$

where, K is 0.451,5; m is 0.741,4; K and m are empirical parameters.

A power law relation accounting for the effect of sediment concentration on Manning roughness is introduced for the preliminary studies. The computed Manning roughness simply decreases with increasing sediment concentration as follows

$$
n = n_0(1 + c_0 - c)^\beta \tag{9}
$$

where, n_0 is initial Manning roughness; c_0 is initial volumetric sediment concentration; β is 3.

3　Numerical case studies

In this section, numerical case studies are conducted to examine the relative importance of roughness reduction and bed erosion on downstream peak discharge increase. For this purpose, the governing equations are solved based on a finite volume method, which can accurately capture shock waves and contact discontinuities (Toro, 2001, 2009; Stecca et al., 2010).

3.1 Numerical cases

Highly silt – laden flow with sediment concentrations of $100 \sim 400$ kg/m^3 occurred in the X—H reach due to turbidity currents flushed from the Xiaolangdi Reservoir during a water – sediment regulating operation (Qi, 2010). A downstream peak discharge increase was observed in some of those man – made hyperconcentrated floods without floodplain inundation. For the flood event that occurred in August 2004, the flood peak was 2,690 m^3/s at the Xiaolangdi station at 08: 36 on 23rd August. About 16 hrs later, the downstream peak discharge was observed to be as high as 3,990 m^3/s at the Huayuankou station (125.8 km downstream of the Xiaolangdi station). In the 125.8 km long reach, the sediment peak happened later than the flood peak, with the time lags increasing downstream from 15.4 hrs at Xiaolangdi to 18 hrs at Huayuankou. The maximal sediment concentrations at both stations are around 350 kg/m^3. This flood was simulated before by a 2D model with YRCC formulae and the observed bathymetry (Jiang et al, 2006). Yet, the relative importance of bed roughness reduction and bed deformation remains unclear, due to the complex topography, the decoupled solution and the many empirical parameters.

Focusing on the essence of the phenomena, we conduct 1 – D numerical experiments for this flood using a fully coupled model. Three cases are considered as presented in Tab. 1. Comparison of Cases 1 and 2 demonstrates the role of roughness reduction, comparison of Cases 1 and 3 illustrates the role of bed erosion.

Tab. 1 Summary of numerical cases

Case no.	Effects of concentration on roughness	Feedback impacts of bed erosion	Remarks
1	Considered by Eq. (9)	Considered, see Eq. (3)	Fully coupled model
2	Ignored, n = constant	Considered, see Eq. (3)	
3	Considered by Eq. (9)	Ignored	Partially coupled model

3.2 Boundaries and parameters

The flood process in the 226.6 km long reach between Xiaolangdi and Jiahetan is investigated. Referring to the background of the Lower Yellow River, it is schematized to be a 1 – D channel with rectangular cross – sections and uniform sediment: channel width = 1,000 m, bed slope = 2.55×10^{-4}, $d_{50} = 0.02$ mm, $p = 0.45$. The bed elevation at Xiaolangdi is set to 132.03 m, according to the Dagu height reference system. Starting from a unit – width discharge of $q_0 = 0.421$ m^2/s, the initial water depth $h_0 = 0.75$ m is obtained from the stage – discharge relationship at Jiahetan and $h_0 = 0.56$ m/s. The initial flow condition is assumed to be at steady and uniform regime, which can be used to back – estimate the initial Manning roughness $n_0 = 0.023,7$. The initial sediment concentration is assumed to be at the capacity state and computed by Wu et al. (2008) formula to $c_0 = 0.003,5$. The measured discharge and sediment concentration at the Xiaolangdi station are taken to be upstream boundary conditions (Fig. 1). The spatial step Δx is set to 100 m and the time step Δt is controlled by CFL condition with the Courant number $C_r = \lambda_{max} \Delta t/\Delta x = 0.45$. The total simulation time is 504 h from 20th August to 10th September 2004.

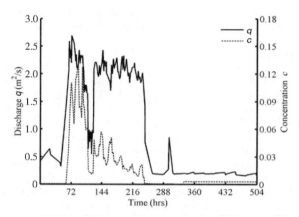

Fig. 1　Flow discharge and volumetric concentration at the upstream boundary (Xiaolangdi)

4　Numerical results and discussions

4.1　Discharge hydrographs and sediment processes

Fig. 2 shows the comparison of the computed discharge hydrographs at distinct cross – sections for the three cases. Fig. 3 shows the comparison of the measured data at Huayuankou ($x =$ 125.8 km) with the computations.

From Figs. 2 and Figs. 3, the feedback impacts of bed deformation appear negligible for the concerned flood process, as no difference can be detected between Cases 1 and 3. In contrast, the effects of sediment concentration on bed roughness are considerable, with the computed peak discharge for Case 2 much smaller than that for Case 1.

The peak discharge decreases gradually downstream for Case 2 when the roughness is set constant, whereas for Case 1 substantial increase of peak discharge in the downstream direction is computed, though limited in the upper reach (i. e. , x < 130 km). The results for Case 1 are consistent with the field observations that downstream peak discharge increase was not observed in the reach between Huayuankou and Jiahetan. Most importantly, for Case 1 the computed time variations of discharge and sediment concentration at Huayuankou compares rather well with the measurement (see Fig. 3), though quantitative differences are evitable. In terms of the peak discharge, the computed value is about 3.6 m^2/s at Huayuankou, while the measured value is about 4 m^2/s. This indicates a relative difference of 11% , which should be within the acceptable range. It is understandable considering the present simplifications for channel width, bed elevation, single – sized sediment, power law formula of roughness, etc. This discrepancy might be improved by using realistic topography, multiple sediment sizes, better calibrated parameters, physically – based formulae, etc.

The above observations are not surprising as seen in Fig. 4, which shows the computed roughness and concentration for Case 1. The bed roughness is computed with 60% ~ 70% reduction by sediment concentration higher than 0.1. Thus later parts of the flood wave experience less friction and can overtake the wave front at some distances. The peak discharge increase is discerned to diminish in the further downstream with weaker effect of bed roughness reduction (e. g, x = 200 km in Fig. 2 and Fig 4).

The current 1 – D modeling for Cases 1 and 3 can satisfactorily reproduce the phenomenon of peak discharge increase for the man – made flood in 2004. This indicates that the effects of sediment concentration on roughness should account for the phenomenon between Xiaolangdi and

Huayuankou in August 2004, which dominates over those of bed erosion.

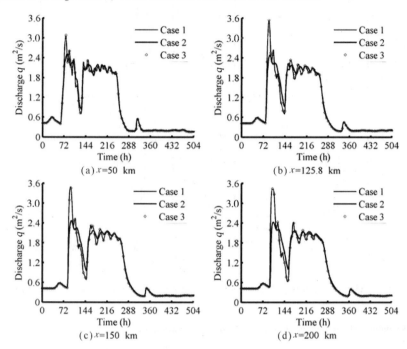

Fig. 2　Discharge hydrographs at distinct cross – sections for the three cases

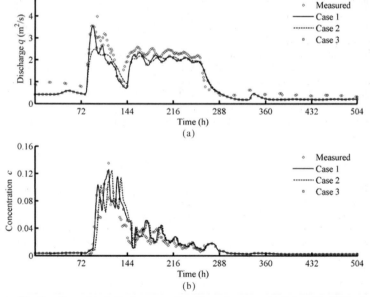

Fig. 3　Computed discharge and sediment concentration with field observations at the Huayuankou station

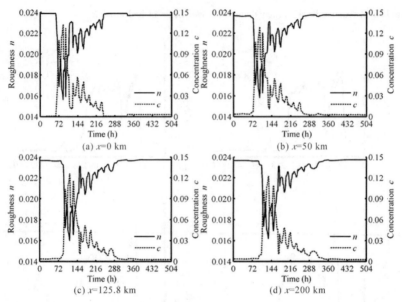

Fig. 4 Time variation of bed roughness and sediment concentration at distinct cross – sections for Case 1

4.2 Influence of downstream stage – discharge relationship

It should be noted that downstream boundary condition could affect the flow and sediment evolution within computational domain. The above used stage – discharge relationship is derived from the limited measured data at the Jiahetan station, which may be lack of validation for the current 1 – D modeling. In this section, a sufficiently long channel (800 km) is considered so that the flood wave does not arrive at the downstream boundary during the simulation time. Case 1 is recalculated in this long channel.

Fig. 5 illustrates the discharge hydrographs and relative changes for water level at two cross – sections for Case 1 in the short reach (226.6 km) and long reach (800 km) respectively. It is seen that the discharge hydrographs in the short reach are almost the same as those in the long reach while the stage has little discrepancy between the two reaches at the location of $x = 226.6$ km. It implies that the stage – discharge relationship at the outlet of the short reach is justified for the 1 – D modeling and the hydrographs in the short reach can reasonably reflect those in the long reach of natural case.

5 Conclusions

We have investigated the mechanisms underlying downstream peak discharge increase in hyperconcentrated flow by analyzing the relative importance of bed roughness reduction and bed erosion with a mathematical model. A coupled shallow water hydrodynamic model using a structured finite volume method and a high resolution of shock – capturing algorithm is presented and applied to the 2004 man – made flood with sediment concentrations from 100 kg/m^3 to 400 kg/m^3. A schematic representation of the Xiaolangdi—Jiahetan reach in the Lower Yellow River is considered.

The current model can satisfactorily reproduce the downstream peak discharge increase of this man – made flood if the influence of sediment concentration on bed roughness is taken into account. If not, the downstream peak discharge decreases continuously. This shows that the roughness reducing effect of high sediment concentrations accounts for the peak discharge increase between Xiaolangdi and Huayuankou in this particular flood. This effect dominates over that of bed erosion. The rapid and considerable bed roughness reduction in hyperconcentrated flow makes later parts of the flood wave experience less friction and overtake the wave front. The computed discharge (Case 1) is about 3.6 m²/s at Huayuankou, which is about 11% less than the measured value. Inevitably, the model bears uncertainty, which may arise from the involvement of the empirical parameters, the simple bed roughness formula, and the simplified channel geometry. Yet, the current study provides for an explanation of the observed downstream peak discharge increase.

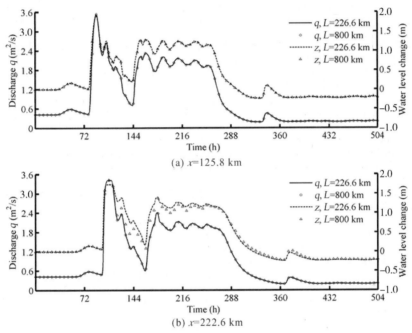

(a) x=125.8 km

(b) x=222.6 km

Fig. 5   Discharge and stage hydrographs for Case 1 at distinct locations in short reach and long reach

References

Cao Z X, Pender G, Carling P. Shallow Water Hydrodynamic Models for Hyperconcentrated Sediment – laden Floods over Erodible Bed[J]. Advances in Water Resources, 2006,29:546 – 557.

Cao Z X, Yue Z Y, Pender G. Landslide Dam Failure and Flood Hydraulics. Part II: Coupled Mathematical Modeling[J]. Natural Hazards, 2011,59(2):1021 – 1045.

Chien N, Wan Z H. Mechanics of Sediment Transport [M]. Beijing: Science Press, 1983.

International Research and Training Center on Erosion and Sedimentation (IRTCES) 2000 ~ 2009, Bulletin of Chinese River Sediment.

Jiang E H, Zhao L J, Wei Z L. Mechanism of Flood Peak Increase Along the Lower Yellow River and Its Verification[J]. Journal of Hydraulic Engineering, 2006,37(12):1454 – 1459.

Qi P. Analysis of Characteristics of '92.8 Hyperconcentrated Flood' Evolution in the Lower Yellow River. Zhengzhou:Yellow River Conservancy Press, 1992.

Qi P, Zhao Y A, Characteristics of Sediment Transport and Channel Formation by Floods at Hyperconcentrations of Sediment in the Yellow River[J]. International Journal of Sediment Research,1993,8(1): 69 – 84.

144

Qi P, Li W X. Evolutional Characteristics of Hyperconcentrated Flow in Braided Channel of the Lower Yellow River[J]. International Journal of Sediment Research, 1996,11(3): 49 -57.

Qi P, Sun Z Y, Qi H H. Flood Discharge and Sediment Transport Potentials of the Lower Yellow River and Development of an Efficient Flood Discharge Channel [M]. Zhengzhou: Yellow River Conservancy Press, 2010.

Richardson J F, Zaki W N. Sedimentation and Fluidization, Part I. Transactions of the Institution of Chemical Engineers[J]. 1954, 32: 35 -53.

Stecca G, Siviglia A, Toro E F. Upwind - biased Force Schemes with Applications to Free - surface Shallow Flows[J]. Journal of Computational Physics, 2010,229(18): 6362 -6380.

Toro E F. Shock - capturing Methods for Free - surface Shallow Flows[M]. Wiley and Sons Ltd., England,2001.

Toro E F. Riemann Solvers and Numerical Methods for Fluid Dynamics: A Practical Introduction [M]. Beijing:Springer - Verlag, 2009.

Wang G Q, Xia J Q. Channel Widening During the Degradation of Alluvial Rivers [J]. International Journal of Sediment Research, 2001,16(2):139 -149.

Wang Z Y, Qi P, Melching C S. Fluvial Hydraulics of Hyperconcentrated Floods in Chinese Rivers [J]. Earth Surface Processes and Landforms, 2009,34: 981 -993.

Winterwerp J C, Lely M, He Q. Sediment - induced Buoyancy Destruction and Drag Reduction in Estuaries[J]. Ocean Dynamics, 2009,59: 781 -791.

Winterwerp J C, de Vriend H J, Wang Z B, Fluid - sediment Interactions in Silt - laden Flow [C]. International Yellow River Forum on River Basin Management, 2003.

Wu B S, van Maren D S, Li L Y. Predictability of Sediment Transport in the Yellow River Using Selected Transport Formulas[J]. International Journal of Sediment Research,2008, 23(4): 283 -298.

Zhang R J, Xie J H. Sedimentation Research in China - systematic Selections[M]. Beijing: China WaterPower Press.

Zhao L J, Zhang H W, Study on the Characteristics of Flow Resistance in the Lower Yellow River [J]. Yellow River, 1997,19(9):17 -20.

Zhu C J, Hao Z C, A Study on the Resistance Reduction of Flows with Hyper - concentration in Open Channel. In: International Workshop on Education Technology and Training & 2008 International Workshop on Geoscience and Remote Sensing, Shanghai, China, 2008, 141 -144.

Study on Key Technology of Xiaolangdi Reservoir Operation for Flood Control and Sediment Reduction during the Late Sediment-retaining Period[①]

Liu Jixiang, *Zhang Houjun*, *An Cuihua*, *Li Shiying* and *Liu Hongzhen*

Yellow River Engineering Consulting Co. , Ltd. , Zhengzhou, 450003, China

Abstract: This paper conducts a comprehensive and deeply analysis and research on key technology of Xiaolangdi Reservoir operation for flood control and sediment reduction during the late sediment-retaining period by theoretical study, filed survey, observed data analysis, mathematical model analysis and physical model tests etc. Based on the study of flow and sediment movement laws in the reservoir during the late sediment-retaining period and characteristics of sediment transport and channel evolution along the lower Yellow River, combining of scheme calculation and analysis, the paper presents regulation indexes of the reservoir such as regulatory and control discharge and duration, regulatory and control capacity, and time to scour by lowering operation water level etc. The paper mainly studies two operation modes for sediment reduction, one of which is "raising water level gradually and retaining coarse sediment and discharging fine sediment" and another is "to regulate sediment for many years and seek chances to scour by lowering operation water level, retaining sediment and adjusting water and sediment", and the paper has studied operational mode for flood control. The paper has conducted analysis and research on different schemes from flood control effect of the reservoir, medium flood control, reservoir sedimentation, the sediment flushing effect, extension of useful life of reservoir sediment capacity, the effect on sediment reduction in the lower River and median water river bed maintaining with the reservoir, benefit of power generation and water supply etc. The paper presents the operation mode and dispatching principle for flood control and sediment reduction during the late sediment-retaining period, with the core idea which is "to regulate sediment for many years and seek chances to scour by lowering operation water level, retaining sediment and regulating water and sediment,". The operation mode for flood control and sediment reduction has the following characteristics: flexibility of reservoir operation, great adaptability with incoming water and sediment condition, long service life of sediment capacity, with great sediment reduction effect in the lower river, and significantly modifying the contradiction of flood control in the lower flood plain.

Key words: late sediment-retaining period, regulatory and control discharge and duration, regulatory and control capacity, time to scour by lowering operation water level, operation mode of sediment reduction, operation mode of flood control

1   Basic situation of Xiaolangdi Reservoir

Xiaolangdi hydraulic complex is located at the last valley exit of the middle reach of the Yellow River. The development goal is mainly for flood control (ice prevention) and sediment reduction, in addition to water supply, irrigation, hydroelectric power generation, to promote the benefit and abolish the harm for comprehensive utilization. It is a key project irreplaceable to solve the problem such as flood control and sediment-reduction of the lower Yellow River. It completed closure in 1997 and the first machine generated electric combined to the power grid in 1999, and all of the project are completed and put into operation in 2001. The normal water level of the Xiaolangdi

① Fund Project: the Special Scientific Research of the Ministry of Water Resources for Public Welfare Industry (200901017)

Reservoir is 275 m and the corresponding total capacity is 12.65×10^9 m^3, and the long-term effective capacity is 5.1×10^9 m^3, and the sediment-retaining capacity is 7.55×10^9 m^3, as well as the capacity for water and sediment regulation is 1×10^9 m^3. More than ten tributaries flow into the reservoir including Zhenshui and the tributaries distributed in the middle and lower section of the reservoir. The sediment deposited in the reservoir is up to 2.772×10^9 m^3 from May 2000 to October 2010, in which 2.194×10^9 m^3 deposited in the main stream accounting for 79.1% of total sedimentation. The whole lower river course was scoured and the scoured amount was 1.848×10^9 t from Baihe to Lijin.

2　Operational phases of the Xiaolangdi Reservoir

The division of the reservoir operational phases should follow the development target of the hydraulic complex, and it especially should follow the principle that is beneficial to flood control and sedimentation reduction for the lower Yellow River. It not only should consider deposition amount, deposition feature of the reservoir, and reservoir capacity maintaining situation, but also should consider the boundary conditions of the downstream river, the conditions of incoming water and sediment, and the requirement of ecological water, irrigation and water supply of the downstream river. The reservoir operation periods are divided into sediment retaining period and normal operation period. According to variation of water storage capacity and deposition amount of the reservoir, the sediment retaining period is divided into initial sediment retaining period and late sediment retaining period. It is initial sediment retaining period before the capacity under the initial operational water level is silted up (sediment deposited in the reservoir is up to 2.1×10^9 m^3 to 2.2×10^9 m^3). The late sediment retaining period is from when the sediment deposited in the reservoir is up to 2.2×10^9 m^3 to the end of the whole sediment retaining period. The late sediment retaining period is divided into three phases, the first phase is from the end of initial sediment retaining period to before the sediment deposited in the reservoir up to 4.2 billon m^3, and the reservoir is mainly on sediment retaining. The second phase is from when the deposition in the reservoir is 4.2×10^9 m^3 to the deposition is up to 7.55×10^9 m^3, and retaining sediment and regulating water and sediment are conducted at the same time in this phase. According to hydrological forecast, to scour by lowering operation water level in short time for capacity recovery when continues great discharge is flowing into the reservoir. Raising water level gradually and retaining coarse sediment and discharging fine sediment when the water and sediment condition is normal. This phase is the key phase to extend the ages of the late sediment retaining period reasonably. The third phase is from the end of the second phase to the end of the whole sediment retaining period (The floodplain elevation is up to 254 m in front of the dam). The capacity for retaining sediment is little in this phase, and actually it is the transition phase from sediment retaining period to normal operation period. The sediment retaining period is the important period for sediment reduction in the lower river course. During the normal operation period, the capacity for retaining sediment is silted up and the reservoir can operate for flood control and water and sediment regulation in a long time by long-period effective capacity with the volume of 5.1×10^9 m^3, of which 1.0×10^9 m^3 is used for water and sediment regulation.

3　Requirement of river course evolution law at the lower Yellow River for the Xiaolangdi Reservoir operation

3.1　Characteristics of scouring and deposition of grouping sediment in flood at the lower Yellow River and the requirement to the reservoir operation

The lower Yellow River is an alluvial channel and the river course performs scouring and deposition adjustment continuously with the variation of the incoming water and sediment conditions. In this process, it is always accompanied by the scouring and deposition adjustment of coarse sediment and fine sediment. According to analysis of characteristics of scouring and deposition of grouping sediment (fine sediment, $d < 0.025$ mm, middle sediment, $d = 0.025 \sim 0.05$ mm, coarse sediment, $d > 0.05$ mm, similarly hereafter) in flood at lower Yellow River, from scouring and deposition situation along the river with different particle size of sediment, the scouring and deposition affected by composition of suspended sediment is mainly in the upper reaches of Gaocun. After scouring and silting adjustment, composition of suspended sediment has little influence on reaches downstream from Gaocun. What plays a major role in river scouring and deposition is the magnitude of discharge and sediment concentration for the lower river especially for the reaches downstream from Gaocun. The influence of composition of suspended sediment on scouring and deposition in the lower river mainly reflected in the case of river deposits seriously, such as hyper-concentrated flood and small flow with high sediment concentration flood. According to the analysis with observed data, the sediment release ratio of coarse sediment with flood is about 60% in the lower river and it has the certain ability of carrying and transporting coarse sediment. Therefore, reservoir operation for retaining coarse sediment and discharging fine sediment is not asking for retaining all the coarse sediment in the reservoir.

3.2　Characteristics of scouring and deposition in flood with different water and sediment condition at lower Yellow River and the requirement to the reservoir operation

There are many factors affecting scouring and deposition in the lower river, but the main factor is magnitude of flood discharge and sediment concentration. The sediment transport capacity is proportional to high order of the discharge in the lower Yellow River, as the sediment transport capacity is great when the discharge is large. For the reservoir management and regulation, it should make full use of the large flow to transport sediment for sediment reduction in the river. According to characteristics of flood routing in the lower Yellow River, if the flood duration is too short then the discharge will attenuate too fast due to the channel storage role of the downstream river and it will not meet the requirements of discharge for scouring in next reach. The sediment transport capacity of the reaches in the lower Yellow River depends not only on magnitude of discharge, but also requires a certain water quantity. Therefore, it needs a continuous duration when the flow is large.

According to analysis of observed flood data in the lower Yellow River, characteristics of scouring and deposition in flood under different water and sediment condition are mainly manifested in the following aspects:

(1) For non-floodplain flood with general sediment concentration in flood season.

①When the sediment concentration is less than 20 kg/m^3, the whole downstream is scoured basically. When the discharge is larger than 2,500 m^3/s, the scouring efficiency increases significantly in the lower river and in the reach from Aishan to Lijin. When the discharge is larger than 3,500 m^3/s, the scouring efficiency increases further in the whole downstream. Seeing from the scouring and deposition in the lower river under different flood duration, it has good scouring effects in the whole downstream and the reaches downstream from Gaocun when the discharge is between 3,500 m^3/s and 4,000 m^3/s and the flood duration is from 4 d to 5 d.

②When the sediment concentration is between 20 kg/m^3 and 60 kg/m^3, it gradually converted deposition to scouring in the lower river. When the discharge is larger than 2,500 m^3/s, the whole downstream and the reaches downstream from Gaocun is scoured basically, and the scouring efficiency in the whole downstream has an increasing trend. Seeing from the scouring and deposition in the lower river under different flood duration, it has good scouring effects in the whole downstream and the reaches downstream from Gaocun when the discharge is between 2,500 m^3/s and 3,500 m^3/s and the flood duration is from 6 d to 7 d. It has good scouring effects in the whole downstream and the reaches downstream from Gaocun when the discharge is between 3,500 m^3/s and 4,000 m^3/s and the flood duration is from 4 d to 5 d.

③When the sediment concentration of flood is between 60 kg/m^3 and 100 kg/m^3 and that is between 100 kg/m^3 and 300 kg/m^3, as the discharge increases, the deposition efficiency and deposition ratio in the whole downstream showing a decreasing trend. The deposition efficiency reduces significantly in the whole downstream when the discharge is between 3,000 m³/s and 3,500 m³/s.

(2) For floodplain flood with general sediment concentration in flood season, the reach from Aishan to Lijin is scoured under overbank flood with each flow rate stage. Seeing from changes of deposition efficiency under different overbank flood with different flow rate stage in the downstream, reaches downstream from Gaocun are scoured when the discharge is between 2,500 m³/s and 3,000 m³/s, and the scouring efficiency is 5.6 kg/m³ and 2.54 kg/m³ respectively. The scouring efficiency is high relatively in the whole downstream river and in reaches from Aishan to Lijin when the discharge is between 3,500 m³/s and 4,000 m³/s.

According to analysis of characteristics of scouring and deposition in flood mentioned above, there will be good sedimentation reduction effect when the discharge is chose between 2,500 m³/s and 3,000 m³/s from the view of sedimentation reduction in the lower Yellow River. The sedimentation reduction effect with the flow rate stage between 3,500 m³/s and 4,000 m³/s is better than that with the flow rate stage between 2,500 m³/s and 3,000 m³/s. From the view of requirement of maintaining normal channel in the downstream river, it needs a certain proportion flood with discharge larger than 3,500 m³/s to maintain the normal channel with about 4,000 m³/s bankfull discharge in the downstream river.

(3) For flood with high sediment concentration.

According to characteristics of scouring and deposition in the downstream river under flood with high sediment concentration, regulation of flood with high sediment concentration under different water and sediment condition should be treated differently. For the non-floodplain flood with high sediment concentration whose average discharge is less than 2,500 m³/s, the main channel of the whole downstream river deposits seriously, and then the reservoir should store runoff to avoid main channel deposited with such flood flowing into downstream. For the non-floodplain flood with high sediment concentration whose average discharge is larger than 2,500 m³/s, deposition of downstream river is still evident, therefore, the reservoir should store runoff appropriately and flush sediment out of the reservoir by low backwater to reduce the deposition in the downstream river. For the overbank flood with high sediment concentration, it has effect on silting floodplain and scouring channel and the reach downstream from Gaocun is not deposited basically, therefore the reservoir will not store runoff.

4 Law of flow and sediment motion in the Xiaolangdi Reservoir

4.1 Flow pattern and sediment flow pattern in the reservoir area

Flow pattern in the reservoir area can be roughly divided into two kinds, firstly, as a result of increasing water level by water retaining structures, the water surface will form backwater curve, water depth gradually increasing and flow velocity gradually reducing along the way, and this flow pattern is called backwater flow pattern. Secondly, water retaining structures can not afford to backwater effect and the water surface curve is close to the natural situation, then the flow pattern is similar to uniform flow and in practical applications it will be treated as uniform flow generally, and this flow pattern is called uniform flow pattern. Due to different flow pattern, the sediment flow pattern is different.

4.1.1 Backwater sediment flow pattern

In backwater sediment flow pattern, the sediment transporting characteristics are different under different water storage volume, water depth, and water and sediment condition. Accordingly the sediment flow pattern is divided into backwater sediment flow pattern, density current sediment flow pattern and sediment flow pattern of turbid water reservoir.

(1) Backwater sediment flow pattern.

The characteristics of the flow pattern is the sediment will spread to the full section of the flow when turbid water flowing into the backwater segment of reservoir area, there is a certain velocity and sediment concentration throughout the flow section and the flow velocity gradually reducing along the way, therefore, the sediment carried by flow is reducing along the way. The sediment is sorted along the way and sediment grain size of depositing matter is coarse at upstream and that is fine at downstream along the way.

(2) Density current sediment flow pattern.

The characteristics of the density current sediment flow pattern is turbid water will not mixed with clear water of the backwater segment when turbid water flowing into the backwater segment, but into below the clear water and continuing to move downstream along the bottom of the reservoir, as the flow entering is sediment carrying flow and content of fine sediment is large. For turbid water layer of density current diving into below the clear water, the flow velocity first increases and then decreases along the water depth from top to bottom. The flow velocity is relatively large at lower position in the turbid water layer, and sediment concentration is greater at position closer to the bottom. Due to different boundary condition of the reservoir, distance of backwater, and water and sediment condition entering, some density current can move farer and reach to the front of the dam to be flushed out of the reservoir, and some density current may stop at midway.

(3) Sediment flow pattern of turbid water reservoir.

The sediment flow pattern of turbid water reservoir is rather special. In most cases, density current in front of the dam can not be flushed out of the reservoir promptly and this cause sluice-stopping. Because the sediment grain size is fine in the density current, the settlement way of sediment in turbid water reservoir is different obviously from that of dispersed sediment particles settlement process in free flow. The settlement characteristic is unique and general performance of the settling velocity is extremely slow.

4.1.2　Sediment flow pattern of uniform flow

Under the sediment flow pattern of uniform flow, the water current is basically equal to the natural situation and water can carry a certain amount of sediment. When the sediment content is greater than sediment content can be carried by the water, the reservoir will deposit and the sediment is sorted along the way. Conversely, the reservoir will scour when the sediment content is less than sediment content can be carried by the water.

The above laws show that during initial sediment retaining period the retention water storage is large and the sediment flow pattern in the reservoir area is mainly density current sediment flow pattern and turbid water reservoir sediment flow pattern. During late sediment retaining period, the retention water storage is relative less than that of initial sediment retaining period, and sediment deposition increases. There is not only backwater sediment flow pattern, and some flow pattern is similar to the uniform flow in the upper reservoir segment out of the backwater influence. Therefore, during late sediment retaining period the flow pattern and sediment flow pattern in the reservoir area is complex and diverse. Therefore, it is also complex and critical to establish the operation mode. The laws also show that riverbed scouring may only occur at the reaches under uniform flow pattern.

4.2　Necessary condition for reservoir to restore storage capacity

4.2.1　Theoretical basis for recovering and maintaining long-term storage capacity

Due to the reservoir retaining water, the water depth increases and the flow velocity decreases, and the longitudinal sediment transporting balance in the reservoir area is destroyed, under automatic adjustment function of the riverbed, the reservoir will improve the river's sediment transport capacity by continuous deformation of deposition and towards to the new equilibrium. Therefore, accompanied by the reservoir deposition, the sediment transport capacity is gradually restored in the reservoir area. When the reservoir deposition reach to a certain level, lowering the water level and gradually emptying water storage, the water surface slope increases and the sediment transport capacity enhance, then flow is likely to change by the saturation as the secondary and the reservoir will occur scouring for restoring capacity.

According to the formula of sediment carrying capacity of flow $S^* = k\left(\dfrac{U^3}{gR\omega}\right)^m$, the sediment carrying capacity is proportional to high order of the flow velocity. During great flood period, the flow velocity is large and sediment transport capacity is great, and it is conducive to the reservoir scouring. At the same time, the sediment carrying capacity is inversely proportional to hydraulic radius. In the backwater case, the size of the hydraulic radius has a considerable extent depends on the storage capacity and the water level. Therefore, the reservoir can take advantage of large discharge lowering the water level in short-term during flood period to scour sediment and restore the capacity, and to maximize the comprehensive benefits of the reservoir.

4.2.2 Main measures for recovering and maintaining long-term storage capacity of built reservoirs

After practical application of built reservoirs for many years such as Sanmenxia, Tianqiao, Qingtongxia, Hengshan and Wangyao, the successful experiences of maintaining long-term storage capacity have been summarized.

The operation mode of Sanmenxia Reservoir is "storing clear and releasing muddy". The water level in front of dam is controlled at 305 m, and the reservoir will basically reach to the balance of scouring and deposition when the runoff is abundant (such as 1974 ~ 1985). The reservoir can not reach to the balance of scouring and deposition when the runoff is low (such as 1986 ~ 1991), and the reservoir will deposit as releasing sediment under low backwater. After the reservoir emptying to scour sediment by lowering water level in flood period, the reservoir will not deposit and will restore some capacity. After Xiaolangdi Reservoir put into operation, Sanmenxia Reservoir is operated combined with Wanjiazhai Reservoir. Using artificial shaping flood process entering, and emptying to scour sediment by lowering water level at the same time (the water level reduced to 289.2 m on July 5,2010), and the capacity is recovery.

The operation mode of Qingtongxia Reservoir is sediment peak passing in flood season combined with flushing sediment at the end of the flood season from 1991. Easing of continuous deposition by the deterioration of water and sediment, and scheduling with the cascade reservoirs upstream and create a large flow with continuous process to scour the reservoir area, it not only be able to discharge sediment entering reservoir in that year, but also gradually recover part of the silt storage capacity.

Tianqiao Reservoir operates in free sediment discharge to restore the capacity in annual flood season during downtime. The characteristics of Hengshan Reservoir operation are: perennial impoundment combining with concentrated sediment discharging, abolishing the harm by sediment discharging combining with promoting the benefit by using sediment and regulating the reservoir sediment for many years. Therefore, Hengshan Reservoir takes the advantages of large longitudinal slope and makes full use of the detention sediment discharging and density current sediment discharging effect, greatly slowing the silting of the reservoir, but the key to restore and maintain the storage capacity is emptying to sediment discharging every three years to four years.

The operation experiences of Wangyao Reservoir management for many years are as follows: the operation under low water level of the backwater to sediment discharging, and it can only slow down the deposition, but can not clear away sediment deposited in the reservoir. Reasonable emptying should be combined with a variety of methods in order to form the ultimate capacity with a relative scouring and deposition balance, and then it should transfer to a reasonable periodic operation mode that storing clear water and flushing muddy combined with emptying for free discharging sediment.

The above data indicate that it is feasible to recover and maintain long-term capacity of the reservoirs in ways of establishing reasonable operation mode according to the law of flow and sediment motion in the reservoir area.

5 Types and key technologies of sedimentation reduction operation of Xiaolangdi Reservoir

5.1 Types of sedimentation reduction operation

The operation for sedimentation reduction is summarized as two types. One type is "raising water level gradually and retaining coarse sediment and discharging fine sediment" and it is called the operation mode one. Another type is "to regulate sediment for many years and seek chances to scour by lowering operation water level for retaining sediment and regulating water and sediment", namely reservoir regulate sediment for many years, retaining sediment combined with discharging sediment, making full use of sediment transport capacity with continuous large flow, reducing deposition in lower river especially in reaches downstream from Gaocun, and it is called the the second operation mode.

5.2   Key technologies of different operation modes

5.2.1   Key technologies of the first operation mode

The core of the first operation mode is based on the sediment transport capacity is greater with finer sediment in the lower Yellow River under large discharge and it has a certain capacity to transport sediment with particle size greater than 0.05 mm. Therefore, the reservoir outflow has polarization in main flood season. The reservoir keep low backwater and the water storage capacity is 0.3×10^9 m^3, controlling reservoir sediment releasing ratio is about 60%. When the sediment releasing ratio of total sediment is 60%, the sediment releasing ratio of fine sediment is 84%, and that of median sediment and coarse sediment is 35.7% and 26.9% respectively. So that reasonable retaining coarse sediment and discharging fine sediment, and reducing the deposition in downstream river. Water level has a small range of changes during the operation process, but the overall trend is gradually increased and deposition surface of swale is gradually increased at the same time. When the deposition surface in front of dam reach to 245 m (corresponding reservoir sedimentation is about 7.86×10^9 m^3), then silt beach and scour channel to form the high beach deep trench, and then use the channel capacity for retaining coarse sediment and discharging fine sediment.

5.2.2   Key technologies of the second operation mode

According to analysis of deposition matter characteristics of built reservoirs such as Guanting and Sanmenxia, the dry density increasing with deposition thickness and deposition body subject to force consolidation for a long time and is not easy to scoured. Therefore, for the capacity recovery, scouring the reservoir after a long-time deposition is not better than scouring and deposition alternately after a certain period operation. On the other hand, the runoff flowing into Xiaolangdi in flood season reduce significantly after Longyangxia and Liujiaxia put into use, and industrial and agricultural water in the upper and middle reaches is growing, and in flood season normal floods appear probability is reducing. Therefore, there is large risk to reduce the water level to scour and recover capacity when the reservoir sedimentation is great. There is an urgent need for a reservoir operation mode for the changed water and sediment conditions and riverbed boundary conditions.

The key indicators of the second operation mode in flood season are the regulation discharge, the regulation capacity and the opportunity to scour by lowering the water level.

(1) Regulation discharge: regulation discharge contains the regulatory minimum discharge and regulation maximum discharge. Regulatory minimum discharge refers to maximum flow in the polarization of small flow and regulatory maximum discharge refers to minimum flow in the polarization of large flow.

Regulatory minimum discharge, according to the erosion and deposition characteristics in flood season normal water period and in non-flood season of the downstream river, great discharge is favorable to scouring in reach upstream from Gaocun but not favorable to sedimentation reduction in reach downstream from Gaocun. Therefore, considering from sedimentation reduction in the downstream river especially in the reach from Gaocun to Lijin, Xiaolangdi Reservoir should control discharge and the discharge of Xiaoheiwu should be less than 800 m^3/s. For meet the requirements of the reservoir generation, discharge should be controlled as 400 m^3/s in flood season.

Regulatory maximum discharge is calculated and analyzed with two schemes of 3,700 m^3/s and 2,600 m^3/s. Due to the regulation capacity is large with 3,700 m^3/s scheme, the deposition is a little greater, but the final service life of the late sediment retention period has little difference for the two schemes. According to analysis of the lower river, regardless of the previous 10 years or the late sediment retaining period, analyzed from the water and sediment conditions, sediment retention sedimentation reduction ratio (sediment retaining amount in the reservoir when the sedimentation reduction is 0.1×10^9 t in the lower river, similarly hereafter) of the whole lower river, sedimentation reduction effect in the whole lower river and in reach downstream from Gaocun, the maintaining role of normal channel of lower river, the scheme of 3,700 m^3/s is better than that of 2,600 m^3/s. According to the above analysis, the regulatory maximum discharge is adopted as 3,700 m^3/s.

(2) Regulation capacity: regulation capacity is the maximum capacity above the initial running water level. It is calculated and analyzed with two schemes of 1.1×10^9 m^3 and 1.3×10^9 m^3. There is little difference for the two schemes from regulation water effect during the whole late sediment retention period. Analyzed from the sediment retention sedimentation reduction ratio of the whole lower river, sedimentation reduction effect in the whole lower river and in reach downstream from Gaocun, the maintaining role of normal channel of lower river, the scheme of 1.3×10^9 m^3 is better than that of 1.1×10^9 m^3. Therefore, the regulatory capacity is adopted as 1.3×10^9 m^3.

(3) Opportunity to scour by lowering the water level is refers to starting time of the reservoir emptying scouring, represented with sedimentation volume when it reaches a certain value. An initial draft of opportunity to reduce the water level to scour is 3.2×10^9 m^3, 4.2×10^9 m^3, 5.8×10^9 m^3, 7.86×10^9 m^3, and calculation of regulatory maximum discharge is 2,600 m^3/s and 3,700 m^3/s. According to analysis of the regulation indicators of the reservoir, it performs basically the same laws with regulatory maximum discharge of 2,600 m^3/s and 3,700 m^3/s:

①Regardless of the previous 10 years or the entire late sediment interception period, service life of the entire late sediment retention period of opportunity to reduce the water level to scour with 7.86×10^9 m^3 is shorter 7 years to 8 years than that of the other three opportunities. Water and sediment at Huayuankou is significantly reducing with discharge over 2,600 m^3/s and with floodplain flood (Discharge of Huayuankou is greater than the 4,000 m^3/s). Days and water at Huayuankou is significantly reducing with discharge over 2,600 m^3/s and over 3,700 m^3/s continuously occur in 4 d, 5 d and 6 d. The reservoir deposition speed is too fast. There is great risk of recovering capacity by reducing water level when the reservoir sedimentation volume is 7.86×10^9 m^3 and the deposition surface in front of the dam is up to 247 ~ 248 m before the start of reducing water level to emptying scouring. Generating capacity is slightly more, but it is not dominant overall with opportunity to scour by reducing water lever of 7.86×10^9 m^3.

②Comparing to times to reduce the water level to scour of 3.2×10^9 m^3, 4.2×10^9 m^3, and 5.8×10^9 m^3, there is a qualitative analysis that the reservoir deposition speed is slower and sediment carried by large discharge is greater when the scouring time is earlier, and the service life of the whole sediment retention period will extend about one year. Quantitative analysis is not very different. But the deposition surface is low when scouring time is 3.2×10^9 m^3 and the elevation is only 200 m, and delta vertex is still away from the dam about 10 km, and the efficiency of reducing water level to scour is low especially that of retrogressive scouring. And the elevation of deposition surface in front of dam is lower than the lowest operation water level 210 m, and there are no conditions of reducing water level to scour and recover capacity. The reservoir deposition is 5.8×10^9 m^3 when time to reduce water lever to scour is 5.8×10^9 m^3, then the deposition matter deposit for too long time and easy to form erosion resistance, and not take full advantage of large discharge have the few opportunity to flush sediment, and there is still a great risk of recovering capacity by reducing water level. More over, the deposition surface in front of the dam reach to 221 ~ 222 m when the time to scour by lowering water level is 4.2×10^9 m^3, and the efficiency of reducing water level to scour can be increased especially that of retrogressive scouring. Therefore, time to reduce water level to scour with 4.2×10^9 m^3 is appropriate according to analysis of reservoir regulation.

According to the sedimentation reduction achievements of downstream river, the argument with regulatory maximum discharge of 3,700 m^3/s and 2,600 m^3/s for time to reduce the water level to scour reflects basically the same law. Seeing from sedimentation reduction of the lower river in the previous 10 years, the amount of sedimentation reduction is increasing in the whole downstream river and in main channel of reaches downstream from Gaocun when the time to scour is later, and reservoir sediment retention sedimentation reduction ratio tended to increase. Seeing from sedimentation reduction in the whole downstream river during the entire late sediment retaining period, the repeated use probability of sediment trapping capacity is less when the time to scour is later. Therefore, the amount of sedimentation reduction is reducing in the whole downstream river and in main channel of reaches downstream from Gaocun, and reservoir sediment retention sedimentation reduction ratio tended to increase. Seeing from bankfull flow changes in the downstream river, the bankfull flow can be maintained in 4,000 m^3/s in the downstream river during the late sediment retention period when the time to reduce water level to scour is 4. 2 × 10^9 m^3.

Time to reduce water level to scour with 4. 2 × 10^9 m^3 is appropriate for reservoir sediment discharging and sedimentation reduction in the downstream river.

6 Xiaolangdi Reservoir operation mode for sediment reduction

6.1 The first operation mode

From July 1 to July 10 reservoir gradually discharging the adjustable amount of water reserved before the end of June to 0. 2 × 10^9 m^3, to meet water supply, irrigation needs in early July. In case of particularly dry years, it is no longer set aside 0. 2 × 10^9 m^3 and keep dischargeing until the adjustable water is finished.

From July 11 to September 30, when the total discharge of inflow, Heishiguan and Wuzhi is less than 4,000 m^3/s, the reservoir operated to regulate water and sediment. When the inflow is less than 2,600 m^3/s and the outflow is 400 m^3/s, when the inflow is equal to or greater than 2,600 m^3/s and the outflow is equal to the inflow, when the water volume reaches to 0. 3 × 10^9 m^3 above the 210 m, discharge water is 0. 1 × 10^9 m^3 and discharge of Xiaoheiwu is 3,700 m^3/s. When the reservoir sedimentation is greater than or equal to 7. 9 × 10^9 m^3, first empty impoundment and then free discharging sediment, to recover the above operation until the deposition amount is less than or equal to 7. 6 × 10^9 m^3. In the discharge process, the discharge of Xiaoheiwu should be not greater than the bankfull discharge of main channel in the lower river. Controlling the operation water level is less than 254 m and the minimum operation water level is 210 m during the sediment retention period. When the total discharge of inflow and Heishiguan and Wuzhi is greater than 4,000 m^3/s, perform the flood control operation.

From October 1 to October 31, when the total discharge of inflow and Heishiguan and Wuzhi is less than 4, 000 m^3/s, the reservoir discharge according to downstream water supply and the demand for irrigation. Otherwise, perform the flood control operation.

From November to next May, the reservoir discharge according to downstream water supply and the demand for irrigation to regulate runoff, and control the operation water level is lower than 275 m.

From June 1 to June 30 according to runoff situation, first to meet downstream water supply and irrigation requirements, as a precondition the reservoir water level does not exceed 254 m to June 30, conditional circumstances to set aside about 0. 8 × 10^9 m^3 of storage capacity (volume of water can adjusted above 210 m). If there is excess water storage capacity, made discharge peak according to downstream bankfull discharge to scour the lower Yellow River .

6.2 The second operation mode

Between July 11 and September 10 Each year：
(1) When the total discharge of inflow and Heishiguan and Wuzhi is less than 4,000 m^3/s, perform water and sediment regulation. ①If impoundment is exceeding or equal to 0. 6 × 10^9 m^3, and predict there is two days whose discharge is greater than or equal to 2,600 m^3/s, short-term to reduce the water level and make peak and the peak flow is 3,700 m^3/s, making peak 5 d later

according to the downstream river bankfull flow, until the water storage capacity is 0.2×10^9 m^3 above 210 m. After that, if the total discharge of inflow and Heishiguan and Wuzhi is greater than 2,600 m^3/s, outflow is equal to inflow; if the total discharge of inflow and Heishiguan and Wuzhi is less than 2,600 m^3/s, reservoir storage water and the outflow is 400 m^3/s. ②If impoundment is less than 0.6×10^9 m^3, and if the reservoir sedimentation is less than 4.2×10^9 m^3, reservoir storage water and the outflow is 400 m^3/s. If the reservoir sedimentation is more than 4.2×10^9 m^3, and predict there is two days whose discharge is greater than or equal to 2,600 m^3/s, reservoir perform short-time empty scouring by reducing water level. ③If impoundment is exceeding or equal to 1.3×10^9 m^3, fill up the reservoir and make peak, reservoir regulation is the same as discharging and making peak. ④When the inflow is greater than 2,600 m^3/s and sediment concentration is equal to or greater than 200 kg/m^3, perform hyperconcentrated flow scheduling operation, for the flood non-floodplain with high sediment concentration, the reservoir storage runoff appropriately. For floodplain flood with high sediment concentration, the reservoir will not storage water.

(2) When the total discharge of inflow and Heishiguan and Wuzhi is greater than 4,000 m^3/s, perform flood control operation.

Between September 11 and September 30 Each year, when the total discharge of inflow and Heishiguan and Wuzhi is greater than 2,600 m^3/s, outflow is equal to inflow. Otherwise, reservoir impoundment ahead , the outflow is 400 m^3/s, to meet the power generation and water supply requirements . Other period operation principle is the same to the first operation mode.

7 Xiaolangdi Reservoir operation mode for flood control during the late sediment retaining period

7.1 Flood control situation and main problems in downstream river recently

Affected by climate change, human activities and a variety of factors, since the 1990s from the last century, measured flood in the middle and lower reaches of the Yellow River changed greatly compared with the 1950s and 1960s. Mainly for flood frequency was reduced, reducing magnitude, and duration has been shortened, base flow is reduced before the peak, proportion of the runoff upstream from Hekouzhen reducing etc. The main channel in lower Yellow River siltation atrophy, "Secondary perched river" situation is grim; Xiaolangdi Reservoir has operated more than 10 years, the minimum beach discharge in downstream river is about 4,000 m^3/s. At the same time, the floodplain is both flood passage and home for about 1.9×10^6 people in the survival. Currently the construction of flood plain is lagging behind, inundation losses is great, social and economic development, livelihood water conservancy request, making the beach area to the focus of conflicts flood control. Normal flood regulation has become the bottlenecks of the flood control operation in downstream river.

The flood control capacity of Xiaolangdi Reservoir of late sediment retention period is reduced from nearly 8.9×10^9 m^3 (capacity in April 2009) to 4.05×10^9 m^3, in original design, it will not control flood less than 8,000 m^3/s in the normal use period at Huayuankou. During the late sediment retention period, how to use the larger flood protection capacity under the premise not affect the downstream flood safety and take into account the flood plain flood control, but also long-term play the role of the flood control and sedimentation reduction is the core problem of the flood control study.

7.2 Study on operation mode for flood control of normal flood

7.2.1 Magnitude of normal flood

According to flow capacity of main channel and dikes in the lower Yellow River, taking the reference criteria for the classification of the national flood magnitude, Combine the research results in the past and the design flood study, determining the flood with peak flow at Huayuankou of

4,000 ~ 10,000 m^3/s as the normal flood in the lower Yellow River. In which 4,000 m^3/s is the long-term flow capacity in main channel of the downstream river; and 10,000 m^3/s is minimum embankment fortification discharge deducted discharge of Changqing and Pingyin added water.

7.2.2 The flood plain inundation losses and control discharge of normal flood in lower Yellow River

Based on the 2009 terrain before flood season, the submerged area and flooded population of the lower Yellow River of different order of magnitude with flood routing and disaster assessment models were calculated. The results show that the floodplain majority have been inundated when the flood of Huayuankou is 8,000 m^3/s. If the the Huayuankou peak flow is controlled to less than 6,000 m^3/s, it can effectively reduce the flooded area of the beach area. By repetitive real often flood control calculated, after the role of the reservoirs in the middle reaches, it is basically able to control often flood peak flow at Huayuankou not more than 6,000 m^3/s. Therefore, from effectively reduce the floodplain inundated loss and analysis of flood magnitude between Xiaolangdi and Huayuankou, the control discharge of normal flood is 4,000 m^3/s to 6,000 m^3/s.

7.2.3 Required storage capacity for normal flood control

After calculation and analysis of design flood and the 100 field actual floods, required storage capacity for different order of magnitude flood at Huayuankou are obtained, as shown in Tab. 1. If the magnitude of normal flood is larger and control discharge is smaller, then the required flood control capacity is larger.

Tab. 1 Required storage capacity for different order of magnitude flood at Huayuankou

Unit: $\times 10^8$ m^3

Flood magnitude(m^3/s)	Control discharge(m^3/s)		
	4,000	5,000	6,000
10,000	18	8.7	6.0
8,000	10	5.5	3.2
7,000	5.7	3.3	2.4

7.2.4 Flood limit level, flood control capacity change and phases division of flood control operation during the late sediment retaining period

According to research results of operation for sedimentation reduction, adopting 1.3×10^9 m^3 as water storage capacity under the flood limit water level to determine the different flood limit water level under different deposition amount. In order to maintain long-term reservoir capacity, the water level should not more than 254 m in flood control operation during the late sediment-retaining period. Flood limit level and flood control capacity are showed in Tab. 2.

Tab. 2 Flood limit level and feature capacity under different deposition volume

Deposition volume($\times 10^8$ m^3)	Flood limit level(m)	Corresponding capacity under flood limit level ($\times 10^8$ m^3)	Capacity under flood limit level 254 m($\times 10^8$ m^3)	Flood control capacity($\times 10^8$ m^3)
42	240	11.9	23.7	70.3
60	250	13.2	6.8	52.6
75.5	254	10.0	0	41.0

According to the results in Tab. 1, for the flood about 10,000 m^3/s at Huayuankou, the required flood control storage capacity of Xiaolangdi is approximately 1.8×10^9 m^3, 0.9×10^9 m^3

and 0.6×10^9 m^3 to control flood at Huayuankou 4,000 m^3/s, and 5,000 m^3/s and 6,000 m^3/s respectively. In preliminary design of Xiaolangdi Reservoir, it prepared to adopt 0.79×10^9 m^3 as beach protection capacity in normal operation. Before sedimentation volume in Xiaolangdi reservoir reached 6×10^9 m^3, the flood control capacity under 254 m and can basically meet the requirement that normal flood flow not more than 6,000 m^3/s at Huayuankou. After deposit volume reached 6×10^9 m^3, Xiaolangdi Reservoir may use capacity above 254 m for normal flood control operation. Therefore, the deposit volume reached 6×10^9 m^3 as a flood control operation cut-off point.

Based on sedimentation reduction in operation phases and flood control storage capacity change of deposition in Xiaolangdi, the flood control operation is divided into three phases by deposition volume if reached 4.2×10^9 m^3 and 6×10^9 m^3. The first phase is the reservoir sedimentation less than 4.2×10^9 m^3 and flood control capacity more than 2×10^9 m^3 under 254 m. The second phase is when the reservoir sedimentation volume is 4.2×10^9 m^3 to 6×10^9 m^3, and reservoir flood control capacity is reduced more and often flood control operation water level is still not more than 254 m. The third phase is when deposition is greater than 6×10^9 m^3 and often flood control operation may use flood control storage capacity above 254 m.

7.2.5 Normal floods operation mode for flood control

The normal flood at Huayuankou occurs about 2 times in one year, and it is the often flood in the actual scheduling. Sediment concentration is great in Yellow River flood, according to recent measured data statistics, flood at Tongguan station is majority of hyperconcentrated flood with discharge more than 6,000 m^3/s. Therefore, adjustment with a long series, sediment accumulated effect for many years is calculated, long-term impact of different flood control operation on the reservoir and channel and floodplain in downstream river is analyzed.

Operation mode of hyperconcentrated flood is the key issue of the Xiaolangdi Reservoir flood control operation. Therefore, the development of the whole control scheme and hyperconcentrated open discharge, non- hyperconcentrated control discharge two major programs operation. The whole control scheme analyzed four schemes to control Huayuankou discharge at 4,000 m^3/s, 5,000 m^3/s, 6,000 m^3/s and 4,000 ~ 6,000 m^3/s. After a comprehensive analysis, that control of 4,000 ~ 6,000 m^3/s is slightly better than the other three schemes, and recommend the program of this control scheme as a recommend fully-controlled manner. Full control recommended programs and hyperconcentrated flood open discharge, non-high sandy controlled discharge program were analyzed, the results show that the sediment retaining period of hyperconcentrated open discharge scheme is about 17 years, significantly better than 4,000 ~ 6,000 m^3/s scheme (the length of the sediment retention period is about 14 years). Sediment retaining sedimentation reduction ratio of the two programs has little difference, the hyperconcentrated open discharge is slightly better; the program the hyperconcentrated open discharge is significantly higher than the control recommended program when discharge at Huayuankou is more than 6,000 m^3/s.

Hyperconcentrated flood program open discharge can not only extend the service years of sediment retaining period of Xiaolangdi Reservoir, and better able to play the role of the brush slot silt beach with hyperconcentrated flood on the downstream river, and it will help to improve the unfavorable situation of the "secondary perched river". The inundation losses caused by hyperconcentrated flood open to discharge can gradually resolve by the use of the the comprehensive management of the beach area safe construction and compensation flooded policy measures. Program for non-hyperconcentrated flood control operate, controlling flow gradually increase with the deposition volume increased in Xiaolangdi Reservoir; it is coordinated with the construction of downstream floodplain security arrangements. With the gradual implementation of the security building in the downstream floodplain, the floodplain bear often flood risk improvement gradually; floodplain flood damage compensation, problem of flood control for floodplain flood of normal flood can be gradually resolved. Therefore, a comprehensive comparison of reservoir and river sedimentation reduction, the length of the reservoir storage period, the downstream floodplain disaster reduction and floodplain comprehensive management plan and other factors, from the long-term role of Xiaolangdi Reservoir and detention settling role of the floodplain, the final recommendation is use the program open discharge of high sediment concentration flood as normal flood operation.

7.3 Study on flood control operation with great flood

According to different characteristics of flood, study Xiaolangdi Reservoir flood control operation mode with the "upper great flood" and "lower great flood". Considering the flood control of often flood, floods from up stage, first operate by often flood control operation, when often flood storage capacity is filled up or the flow in Xiaohua section exceed the often flood control standard, Xiaolangdi Reservoir transferred into great flood control mode operation. And consider the Sanmenxia Reservoir joint operation and the downstream Dongpinghu detention basin operation, the flood control operation mode of Xiaolangdi Reservoir is:

(1) Forecast of Huayuankou flood peak flow is $4,000 \sim 8,000$ m^3/s.

① For flood mainly of incoming water from upper river to the Tongguan, if the medium-term forecasts of there is heavy rainfall in the middle reaches of the Yellow River or threre is flood at Tongguan station with sediment concentration greater than or equal to 200 kg/m^3, Xiaolangdi Reservoir operates in the way open to discharge flood detention. Otherwise, according to operation phase of the Xiaolangdi Reservoir for flood control, in accordance with controling discharge at Huayuankou not more than $4,000$ m^3/s to $6,000$ m^3/s, control the operation water level of reservoir for flood control not exceeding 254 m as much as possible.

②For the flood mainly of incoming water from reach between Sanmenxia and Huayuankou, Xiaolangdi Reservoir operates in the way controlling the discharge at Huayuankou no more than the control targets $4,000$ m^3/s to $6,000$ m^3/s, when the discharge between Xiaolangdi and Huayuankou is up to $4,000$ m^3/s to $6,000$ m^3/s, the reservoir in accordance with controlling the maximum discharge not exceeding the power generation discharge, and controlling the operation water level for flood not exceeding 254 m control as far as possible.

(2) Forecast of Huayuankou flood peak flow is $8,000 \sim 10,000$ m^3/s.

According to the flood source, sediment concentration and reservoir sedimentation, Xiaolangdi Reservoir operate in open discharge or control discharge (if the flood comes mainly from the river over Tongguan, in principle in accordance with open discharge operation; if the flood is mainly derived from Sanmenxia to Huayuankou, depending on the sediment concentration in flood, flood process, Xiaolangdi Reservoir sedimentation volume, control operation as appropriate).

(3) Forecast of Huayuankou flood peak flow is greater than 10,000 m^3/s, Xiaolangdi Reservoir operate in accordance with controlling discharge of Huayuankou not exceeding 10,000 m^3/s, in the control operation process, if the forecast of flow between Xiaolangdi and Huayuankou is greater than or equal to 9,000 m^3/s, discharge no greater than 1000 m^3/s. "Upper great flood" water level is up to 263 m to 266.6 m, Xiaolangdi Reservoir operate to increase discharge volume to open to discharge or operate with maintaining the water level. Forecast of flood volume over ten thousand at Huayuankou is 2×10^9 m^3, Xiaolangdi Reservoir recover controlling the discharge of Huayuankou 10,000 m^3/s. "Lower great flood" water level is up to 263 m to 269.3 m, Sanmenxia Reservoir control operation.

(4) Forecast of the discharge of Huayuankou flow down to less than 10,000 m^3/s, operate in controlling the discharge of Huayuankou not exceeding 10,000 m^3/s for flood discharge, until the water level down to the flood control level in Xiaolangdi Reservoir.

8 Study on Xiaolangdi Reservoir operation plan for sediment reduction during the late sediment retention period.

8.1 Design water and sediment conditions and the initial topography for calculation

The water and sediment conditions adopted is on 2020 design level in research of Xiaolangdi Reservoir operation plan for sediment reduction during the late sediment retention period and annual average sediment reduction amount is 0.5×10^9 t. For the water volume and sediment amount entering Xiaolangdi Reservoir under design water and sediment conditions, the average water volume is 13.77×10^9 m³ and the average sediment amount is 1.078×10^9 t in flood season of the first decade, and the annual average water volume is 26.795×10^9 m³ and the average sediment amount is 1.098×10^9 t. The average water volume is 13.203×10^9 m³ and the average sediment amount is 0.983×10^9 t in flood season of the first 20 years, and the annual average water volume is 26.184×10^9 m³ and the average sediment amount is 1.006×10^9 t.

The initial terrain adopted in scouring and deposition calculation of the reservoir area and the downstream river and that adopted in physical model test is the terrain measured in October 2007.

8.2 Calculation results and comparison

8.2.1 Calculation results of the first decade
(1) Reservoir regulation situation.

When the discharge of Huayuankou is larger than 3,700 m³/s, for the first operation mode, the annual average number of days is 8.8 d, and the corresponding water volume flowing out of the reservoir is 2.773×10^9 m³ and the average sediment amount is 0.256×10^9 t. For the second operation mode, the annual average number of days is 13.2 d, and the corresponding water volume flowing out of the reservoir is 4.424×10^9 m³ and the average sediment amount is 0.41×10^9 t. The second operation mode is better to carrying more sediment with larger discharge. Both the number of days continuously appear more than 5 d in main flood season and the corresponding water volume of the second operation mode is 2.2 times to that of the first operation mode. The water and sediment rate is favorable to sediment reduction in the reach downstream from Gaocun.

(2) Reservoir sedimentation amount and sediment release situation.

The sediment release ratio of the reservoir is respectively 43.48% and 58.50% for the first operation mode and the second operation mode, and accumulative sedimentation amount (including measured sedimentation amount in practical operation) is 7.165×10^9 m³ and 5.897×10^9 m³ respectively. Because the second operation mode makes flow peak by filling the reservoir or by discharging when the incoming water is large, and it seeks chance to scour by lowering operation water level, causing the reservoir sedimentation rate is relatively slow. Analyzing from effects of retaining coarse sediment and discharging fine sediment, the proportion of fine sediment in deposition matter is 46.53% and 50.53% respectively for the first and the second operation mode, and there is only a little difference. The effect of retaining coarse sediment and discharging fine sediment of the first operation mode is slightly better.

(3) Effect of sediment reduction in the lower river.

The accumulative sedimentation reduction amount of full section in the lower river is 4.214×10^9 t and 3.359×10^9 t respectively for the first and the second operation mode in the first decade, among which sedimentation reduction amount in the reach downstream from Gaocun is 1.652×10^9 t and 1.377×10^9 t, accounting 39.2% and 41.0% for sedimentation reduction amount in the whole lower river respectively. The accumulative sedimentation reduction amount of main channel in the lower river is 1.343×10^9 t and 0.899×10^9 t respectively for the first and the second operation mode, among which sedimentation reduction amount of main channel in the reach downstream from Gaocun is 0.6×10^9 t and 0.375×10^9 t, accounting 44.7% and 41.8% for sedimentation reduction amount of main channel in the lower river respectively. The result is shown in Tab.3. The reservoir sedimentation of the first operation mode is more 1.268×10^9 m³ than that of the second operation mode. Therefore, the sedimentation reduction amount of the second operation mode is less than that of the first operation mode in the lower river. Sediment retaining sedimentation reduction ratio (sediment retaining amount in the reservoir when the sedimentation reduction is 0.1×10^9 t in the lower river) is smaller, showing that the sedimentation reduction effect in the downstream river of reservoir is better. The reservoir sediment retaining sedimentation reduction ratio of the first and the second operation mode is 1.47 and 1.36 respectively in the first decade. Therefore, from efficiency of reservoir sediment retaining and sedimentation reduction, the second operation mode is better than the first operation mode.

Tab. 3 Effects of sedimentation reduction in the downstream river of the first and the second operation mode in the first decade

Item	Operation mode	Upper river to Huayuankou	Huayuankou—Gaocun	Gaocun—Aishan	Aishan—Lijin	Upper river to Lijin
Accumulative sedimentation reduction amount of the main channel(1×10^8t)	The first operation mode	0.77	6.66	3.29	2.71	13.43
	The second operation mode	0.24	4.99	2.06	1.69	8.99
Accumulative sedimentation reduction amount of the full section($\times 10^8$t)	The first operation mode	1.89	23.73	11.82	4.7	42.14
The second operation mode		0.66	19.16	10.39	3.38	33.59

(4) Change of bankfull discharge in the downstream river.

The average bankfull discharge in the whole downstream river is $4,666 \ \mathrm{m^3/s}$ and $4,824 \ \mathrm{m^3/s}$ respectively for the first and the second operation mode in the first decade. From the normal channel shaping in the downstream river and the role maintaining, the second operation mode is better than the first operation mode.

8.2.2 Analysis of calculation result of the late sediment retaining period

(1) Service life of the late sediment retaining period.

Service life of the late sediment retaining period is 11 years and 18 years for the first and the second operation mode respectively. The service life of the second operation mode is longer 7 years than that of the first operation mode. If coupled with eight years the reservoir actual operation, the service life of the entire late sediment retaining period is 19 years and 26 years for the first and the second operation mode respectively. Considering from the perspective of extending the sediment retaining capacity service life, the second operation mode is significantly better than the first operation mode.

(2) Reservoir regulation situation.

When the discharge of Huayuankou is larger than $3,700 \ \mathrm{m^3/s}$ in main flood season, for the first operation mode, the annual average number of days is 7.33 d, and the corresponding water volume flowing out of the reservoir is $2.346 \times 10^9 \ \mathrm{m^3}$ and the average sediment amount is 0.256×10^9 t. For the second operation mode, the annual average number of days is 12.22 d, and the corresponding water volume flowing out of the reservoir is $4.072 \times 10^9 \ \mathrm{m^3}$ and the average sediment amount is 0.373×10^9 t. The second operation mode is better to carrying more sediment with larger discharge. Both the number of days continuously appear more than 5 d in main flood season and the corresponding water volume of the second operation mode is 2.9 times to that of the first operation mode. The water and sediment rate is favorable to sediment reduction in the reach downstream from Gaocun.

(3) Reservoir sedimentation amount and sediment release situation.

The reservoir is basically filled up by the sediment retaining period for the first and the second operation mode, and the accumulative sedimentation amount and annual sediment release ratio has little difference for the two operation modes. The sediment release ratio of the reservoir is respectively 61.48% and 62.29% for the first operation mode and the second operation mode, and accumulative sedimentation amount (including measured sedimentation amount in practical operation) is $7.780 \times 10^9 \ \mathrm{m^3}$ and $7.668 \times 10^9 \ \mathrm{m^3}$ respectively. Analyzing from effects of retaining coarse sediment and discharging fine sediment, the proportion of fine sediment in deposition matter is 44.91% and 49.72% respectively for the first and the second operation mode, and the first operation mode is slightly better.

(4) Sedimentation morphology in the reservoir area.

Sedimentation morphology of the main stream: for the first operation mode, by the tenth year the deposition surface of main stream form high beach and high channel, and the elevation of beach face is about 250 m in front of the dam, and the elevation of the channel bottom is about 247 m. When the reservoir sediment retaining amount is 7.9×10^9 m^3, begin to lower the operation water level to form swale. By the eighteenth year the elevation of beach face is about 254 m in front of the dam, and the elevation of the channel bottom is about 230 m. For the second operation mode, by the tenth year the elevation of beach face is about 240 m in front of the dam, and the elevation of the channel bottom is about 225 m. By the eighteenth year the elevation of beach face is about 254 m in front of the dam, and the elevation of the channel bottom is about 226 m. Operate reservoir in the second operation mode, the reservoir beach and slot formed simultaneously, the formation process of swale is significantly different from that of the first operation mode.

Sedimentation morphology of the tributary: the beach elevation of tributary outlet gradually raise with that of the main stream, for the two operation mode, the sedimentation morphology of tributaries is the basically same as inverted cone, and there formed a sediment dam with height of about 4m at the tributaries outlet, capacity of tributary downstream from the sediment dam is fill with sediment.

(5) Effect of sediment reduction in the lower river.

By the late sediment retaining period, the accumulative sedimentation reduction amount of full section in the lower river is 5.127×10^9 t and 5.465×10^9 t respectively for the first and the second operation mode, among which sedimentation reduction amount in the reach downstream from Gaocun is 1.812×10^9 t and 2.075×10^9 t, accounting 35.3% and 38.0% for sedimentation reduction amount in the whole lower river respectively. The accumulative sedimentation reduction amount of main channel in the lower river is 1.381×10^9 t and 1.617×10^9 t respectively for the first and the second operation mode, among which sedimentation reduction amount of main channel in the reach downstream from Gaocun is 0.643×10^9 t and 0.781×10^9 t, accounting 46.5% and 48.3% for sedimentation reduction amount of main channel in the lower river respectively. The result is shown in Tab. 4. The sedimentation reduction effect of the second operation mode in the downstream river of reservoir is better than that of the first operation mode during the longest sediment retaining period. The reservoir sediment retaining sedimentation reduction ratio of the first and the second operation mode is 1.37 and 1.25 respectively during the late sediment retaining period. From efficiency of reservoir sediment retaining and sedimentation reduction, the second operation mode is better than the first operation mode.

Tab. 4 Effects of sedimentation reduction in the downstream river of the first and the second operation mode during the late sediment retaining period

Item	Operation mode	Upper river to Huayuankou	Huayuankou— Gaocun	Gaocun— Aishan	Aishan— Lijin	Upper river to Lijin
Accumulative sedimentation reduction amount of the main channel($\times 10^8$t)	The first operation mode	0.78	6.61	3.48	2.94	13.81
	The second operation mode	0.44	7.91	4.01	3.8	16.17
Accumulative sedimentation reduction amount of the full section($\times 10^8$t)	The first operation mode	2.39	27.97	13.96	4.16	51.27
	The second operation mode	2.8	31.1	15.64	5.11	54.65

(6) Change of bankfull discharge in the downstream river. The average bankfull discharge in the whole downstream river is 4,537 m^3/s and 4,723 m^3/s respectively for the first and the second operation mode during the late sediment retaining period. From the normal channel shaping in the

downstream river and the role maintaining, the second operation mode is better than the first operation mode.

In a word, the second operation mode compared with the first operation mode, water and sediment process out of the reservoir under reservoir regulation is better, depositing slower, and the late sediment retaining period is longer, all aspects show that the second operation mode is superior. Analyzing from effects of retaining coarse sediment and discharging fine sediment, the sediment retaining amount of the first operation mode is slightly more than that of the first operation mode. From efficiency of sedimentation reduction in the downstream river and the normal channel shaping and maintaining, the second operation mode is both better than the first operation mode.

8.2.3 Physical model test results

(1) Physical model test results in the reservoir area.

① Reservoir sedimentation amount.

By the end of the sediment retaining period, the total sedimentation amount (including measured sedimentation amount in practical operation) is 7.250×10^9 m^3 and 7.471×10^9 m^3 respectively for the first and the second operation mode, among which the sedimentation amount in main stream is 5.791×10^9 m^3 and 5.737×10^9 m^3 respectively, and the sedimentation amount in tributary is 1.459×10^9 m^3 and 1.734×10^9 m^3 respectively. The sedimentation amount of the second operation mode is more 0.221×10^9 m^3 than that of the first operation mode, among which main stream deposit less 0.054×10^9 m^3 and tributary deposit more 0.275×10^9 m^3. It shows that sedimentation distribution of the second operation mode is more reasonable.

②Reservoir capacity.

By the end of the sediment retaining period, the total reservoir capacity is 5.504×10^9 m^3 and 5.283×10^9 m^3 respectively for the first and the second operation mode, and the effective capacity above 254 m is 4.018×10^9 m^3 and 4.053×10^9 m^3 respectively, and both can meet the basic requirements for flood control.

(2) Physical model test results in the downstream river.

The physical model test with 15 years series is performed for the first and the second operation mode, and the test reach is between Tiexie and Taochengpu.

The total sedimentation volume with section method is 1.8×10^9 t and 0.755×10^9 t respectively for the first and the second operation mode, and the sedimentation volume of the second operation mode is significantly less than the first operation mode.

For bankfull discharge changes in the test reach, 15 years later from the beginning, the minimum bankfull discharge is 3,140 m^3/s and 3,360 m^3/s respectively for the first and the second operation mode, and the value of the second operation mode is more 220 m^3/s than that of the first operation mode. From the role of maintaining normal channel in the downstream river, the second operation mode is better than the first operation mode.

From changes in river regime, the adaptability to the project of the second operation mode is stronger than the first operation mode. Because large discharge duration of the second operation mode is longer, so river regime adjustment of the second operation mode is quicker than the first operation mode, and project relying on slip conditions of the second operation mode is better than the first operation mode in the late test.

According to the preceding analysis, from extending service life of reservoir sediment retaining capacity, coordination of the water and sediment relations in the lower Yellow River, effect of sedimentation reduction by reservoir in the downstream river, the role of restoring and maintaining normal channels etc., the second operation mode is favorable than the first operation mode. Therefore, operation mode for flood control and sediment reduction with the core idea "to regulate sediment for many years and seek chances to scour by lowering operation water level for retaining sediment and regulating water and sediment" is adopted during the late sediment retaining period.

9　Conclusions

（1）During the late sediment retaining period of Xiaolangdi Reservoir, in addition to completing the task of design flood control, the reservoir also bear task for flood control of the normal flood. For three phases of flood control operation during the late sediment retaining period, control discharge for flood control of normal flood is $4,000$ m^3/s, $5,000$ m^3/s and $6,000$ m^3/s respectively, and the maximum operation water level is no more than 254 m for flood control of normal flood. To play role of Xiaolangdi Reservoir in long-term, for normal flood the operation mode is opening discharge for hyperconcentrated flood, and controlling discharge for non-high sediment concentration flood.

（2）According to characteristics of the deposition and scouring in the downstream river and the regulation ability of the reservoir, combined with analysis and calculation, to determine the time to scour by lowering water level is when the sedimentation volume in the reservoir reached 4.2×10^9 m^3, the upper limit regulation discharge is $3,700$ m^3/s, duration is not less than 5 d, and the regulation capacity is 1.3×10^9 m^3 during the late sediment retaining period of Xiaolangdi Reservoir.

（3）Integrated analyze and demonstrate from flood control function of reservoir for the downstream river, extending service life of reservoir sediment retaining capacity, coordination of the water and sediment relations in the lower Yellow River, effect of sedimentation reduction by reservoir in the downstream river, the role of restoring and maintaining normal channels etc. , operation mode for flood control and sediment reduction with the core idea "to regulate sediment for many years and seek chances to scour by lowering operation water level, retaining sediment and regulating water and sediment" is presented during the late sediment retaining period, and it meet the requirements for reservoir operation with changes of water and sediment condition and riverbed boundary of the Yellow River as well as the economic and social development.

References

Liu Jixiang. Study and Practice of Reservoir Operation Mode[M]. Beijing: China Water Power Press; Zhengzhou: Yellow River Conservancy Press, 2008.

Xia Zhenhuan, Cao Ruxuan et, al. Reservoir Sediment [M]. Beijing: Water Resources and Electric Power Press, 1979.

A Global Sediment Data and Information Sharing Platform and Global Land – ocean Sediment Flux

Liu Cheng[1,2], *Hu Chunhong*[1,2], *Wang Yangui*[1,2], *Zhang Yanjing*[1,2] and *Shi Hongling*[1,2]

1. International Research and Training Center on Erosion and Sedimentation,
Beijing, 100048, China
2. China Institute of Water Resources and Hydropower Research, Beijing, 100038, China

Abstract: Sediment data of rivers, as one of the basic scientific data resources, are likely to have a significant importance for studies on river channel morphology, river management, river ecology, river nutrient flux, navigation, and erosion and sedimentation in the delta. To provide basic sediment data of world rivers for scientists, engineers and managers in related fields, a sediment data and information sharing platform, Global Data of Erosion and Sedimentation (GDES), which allows free access for public and scientific research communities, has been established by the International Research and Training Center on Erosion and Sedimentation (IRTCES) under support of Chinese government as one of contributions to the UNESCO IHP – International Sediment Initiative (ISI). The database and its architecture, key technologies and main contents are introduced in the paper. The GDES is designed and developed by adopting the B/S technology, using a fat server and thin – client model. Web applications with GIS function have been achieved by using Java language, JSP development tools and component technology based on the J2EE environment. It includes 6 sections, i. e. Basic Data, Literature, Simulation Technology, Major Projects, Institutes and Experts, as well as Multimedia. Management services for the GDES data enquiry, publishing and information sharing based on the spatial distribution are supplied. Users can browse and download all relevant information, data and documents, visually navigate hydrology gauge stations by using WebGIS and Google Map, and access sediment attribute data displayed with distribution lines, column charts, bar charts and two – dimensional tables. Referring the available sediment load data of large rivers of the world obtained from the GDES and using the contemporary land – ocean flux in the absence of reservoir trapping, 16.2 Gt/a, as a baseline, we roughly estimate the recent global land – ocean sediment flux is 9.3 ~ 10.5 Gt/a.

Key words: database, WebGIS, erosion and sedimentation, sediment load, land – ocean sediment flux

1 Introduction

Sediment data of rivers, as one of the basic scientific data resources, are likely to have a significant importance for studies on river channel morphology, river management, river ecology, river nutrient flux, navigation, and erosion and sedimentation in the delta. To provide basic sediment data of world rivers for scientists, engineers and managers in related fields, a sediment database, Global Data of Erosion and Sedimentation (GDES), has been constructed by the International Research and Training Center on Erosion and Sedimentation (IRTCES) under supports of the Ministries of Finance and Water Resources of China. It was intended to be constructed as a sediment data and information sharing platform with free access for public and scientific research communities. The GDES, including 6 sections of Basic Data, Literature, Simulation Technology, Major Projects, Institutions and Experts, Multimedia, was set up based on modern technologies of information collection, data mining, data warehouse and Geographic Information System (GIS). The GDES was established as one of contributions to the UNESCO IHP – International Sediment Initiative (ISI) with aims to strengthen, at global level, awareness about the importance of erosion and sediment processes and their impacts, to promote exchange of information on relevant data, and to foster cooperation in erosion and sediment – related research and education.

2 Architecture and development techniques

2.1 Architecture

Fig. 1 shows the GDES architecture. The bottom layer is the basic data layer supported by the database, used for storage, access and management of spatial and non – spatial data, as well as for data services for the application system. The middle layer is the business layer by which the functions of data query and search, information dissemination, dynamic data update and information service can be achieved. The top layer is the presentation layer which is the application service layer for different users.

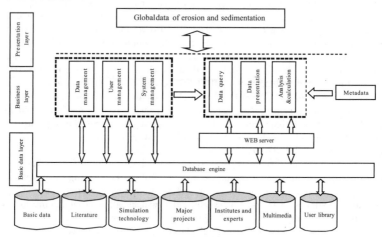

Fig. 1　GDES Architecture

The GDES is designed and developed by adopting the B/S technology, using fat server and thin – client model (Shapiro, 2006). A thin client makes it possible for all business processes to be performed in a controlled environment on a managed server (fat server). It also eliminates the requirement that clients download additional software in order to use the site. This distributed computing architecture enables large number of users to easily and efficiently access information from all over the world. A standard web browser is used to access and manipulate dynamic information over the Internet. The main components of the GDES include: Internet access lines and the corresponding equipments, safety and security equipments, information dissemination and storage equipments, information management and maintenance equipments, as well as the corresponding supporting software. The GDES network topology is shown as Fig. 2.

The server – side includes: ①WebGIS server: it mainly provides management services for GIS data resource, such as release, query and analysis of electronic map, to realize the ArcGIS – based storage and management of the spatial information. ②Web server: it is supported by the database server and the GIS server, to provide users with information dissemination and inquiry service in the form of dynamic WebPages. ③Database server: it is mainly used for storage and management of attribute data, to provide support for user query and retrieval.

2.2　Java technology

Java language is an object – oriented and platform – independent programming language with advantages of object – oriented nature, portability, platform independence and security (Smith, 1999). ①Applet: it is a small application program that can be embedded into HTML for providing users with nice interface for administration. It is interactive well that it can control the Applet to do

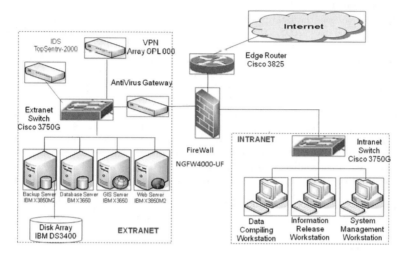

Fig. 2   GDES Network Topology

a lot of extra work, e. g. data checks to detect errors, before sending data back to the Web server. By this the network transmission cost can be reduced and the response time can be improved. ②Servlet: it is a server – side program written in Java. The main function is browsing and modifying data interactively to generate dynamic Web content, based on request/answer mode. ③JSP technology: JSP pages contain Java code that can generate dynamic content in the HTML page. It is more convenient to write a static HTML in the JSP, which fetch up the shortcomings of Sevrlet program such as no separation of direct web page presentation control, transaction logic and design logic. ④ JavaBeans: JavaBeans is a portable, platform – independent component model written in the Java programming language conforming to a particular convention. They are used to encapsulate many objects into a single object (the bean), so that they can be passed around as a single bean object instead of as multiple individual objects. Combined with JavaBeans, JSP and Servlet expand functions of the program in web pages (Robert, 1997).

2.3   Key technology

2.3.1   WebGIS technology

WebGIS is a product combined by GIS and Internet technology, which has become the important direction of GIS development at the present time. It has become the most effective way of GIS' socialization and popularization particularly as the increasing of user' demands that is using Internet technology to express information on Web and providing user browsing. The GDES uses the U. S. WebGIS development platform, ArcIMS, provided by ESRI. ArcIMS has the following advantages:①It is an Internet – based GIS, which allows the centralized establishment of a wide range of GIS map database, and provides the information to a wide range of users in the Internet/Intranet. ②It extends the general site' s capacity to provide GIS data and application services. ArcIMS provides free HTML and Java browsing tools, as well as supports other users. ③ArcIMS establishes a public platform for releasing data and services in the Web. It is more than an Internet mapping solution, but also a framework to release GIS functions through the Internet. ④ArcIMS supports the integration of network data and local data, but also supports raster data and vector data formats. ⑤The provided ArcIMS Manager can easily edit and modify web publish maps, so that secondary development has become very easy. It supports a full range of operating systems (ESRI, 2004).

2.3.2 Ajax technology

Ajax (Asynchronous Java Script and XML) was put forward by Jesse James Garrett of the Adaptive Path in February 2005. Strictly speaking, Ajax isn't a new technology. It's really several technologies, each flourishing in its own right, coming together in powerful new ways. Ajax incorporates: standards – based presentation using XHTML and CSS; dynamic display and interaction using the Document Object Model; data interchange and manipulation using XML and XSLT; asynchronous data retrieval using XMLHttp Request; and JavaScript binding everything together. The core of Ajax is based on XMLHttp Request asynchronous request mechanism, which had been applied before the word Ajax appeared but had not been given sufficient attention. Currently, Ajax – based Web geographic information services are emerging, such as Google Maps, Baidu Map, Go2map etc. These are rather typical successful application (McLellan, 2005).

Google is the first successfully using Ajax technology into commercial products, and Google Map services have been familiar. Google launched its own API program to enable world – wide who are interested in Google Maps can develop their own Google Maps service in June 2005, and Google can also manage these services through API (Google, 2005). Developers can easily interface to the Google Maps service to their own web pages by simply using the JavaScript scripting language. The map services provided by Google Maps are used in the GDES for publishing the world geographic information, and secondary development was carried out using Google Maps API to realize the query and share services for world hydrology gauge stations.

2.3.3 River sediment attribute data visualization

There are various types of data in the GDES. It would be more convenient for both professional and non – professional users to access data with visual information and knowledge. So it is necessary to carry out study on data visualization and the corresponding software development. In this system, the river sediment attribute data, including water level, flow rate, runoff, sediment concentration, sediment transport rate and sediment median size, are all numerical. Different visualization methods, such as distribution lines, column charts, bar charts and two – dimensional tables, were developed for describing the time series data. In particular, comparative charts of some parameters between two different stations or two different periods were developed for easy data analysis. The "comparative chart of related parameters between two stations" and "comparative chart of related parameters between two periods" can draw comparative chart of river sediment parameters such as runoff, sediment transport rate, and sediment concentration by collecting data from the database. Fig. 3(a) shows comparative chart of annual runoff between Datong station of the Yangtze River and "Tarbert RR + Landings" stations of the Mississippi River during 1950 to 1979, and Fig. 3(b) shows comparative chart of annual sediment load on Lijin Station of the Yellow River between two periods of 1952 ~ 1971 and 1986 ~ 2005.

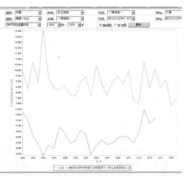
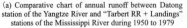
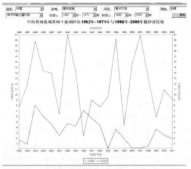

(a) Comparative chart of annual runoff between Datong station of the Yangtze River and "Tarbert RR + Landings" stations of the Mississippi River during 1950 to 1979

(b) Comparative chart of annual sediment load on Lijin Station of the Yellow River between two periods of 1952~1971 and 1986~2005

Fig. 3　Comparative charts generated by the database

3　GDES main sections

The GDES includes 6 sections of Basic Data, Literature, Simulation Technology, Major Projects, Institutes and Experts, as well as Multimedia. It provides internet – based shared information in 3 categories, i. e. public browse, ordinary membership and senior membership.

3.1　GDES homepage

The GDES Homepage was designed taking into account of good appearance, nice arrangement, and good guidance function. Major sections of the database, including Basic Data, Literature, Simulation Technology, Major Projects, Institutions and Experts, Multi – Media, are listed by navigation bands at the page top, and the navigation bands are appeared on each page to facilitate the users ready to access to any of the major sections.

There are columns of News, Search, World Map and General Charts established in the middle part of homepage. The column of "News" is used to publish related news and information; "Search" is used to search text files in the database with keywords; The World Map directly lead users to the data retrieval page of the section of "Basic Data"; and "General Charts" publishes charts and tables of general information related to the erosion, runoff and sediment in rivers at global and national scale, such as "Dynamic presentations of monthly variations of runoff and sediment load for global rivers", "Distribution of suspended sediment loads of the world's rivers", "Total suspended solid loads for Asian rivers", "Variations in river runoff by continent through most of the 20th century", etc. On both sides of the homepage, columns of "Member Login", "Recent Released", "Forum", "WebGIS System", "Important Researches", "Related Links" and others are listed for users' easy browsing.

3.2　Main sections

3.2.1　"Basic Data" section

This section includes five sub – sections of "River Data", "Soil Erosion", "Engineering Sediment", "River Erosion & Deposition" and "Gazette".

"River Data" sub – section is used to release numerical data related to sediment of global rivers, mainly including runoff, sediment load, sediment concentration, sediment median size etc. The time series parameters of a hydrology gauge station are showed in distribution lines, column charts or two – dimensional tables. Fig. 4 presents daily sediment concentration of Columbus Station on the Colorado River during 1960 ~ 1974. Each gauge station can be found by three methods. One is to find the gauge station with navigation boxes by selecting country – river basin – river – gauge station in sequence. The second is to find a gauge station by zooming in level by level in the world map of the WebGIS System. The third is to find a gauge station in the world map with gauge stations developed by map service provided by Google Maps (Fig. 5).

"Soil Erosion", "Engineering Sediment" and "River Erosion & Deposition" sub – sections publishes data, information, charts and pictures related to the soil erosion, erosion and deposition caused by hydraulic projects, and erosion and deposition in river channels, respectively. In the "Gazette" sub – section, all issues of the China Gazette on River Erosion and Sedimentation are released in electronic version.

3.2.2　Other sections

"Literature" Section: This section includes "Research Report", "Journals", "Proceedings", "Bibliography" and "Three Gorges Project" sub – sections. The "Research Report", "Journals" and "Proceedings" sub – sections release full – length sediment related to research reports, journal papers and academic conference proceedings. "Bibliography" sub – section releases sediment – related bibliographies and brief monograph introductions. In the "Three Gorges Project" sub – section, the research results of sediment issues of the Three Gorges Project are released.

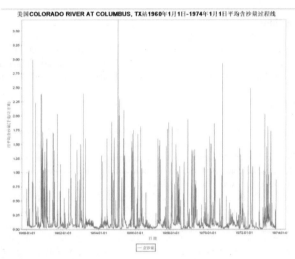

Fig. 4 Daily sediment concentration of columbus station on the Colorado River during 1960 ~ 1974

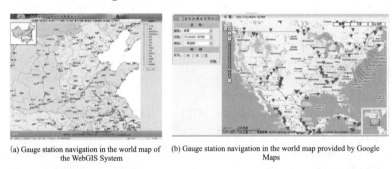

(a) Gauge station navigation in the world map of   (b) Gauge station navigation in the world map provided by Google
the WebGIS System   Maps

Fig. 5 Gauge station navigation by using the WebGIS system and google maps

"Simulation Technology" Section: this section is used to release introductions of sediment related numerical model and physical model, with two sub – section of "Numerical Models" and "Entity Models". "Major Projects" Section: Major hydraulic projects are introduced in this section. "Institutions and Experts" Section: this section is used to release brief introductions and home page linkages of sediment related research, coordination and management institutes and organizations, as well as to publish homepages for experts with interest of sediment and erosion research. "Multimedia" Section: this section publishes related images and videos.

4 Global Land – ocean Sediment Flux

It is difficult to quantify a reliable global land – ocean sediment flux due to a number of important problems, including availability and reliability of data on sediment loads data for rivers, the existence of many ungauged rivers, and only short period of sediment loads data for many gauged rivers. Despite these difficulties, the fluvial land – ocean sediment flux has been estimated using different methods by many researchers, with wide values ranging from 8.3 Gt/a to 51.1 Gt/a (Walling, 2006). However, recent work based on more available data and newly developed numerical models has provided more consistent results. The estimates of global land – ocean sediment flux have tended to converge on a mean annual flux in the region of 15 ~ 20 Gt/a. Syvitski et al. (2003, 2005) provided global estimations of the seasonal flux of sediment under modern and prehuman conditions making use of new global databases to represent land surface configuration and properties and hydrometeorological conditions at the global scale with a numerical model trained on a global database of 340 rivers. Syvitski et al. (2005) estimated that the contemporary land – ocean sediment flux is 12. 6 Gt/a, and that the contemporary flux in the absence of reservoir trapping would be 16. 2 Gt/a. They also provide an estimate of the pristine flux for the pre – human period of 14 Gt/a. Walling (2006) suggests that the reduction in the land – ocean sediment flux associated with reservoir trapping could be as much as 10 Gt/a and pre – human land – ocean sediment flux be 12 Gt/a, using the estimate of the contemporary land – ocean sediment flux of 12. 6 Gt/a provided by Syvistski et al. (2005).

Referring the available sediment load data of large rivers in the world in the GDES and using the contemporary land – ocean flux in the absence of reservoir trapping, 16. 2 Gt/a (Syvitski et al. , 2005), as a baseline, we try to roughly estimate the changing trend of the global land – ocean sediment flux in this paper.

Tab. 1 lists available sediment load data of the most downstream reaches in the large rivers of the world with high sediment loads obtained from the GDES and other literatures. The mean annual sediment loads of history or pre – dam values and recent or post – dam values are presented. The total value of history or pre – dam annual sediment loads of rivers listed in the Tab. 1 is 5. 4 Gt/a that accounts for 34% contemporary land – ocean flux in the absence of reservoir trapping. The table also shows that marked reductions in sediment load occurred in most of large rivers with total sediment reduction rate of 58%.

Tab. 1   Changes in annual sediment loads of large rivers in the world

	River	Mean annual sediment load ($\times 10^6$ t/a)		Sediment load reduction rate (%)	Source
		History or pre – dam value	Recent or post – dam value		
Asia	Chinese rivers	2090	580	72	Liu et al. (2009, 2011) ; GDES
	Euphrates	100	14	86	Al – Ansari et al. (1988)
	Indus	240	50	79	Walling (2009) ; GDES
	Ganges	520	262	50	Rice(2007)
	Brahmaputra	540	387	28	Rice(2007)
	Godavari	145	57	61	Malini & Rao (2004)
	Irrawaddy	364	379	– 4	Robinson (2007) ; Furuichi et al. (2009)
	Mekong	160	145	9	Wang et al. (2011)
	Song Hong (Red)	130	40	69	Le et al. (2007)
North America	Mississippi	497	133	73	Heimann et al. (2011) ; GDES
	Red	178	64.9	64	Heimann et al. (2011)
	Colorado	146.2	5.7	96	GDES
Europe	Danube	65	15	77	Walling (2009) ; GDES
	Rhine	4.3	2.1	51	Walling (2009) ; GDES
Africa	Nile	120	0.2	100	Milliman et al 2011
South America	Magdalena	144	174	– 21	Restrepo and Kjerfve (2000) ; Walling (2009)
Total		5,443.5	2,308.9	58	

In Tab. 2 the contemporary land – ocean sediment fluxes in the absence of reservoir trapping are listed in different land masses according to Syvitski et al. (2005) , and the sediment flux reduction rate are roughly decided for each land masses referencing those real values listed in the Tab. 1 and considering some key drivers, namely the impact of dams in reducing sediment loads as a result of sediment trapping and the various anthropogenic impacts, such as land clearance and catchment disturbance, leading to increased sediment loads. The estimated recent global land – ocean sediment flux is 9. 3 ~ 10. 5 Gt/a.

Tab. 2   Estimate of changing trend in the global land – ocean sediment flux

Land mass	Contemporary sediment flux in the absence of reservoir trapping ($\times 10^6$ t/a)	Sediment flux reduction rate (%)	sediment flux reduction ($\times 10^6$ t/a)	Estimate sediment flux ($\times 10^6$ t/a)
Africa	1,200	40 ~ 50	480 ~ 600	600 ~ 720
Asia	6,750	55 ~ 60	3,712 ~ 4,050	2,700 ~ 3,038
Europe	840	50 ~ 60	420 ~ 504	336 ~ 420
North America	2,340	45 ~ 55	1,053 ~ 1,287	1,053 ~ 1,287
Australasia	440	0 ~ 10	0 ~ 44	396 ~ 440
Indonesia	1,680		0 ~ 168	1,512 ~ 1,680
South America	2,950		0 ~ 295	2,655 ~ 2,950
Global	16,200		5,665 ~ 6,948	9,252 ~ 10,535

5   Conclusions

The database, Global Data of Erosion and Sedimentation (GDES) , has been constructed as sediment related data and information sharing platform with free access for public and scientific research communities. It is small, but rich in content and coverage. It includes not only the basic data of erosion and sedimentation, but also related literatures, documents, pictures and videos. The basic data of each hydrology gauging stations can be visually navigated by using WebGIS and Google Map. Currently a large number of available data on river sediment, related documents and information has been collected and included in the GDES. The database is ready for access through the Internet (http://data. irtces. org) , which provides online resources query, browsing, and downloading. Considering data collection, database updating and system improvement for the GDES require long and continuous efforts, any comments and suggestions, data and material contributions are welcomed.

Referring available sediment load data of large rivers of the world obtained from the GDES and using the contemporary land – ocean flux in the absence of reservoir trapping as a baseline, we roughly estimate the recent global land – ocean sediment flux is 9. 3 ~ 10. 5 Gt/a.

Acknowledgements
The work reported was supported by the China Institute of Water Resources and Hydropower Research (No. Shaji – 1230).

References

Al – Ansari N A, Asaad N M, Walling D E, et al. The Suspended Sediment Discharge of the River Euphrates at Haditha, Iraq: An Assessment of the Potential for Establishing Sediment Rating Curves. Geografiska Annaler[J]. Series A, Physical Geography, 1988,70(3): 203 – 213

ESRI. Getting Started with ArcIMS[M]. California: ESRI Press, 2004:66.

Furuichi T, Win Z, Wasson R J. Discharge and Suspended Sediment Transport in the Ayeyarwady River, Myanmar: Centennial and Decadal Changes[J]. Hydrol. Processes, 2009,23 (11): 1631 – 1641.

Garret J J. Ajax: A New Approach to Web Applications. http://www. adaptivepath. com/, 2005.

Google (2005). Google Maps API. http://code. google. com/

Heimann D C, Sprague L A, Blevins D W. Trends in Suspended – sediment Loads and Concentrations in the Mississippi River Basin, 1950-2009: U. S. Geological Survey Scientific Investigations Report 2011: 33.

Le TPQ, Garnier J, Gilles B, et al. The Changing Flow Regime and Sediment Load of the Red River, Viet Nam[J]. Journal of Hydrology. 2007,334(1 – 2):199 – 214

Liu C, Hu C H, Shi H L. Changes of Runoff and Sediment Fluxes of Rivers in Mainland of China Discharged into Pacific Ocean. Sediment Research, 2011(1):70 – 75.

Liu C, Wang Z Y, Souza F. Variations of Runoff and Sediment Fluxes into the Pacific Ocean from the Main Rivers of China[C]// Proceedings of the Third International Conference on Estuaries and Coasts. 2009(1):94 – 100

Malini B H, Rao K N. Coastal Erosion and Habitat Loss along the Godavari Delta Front a Fallout of Dam Construction[J]. Curr. Sci. , 2004(87):1232 – 1236.

McLellan D. Very Dynamic Web Interfaces. http://www. xml. com/, 2005.

Milliman J D, Farnsworth K L. River Discharge to the Coastal Ocean: A Global Synthesis[M]. Cambridge: Cambridge University Press, 2011.

Restrepo J D, Kjerfve B. Magdalena River: Interannual Variability (1975 – 1995) and Revised Water Discharge and Sediment Load Estimates[J]. Journal of Hydrology, 2000(235): 137 – 149.

Rice S K. Suspended Sediment Transport in the Ganges – Brahmaputra River System, Bangladesh [D]. Texas: Texas A&M University, USA. , 2007, 81(8).

Robert E. Developing Java Beans. O'Reilly Media, 1997:316.

Robinson R A J, Bird M I, Nay W O, et al. The Irrawaddy River Sediment Flux to the Indian Ocean: the Original Nineteenth – century Data Revisited[J]. J. Geol. , 2007(115): 629 – 640.

Shapiro J R. Microsoft SQL Server 2005: The Complete Reference: Full Coverage of all New and Improved Features[M]. McGraw – Hill Osborne Media. 2006.

Smith M A. Java: An Object – Oriented Language[M]. McGraw – Hill College, 1999.

Syvitski J P M, Vörösmarty C J, Kettner A J , et al. Impact of Humans on the Flux of Terrestrial Sediment to the Global Coastal Ocean[J]. Science, 2005(308):376 – 380.

Walling D E. Human Impact on Land – ocean Sediment Transfer by the World's Rivers[J]. Geomorphology, 2006 (79):192 – 216.

Walling D E, Fang D. Recent Trends in the Suspended Sediment Loads of the World's Rivers [J]. Global and Planetary Change, 2003, 39(1 – 2):111 – 126.

Walling D. E. The Impact of Global Change on Erosion and Sediment Transport by Rivers: Current Progress and Future Challenges [R]. The United Nations World Water Assessment Programme Scientific Side Paper Series. UNESCO, Paris, 2009:26.

Wang J J, Lu X X, Kummu M. Sediment Load Estimates and Variations in the Lower Mekong River[J]. River Research and Applications, 2011,27(1):33 – 46.

Bedform and Sediment Transport Mode in the Lower Yellow River

Zhang Yuanfeng[1] and *Lv Xiuhuan*[2]

1. Yellow River Institute of Hydraulic Research, Zhengzhou, 450003, China
2. Department of International Cooperation, Science and Technology, Yellow River
Conservancy Commission, Zhengzhou, 450003, China

Abstract: Bedform is a very important factor of ones which influence significantly resistance – to – flow and sediment transport in alluvial systems. With a control factor m which can reflect the bed forms developments, bedform features of the lower Yellow River are analyzed and discrimination method to bedform types of the lower Yellow River is established in this paper. In addition, based on Bagnold's resuspension efficiency concept (η_{sl}), we derive a transport capacity equation for bed – material load. It is suggested that, for a given alluvial system, $\eta_{sl} C_*$ in the equation can be deduced from the knowledge of m. In particular, the relationship between $\eta_{sl} C_*$ and control factor m is tested using observations collected in the the lower Yellow River. The sediment concentrations predicted by the formula in which m has been introduced, agree fairly well with observed ones during the lower Yellow River 'silt – flush experiment'. The use of an explicit bedform descriptor in a transport capacity formulation is novel and seems very promising.
Key words: the lower Yellow River, control factor m, bedform discrimination, sediment transport capacity

1   Introduction

Research on bed form and sediment transport has been carried for ages. Although numerous well known theories and formulae have been developed, the problem is still unsolved very well due to its complication. Comparison of transport formulae proposed in recent years for sediment concentration prediction shows that the discrepancy, ratios of computed values to be measured, still far away from 1 even for flume experiment data(Albert Molinas, 2001; Yang Shuqing, 2004; Guo Qingchao, 2006). Up to now it is realized that in a dynamic river such as the Yellow River, bedform always adapts according to flow and incoming sediment sollicitation, ultimately conditioning the extent of sediment transport in the alluvial system. But how do bed forms influence the sediment transport capacity? The bed form control factor m(Verbank,2004) proposed by Verbanck according the turbulent flow theory, provided us a opportunity to introduce bed form effects to sediment transport formula directly and research the bed form types with a new way. Flume experiments of flowing over different bed form types were conducted and summed by H. P. Guy et al (1956 ~ 1961), Williams, Praat, WlS, Wang Shiqiang et al. In this paper the classification of bed forms in the Yellow River are researched with Guy and Williams data. Then based on energy dissipation theory deduced by RA Bagnold and the concept of bed form control factor m, a new sediment transport mode is analyzed with aforementioned flume data. Finally a new equation proposed in this paper is improved by the Yellow River data and applied in Yellow River.

2   Bed form types

2.1   Criteria of bed form types

Bed forms types follow the classical sequence depicted by Simons and Richardson (1966): lower stage plane(Bed form code 1,), ripples(2), dunes(3), dunes in transitional state(4), high stage plane(5), in phase wave(anti – dune standing waves,6), breaking waves(7), chutes and pools(8). Some criterions on lower regime and upper regime bed form types, with parameters such as Fr (Froude Number), $U_* D/\nu$ (U_* is shear velocity, D is flow depth, ν is viscosity coefficient), U_*/ω (ω is settling velocity) and θ ($\theta = \dfrac{\tau_0}{(\gamma_s - \gamma)D}$, τ_0 is shear stress, γ_s, γ is sediment and flow density respectively) et al., were summarized by Qianning and WanZhaohui (1983). Leo C. van Rijn(1985) proposed the classification of bed forms for lower and transitional flow regime stages by a sediment transport parameter and a particle parameter. But the method developed from bed form control factor m and Fr is more visible and easier to be applied in natural rivers. Control factor m can be calculated as following(Verbank,2004):

$$\frac{m}{2\pi} = s^{0.3} Fr_g^{-1} \qquad (1)$$

where, Fr_g is general Froude Number, $Fr_g = U/C$, U is flow velocity, C is gravitational wave celerity.

$$C = \sqrt{\frac{g}{2\pi/\lambda_{BP}} \arctan \frac{2\pi D}{\lambda_{BP}}} \qquad (2)$$

where, λ_{BP} is wave length; g is acceleration of gravity.

For very shallow configuration of the flow C can be approximated as $\sqrt{gD}$. So the simple Froude Number Fr_s can be computed as $Fr_s = U/\sqrt{gD}$. Theoretically the bed form types of ripples and in phrase wave correspond to $m = 1$ and fully deve-loped dunes correspond $m = 2$. Flow intensities are very different from ripples to in phrase wave. Based on the aforementioned classifications, we choose m and Fr_s to distinguish bed form types. In order to avoid the disturbance of sediment size the normalized Froude Number Fr_{dd} ($Fr_{dd} = F_{dd}/Fr_s$) is introduced here. F_{dd}, Froude Number, corresponding to fully developed dunes, can be computed as following formula (G. V Luong, 2007):

$$F_{dd} = 0.044,6 \, A_r^{0.2} + 0.251,5$$

where, A_r is Archimedes Number, $A_r = g\Delta D/\nu$, Δ is relative excess density.

The applicability of m and Fr_{dd} in determining the bed form types with the flume data of Williams and Guy is shown in Fig. 1.

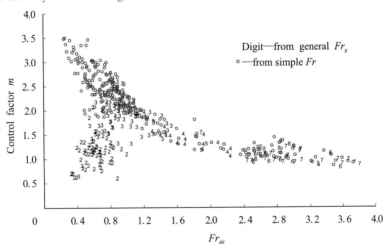

Fig. 1 Bedform changes with the relationship between F_{dd} and m

The digits in Fig. 1 represent bed form codes. It is very clear that different bed form codes are grouped into different zones. We can see that ripples are in the zone of $m < 2.5$(most of points around 1) and $F_{dd} < 0.8$, dunes in the zone of $1.5 < m < 2.5$ and $0.8 < F_{dd} < 1.6$, transitional dunes in the zone of $1 < m < 1.5$ and $1.6 < F_{dd} < 2.4$, in phase waves and breaking waves in the zone of m around 1 and $F_{dd} > 2.4$.

Usually bed form wave length is difficult to measure especial for natural rivers. Thus it is necessary to compute m by F_{rs} in shallow flow condition. But some times this simplification can produce some deviation. From the comparison of general m simple m the data points distributions agree with each other well except the zone of $F_{dd} < 0.8$ (Fig. 1). When Fdd is bigger than 0.8, the trend of simple m data points is opposite to that of general m. Ripples are usually in the zone $m > 2$ and $F_{dd} < 0.8$ for simple m.

2.2 Bed form types in the Yellow River

The Yellow River is very heavy sediment laden flow with intensive fluvial process. The sediment transport depends not only on incoming flow and sediment load but also on bed form adaptation. It is researched that incoming sediment load can be almost transport into downstream in the lower parts of the lower Yellow River when the bed form in the state of upper regime plane (Qi Pu, 2005). So it is very significant to research bed form types for flood control. But because the measurement of bed forms dimensions in the Yellow River is very difficult, research on bed form types rare. With aforementioned method and observed data the bed form types in the Yellow River can be analyzed. From Fig. 2 we can see that the Yellow River data points distributing tendency agree with that flume data but the values are about 0.8 lower at the same F_{dd}. One reason of the differences is the m computation without sediment size properties. The flume experiment sand sizes are much coarser than that of in the Yellow River. But the finer the sediment size is the smaller m is at the same flow condition according the definition (Verbank, 2004).

Combing Fig. 1 and Fig. 2 we suggest that bed forms mainly represent the type of ripples with $m > 2.0$ and $F_{dd} < 0.8$, dunes with $0.7 < m < 2$ and $0.8 < F_{dd} < 1.6$, transitional dunes with m around 1 and $1.6 < F_{dd} < 2.4$ and in phase waves with $F_{dd} > 2.4$. So the Yellow River usually is in the cases of ripples and dunes and less in the transitional dunes and in phrase wave.

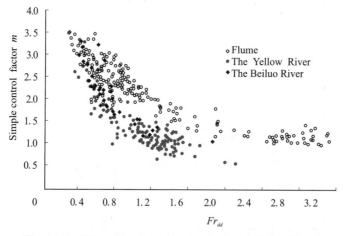

Fig. 2　Bedforms Discrimination to the lower Yellow River

3　Sediment transport mode

The concept of stream power for sediment transport proposed by Bagnold is very clear and reasonable. Based on the concept, it is suggested to study the value of $\eta_{sl} C_*$ in the following power balance equation $[\,W/m^2\,$ of riverbed area$]$:

$$(\rho_s - \rho_l)\,g\omega_s C_v R = \rho_l \eta_{sl} C_* \, u_* \, (u_*^2 - u_{*c}^2) \tag{3}$$

where, ρ_s, ρ_l stands for density of sediment and fluid respectively; R is hydraulic radius; ω_s is effective settling velocity; C_v is volume concentration; C_* is non – dimensional Chezy coefficient (the one divided by $\sqrt{g}$); η_{sl} is suspension efficiency; u_{*c} shear velocity at the onset of motion for sediments forming the channel bed.

According Bagnold's definition $\eta_{sl} C_*$ can be represented as $\eta_{sl} C_* = 0.266\left(\dfrac{\bar{v}'_{max}}{U_*}\right)^3$ ($\bar{v}'_{max}$ is

maximum root – mean – square vertical velocity). And he suggested $\left(\dfrac{\bar{v}'^3_{max}}{U_*}\right)$ appears to be in the

neighbourhood of unity (Bagnold, 1966). But it is shown by more experiment data that considering $\eta_{sl} C_*$ as invariant is not sufficient and it is expected to be associated with control factor m. With Guy's data the inverse of m seems to change with $Ec\,(Ec = (1.5\pi\eta_{sl} C_*)^{0.6}$ as sine curve mode which is given by Fig. 3. According the definition m should range from 1 to 2. But computed m is usually beyond this range due to measurement error (such wave length, energy slope etc) and adoption of the simple m (natural river, some flume data). For the simplicity sediment transport only for m ranging from 1 to 2 is discussed in this paper. For ripples flow intensity is very weak and a few sediment particles occur. With bed form developing from ripples to dunes flow resistance increase continuously until the fully developed dunes formed. Then dunes develop to upper regime flat or in phrase wave and resistance decrease (Qianning, 1983). This process can be reproduced clearly in Fig. 3.

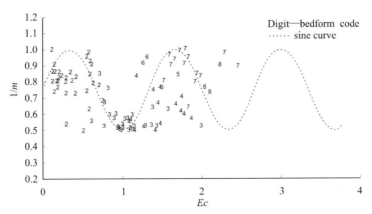

Fig. 3　Relation between $1/m$ and Ec

We can see that ripples occur with $Ec < 0.8$. When bed forms change from ripples to dunes, $1/m$ will decrease with increasing from 0.6 to 1.2. On the contrary, $1/m$ increases with increasing from 1.2 to 1.8 when bed forms change from dunes to breaking waves. With bed forms changing to dunes from breaking waves, $1/m$ decreases again with increasing. The sine curve circulation law is also proved by other data groups. Although data points are little bit scattering, the tendency of sine curve law is very clear. The sine curve can approximately be written as following equation:

$$\frac{1}{m} = 0.75 + 0.25 \sin (k_1 \pi \cdot Ec) \tag{4}$$

where, $k_1 \pi$ is a parameter on the period of sine curve.

According the experiment data shown in Tab. 1 $k_1 \pi$ is suggested as 1.5π. The hydraulic conditions of these flume data are summarized in Tab. 1.

Tab. 1 Summary of hydraulic conditions in Fig. 3

Researchers	Discharge (m^3/s)	Width/ depth(m)	Depth (m)	Slope (‰)	D_{50} (mm)	Concentration (kg/m^3)	Runs
Guy et al.	0.09 ~ 0.63	7 ~ 18	0.13 ~ 0.33	0.15 ~ 3.00	0.19	0 ~ 9.20	27
Guy et al.	0.15 ~ 0.62	7 ~ 18	0.14 ~ 0.34	0.18 ~ 9.50	0.27	0.001 ~ 35.60	16
Guy et al.	0.12 ~ 0.62	7 ~ 27	0.09 ~ 0.33	0.41 ~ 9.30	0.28	0.01 ~ 42.40	25
Guy et al.	0.03 ~ 0.15	4	0.15	0.29 ~ 8.00	0.33	0.004 ~ 15.00	6
Guy et al.	0.05 ~ 0.61	8 ~ 42	0.06 ~ 0.31	0.20 ~ 6.20	0.45	0.001 ~ 11.50	41
Guy et al.	0.20 ~ 0.60	6 ~ 22	0.11 ~ 0.41	0.47 ~ 9.60	0.47	0.002 ~ 17.70	47
Guy et al.	0.20 ~ 0.64	8 ~ 17	0.14 ~ 0.33	0.37 ~ 9.40	0.93	0.02 ~ 6.10	24

4   Sediment transport in the Yellow River

Since 2002 experiments of regulation for sediment and flow in the Yellow River have been conducted in flood season in order to increase the bank full discharge for flood control. The basic regulating rule is to control discharge and concentration about 3,000 m^3/s and 20 kg/m^3 at Huanyuankou Hydrometric Station respectively by Xiaolangdi Reservoir. During experiment period erosion occurred along the whole course of the lower Yellow River and the erosion is much more in upstream of Gaocun Station than that in downstream. Here the flush experiment data observed at stations downstream of Gaocun in years from 2000 to 2004 are used to test the above sediment transport mode. Plotting Ecagainst $1/m$ in Fig. 4 shows that the sine curve law is still very clear.

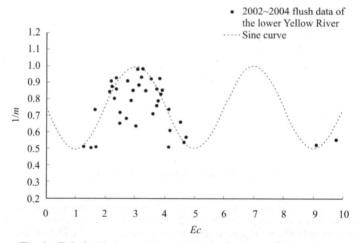

Fig. 4   Relation between $1/m$ and Ec in the lower Yellow River

The period of sine curve is different from that of flume data. According to the Yellow River data $k_1 \pi$ is suggested as 0.5π. So Fig. 4 can be expressed approximately as the following Equation:

$$\frac{1}{m} = 0.75 + 0.25 \sin(0.5\pi \cdot Ec) \qquad (5)$$

From Eq. (5) we can have Ec:

$$Ec = \frac{\left| \arcsin(\frac{4}{m} - 3) + k_2 \pi \right|}{0.5\pi} \tag{6}$$

Applying Eq. (6) into Eq. (3) we have:

$$C_v = \frac{3}{2\pi} \cdot \frac{\rho_l u_* (u_*^2 - u_{*c}^2)}{(\rho_s - \rho_l) g\omega_s R} \left(\frac{\left| \arcsin(\frac{4}{m} - 3) + k_2 \pi \right|}{0.5\pi} \right)^{5/3} \tag{7}$$

Parameter k_2 is determined by period $k_1 \pi$ and Ec. $k_2 = 0, 1, -2, 3, -4, 5 \cdots$ In order to test Eq. (7) the data sets observed in 1980s and early 1960s in the lower Yellow River and the data observed at tributaries of the Yellow River, the Weihe River and the Beiluohe River, are used. From 1960 to 1963 erosion also took place all the way in the lower Yellow River. The data selections in 1980s (including 1950s data at Tcz station) were carried out by judging erosion and erosion with the variation of water surface elevations under the same discharge. So the data can be regarded as relative equilibrium data. The Weihe River and the Beiluohe River data were not selected carefully. Comparison of computed concentration and observed values are shown in Fig. 5. Discrepancy ratio(R_i), average discrepancy ration($\overline{R}$) and mean normalized error(MNE) are tabulated in Tab. 2. R_i, $\overline{R}$, MNE is defined as following:

$$R_i = \frac{C_{ci}}{C_{mi}} \qquad \overline{R} = \frac{1}{N} \sum_{i=1}^{N} R_i \qquad MNE = \frac{100}{N} \sum_{i=1}^{N} \left| \frac{C_{ci} - C_{mi}}{C_{mi}} \right|$$

where, C_{mi}, C_{ci} are observed and computed sediment concentration.

Molinas A. and WU B. S(2000) also made a comparison(shown in Tab. 2) among several well-known equation with the data from Amazon and Orinoco River systems, Mississippi River system, Atchafalaya River and Red River.

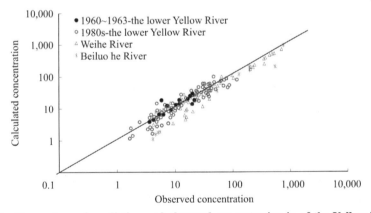

Fig. 5　Comparison of prediction and observed concentration in of the Yellow River

It can be seen that R_i, $\overline{R}$, MNE for Eq. (7) are more reasonable comparing with others though the No. of data sets are relative less. For the Yellow River data different data groups the predictions by formula also represent some difference. For the lower Yellow River data in 1980s and 1960s, the Weihe River, and the Beiluohe River the percentage of data sets with R_i between 0.75 ~ 1.25 is 64%, 50%, 63% and 47% respectively. But even for Beiluohe River the result is still reasonable. So up to we can see Eq. (7) is very potential for application in the Yellow River.

Tab. 2　Verification of new sediment transport equation

Author of equation	Data in rang of discrepancy ratio R_i (%)				MNE	Data source & No.
	0.75 ~ 1.25	0.5 ~ 1.5	0.25 ~ 1.75	$\overline{R}$		
Eq. (7)	53.9	77.8	94.4	1.05	30.6	YR 180
Molinas A. and WU B. S	41.6	70.3	85.0	1.14	50.3	
Engelund and Hansen	23.0	46.9	68.4	2.21	164.9	USA
Ackers and White	30.4	52.9	71.3	1.65	108.0	Rivers
Yang C. T	10.6	26.2	55.2	0.63	84.3	414
Toffaleti	33.1	62.8	80.2	1.2	62.2	

5　Discussions

Sediment transport issue is very challenging even for flume experiments cases. Up to now we still can't say it has been solved well. The sediment transport mode proposed in this paper is a new way to the problem and also there are some aspects need to be improved. For this mode the sine curve period of $k_1\pi$ is most important. For flume data shown in Tab. 1 $k_1\pi$ is about 1.5π though sediment sizes range from 0.078 mm to 0.93 mm. But for the Yellow River data given in Tab. 2 with much finer sediment sizes of ranging from 0.005 mm to 0.05 mm 0.5π seems reasonable for $k_1\pi$. It is expected that $k_1\pi$ is related with effective settling velocity but not very sensitive. The dramatic difference experienced in flow depths, Reynolds Number, Froude Number, and water surfaces slope between the Yellow River and laboratory flumes affect the resistance to flow as well as sediment wave movement and suspension, and consequently, the sediment transport. So Molinas A. and WU B. S(2000) think that it is not reasonable to extrapolate sediment transport equation developed from laboratory data to prototype conditions without any adjustment and modification. In addition most of flume experiment materials are non – cohesive and non – uniform particles. But the Yellow River sediment is very fine and sediment sizes gradations are very large. How do these differences influence $k_1\pi$? It is not very clear and $k_1\pi$ should be improved in the future. Thus it is suggested that should be calibrated with local data before applying Eq. (7) into natural rivers at present.

6　Conclusions

In this study the method of distinguishing bed form types and new sediment transport equation are proposed by introducing the concept of bed form control factor m. After comparison with flumes and the Yellow River data the following conclusions can be reached.

(1) With control factor m and normalizing Froude Number the bed form types of natural rivers can be determined.

(2) The sine curve law of $1/m$ againstis confirmed by a lot of flume data and the Yellow River data. But the sine curve period of the Yellow River is different from laboratory flumes.

(3) Predictions of sediment concentration in the Yellow River by sediment transport formula proposed in this paper agree well with observed values. The percentage of data sets with discrepancy ratios between 0.5 ~ 1.5 is about 80% for the data of the lower Yellow River.

(4) The predictions are carried in the case of control factor m ranging from 1 to 2. Beyond this range the sediment transport equation should be modified and improved.

References

Bagnold R A. An Approach to the Sediment Transport from General Physics, Physiographic and

Hydraulic Studies of Rivers, Geological Survey Professional Paper, 422 – I, 1996.

Guo Q C. Sediment – carrying Capacity in Natural Rivers[J]. Journal of Sediment Research, 2006, 5: 45 – 51.

Luong G V, Verbank M A. Froude Conditions Associated with Full Development of 2D Bed Forms in Flumes, 2007.

Molinas A, Wu B S. Transport of Sediment in Large Sand – bed Rivers[J]. Journal of Hydraulic Research, 2000, 39(2): 135 – 146.

Qian Ning, Wan Z H. Mechanics of Sediment Transport, Beijing, 168.

Simons D B, Richardson E V. Resistance to Flow in Alluvial Channel[J]. USGS prof. , 1966, 422 – j.

Qi P, Sun Z Y, et al. Influence of Unsteady Floods on Sediment Transport and River Bed Evolution in Yellow River[J]. Journal of Hydraulic Engineering, 2005, 36(6): 1 – 9.

Verbanck M A. Sand Transport at High Stream Power: Towards a New Generation of 1D River Models [C] // Proceedings of Ninth International Symposium on River Sedimentation, Yichang, 2004.

Van Rijn L C. Sediment Transport, Part III: Bed Forms and Alluvial Roughness[J]. Journal of Hydraulic Engineering, 1985, 110(12): 1733 – 1755.

Yang S Q. Prediction of Total Bed Material Discharge[J]. Journal of Hydraulic Research, 2004, 43(1): 12 – 22.

Experimental Study on Sediment Settling of Xiaolangdi Reservoir

Jiang Siqi[1] , *Wang Hongwei*[2] , *Chen Shukui*[1] , *Li Tao*[1] and *Li Pin*[1]

1. Yellow River Institute of Hydraulic Research, YRCC, Key Laboratory of Yellow River
Sediment Research, Zhengzhou, 450003, China
2. Yellow River Engineering Consulting Co. , Ltd. , Zhengzhou, 450003, China

Abstract: Sediment settling velocity is an important parameter in particle suspension and indispensable link of hyperconcentrated flow as well as basic theory support for efficient sediment transport of Xiaolangdi Reservoir. This paper has adopted current sediment sample in Xiaolangdi Reservoir to take still water settling experiment, observed carefully silt sediment distribution during different settling duration under various initial concentrations, analyzed common settling velocity calculation method, and adopted settling duration curve method as calculation method to obtain a constant settling velocity of free settling section. Typical settling velocity that is obtained from experiments makes supplement and fitness to settling velocity formula of the existing groups of sediment sample, amends the formula coefficient, further explore the settling characteristics and settling velocity of fine particle mixed viscous sediment with water concentration change, and meanwhile this paper researched settling distance impact on settling velocity, existing sediment group settling transition to mesh settling critical transition silt concentration of Xiaolangdi Reservoir. This paper has analyzed the relationship between settling velocity of muddy surface and settling velocity of mixed sediment group, and discussed silt concentration change trend with time of the muddy water below the muddy surface after settling.

Key words: settling duration curve method, settling velocity, mixed sediment, critical silt concentration, muddy surface

1   Experiment Overview

Still natural settling experiment of Xiaolangdi Dam sediment sample adopts organic glass cylinder, which is 2 m in height, 0. 2 m in outer diameter and 0. 19 m inner diameter. Settling cylinder is put vertically, with one sampling outlet hole every 10 cm on the cylinder wall and 14 outlet holes on every cylinder (Fig. 1 and Fig. 2).

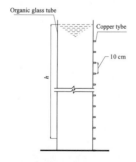

Fig. 1   Settling cylinder sketch map   **Fig. 2   Settling cylinder physical map**

The experiment uses distilled water to substitute tap water so as to reduce experiment error resulted from water quality influence.

Settling experiment adopts sediment sample with silt concentration from 800 kg/m³ to 50 kg/m³, reduced by 50 kg/m³ each time, to take still water settling experiment of two groups. Sediment particle distribution curve of the eperiment refers to Fig. 3.

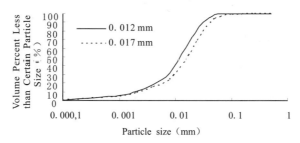

Fig. 3   Sediment prticle dstribution crve of the eperiment

2   Experiment anysis mthod and rsult

Experiment data analysis adopts settling duration curve method.

Setting velocity ω at h water depth after t is obtained from settling duration curve $h \sim t$ under different silt concentration percent (S/S_0) with various initial concentrations:

$$\omega = \frac{dh}{dt} \tag{1}$$

Silt concentration percent P ($= S_t/S_0$) is actually the same as weight percent P of less than certain particle size d in initial distribution curve of mixed sediment, for example, $P = 0.5$ refers to half concentration with half particles settled in tested section. Because measured concentration S at certain settling duration is concentration after distributed settling at h water depth, while ($S_0 - S_t$) concentration coarse particle has been settled and silt concentration percent ($S_t/S_0 = P$) reflects particle size of initial distribution. Settling velocity under various silt concentration percent is measured settling velocity of d particle.

Take instant settling velocity ω (cm/s) as x – coordinate and silt concentration percent P (%) less than the settling velocity as y – coordinate to draw $P \sim \omega$ curve chart (refers to Fig. 4 and Fig. 5). Among which, (a) refers to $10^{\#}$ sampling hole, settling distance 1.05 (b) refers to $8^{\#}$ sampling hole, and settling distance is 0.85 .

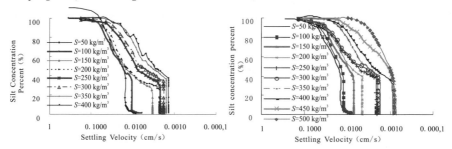

Fig. 4   Sling eperiment of Xiaolangdi Dam sdiment smple ($d_{50} = 0.017$ mm)

Seen from a series of settling duration curve chart ($P \sim \omega$ curve), $P \sim \omega$ curve of $d_{50} = 0.017$ mm fine sediment in Xiaolangdi Reservoir has obvious turning point when silt concentration is more than 150 kg/m^3, while $P \sim \omega$ curve of $d_{50} = 0.012$ mm fine sediment in Xiaolangdi Reservoir has obvious turning point when silt concentration is more than 100 kg/m^3, and is the same smooth curve as Yellow River Coarse $P \sim \omega$ curve and clean water $P \sim \omega_0$ curve when silt concentration is 10 kg/m^3. Physical meaning of the curve refers to particle percent taking part in group settling under certain silt concentration of certain sediment sample. Turning point refers to that the corresponding particles behind will be settled evenly as per certain settling velocity for a long time, and the relative smooth curve and turning point have restricted particle settling velocity.

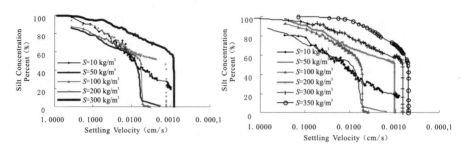

Fig. 5 Settling experiment of Xiaolangdi Dam sediment sample (d_{50} = 0.012 mm)

3 Determination of critical silt concentration

We can take measured middle settling velocity of viscous sediment as typical settling velocity of the group, dot the relation of initial volume concentration S_{V0}, and then get linear relation of one turning point (Fig. 6). Seen from Fig. 6, corresponding volume concentration of Xiaolangdi sediment sample at muddy water turning point when d_{50} = 0.017 mm, d_{50} = 0.012 mm is 0.057 and 0.038, namely critical silt concentration of the sediment sample transiting from Newtonian fluid to non – Newtonian fluid is 150 kg/m³ and 100 kg/m³.

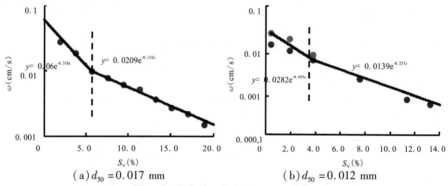

(a) d_{50} = 0.017 mm   (b) d_{50} = 0.012 mm

Fig. 6 Relation between ω and S_V

4 Discussion of Muddy Surface Settling Formula

One of main characteristics of muddy water settling to certain concentration is obvious joint surface between clean water and muddy water during settling period. It has been shown in various experiments that joint surface between clean water and muddy water only occurs when initial concentration is up to certain value, namely muddy surface occurs settling.

Fei Xiangjun pointed out in slurry research analysis that uneven particle group settling velocity can be substituted by muddy surface settling to a certain degree in fine particle slurry and muddy water clarifying issues, while muddy surface settling velocity reflects settling velocity of flock made of fine particles.

According to slurry settling experiment, we have analyzed and deduced muddy surface settling velocity relation formula as follows:

$$\omega = \frac{a_s(1 - S_V)(\gamma_s - \gamma)}{\mu_0} \tag{2}$$

Among which,
$$S_* = 6S_V \sum \frac{\Delta p_i}{d_i} \quad (\text{mm}^{-1}) \tag{3}$$

When $S_* < 170$ (mm^{-1}),
$$a_* = 0.017,5e^{-0.0155*} \times 10^{-5} \quad (\text{cm}^2) \tag{4}$$

When $S_* > 170$ (mm^{-1}),
$$a_* = 0.002,75e^{-0.004S*} \times 10^{-5} \quad (\text{cm}^2) \tag{5}$$

In the above equation, μ_0 refers to viscosity coefficient of clean water, d refers to particle size, $6\sum \frac{\Delta p_i}{d_i}$ refers to specific surface area consisting of particle size, while S_V refers to initial volume concentration.

As to two particle size of Xiaolangdi Reservoir sediment sample adopted in this experiment, we have made revision for Fei Xiangjun muddy surface settling formula under the precondition that critical silt concentration has been determined in the former experiment (Fig. 7).

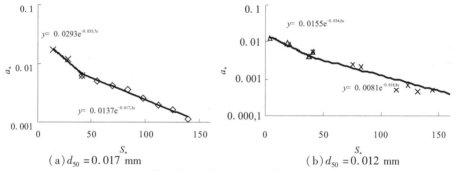

Fig. 7 a_* **Formula revision**

(1) Xiaolangdi Dam Sediment Sample ($d_{50} = 0.017$ mm).

When $S_* < 41.94$ (mm^{-1}),
$$a_* = 0.002,93e^{-0.035,7S*} \times 10^{-5} \quad (\text{cm}^2) \tag{6}$$

When $S_* > 41.94$ (mm^{-1}),
$$a_* = 0.013,7e^{-0.017,3S*} \times 10^{-5} \quad (\text{cm}^2) \tag{7}$$

(2) Xiaolangdi Dam Sediment Sample ($d_{50} = 0.012$ mm)

When $S_* < 37.67$ (mm^{-1}),
$$a_* = 0.015,5e^{-0.003,46S*} \times 10^{-5} \quad (\text{cm}^2) \tag{8}$$

When $S_* > 37.76$ (mm^{-1}),
$$a_* = 0.008,1e^{-0.001,88S*} \times 10^{-5} \quad (\text{cm}^2) \tag{9}$$

Put effective data into Eq. (2) to get group settling velocity, and compare the results with measured data, referring to Fig. 9. Seen from Fig. 9, measured values fit better than calculated values; use Eq. (2) to calculate settling velocity according to settling change process with muddy water duration, on the other hand to calculate d_{50} sediment particle in primary water level.

5　Conclusions

(1) Still settling experiment of Xiaolangdi Dam sediment adopts instant settling velocity in settling duration curve as typical settling velocity of muddy sediment and determines turning point as critical concentration of transition from flock settling stage to flock mesh settling stage in silt concentration ~ settling velocity chart.

(2) According to existing two types of Xiaolangdi Dam sediment samples as critical concentration, muddy surface settling velocity formula of Fei Xiangjun is made revision and

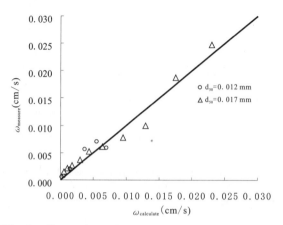

Fig. 8   Comparison of experiment data and Eq. (2)

determined to calculate group settling velocity of Xiaolangdi Dam existing sediment sample.

(3) Under the precondition that even silt concentration of muddy water is considered as silt concentration below muddy surface, experiment value is adopted to deduce the change formula of muddy water silt concentration below muddy surface with time and accordingly to predict even silt concentration of muddy water under certain initial silt concentration and at certain time.

Acknowledgements

The research was supported by public welfare industry research special fund of the Ministry of Water Resources (MWR) (No. 200901015, No. 200801024), the National Natural Science Foundation of China (NSFC) (NO. 51179072) and central-level nonprofit research institute fund (HKY – JBYW – 2009 – 6).

References

Jin Tonggui. Precipitation Separation[D]. Xi'an: Xi'an University, 1999.
Zhang Shude. Research on Sediment Settling of Yellow River and Particle Dispersion Characteristics[D]. Xi'an: Xi. 'an University, 2002.
Qian Ning. Hyperconcentrated Flow Motion [M]. Beijing:Tsinghua University Press, 1989.
Fei Xiangjun. Transportation Hydraulics of Slurry and Grainy Material [M]. Beijing: Tsinghua University Press,1994.
Han Qiwei. Reservoir Sedimentation [M]. Beijing: Science Press, 2003.
Jiang Naiqian. Experimental Research on Ball Deposition in Turbulent Muddy Water [R]. Zhengzhou:Hydraulic Scientific Research Institute of Yellow River, 1989.
Yang Meiqing. Flocculated Micro – Structure and Influence on Flocculated Structure of Flocculation [R]. Beijing: Research Report of Sediment Research Institute of Department of Water Conservancy of Tsinghua University, 1986.
Tang Maoguan, Tan Guangming. Some Analysis, Comparison, and Remark on Turbulence Measure – ments, Flow Modeling and Turbulence Measurements [M]. Hemisphere Publishing Corporation, 1991.
Qian Ning, Wan Zhaohui. Sediment Transport Mechanics [M]. Beijing:Science Press, 1982.
Lee S L,Durst F. On the Motion of Particles in Turbulent Duct Flows[J]. Multi – phase Flow, 1982 (8):125 – 146.
Tsuji Y,Morikawa Y LDV Measurements of an Airsolid Two – phase Dlow in a Horizontal Pipe[J]. Fluid Mechanics 120,1982(12):385 – 409.
Tsuji Y,Morikawa Y,Shiomi H. LDV Measurements of an Airsolid Two – phase Flow in a Vertical Pipe[J]. Fluid Mechanics,1984(139):417 – 434.

Liu Qingquan, Chen Li. Analysis of Sediment Influence on Flow Flocculation [J]. Journal of Hydraulic Engineering, 1997(9):77 – 82.

Song Genpei. Experimental Research on Deposition Characteristics of Mixed Sediment [J]. Sediment Research, 1985(6):40 – 50.

Huang Jianwei. Discussion on Memorial Sediment Deposition Characteristics in Still Deposition Test of Lianyungang Sediment (Stage II Report) [R]. Nanjing: Nanjing Hydraulic Research Institute, 1980.

Wu Chigong. Hydraulics (I) [M]. Beijing: Higher Education Press, 1982.

Discussion on the Construction of Sediment Retention Works in the Concentrated Coarse Sediment Source Areas of the Yellow River

Hu Jianjun, *Liu Xuan' e* and *Cheng Kun*

Upper and Middle Yellow River Bureau, YRCC, Xi' an, 710021, China

Abstract: The sticking point for the regulation of the Yellow River lies in sediment, where coarse sediment is most critical. Ever since a long time ago, a large quantity of sediment moves to the lower Yellow River due to severe soil loss from the loess plateau, particularly the coarse sediment fraction which silted in the lower course, directly threatening the safety of the people and properties on both banks of the lower Yellow River. Trapping of the sediment to the Yellom River in the loess plateau region of the Yellow River Basin is the most effective solution to the Yellow River sediment problem. There are 454,000 km^2 areas subject to soil erosion in the loess plateau region of the Yellow River Basin, which mean a very high difficulty in harnessing, and that long – term unremitting efforts are needed. In order to make the limited investments give play to the effect of sedimentation reduction as early as possible, the coarse sediment that make the biggest contribution to the sedimentation of the lower Yellow River must be controlled. Years of research shows that the 18,800 km^2 concentrated coarse sediment source areas of the middle Yellow River are the key to stopping the coarse sediment from flowing into the Yellow River. Construction of sediment retention works in these areas can play an enormous role in balancing runoff – sediment relationship, containing the uninterrupted development of course shrinking and "secondary suspended river", reducing the sedimentation of the lower course, and extending the service life of the large reservoirs for sediment retention. This paper analyzes the influence of the coarse sediment on the rise of the lower riverbed, discusses the construction measures, layout and scale of the sediment retention works, predicts the effect of construction of the sediment retention works in the concentrated coarse sediment source areas, and addresses the safety issues to which attention must be given in the engineering construction.

Key words: coarse sediment, concentrated source areas, sediment retention works, construction

1   Profile of the concentrated coarse sediment source areas

The concentrated coarse sediment source areas are found mainly in the right bank of the middle Yellow River around nine main tributaries, i. e. Huangfu, Qingshui, Gushan, Kuye, Tuwei, Jialu, Wuding, Qingjian and Yan with total area of 18,800 km^2. Twelve counties (Fugu, Shenmu, Yuyang, Jiaxian, Mizhi, Suide, Zizhou, Hengshan, Jingbian, Qingjian, Ansai and Zichang) of Shaanxi Province and three counties (Da, Yi and Zhun) of Inner Mongolia are involved, where total population is 1,264,100, and population density is 61 persons/km^2.

A majority of the areas are located in the first sub – zone of the loess hilly region with numerous ridges and ravines, where gully density reaches 3. 0 ~ 8. 0 km/km^2. In the southern part are mainly mound – like loess hills; in northern part are mainly beam – like ones. Surface soil is mainly loess, accounting for 77. 83% of total area, followed by 15. 68% sandstone and 6. 49% aeolian sandy soil. From southeast to northwest, vegetation transits from open forests, shrubs and grasslands to typical steppes and desert steppes. The higher coverage areas are 5. 8 km^2 only, the rest areas are less covered.

In the areas, mean annual precipitation is 350 ~ 500 mm with an uneven spatial and temporal distribution, which decreases from southeast to northwest. The precipitation in rainy season (June to September) accounts for 60% ~ 90% of yearly precipitation. Mean annual evaporation is 1,030 mm.

Maximum evaporation occurs in May, June and July. The monthly evaporation of these months accounts for about 15% of total evaporation. Mean annual runoff is 1.019×10^9 m^3 with a very uneven distribution in a year, and mainly formed by heavy rains with the features of ephemeral stream. Flood generally occurs in July to September, especially in July and August, and is characterized by high peak and low volume, sharp rise and fall, and dramatic variation. In the coarse sediment source areas, mean annual quantity of the sediment to the Yellow River is 4.08×10^8 t, and erosion modulus is as high as 21,700 t/(km^2 · a), and reaches up to 30,000 t/(km^2 · a) or higher in parts of the areas. It is the most intensive erosion area in loess plateau region. The intensive erosion area accounts for 51.2% of the intensive erosion area (36,700 km^2) of the loess plateau. Erosive sediment yield is mainly concentrated in June to September, which accounts for 93% of yearly sediment yield, and even mostly comes from a few storm floods. As per the measured data of Gaoshiya hydrological station on Gushan River, the maximum annual sediment runoff is 100 times the minimum one.

2 The rise of the lower riverbed is most affected by the coarse sediment from the concentrated source areas

It is well known that the Yellow River is characterized by "less water and more sediment, and uncoordinated runoff – sediment relationship". The "suspended river" in the lower reaches frequently causes the occurrence of complex and refractory flood disasters. Analysis showed that the coarse sediment originating from the region of the middle Yellow River is the key to the rise of the lower riverbed, especially those coming from the concentrated coarse sediment source areas.

According to the 1952 ~ 1998 data, the total quantity of the sediment to the lower course is 5.66×10^{10} t, of which, the sediment with particle size less than 0.05 mm is 4.54×10^{10} t, accounting for 80%; those (coarse) greater than 0.05 mm is 1.12×10^{10} t, accounting for 20%; those (coarse) greater than 0.1 mm is 2.3×10^9 t, accounting for 4%. During the same period, a total of 9.2×10^9 t sediment deposited in the lower course, accounting for 16.3% of total incoming sediment, of which, the sediment with particle size less than 0.05 mm is 4.71×10^9 t; those (coarse) greater than 0.05 mm is 4.49×10^9 t; those (coarse) greater than 0.1 mm is 1.78×10^9 t, respectively accounting for 51%, 49% and 19% of total sediment deposition, and 10.3%, 40.1% and 77.4% of the incoming sediment of respective fractions, and the respective incoming sediment is 1/8, 2 and 19 times of the total incoming sediment. The 1990 ~ 1998 data show that 90% or more of the coarse sediment greater than 0.1 mm deposited between Xiaolangdi Dam and Lijin Section. The three water and sediment regulation tests performed between 2002 and 2004 washed away totally 1.455×10^9 t sediment to the sea, of which, the coarse sediment greater than 0.1 mm is about 5×10^6 t, accounting for 3.6% only. Analysis of the experimental data shows that most of the sediment brought into the sea is fine sediment, and the coarse sediment greater than 0.1 mm is difficult to carry into the sea through water and sediment regulation. While the coarse sediment is a smaller fraction of the total incoming sediment, a majority of them is trapped in the course. It is the very crux of the severe siltation problem of the lower Yellow River. The lower course will continue to be raised with the deposition of coarse sediment, thus further increasing the threat to the safety of the people and properties on both banks of the lower Yellow River.

According to the 1954 ~ 1969 data (Tab. 1), in the rich coarse sediment areas, the soil erosion area accounts for 17.3% in comparison with the region of the middle Yellow River. For sediment runoff in the rich coarse sediment areas, the total sediment, the coarse sediment greater than 0.05 mm and the coarse sediment greater than 0.10 mm respectively account for 62.8%, 72.5% and 78.8% of the sediment runoffs of respective fractions of the middle Yellow River region. For erosion modulus, the total sediment, the coarse sediment greater than 0.05 mm and the coarse sediment greater than 0.10 mm are respectively 3.6, 4.2 and 4.6 times the average value of the middle Yellow River region. The ratio of the coarse sediment greater than 0.05 mm to the total sediment and of the coarse sediment greater than 0.10 mm to the total sediment are almost the same as those in the middle Yellow River region. Comparing with the region of the middle Yellow River, in the concentrated coarse sediment source areas, the soil erosion area accounts for 4. 1%. For sediment runoff in the concentrated coarse sediment source areas, the total sediment, the coarse sediment greater than 0. 05 mm and the coarse sediment greater than 0. 10 mm respectively account for 21. 7%, 34. 5% and 54% of the sediment runoffs of respective fractions of the middle Yellow River region . For erosion modulus, the total sediment, the coarse sediment greater than 0. 05 mm and the coarse sediment greater than 0. 10 mm are respectively 5. 4, 8. 4 and 13. 2 times the average value of the middle Yellow River region. The ratio of the coarse sediment greater than 0. 05 mm to the total sediment and of the coarse sediment greater than 0. 10 mm to the total sediment are respectively 1. 6 and 2. 5 times of the average value of all soil erosion areas. Comparing with the rich coarse sediment areas, in the concentrated coarse sediment, the soil erosion area accounts for 23. 9%. For sediment runoff, the total sediment, the coarse sediment greater than 0. 05 mm and the coarse sediment greater than 0. 10 mm respectively account for 34. 5%, 47. 7% and 68. 5% of the sediment runoffs of respective fractions of the rich coarse sediment areas. For erosion modulus, the total sediment, the coarse sediment greater than 0. 05 mm and the coarse sediment greater than 0. 10 mm are respectively 1. 4, 2. 0 and 2. 9 times the average value of the rich coarse sediment areas.

Tab. 1　The sediments of different grain size fractions to the Yellow River in different areas between 1954 and 1969

Area	Middle Yellow River region	Rich coarse sediment areas		Concentrated coarse sediment source areas	
		Qty	Percentage of Middle Yellow River region	Qty	Percentage of rich coarse sediment areas
				Percentage of middle Yellow River region	
Soil erosion area ($\times 10^4$ km^2)	45. 4	7. 86	17. 3	1. 88	4. 1	23. 9
Total sediment yield — Sediment yield ($\times 10^8$ t)	18. 81	11. 82	62. 8	4. 08	21. 7	34. 5
Total sediment yield — Sediment transport modulus (t/(km^2 · a))	4 ,143	15 ,038	363	21 ,702. 1	523. 8	144. 3
Coarse sediment ≥0. 05 mm — Sediment yield ($\times 10^8$ t)	4. 4	3. 19	72. 5	1. 52	34. 5	47. 6
Coarse sediment ≥0. 05 mm — Sediment transport modulus (t/(km^2 · a))	969	4 ,059	418. 8	8 ,085. 11	834. 2	199. 2
Coarse sediment ≥0. 1 mm — Sediment yield ($\times 10^8$ t)	1. 13	0. 89	78. 8	0. 61	54	68. 5
Coarse sediment ≥0. 1 mm — Sediment transport modulus (t/(km^2 · a))	249	1 ,132	454. 9	3 ,244. 68	1 ,303. 6	286. 6

It can be seen from the Tab. 1 that sedimentation and rise of lower riverbed is mainly a result of the coarse sediment from the concentrated coarse sediment source areas of the Yellow River.

3　Thought on the construction of sediment retention works

The general ideas for the construction of sediment retention works are as follows: build medium and small check dams on individual small gullies of each tributary stream, and build large check dams on larger gullies as appropriate. Use the numerous dams to retain the sediment generated from slopes and grooves in numerous gullies and ravines so as to reduce the entry of coarse sediment to the Yellow River. In addition, take measures to control the loss of sediment from slopes and provide necessary conditions to reduce erosion.

3.1 Construction measures

Years of practice proves that construction of check dam is an effective means for retention of sediment in the loess plateau. Build large, medium and small check dams at various junctions and passages of sediment to form an artificial barrier. Within the service life of the check dams, not only is the sediment retained in gullies, but the base level of erosion is raised, the depth of the gullies is reduced, gully slopes are stabilized, gully expansion and incision are contained, and gully erosion is alleviated. Moreover, this method can build farmlands by retention of the sediment and provide impoundment in early operation stage, helpful for improving the production and living conditions of local people. It also can create a suitable environment for the growth of trees and grasses, thus consolidating the achievements of "returning farmland to forest and grassland" and favorable for improving the ecological environment. Therefore, the construction of sediment retention works shall focus on building of large, medium and small check dams.

3.2 Construction layout

General layout: Firstly to take the tributaries as framework and the minor watersheds as units, and properly coordinate the relations between mainstreams and its tributaries and between the upper and lower reaches to make the large, medium and small check dams form a complete set in a reasonable layout by optimizing the runoff − sediment relationship. Secondly the dams shall be located in a place with "small mouth and large stomach", easy access, less reservoir inundation losses and construction work, and space for placing outlet structures.

The large check dams mean that the control area of a single dam is greater than 8 km^2, and its storage capacity is greater than $5 \times 10^6 \text{ m}^3$. They generally apply to large gullies with larger control area and storage capacity and significant sediment trapping capability. The large check dams shall be placed where inundation losses are low, good construction conditions are available, and medium and small ones are incompetent. The large check dams can detain the coarse sediment within a short period of time.

The large check dams shall be different from reservoirs. firstly they generally address sediment retention, while the reservoirs focus on impoundment. Secondly, their operation mode is different. The large check dams are so operated as to retain the flood and release the clear water, so they are generally empty and ready for flood retention, and seldom kept at high water level. The reservoirs are designed for flood storage and sediment flushing. They often operate at high water level with frequent fluctuation of water level. Therefore, the large check dams are designed differently from the reservoirs. A large part of the reservoirs built in the concentrated coarse sediment source areas has become the large check dams.

The medium check dams mean that the control area of a single dam is around 3 km^2, and its storage capacity ranges from $0.5 \times 10^6 \text{ m}^3$ to $5 \times 10^6 \text{ m}^3$. They, arranged in medium gullies, are characterized by small size, low inundation losses, and easy construction and operation. Construction sites and dammed lands are also closer to the residential areas. With high and stable yield, the dammed lands can offer better livelihood to local farmers, promote the returning of steeply sloped croplands, move forward with the recovery of vegetation, and improve the ecological environment. In early operation stage, the medium check dams can also increase the utilization of water resources by accumulating surface runoff to provide water for the life and production of people in the areas that are short of water recourses. Therefore, the medium check dams are not only fast and efficient for stopping sediment yield by erosion, but fit the demand from local natural and social development, so welcome by local people.

The small check dams mean that the control area of a single dam is less than 3 km^2, and its storage capacity is less than $0.5 \times 10^6 \text{ m}^3$. They are generally applicable to small gullies with the functions of both sediment retention and building farmland, helpful for increasing production.

3.3 Construction scale

For the large check dams, their construction scale is determined by taking the tributaries as units on the base of detailed field survey on each dam site in combination with the study of 1:10,000 topographic maps. Initially, 86 large check dams are chosen to build by studying the topographic maps. After field survey, it is preliminarily determined to build 79 dams. Finally, the construction scale is determined at 35 large check dams through further research and analysis of construction conditions, inundation losses, gully features, water and sediment control effect, and the coordination with the medium and small check dams in the minor watersheds.

For the medium check dams, their construction scale is determined by taking the minor watersheds as units on the base of detailed field survey on each dam site in combination with the study of 1:10,000 topographic maps. Initially, 2,324 medium check dams are chosen to build by studying the topographic maps. After field survey, it is preliminarily determined to build 2,159 dams. Finally, the construction scale is determined at 2,014 medium check dams through further research and analysis of construction conditions, inundation losses, gully features, water and sediment control effect, and the coordination with the small check dams in the minor watersheds.

For the small check dams, 18 typical minor watersheds are selected for demonstration of dam layout in the minor watersheds with a proportion of the allocation of medium and small check dams obtained, which is then used to ascertain that the construction quantity for the small check dams is 8,204 dams.

Tab. 2 Number and control area of check dams on the tributaries

Tributary	Number of check dams (pcs)				Control area of check dams(km^2)			
	Large	Medium	Small	Total	Large	Medium	Small	Total
Huangfu	3	378	1,134	1,515	171.6	1,065.6	654.62	1,442.6
Qingshui		103	394	497		350.2	180.52	446.26
Gushan	2	87	307	396	89.5	246.03	132.78	342.51
Kuye	12	327	1,084	1,423	1,067.1	997.39	384.67	1,821.2
Tuwei	8	170	850	1,028	402.6	463.35	450.5	886.35
Jialu	3	100	500	603	383.9	340	214	687.9
Wuding	7	687	3,449	4,143	556	2,335.8	1,524.8	3,315
Qingjian		2	6	8		6.8	2.36	7.64
Yan		32	96	128		108.8	43.52	128.64
Others		128	384	512		323	190.78	397.42
Total	35	2,014	8,204	10,253	2,670.7	6,236.97	3,778.55	9,475.5

Note: Repeated area is not included in the total control area.

4 Construction of sediment retention works in the concentrated coarse sediment source areas is most significant

For the construction of sediment retention works in the middle Yellow River region, in the context of the same quantity of sediment to be trapped, firstly, the sedimentation reduction benefits are higher in the areas with more coarse sediment than in those with less coarse sediment. The Yellow River has the mean annual sediment runoff of 1.6×10^9 t. About 4×10^8 t of them was deposited in the lower reach. That is to say, if 1×10^8 t sediment is headed off in the upper and middle reaches, the sedimentation in the lower reach will be reduced by 0.25×10^8 t. If the 1×10^8 t is detained in the concentrated coarse sediment source areas, the reduction will be increased to 0.33×10^8 t, i.e. being 1.4 times of the upper and middle reaches. Secondly, in the areas of high erosion modulus, the storage capacity would be filled up at a faster speed so that the effect of reducing the sedimentation in the lower reach can be achieved sooner. In the concentrated coarse sediment source areas, the ratios of the coarse sediment greater than 0.05 mm and greater than 0.1 mm to the total sediment are respectively 1.5 and 2.5 times of the average value of all soil erosion areas, and the erosion moduluses of the total sediment, the coarse sediment greater than 0.05 mm and the coarse sediment greater than 0.10 mm are respectively 5.4, 8.4 and 13.2 times of that of all soil erosion areas. Seeing either from the sedimentation reduction benefits or from the time that sedimentation reduction effect are achieved, subject to equal storage capacity, it is better to select the construction of sediment retention works in the concentrated coarse sediment areas than in other areas. It can improve the production and living conditions of local people and local ecological environment.

4.1 Ameliorate the runoff – sediment relationship of the Yellow River, mitigate the sedimentation of the course; extend the sediment retention life of major eservoirs (Xiaolangdi, etc)

At present, the water and soil conservation measures averagely reduce 4×10^8 t sediment into the Yellow River each year, which is about 25% of the mean annual incoming sediment (1.6×10^9 t) in natural conditions. On the other hand, water consumed by socio – economic development reaches $3.5 \times 10^{10} \sim 4.0 \times 10^{10}$ m^3, which is about 70% of the natural water resources (0.535×10^{10} m^3). The regulation and storage of the large reservoirs (Longyangxia, Liujiaxia, etc.) on the upper reach leads to the reduction of the percentage of the runoff during flood season to that in a whole year from 60% to 40%, while the percentage of the sediment runoff during flood season to that in a whole year remains 90%. The normal medium floods in the lower Yellow River were weakened greatly while sediment reduction is insignificant. As a result, the normal medium floods become the small floods with hyper – concentration of sediment in the lower Yellow River. For both quantity and process, current runoff – sediment relationship of the lower Yellow River gets worse than before (in natural conditions). If no effective measures are taken, the runoff – sediment relationship would be further worsened with the increase of water consumption by society and economy.

Not only does the deterioration of the runoff – sediment relationship aggravate the sedimentation of the lower course, but cause the sharp shrinking of the medium – flow channel, so that the "secondary suspended river" develops rapidly, and the normal medium floods are much more likely to threaten the embankments than before. Flood control situation is severe.

In order to reduce the sedimentation of the lower course and alleviate the severe flood control situation, the Xiaolangdi Reservoir was built in 2000, which plays an extremely important role in reducing the sedimentation of the lower course and restoring the flood discharge and sediment transport capabilities of the medium – flow channel.

The Xiaolangdi Reservoir can cut down the 100 and 1,000 – year flood peak flow at the Huayuankou Section respectively from 29,200 m^3/s to 15,700 m^3/s and from 42,100 m^3/s to 22,600 m^3/s through coordinated operation with Sanmenxia, Luhun and Guxian reservoirs, and can raise the flood control standard of the lower Yellow River embankment works from less than 60 – year flood to 1,000 – year flood. Through sediment trapping and water and sediment regulation, the reservoirs play an important role in reducing the sedimentation of the lower course and restoring the discharge capability of the medium – flow channel. However, the Xiaolangdi Reservoir only has a sediment retention capacity of 7.5×10^9 m^3 (totally 1×10^{10} t), which is limited in comparison with unlimited incoming sediment. Its sediment retention function can only work in a limited period of time. It is estimated that the Xiaolangdi Reservoir will reach a deposition – erosion equilibrium state by 2020, when its sediment retention function will lost, and the water and sediment regulation alone can not coordinate well the runoff – sediment relationship of the Yellow River any longer. The lower course will enter the period of sustained siltation and elevation again. The flood control standard of the lower embankment works will gradually decline as well. Moreover, after the Xiaolangdi Reservoir enters the period of floodplain sedimentation and channel scouring, the small flow with high sediment concentration will be more likely to occur in the lower reach as an effective runoff and sediment regulation can not be achieved due to the limited regulation storage of the reservoir, when the discharge capability of the medium – flow channel will be difficult to maintain for a long period of time, very unfavorable for the flood control of the lower reach.

Construction of the sediment retention works in the concentrated coarse sediment source areas is conducive to improving the runoff – sediment relationship of the Yellow River. The main problem of the Yellow River lies in less water and more sediment as well as lack of balance between runoff and sediment load. After the planned sediment retention works are all built and put into operation in the concentrated coarse sediment source areas, the sediment into the Yellow River will be reduced by additional 1.98×10^8 t/a per year within 20 years, which is about 12.4% of the mean annual incoming sediment (1.6×10^9 t), while the maximum water consumption of the sediment retention works is 3.16×10^8 m^3, which is only 0.59% of the mean annual natural runoff (5.35×10^{10} m^3). From this perspective, the sediment retention works can reduce much more sediment load than they consume water, very beneficial for improving the runoff – sediment relationship of the Yellow River.

Construction of the sediment retention works in the concentrated coarse sediment source areas will effectively reduce the sedimentation of the riverbed and large reservoirs of the lower Yellow River. After the planned sediment retention works are all built and put into operation, they can trap 3.963×10^9 t sediment, thus reducing the sedimentation of the riverbed and large reservoirs of the lower Yellow River by 1.308×10^9 t, and reducing the rise of the lower riverbed by 0.42 m accordingly. Since the base level of erosion is raised when the sediment retention works are filled up with sediment, they can still reduce the sediment into the Yellow River by averagely 0.5×10^8 t per year for a long period of time without taking into account the raising of them. Therefore, construction of the sediment retention works in the concentrated coarse sediment source areas is very useful for sediment reduction of the lower course.

4.2　Improve the production and living conditions of local people and local ecological environment

A majority of the concentrated coarse sediment source areas are located in the first sub – zone of the loess hilly region with numerous ridges and ravines, where gully density generally ranges from 3.0 km/km^2 to 8.0 km/km^2. Mound – like and beam – like loess hills are present with a broken terrain and a very fragile ecological environment. Water and soil losses are very serious. Farming production is largely performed on steep hillside land with very low yield. Local people still live at the mercy of the weather. In addition to poor production conditions, the areas are also short of drinking water and difficult of access. In order to make a living, farmers have to push up the hillsides, which caused massive destruction to the fragile ecological environment, further exacerbating soil and water loss.

Construction of the sediment retention works in the concentrated coarse sediment source areas can improve the production and living conditions of local people and local ecological environment. The dammed lands by sediment retention are high and stable yield farmlands with ample moisture and high fertility due to higher ground water table by rainwater infiltration, and the high yields can be ensured irrespective of drought or water logging. According to years of statistics, the yields of the dammed land per mu are 5 ~ 8 times of the hillside land. The land can be used in the year after 3 to 5 years of the dams built. A large area of dammed lands can be formed after the storage capacities are filled up. At that time, local people can move to the vicinity of the dammed lands in a planned way to live and farm there in stead of hard farming on hillside land, while the people can get drinking water in the dammed land. By then, the hillside land can be returned to forests or grasslands. After the planned sediment retention works are all built and put into operation in the concentrated coarse sediment source areas, 37,986.67 hm^2 of dammed land can be formed, and 265,906.67 hm^2 of hillside lands can be permanently returned to forests or grasslands, greatly improving the production and living conditions of local people and local ecological environment. To the local people, retention of sediment in the concentrated coarse sediment source areas, so to speak, is an infrastructure for attaining the objective of building a well – off society in an all – round way.

5 Great importance shall be attached to security issues in engineering construction

The construction of the sediment retention works covers a wide scope with lots of dams and huge working quantity to be completed. Great importance must be attached to security issues when the construction is initiated. To be absolutely safe, precautionary measures shall be taken against potential safety hazards that are present in engineering design, flood control during construction period, construction quality, and operational management.

Engineering design must be entrusted to the water conservation design institutes with corresponding qualifications. The designers must strictly observe the related hydraulic engineering specifications, such as Technical Code of Key Dam for Soil and Water Conservation, Design Code for Rolled Earth – Rock Fill Dams, etc. For the dams in series, the impact of flood discharge from upper dams on lower dams must be considered. A joint flood regulation shall be carried out to address the potential safety hazards in flood control.

For flood control during construction period, the net construction period of a check dam is generally six months, which shall be arranged after the flood season of a year and before the flood season of next year to avoid construction in flood season. When inevitable, construction cannot be carried out before the earth dams have been high enough to resist flood and the water release structures have been tested to be qualified so as to evade the flood risk during construction.

For construction quality, firstly, a Trinitarian quality assurance system, i. e. constructors are responsible for quality control; developers and supervisors are responsible for quality surveillance; government quality inspection departments are responsible for holding the pass, shall be established. The constructors shall designate a special quality safety supervisor to each dam. The developers and supervisors shall separately have a technical staff and a supervising engineer in charge of inspection and acceptance after completion of a concealed work or a phase of work with interim and final inspection reports presented. Secondly, related national technical codes and guidelines must be strictly observed during construction. The quality responsibilities of each relevant personnel shall be clarified. Thirdly, construction design and arrangement shall be meticulously prepared in light of local conditions to expedite construction progress to meet the construction schedule.

For operational management, firstly, the operation period of check dam is generally ten to thirty years. The operation and maintenance facilities, such as access road, communication, land and housing needed for water release structures and management, shall all be completed in construction period. The units or individuals responsible for the management and maintenance thereof shall be appointed with a given cost appropriated after final acceptance and take – over. Secondly, most of check dams consist of three major parts. When a sediment retention works reaches its service life, its sediment retention capacity would be filled up, when its flood control function is completely undertaken by flood control capacity and spillway. The safety of the works depends on the spillway. So the spillway should be frequently inspected to remove the potential problems and strengthen the works in time. Thirdly, the flood control and safety management regulations shall be established. Local governments at all levels should incorporate the flood control of the sediment retention work into local flood prevention management system. According to the Flood Control Law, the administrative heads of governments at all levels shall assume overall responsibility for the work of flood control, with different levels responsible for respective part of flood control work.

6 Conclusions

(1) In the coarse sediment source areas, mean annual quantity of the sediment to the Yellow River is 4.08×10^8 t, and erosion modulus is as high as 21,700 t/(km^2 · a), and reaches up to 30,000 t/(km^2 · a) or higher in parts of the areas. It is the most intensive erosion area in loess

plateau region. The sediment coming from the area has the greatest impact on the rise of the lower course of the Yellow River. Construction of the sediment retention works in this area is most effective.

(2) The construction of the sediment retention works shall focuses on large, medium and small check dams. General layout shall take the tributaries as framework and the minor watersheds as units, and properly coordinate the relations between mainstreams and its tributaries and between the upper and lower reaches to make the large, medium and small check dams form a complete set in a reasonable layout by optimizing the runoff – sediment relationship. Generally, build large check dams on large gullies; build medium check dams on medium gullies; and build small check dams on small gullies.

(3) Through investigation and analysis, it is planned to build 35 large check dams, 2,014 medium check dams and 8,204 small check dams in the coarse sediment source areas. After the planned sediment retention works are all built and put into operation, they can trap 3.963×10^9 t sediment, thus reducing the sedimentation of the riverbed and large reservoirs of the lower Yellow River by 1.308×10^9 t, and reducing the rise of the lower riverbed by 0.42 m accordingly.

(4) The construction of the sediment retention works covers a wide scope with lots of dams and huge working quantity to be completed. Great importance must be attached to security issues when the construction is initiated. To be absolutely safe, precautionary measures shall be taken against potential safety hazards that are present in engineering design, flood control during construction period, construction quality, and operational management.

References

Yellow River Conservancy Commission, MWR. Yellow River Short – Term Keynote Development Plan[M]. Zhengzhou: Yellow River Conservancy Press, 2002.

Xu Jianhua, Lin Yinping, et al. Definition on Source Area of Centralized Coarse Sediment in Middle Yellow River[M]. Zhengzhou: Yellow River Water Conservancy Press, 2006.

Qian Ning, Wang Kechin, Yan Linde, et al. The Concentrated Coarse Sediment Source Areas of the Middle Yellow River and Its Influence on the Erosion and Deposition of the Lower Yellow River[M]. Bejing:Guanghua Press, 1980.

Zhang Ren, Cheng Xiuwen, Xiong Guishu, et al. Impact of Coarse Sediment Reduction on the Erosion and Deposition of the Lower Yellow River[M]. Zhengzhou: Yellow River Conservancy Press, 1998.

Warping for Sediment Utilization of the Yellow River

Shi Hongling[1] , *Tian Qingqi*[2] , *Wang Yangui*[1] and *Liu Cheng*[1]

1. International Research and Training Center on Erosion and Sedimentation,
Beijing, 100048, China
2. Ministry of Water Resources, P. R. China,Beijing, 100053, China

Abstract: The Yellow River is a well – known heavily sediment – laden river in the world. The sediment issues caused by redundant sediment yield from catchment and sediment deposition in the river channel, such as serious soil erosion, aggravated river bed of the lower Yellow River (the LYR), decrease of reservoirs storage capacity, irrigation canal obstruction, etc. are so harmful to the harness of the Yellow River. While, considering the sediment as one of nature resources, sediment issues would be mitigated with sediment rational uses. By sediment utilization, the excessive useless sediment will be changed to helpful resources. Warping is one of the most efficient measures for sediment utilization. In this paper, depending on the sediment features and flood control conditions of the LYR, some practices of warping for sediment utilization in the Yellow River are introduced. These include warping for irrigation, warping for consolidating levees, warping for heightening and widening the ground of levees which is used to construct the relatively normal river. Warping for sediment utilization would be of significant for the integrated management of other heavily sediment – laden rivers.
Key words: the Yellow River, sediment issues, sediment utilization, warping

1 Introduction

The Yellow River is well – known as one of the most heavily sediment – laden rivers in the world. From 1919 to 1960 the mean annual runoff of the Yellow River was 58×10^9 m³; the mean annual sediment load reached 1.63×10^9 t; and the mean annual sediment concentration was as high as 35 kg/m³, which is on the top of large rivers worldwide, as shown in Tab. 1.

The redundant sediment yield in upper and middle catchment and sediment deposit in lower reaches, which makes the sediment issues serious all around the Yellow River Basin, such as soil erosion, a perched river channel, loss of the storage capacity of reservoirs, irrigation canal obstruction, etc. The sediment issues are disadvantageous to the Yellow River improvement, and how to resolve or mitigate sediment issues has always been a concern of the integrated management of the Yellow River Basin.

Tab. 1   Water and sediment data of some large rivers worldwide

Country	River	Catchment area ($\times 10^3$ km²)	Annual runoff ($\times 10^9$ m³)	Annual sediment load ($\times 10^9$ t)	Annual sediment concentration (kg/m³)	Sediment load / catchment area (t/(km² · a))
China	Yellow	752.4	47.4	1.63	35.0	2,126
US	Colorado	637	4.9	0.135	27.5	211.9
India, Bangladesh	Ganges	955	371.0	1.451	3.92	1,519
Egypt, Sudan	Nile	2,978	89.2	0.111	1.25	37.3
US	Mississippi	3,230	564.5	0.312	0.55	96.6

1.1   Flood conditions of the lower Yellow River

Serious soil erosion occurs when the Yellow River flows through the upper and middle catchment, especially through the Loess Plateau; it not only deteriorates the local eco – environment but also aggravates the lower Yellow River to be a "perched river channel" (as shown in Fig. 1) , and inevitably roots in the problem of flood control.

From of ancient, the Yellow River is a heavily sediment – laden river, with the sign of frequent disaster divagation in history. Since 1855, the year the latest distinct divagation of the Yellow River stem channel at Tongwaxiang, the serious sediment deposition has made the lower Yellow River channel rose 0. 05 ~ 0. 10 m annually, and made the flood plain 3 ~ 5 m higher than the outside ground, in some places, it even goes more than 7 m higher. Therefore, the lower Yellow River channel, had been being catastrophe in flood overflow all the time, and some epigrams like "two times levee burst in every three years", "once divagation in hundred years" were prevailed in history.

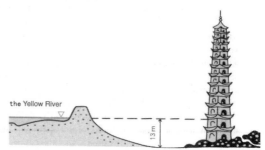

Fig. 1   Sketch map of perched river

1.2   Property of sediment resources

In spite of the sediment disaster was always reminded whenever mentioned the Yellow River, it should be restated that sediment is one of natural resources (Wang Yangui). As the main erosion area of the Yellow River Basin, the loess plateau contributes 90% of the sediment load into the low reach of the Yellow River. It is statistic that, in average, the annual soil loss is 5, 000 ~ 11,000 t/(km^2 · a) on loess plateau, which carries nutrient more than 1,600 kg/(km^2 · a) , among that the organic accounts for 95% and the rest (5%) comprises some fertility as Nitrogen, Phosphor, and Kalium, etc. Corresponding with the loess plateau soil features, the lower Yellow River sediment load is nutrient and fine.

Sediment is absolutely endowed with the natural resources basic properties: the first property is resources validity, meaning of which is good for the society and economy development and environment protection; the second property is resources reconcilability, meaning of which could be assigned or allocated by artificial tools or ways; the third and last property is resources finiteness, meaning of which is the finite in quantity and could be exhausted.

The sediment about the Yellow River is implicated disaster cause of it excessive in amount, while it is definitely a one of natural resources. Under the condition of high technology and economic development, the redundant sediment in the Yellow River could be changed to basic material and to be utilized, which will contribute to produce more social and economic efficiency.

Therefore, sediment issues consequentially make adverse impacts on flood control and environment protection in the Yellow River Basin. However considering the sediment as a type of valuable soil resource, the sediment rational utilization would resolve the sediment issues.

1.3   Types of sediment utilization in the Yellow River Basin

Historically, the main types of sediment utilization were warping for irrigation and warping for levee consolidation. The ancient people recognized the sediment characteristics and used it rationally in their production practice. Based on the historical record, the conception of warping appeared in the early Qin dynasty (221 B. C.). Warping for irrigation was planned in force for increase in crop production in the Tang and Song dynasties (618 ~ 1279 A. D.), while consolidating levees started in late Ming dynasty (1368 ~ 1644 A. D.), and prevailed in the Yellow River and the Haihe River at the epochs of Qianlong and Jiaqing (1736 ~ 1820 A. D.) of Qing dynasty. These measures mentioned above have been widely used in the Yellow River Basin, and get new successful development.

Besides the above – mentioned main types of sediment utilization, there are other types like farmland formation and new land creation, wetlands figuration and production constructing martial, etc.

Taking advantage of the excessive sediment from the Yellow River, the fluxed sediment not only creates new land in a speed of 20 ~ 30 km^2 annually in the Yellow River estuary, but also promotes the farmland, the oil field, the fishery and modern cities. In this case, the sediment is looked as resources prefer to the waste by local resident, and the sediment is far from to destroy the channel and to pollute the environment.

Sedimentation contributes to new land and inevitably derives new wetland. The forming and vanishing of coastal wetlands are related closely in sedimentation. At estuary, the sediment carried by the river runoff will be suspended by tidal wave and re – deposit along coastwise by diffusion effect, and finally forms new wetlands periodically submerged by tidal waves. Thanks to abundant incoming sediment load deposited in the Yellow River estuary, the YRD is one of the quickest continent – building deltas currently. The continually increasing new land supplies a wide space for wetlands development.

Moreover, in history, the ancient people had found the way to fire the dredged sediment to the building material. The Yellow River sediment, due to the fine particles, is difficult to be use as building material directly instead of changing to building material indirectly. Currently, some successful practices of changing the Yellow River sediment into building material have been achieved, for example, the sediment mixes with certain ratio cement and make building brick in Liuzhuang irrigation district.

1.4   Warping for sediment utilization

Warping is one of the most efficient measures for sediment utilization in the Yellow River Basin. Warping has so many virtues, such as more sediment diversion and wide distribution, available diversion infrastructure, simple project demand, short operation period and less economic investment, etc. Therefore, it is widely used in the Lower Yellow River.

As the main type of sediment treatment in the Yellow River, Warping for irrigation means that after muddy irrigation, the farmland was developed by the deposition of fertile sediment. In the regions near the lower reaches, muddy warping has ameliorated some lean sandy or salina land to the fertile farmland; the hyper – concentrated overflowing warping for land – making is also practiced both in the middle and lower reaches; and the storm flood warping is applied in the area of middle reaches through constructing check dams to block silt and develop farmland; It is very significant for creating efficient land by warping for irrigation.

The warping for consolidating levees means that the levees are stabilized and strengthened by sediment deposition near the foot of the levees in the chance of the heavy sediment load in flood seasons. Warping for consolidating the levees not only restrains the levees from piping, seeping and collapse, but also holds up some sediment from the stem river, alleviating the burden of the lower channel deposition. That will improve the situation of flood control, and achieve the target of the Lower Yellow River management: mitigation the perched river to the flat river, which called as the relatively normal river in this paper.

2   Warping for irrigation

As mentioned above, regarding to the sediment load of the lower Yellow River being fine and carrying rich nutrient, warping could be use in rational way to irrigation. As viewed from farmland, warping for irrigation endows with many advantages, such as alleviating alkali, tidying the cropland surface, ameliorating soil structure, enhancing the fertility, etc.

Thanks to the full – developed irrigation district along the double sides of the Lower Yellow River, warping for irrigation can be carried out smoothly combined by the constructed reticular irrigation canals. Through warping, not only the visage of agriculture and environment of the wide irrigation area has been improved, but also the great economy benefit and outstanding society effect has been achieved.

The main type of warping for irrigation in the Yellow River Basin are warping for soil melioration, warping for paddy planting; warping by hyper – concentrated sediment flows and storm flood warping in the middle reaches.

2.1 Warping for soil melioration

From the long time practice, the people has recognized gradually that the Yellow River sediment load is a potential resource depending on the high sediment concentration and fine size in floods, which also carries much nutrient, favourable to the crop.

Based on the sand size observation, the sediment caught by the Lower Yellow River flow belongs to silt, with more than 60% particles finer than 0.05 mm. Except some coarse particles, most sediment particle is fine, which is beneficially to increase the soil nutrient and develop the soil structure. According to the nutrition evaluation of the warped soil, it is found that the organic matter and nitrogen increase 0.3% and 0.03% respectively comparing with the soil non – warped in same region. The warped soil also reduces the soil salinity, and makes the desalination ratio up to 50% ~80%.

In the process of warping planning and practice, three demands should be considered: ①the feasibility of warping and minimum canal deposition; ②the silting area should be warped to "flat, thick and even"; ③the investment and labor should be reasonable.

2.2 Warping for paddy planting

In some irrigation districts, there are a lot of salina depressions near the river banks. In order to change the salina depressions to arable land, the research of the paddy planting was launched in 1950's successfully. The warping for paddy planting would not only make both paddy and wheat get high yield, but also ameliorate the soil by translating the salinity below. The paddy growth phase prolongs the total flood season. Based on the investigation, the diversion sediment concentration usually was $20 \sim 40$ kg/m^3, and the water demand was $15,000 \sim 22,500$ m^3/hm^2, then the diversion sediment load was $30 \sim 45$ t, and the paddy land will be covered by sediment around 4 cm thick annually. It was really a large amount of sediment and efficient for sediment treatment.

Due to limited water resources of the lower Yellow River in recent years, especially in May, June and July of dry years, the paddy planting has not been so prevailing since 1980's. Nowadays the paddy planting is only implemented at RenminShengli irrigation district, Dongying city, Shandong province, located in the Yellow River delta.

2.3 Warping by hyper – concentrated sediment flows

Considering the risk of deposition and blockage of the irrigation canal, the hyper – concentrated warping is usually forbidden in flood seasons when the sediment concentration is lager than a critical level. However, in the case of urgent demand for water resources, the Luohui irrigation district (located in the middle reaches of the Yellow River) broke the sediment concentration limit and diverted the flow with a sediment concentration of 165 kg/m^3 in 1974. The test took full advantage of the hyper – concentrated flood, and the warping succeeded unexpectedly. Thenceforth, the extensive trials developed in other irrigation districts in the years of 1976 ~ 1985, and made a significant progress of the hyper – concentrated warping techniques.

It should be emphasized that the deposition in canals, especially in the ditches, is inevitable and normal during the hyper – concentrated irrigation. But the demand of the balance between aggregation and degradation should be achieved in a definite period, such as a quarter or a year, in order to guarantee the routine irrigation.

3 Warping for levee consolidation

The measure of warping for levee consolidation would divert the sediment – laden flow to the foot of two sides levee wall; leave the sediment to consolidate the levee, and drain the up – layer clear water out for irrigation or industry or town water supply. By this way, on the one hand, the sediment diversion from the Yellow River will alleviate the burden of sediment transportation in the lower reaches; on the other hand, sediment deposition for consolidating the levee will diminish the risk of levee break.

The Yellow River management department puts forward warping for levee consolidation firstly with the purpose of flood control. The gravity flow warping started in 1950's, then the pumping warping experimented in 1960's and fully practiced in 1970's, while the excavating warping appeared in 1990's. Meanwhile, based on the demands of integrated harnessing and development the Lower Yellow River, the conception of the warping for construction of the relatively normal river was formed.

The levees consolidated by warping endured several floods in the Yellow River in the past decades, and the prominent effect of warping has been proved. Until 1993, warping had consolidated 734 km levees with 0.36×10^9 m^3 deposited sediment consumed.

Depending on different tools of diversion, the warping for levee consolidation has three modes: by gravity flow warping, by pumping warping and by excavating warping.

3.1 Warping for levee consolidation by gravity flowing

The gravity flow diverted from the river is usually by culverts or sluice gates. Depending on the slope of the local levee ground, the schema length of warping plot for levee consolidation is controlled in 1 ~ 10 km, and the schema height of warping plot usually is 2 ~ 4 m, but for the deep pool warping, the height could be designed as 11 m as the Huayuankou burst pool. The deposition area and junction canal should be laid out parallel with the levee, but the border dike of warping plot should be built a distance from the toe of the levee. For the type of warping with several sluice gates, the corresponding canal to each sluice gate should be connected in order to control the warping in plan.

Another diversion of gravity flow is by siphon pipelines. The siphon pipeline warping can divert hyper – concentrated floods and warping the depressions. The warping area locations of this type are nearly the same as the gravity flow by culverts or sluice gates, except smaller in scale and more flexible in the warping height.

3.2 Warping for levee consolidation by pumping

This mode of warping is usually applied in the case of local energy resources rich, city water supply demand and gravity flowing is impossible. Combining with the fixed or movable pump station for water supply, the pumping warping is easily implemented. The typical example is the Luojiawuzi pumping station in Dongying city, Shandong province, which pumping water is supplied to the oil field in the Yellow River estuary and for warping to consolidate the Yellow River north levees.

The construction of a pumping station for warping is so simple and flexible that many pumping stations are removed after finishing the mission of warping along the lower Yellow River. But this style of warping invests higher and costs more for per m^3 deposition, owing to the demand of larger discharge and higher water head. The warping by pumping is difficult to be widely applied under present economic and electric power condition in the lower Yellow River catchment, except it combines with the city and mine water supply in some places.

3.3　Warping for levee construction by excavation

After summarizing and learning the good experiences of river dredging by excavation worldwide, warping for levee construction combining with channel excavation was put forward in 1990's. This type of warping technique makes the Yellow River channel desiltation efficiently, strengthens the river bank consistently, and is beneficial to the river regime stabilization.

The process of excavating and warping starts firstly to dredge the muddy deposition from the channel or point bars by motor boats or pumps, then to transport it by pipelines to the warping area located at both sides of the levees. The channel excavation range, construction of pipe line, and warping location should be planed focus on the position of levees, channel or flood plain and the capacity of pump. The excavation warping influences on the environment less and is easy to be operated; thanks to the operation process only disturbs one river side.

From 1970's up to now, more than 220 motor boats were used and hundreds km of levees have been consolidated, which is really significant to increase the levee safety, decrease the levee construction investment, and save farmland from occupied by dredged sediment.

4　Achieving the vision of the relatively normal river by warping in the LYR

With the development of diversion irrigation in the area along the lower Yellow River, the more demands of the diversion water, the more abundant sediment needs to be dealt with before irrigation. Meanwhile, after several years practice, the technique of warping for consolidation of the levees has got rich achievements. The assumption that combining warping for consolidation of the levees with depositing the excessive irrigation imported sediment was put forward.

4.1　Conception of the warping for construction of the relative normal river

According to recent researches, the volume and range of warping along the outside regions of the lower Yellow River levees could be in a huge scale. So, the constructed levees will convert the lower Yellow River "perched channel" to a "normal channel" relative to the widened and heightened levees ground. This prospective vision of the Lower Yellow River channel (as shown in Fig.2) is called "Relative Normal River". Eyes on the river regime, the Lower Yellow River channel is same as other normal rivers' channel, river bed is lower than near ground.

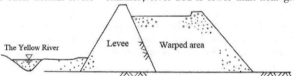

Fig. 2　Cross – section sketch map of the warping for Construction of the Relative Normal River

The purpose of the warping for converting the perched river to the relatively normal river is to build two man – made highlands at both side of levees back to well controlling the flood and dealing with the sediment. The idea of warping for construction of the relatively normal river has been highly appraised for the significance for flood control, and it is a unique action of combining irrigation with the consolidation of the levees. It can be elaborated through warping and raising the ground near the levees, widening the levee base and enhancing the flood control. The finished levees will achieve a standard height and width and the Yellow River will become a "normal river" with the channel bed elevation lower than the levees basement ground.

4.2 Overall management of the warping for construction the relative normal river

The planning of warping for the relatively normal river was made in the years of 1985 - 1990, and the layout of warping area of the lower Yellow River are showed in Fig. 3.

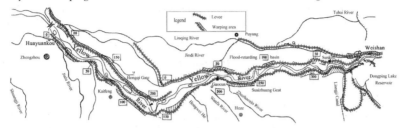

(a) Reach of Huayuankou—Weishan in the Lower Yellow River

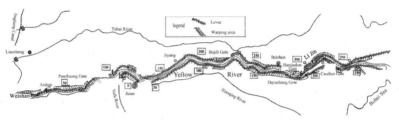

(b) Reach of Weishan—Lijin in the Lower Yellow River

Fig. 3   Layout of construction of the relative normal river in the lower Yellow River

According to the planning, after 10 ~ 20 years construction, the total length of consolidated levees will reach 797.8 km (south bank 370.7 km, north bank 427.1 km), accounting for 61.3% of the total (south levee 62.8%, north levee 60%) and form a constant relatively normal river. Also by this planning, 70% of the diverted sediment (nearly 0.09×10^9 t) will be transported and deposited at both sides of the levees, meanwhile the diameter of the diverted sediment reduces to less than 0.01 mm, which will radically resolve the difficulty of sediment treatment in the irrigation area. With other warping measures and after the arrangement is carried out, the levees of the lower Yellow River will be man – made highlands of 100 ~ 200 m wide and 7 ~ 10 m high. At that time, not only the irrigation sediment will be treated efficiently and the irrigation career will develop healthy, but also the prospect of flood control will be improved greatly.

5   Conclusions

Sediment should be treated as valuable resources and to be used efficiently. Sediment issues of the Yellow River could be alleviated by rational sediment utilization, which changes the excessive useless sediment to helpful resources. Among all the sediment utilization styles, warping is an efficient one.

Based on the inadequate statistics, from the beginning of the diversion irrigation (1970) until 1990, warping diverted 3.865×10^9 t sediment, equal to 0.133×10^9 t annually, and ameliorated 2,300 km^2 and, developed 1,200 km^2 paddy field and consolidated 734 km levees. As far as warping for converting perched river to the relatively normal river was concerned, 70% of annual diverted sediment (nearly 0.09×10^9 t annually) were transported and deposited at both sides of the levees.

The achievement of sediment treatment, especially warping in the Yellow River should be significant for sediment utilization in other heavily sediment – laden rivers.

Acknowledgments

The work is sponsored by the Special Research for Social Commonwealth in Water Resources of P. R. China (research item No: 200901021), the Special Foundation from China Institute of Water Resources and Hydropower Research (research item No: Shaji – 1230) and PHD candidate Foundation.

References

Shi Hongling, et al. Sediment Yield and Sediment Budget of the Yellow River [C]//The Proceeding of 10#ISRS, August, 2007, Moscow, Russia.

Ministry of Water Resources, P. R. China. Gazette of China River Sediment,2002.

Qing Mingzhou, et al. Effect of Using Suspended Sediment Load in the Yellow River on Land Quality and Its Evolution on the Lower Reaches[J]. Journal of Soil and Water Conservation, 2001, 15, (4):107 – 109.

Water Conservancy Encyclopaedia China Vol. 4, Dec. Beijing:Hydropower Press, 1990.

Hong Shangchi, et al. Research on Warping for Constructing the Relative Normal River Combined with Irrigation Sediment Settling[M]. Zhengzhou: Yellow River Conservancy Press.

Jiang Ruqin, et al. Sediment Utility in the Lower Yellow River Irrigation Districts [M]. Zhengzhou: Yellow River Conservancy Press, 1998.

Wang Yangui, et al. Allocation and Utilization for Sediemnt Resource in the Yellow River Basin, Sciene of Soul and Water Conservation, 2010(4): 20 – 26.

The Yellow River Conservancy Committee. The Yellow River Management and Water Resources Development (Integrated Volume) [M]. Zhengzhou: Yellow River Conservancy Press.

Li Guoying. Ponderation and Practice of the Yellow River Control[M]. Zhengzhou: Yellow River Conservancy Press, 2003.

Relationship between Water and Sediment Regulation and Water Supply from the Yellow River

Han Tao[1], *Mi Xiao*[2], *Bai Bo*[1] and *Yang Jianshun*[1]

1. The Administration Office of Yellow River Conservancy Commission, Zhengzhou, 450003, China
2. The North China University of Water Conversancy and Electric Power, Zhengzhou, 450045, China

Abstract: Although the giant achievement has been made according to the water and sediment regulation, people still have different viewpoint about the influence to water supply in the lower Yellow River because of storing water by reservoirs, the river channel being unstable, the river channel section being scoured and so on. This paper analyzes datum of the water and sediment regulation from 2002 to 2010. The outcome shows that: ① the reliability of water supply in the lower Yellow River increased because of the runoff allocation function of the Xiaolangdi Reservoir; ② the application of the water and sediment regulation and the improvement of river training work in the lower Yellow River are beneficial to steady river channel, decrease the degree of river channel wandering and improve the probability of being close between the water supply construction and river channel; ③ the entire lower Yellow River has been successfully scoured because of the water and sediment regulation, which definitely do the negative influence to water supply from the Yellow River. However, in recent years, the average water level of the lower Yellow River has meet the designing demand about water supply of sluice. All the negative influence can be eliminated by the method of dredging the channel in the front of sluice and the sluice area in the lower Yellow River.

Key words: water and sediment regulation, the Yellow River water supply, effect, influence

The Yellow River professionals have explored and developed for decades the theory of the Yellow River water and sediment regulation. With the completion of the Xiaolangdi Reservoir construction, the idea is put into practice. From 2002 to 2010, the entire lower reach of the Yellow River has been successfully scoured after three times of water and sediment regulation tests and seven times of productive practices, which remarkably improves the ability of flood release and sediment discharge in the main channel of the lower reach, and modulates the channel morphology. Great achievements have been made from these exploration and practices, and plenty of inspiration has been obtained for the treatment, development and management of the Yellow River and the reservoir allocations on the muddy water rivers. However, people still have different viewpoints on the influence to water supply in the lower reach because of storing water by reservoirs, river wandering, channel downcutting and so on. Hence, the thesis analyzes the influence to water supply on the lower reach caused by the water and sediment regulation in the past decade, then proposes the preliminarily measures that can eliminate or mitigate the influence, and provides fundamental support and basis for formulating the water drawing measures for the lower reach of the Yellow River.

1 Relationship between water storage by Xiaolangdi Reservoir and water drawn from lower reach of the Yellow River

In the 1990s, the Yellow River experienced continuous dry years accompanying with sharp decrease of inflow, rapid economic and social development in the river basin and great increase of water consumption gave rise to increasingly acute imbalance between water demand and supply, and frequent zero − flow phenomena appeared. The total annual water supply from the lower reach exceeded $10,000,000,000$ m^3 (including inter – basin water transfer), 85% of which was irrigation water supply, and 52% of which was concentrated in the period from March to June which is low – flow period of the Yellow River with little rainfall and inflow. As a result, the water supply from the lower reach and the natural inflow of the Yellow River were quite uncoordinatedly distributed in terms of the time, which greatly influenced the water supply from the Yellow River. In October 1999, the Xiaolangdi Reservoir was put into operation, its controllable annual allocation and storage capacity was fully utilized, and the Yellow River water quantity was managed and regulated in a unified way. These measures optimized the water supply, mitigated the imbalance between water supply and demand in the lower reach in terms of the time, changed the coexistence situation of zero flow and water wasting in the lower reach, and eased the contradiction between the water supply and demand in the lower reach to a certain extent. According to the statistics, with relatively less water inflow in the upper reach of the Yellow River, an allocated amount of $2,000,000,000$ m^3 has been increased annually on average since the year of 2000 by optimizing the Xiaolangdi Reservoir allocation, and the reliability of water supply in lower reach has risen by 43%. At the same time, by applying sediment prevention through water storage and implementing water and sediment regulation, the Xiaolangdi Reservoir decreases the sediment content especially the coarse sediment content that entering the lower river, coordinates the relationship between the water and sediment, lessens the pressure of sedimentation in the water drawing channel, and plays an positive role in the improvement of water supply from the Yellow River.

2   Analysis of influence of river regime change on water supply

The river channel in the lower reach of the Yellow River is wandering because of the wide and shallow river channel, sandy river bed and special water and sediment conditions. Before the operation of the Xiaolangdi Reservoir, the natural flood rose and fell sharply, the sediment content changed greatly, and the river channel was not well controlled, as a result the river channel in the lower reach wanders frequently. Great changes have taken place in the water and sediment supply of the lower reach since the water and sediment regulation: water supply in the lower river is mainly of medium and small flow, the flood is reduced, the flood peak drops, and the fluctuation range of flow becomes controllable. At the same time, the construction of the standard levee and river channel control and guide project for the Yellow River further regulates the wandering of the flow, decreases the degree of river channel wandering, makes the fluctuation range of the main flow show a weakening trend, stabilizes the river regime, and improves the status of the water drawing construction closing to the river channel.

According to the statistics, since the year 2000, there are six ones out of control among the 100 water drawing sluices in the lower reach of the Yellow River. The six sluices respectively are the Liuyuankou water drawing sluice in 2002, the Liuzhuang water drawing sluice in 2002, the Qucun, Nanxiaodi, Liyuan and Wangchenggu water drawing sluices (these four sluices are in the same river reach) in 2010, and they function well after excavation and water drawing. With the refinement of the standard levee and control and guide project for the Yellow River, the wandering of the flow is further regulated, the out – of – control phenomena of the water drawing project will be decreased gradually, and the resulting skepticism about the support ability of water supply from the Yellow River will be vanished.

3   Analysis of influence of river scouring and river downcutting on water supply

The water drawing sluices in the lower reach of the Yellow River were mainly built in 1960s ~ 1980s. At that time, the sediment deposition in river channel of the lower reach rose year by year, therefore, when designing the bottom height of the culvert sluice, the sediment factor was taken into consideration to set a relatively high sluice bottom. The comparison table of water level and sluice bottom height before and after water and sediment regulation in Shandong reach of the Yellow River (Tab. 1) shows that, after the water and sediment regulation in 2010, the upstream water level of 54 sluices of Shandong Province at the river flow of 300 m^3/s was 1.42 m lower than that in 2002 on average, and the upstream water level of the sluices at the river flow of 500 m^3/s was 1.34 m lower than that in 2002. The facts indicate that, since the water and sediment regulation in 2002, the entire main channel of the lower reach has been downcut, the water levels of similar flow have dropped obviously, the bottom height of the Yellow River sluices has risen relatively, and the water drawing of the sluices has been influenced to a certain extent. However, except the Caodian Sluice whose water level at the river flow of 300 m^3/s is a little lower than the bottom height of the sluice, all other sluices are higher than the bottom height of the sluices. Besides, the mean height difference in front of and behind the Yellow River water drawing sluice at the river flow of 300 m^3/s was 1.18 m in 2002, and such height difference was only 0.48 m in 2010, which suggests that reduction of gradient ratio in front of and behind the Yellow River water drawing sluice is also one of the factors that influence the water drawing efficiency of the lower river reach.

Tab. 1 **Comparison of Wwater level and sluice bottom height before and after water and sediment regulation in Shandong reach of the Yellow River**

Water drawing gate	River flow of 300 m^3/s				River flow of 500 m^3/s				Bottom height of sluice (m)
	Water level in front of sluice in 2002 (m)	Water level behind sluice in 2002 (m)	Water level in front of sluice in 2010 (m)	Water level behind sluice in 2010 (m)	Water level in front of sluice in 2002 (m)	Water level behind sluice in 2002 (m)	Water level in front of sluice in 2010 (m)	Water level behind sluice in 2010 (m)	
Yantan	65.50	65.47	64.28	63.74	65.56	65.52	64.71	64.14	62.99
New Xiezhan	63.21	62.99	62.72	62.38	63.51	63.20	62.60	62.45	60.91
Old Xiezhan	63.30	63.00	62.70	62.36	63.60	63.35	62.62	62.40	60.91
Gaocun	61.08	59.20	58.36	58.14	61.40	59.10	58.38	58.27	57.11
Liuzhuang	60.01	57.63	57.41	57.38	60.26	58.85	57.75	57.46	55.20
Susi	56.68	56.06	54.53	54.36	56.90	56.26	54.75	54.58	53.13
Jiucheng	52.64	52.50	50.55	50.52	52.85	52.70	50.75	50.63	49.13
Suge	50.39	49.33	49.30	48.92	50.89	49.45	49.78	49.22	46.08
Yangji	49.32	47.42	47.69	47.32	49.66	47.71	47.87	47.55	44.48
Chengai			45.20	45.00	45.10	44.90	46.10	45.80	42.45
Guonali			0		40.50	39.90	40.56	39.90	38.80
Toachengpu	43.27	41.06	41.20	40.95	0	0	41.39	41.16	36.42
Weishan	42.20	42.00	40.70	40.50	42.30	42.10	41.00	40.80	37.04
Guokou	36.95	35.45	36.80	35.22	37.23	35.36	37.17	35.53	33.87
Panzhuang	35.99	34.60	35.05	34.16	36.60	34.60	34.55	33.22	30.10
Lijia'an	30.13	27.19	27.65	27.06	30.44	26.90	28.26		24.50
Hanliu	34.19	32.92	33.14	32.55	34.93	33.02			28.98
Doufuwo	30.10	28.29	29.88	29.43			29.88	29.43	26.50
Beidianzi			28.37	26.00			28.60	26.00	24.62
Yangzhuang			27.95	Zero flow			28.15	28.10	

Continued Tab. 1

Water drawing gate	River flow of 300 m³/s				River flow of 500 m³/s				Bottom height of sluice (m)
	Water level in front of sluice in 2002 (m)	Water level behind sluice in 2002 (m)	Water level in front of sluice in 2010 (m)	Water level behind sluice in 2010 (m)	Water level in front of sluice in 2002 (m)	Water level behind sluice in 2002 (m)	Water level in front of sluice in 2010 (m)	Water level behind sluice in 2010 (m)	
Laoxuzhuang			No water release	No water release			No water release	No water release	23.61
Dawangmiao			26.60	26.53			27.26	27.03	23.61
Huojialiu	27.65	25.32	24.78	24.32	27.80	25.56	24.98	24.65	21.24
Hujia'an	26.59	25.98	23.63	Zero flow	26.74	26.12	23.85	23.65	20.55
Tuchengzi	25.46	24.29	22.15	21.75	25.62	24.41	22.35	22.15	18.90
Gedian			22.85	21.60			23.05	21.75	17.69
Zhangxin			22.05	20.45			22.26	20.47	17.00
Gouyang			23.50	23.23			24.76	24.04	19.75
Xingjiadu	26.00	25.50	25.12	24.95			26.05	25.82	21.59
Mazhazi			18.80	18.60	15.60		19.02	18.82	
Liuchunjia			15.78	15.44			16.08	15.92	11.40
Zhangqiao	21.50	20.93	20.66	20.36					17.10
Hulou	20.50	20.42	20.36	20.23			20.42	20.27	16.71
West Bojili			20.35	19.73			22.60	19.90	15.10
East Bojili	21.03	19.45	Closed						
Guiren	19.20	17.70	18.30	17.88					15.13
Bailongwan	19.02	17.60	17.50	16.60			17.78	17.67	14.15
Dacui	18.10	16.25	16.50	16.00					13.85
Lanjia	16.85	15.90	15.95	14.96					12.43
Xiaokaihe	17.15	16.50	16.10	15.95					12.63
Dadaowang	16.08	15.02	14.30	13.98					10.27
Daoxu	15.64	14.07	13.35	13.20					9.03
Handun	12.70	12.55	13.50	13.10					9.08
Zhangxiaotang	16.08	14.24	14.68	14.53					11.72
Dayuzhang	14.10	13.10	12.58	12.65					9.04
Mawan	11.50	10.60	9.30	8.90	13.50	12.80	12.50	12.20	8.05
Caodian	8.50	8.20	7.60	7.50	12.50	12.00	8.80	8.50	7.88

Continued Tab. 1

Water drawing gate	River flow of 300 m³/s				River flow of 500 m³/s				Bottom height of sluice (m)
	Water level in front of sluice in 2002 (m)	Water level behind sluice in 2002 (m)	Water level in front of sluice in 2010 (m)	Water level behind sluice in 2010 (m)	Water level in front of sluice in 2002 (m)	Water level behind sluice in 2002 (m)	Water level in front of sluice in 2010 (m)	Water level behind sluice in 2010 (m)	
Shengli	12.25	10.32	10.60	10.41	No water release	No water release	11.29	11.10	7.41
Luzhuang	No water release	No water release	10.12	10.02	No water release	No water release	10.71	10.53	6.41
Sluice No.1	No water release	No water release	7.89	7.82	No water release	No water release	No water release	No water release	4.36
Shibahu Sluice	No water release	No water release	5.65	5.52	6.80	6.30	6.25	6.15	4.42
Wuqi Sluice	No water release	No water release	3.80	3.70	No water release	No water release	No water release	No water release	3.30
Gongjia	13.45	11.50	12.27	10.97	No water release	No water release	12.31	11.01	7.85
Wangzhuang	11.18	10.10	10.17	8.87	No water release	No water release	10.21	8.91	6.63

The comparison analysis shows that, the water and sediment regulation causes the downcutting of entire lower reach, but at perennial average flow, the water level of the lower Yellow River can also meet the designing demand of sluice on water drawing in recent years, and all the negative influence of channel downcutting on water drawing can be eliminated by timely dredging in the front of sluices and drawing channels in the lower reach.

The water and sediment regulation is not simply about pursuing and continuing the river channel descent, it is about building and stabilizing the medium – sized river channel with discharge capacity of 4,000 m³/s, ensuring the unblocked status of water and sediment conveying channels, and guaranteeing the flood control safety of the lower Yellow River. After the water and sediment regulation in 2010, the minimum discharge capacity of the main channel in the lower Yellow River comes back to 4,000 m³/s from 1,800 m³/s (the figure in 2002), which is beneficial to the formation of medium – sized river channel for conveying water and sediment. Therefore, the lower Yellow River will no longer be greatly flushed when regulating water and sediment, and the influence of channel downcutting on water drawing will be remarkably reduced.

4 Conclusions

Multiple water and sediment regulation modes that fit the water and sediment regime of the Yellow River have been successfully explored through ten years practices. With the on – going improvement of productive practice, these modes play a significant role in ensuring the flood control safety in the lower reach of the Yellow River, and obtain great social, economic and ecological benefit. The Yellow River treatment and development practices show that, the water and sediment regulation has become an integrated part both for the safety guarantee and the healthy life maintenance of the Yellow River. We will take account of the relationship between the water and sediment regulation and water supply from the lower reach of the Yellow River in a comprehensive way and set a linking mechanism from now on when meeting the technical demands and continuing the adjustment and improvement of water and sediment regulation implementation. We will also timely adjust and rebuild the Yellow River Drawing Project with engineering technologies according to the characteristics of the water and sediment regulation. For example, the dredging strength of the channels in front of and behind sluices can be enhanced, and allocation and storage projects can be constructed. Based on the practical working scheme of the water and sediment regulation, we will further make best possible use of the allocation and storage projects, regulate water drawing amount in different areas timely and properly, minimize the influence of water and sediment regulation on water supply from the lower reach of the Yellow River, maximize the utilization of the Yellow River water resources, and provide support for the economic and social development in the Yellow River basin.

Key Technology Research of Silting and Scouring Sedimentation Reduction in the Coastal Tidal Gate

Guo Ning[1] , *Sun Hongbin*[2] and *Zhou Heping*[3]

1. Water Resources Department of Jiangsu Province, Nanjing, 210049, China
2. The Authority of Waterway Works to the Sea of Huaihe of Jiangsu Province, Huai'an, 223200, China
3. Luoyun Hydraulic Engineering Management Division of Jiangsu Province, Suqian,223800, China

Abstract: A lot of thetidal gate were built since the establishment of our country, making the most of the estuaries under control. The tidal gate plays an important role in flood control and drainage halogen block tide and irrigation. In the meantime, it also changes the hydrodynamic conditions of the estuary, causes the gate port way siltation, raises the riverbed, reduce the discharge area and the flood discharge capacity of the tidal gate sharply. For many years, a lot of experiments and exploration work are made to reduce the gate port way siltation, such as using upstream water, tidal power and different monitoring methods. Hydraulic silting and scouring are the most direct, cost-effective method by using upstream water, which brought out more sediment by increasing the amount and velocity of low tide, reduced the tide out of sediment by reducing the amount and velocity of high tide. The water source should be made full use of water saving silting and scouring, keeping the main channel unobstructed in nonflood season. It should be use every chance of the silting and scouring to expand the river section in flood season. At the end of the flood season, it should make full use of tail water and increase the silting and scouring strength to improve the silting and scouring effect. The silting and scouring between the brake group combining with actual conditions, should be in line with the principle that "the scheduling concentration and external water rushed down, take turns the high water rushed, flood season clear from the top and discharging water continuous blunt" to formulate feasible operating scheduling solutions for the best silting and scouring effect. Regulating the brake orifice discharge is playing an important role in scouring the silt and protecting the port.

Key words: the tidal gate, scouring the silt and protecting the port, regulate, monitoring methods

Jiangsu Province, east of the Yellow Sea, and the coastline is 953.9 km, Huaihe, Yi, Shu, Si rivers and many such medium and small rivers are all from here into the sea, and more than 60 is larger among those 100 sea-entering rivers. Those rivers which belong to the backbone Spillway river include North Jiangsu Irrigation Canal, Huaihe River sea-entering channel, the old course of the Yellow River, new Yi River and new Shu River, draining nearly 200,000 km² of river basin flood into the sea. Among those, the Doulong Port, Xinyang Port, Huangsha Port, Sheyang River, Liutang River, Chaimi River, Shanhou River and Qiangwei River are part of the regional backbone drainage channel, and discharging abdominal area nearly 20,000 km² area flood into the sea. The main drainage channel of Reclamation area include Jiuyu Port, Tonglv River, Tongqi River, Limin River, Chuandong Port, Wang Port, Nanchao River, Shaoxiang River and Qingkou River, which discharging 25,000 km² of rainfall runoff into the sea. All those sea-entering rivers expect for the irrigation rivers have been controlled by gates. Channel deposition lead to riverbed rising and water section decreasing which dramatically reduce the tidal gate drainage capacity result from tidal wave deformation, discharging runoff reduction and low tide lasted extend after the sluice construction.

In the 1960s, the domestic research about the gate channel siltation began with the analysis about the sediment hydrological change of the Sheyang River after the sluice construction. Since then, with the siltation problem tending to seriously and frequently, many related research institutes and coastal tidal gate? management unit have started to explore and practice about the silting cause and protection. Some of the main protection measures are as follows: ①breaking the dam to prevent from silting; ②clearing the silt by manual or mechanical; ③scouring the silt by waterpower; ④dragging the silt by motor vessel; ⑤carrying water to scouring the silt; ⑥scouring the silt by tide and so on.

Scouring the silt by waterpower is the most direct, effective and economic method among all those protection measures. With the development of national economy, water becomes more and more precious, so how to make scouring the silt by waterpower more effective is a big problem to us. Starting from the 1950s, many experiments and researches have been done through changing the way to open the gate, which has accumulated a lot of practical experiences and valuable information for us.

1 Summarizes of scour the silt by waterpower

Opening the gate to scour the silt by waterpower is very effective. Because the release of the runoff from upstream play two roles: on one hand, to increase the ebb tide water yield and flow velocity which leading ebb tide water bringing out more silts; on the other hand, to decrease tide water yield and flow velocity which reducing flood tide with silts. Through the sluice gates control to carry out the scouring the silt by waterpower, which make full use of limited water resources, control the sluice gates scientifically and? improve? the? protective effect.

The basic principle of exploiting the sluice gates control to improve the scour slits effect is maximizing the transport silts capacity of the gate river ebb flow (the transport silt capacity $S = kv^3/gh\omega$) , in this equation, v is section average flow velocity, g is gravity acceleration, h is water depth, ω is silts sink speed and k is coefficient), which need increase flow velocity v and reduce water depth h. However, h is difficult to control, so it is more commonly to increase v through sluice gates control.

There are two ways to control the sluice gates: one is scouring the silts by single gate, the other is by two or more gates (also known as "Gate Group") control and their control applications with each other.

2 Scouring the silts by single-gate

The mostly coastal scouring water comes from rainfall, and the other is external water. At the non-flood season, the pre-precipitation and diversion should be made full use of preserving the unobstructed of the discharge passage and main channel. When not at the tension flood season, expanding the river section should take every opportunity as far as possible to scouring silts. Finally, at the end of the flood season, in order to improve the scouring effect, we should take full advantage of the last water.

2.1 Open the gate with water level difference

Under the premise of gate body safety, exploiting the maximum allowable water level difference to scouring silts, which can increase flow velocity v and at the same time means that actually improve the transport silts capacity S. Also, practices have proved this method to be effective.

In 1996, the Fangtang River Sluice had conducted this experiment and the result was: August 23 to 25 scouring 3 times, lasting 9 h, draining 2.165,4 × 10⁶ m³ and scouring 4.851 × 10⁶ m³; November 19 to 20 scouring 2 times, lasting 6 h, draining 228.25 × 10⁶ m³, scouring 4.72 × 10⁶ m³.

In 1970, the Niantuo Sluice in Rugao employed this way, and the consequences are as follows: February 6 to 25 opening the gate to scouring 33 times, draining almost 40 million m³, which make the downstream river bottom elevation deep into − 1.3 m from − 0.3 ~ 0 m to guarantee the drainage capacity of the diversion canal.

According to those many years? practical experiences in all aspects of Jiangsu Province, with a 0. 8 ~ 1. 0 m water level difference actually be the best, because it can wash away the 1. 0 km range of the silts, and also able to ensure the tidal gate operate safely. If using the water level of 2. 5 m can actually get the scouring distance up to 3. 5 km.

2.2 Open orifice flow or part of the gate

When the coastal tidal gate is opening, the speed and quantity of the flow turn into variable, and the tide level change constantly, causing the water level difference has also been changed. Opening the gate normally generate the upstream water level fell back quickly, which make a bigger water level difference between upstream and downstream last shorter. However, opening orifice flow allows the upstream water level not fall too fast to keep the bigger difference for a long time, thus, the high speed scouring time become longer and the scouring effect enhanced.

Most of the coastal tidal gates are porous gates. If the near gate segment in the downstream get silted, it can just open part of the gate in the security permission range to scour silts, and this way can save water, extend the high speed scouring time and increase scouring effect.

The Liuduo South Sluice is the coastal escape sluice of irrigation canal, and the water level difference is much bigger between upstream and downstream. Employing the orifice flow not only save water, but also achieve a remarkable scouring effect.

When the Xinyang Port in the water shortage season or the deposition of the upstream gate becomes serious, the orifice flow should be opened under the design permission, which will increase the bottom water flow velocity. Through this way the silts near the sluice can be washed out, and it also can take away the silts on the sluice due to the gate leakage.

The Yunmian River Sluice adopted opening the left or right part of the gate in alternative to scouring silts. This method takes advantage of the water swing to improve the diversion canal scouring effect and save water at the same time. And the Xinyang Port also used the same approach and achieved a good effect.

2.3 Open the gate in spring tide

On the basis of spring tide make big siltation while neap tide make small siltation, opening the gate in spring tide can increase the effect of silting up prevention and reduction. Therefore, in the non-flood season and water shortage area, the proportion of opening gate in spring tide should be send up.

In August 25, 1996, a shut gate deposition test was conducted in Fangtang Sluice, and the result showed that the quantity and speed of silting in the spring tide were much larger and faster.

In the mean time, the other experiment was conducted in Sheyang Sluice also told us that the silts of shutting sluice in spring tide were more than 4 ~ 5 times to neap tide while the drainage is 2 ~ 3 times than shutting sluice in neap tide.

2.4 Open gate after high and low tide

The coastal tide is almost irregular semidiurnal tide, which means a lunar days have two up and two fall, and one is higher while the other is lower. The low tide after the higher tide is high too, named high tide and low tide, while the low tide after the low and high tide is lower, named lower low tide. The water level after lower low tide is the most shallow, and the water level difference between upstream and downstream reaches maximum at this moment. A contrast test had been done on Sheyang River Sluice in the early 1960s. Draining after higher high tide, the average flow is 800 m^3/s (the total drain volume is 14.4×10^6 m^3), the scouring quantity is 18,000 t in the test day with two tides. Moreover, draining after low tide and high tide, the average flow is 800 m^3/s (the total drain volume is 14.4×10^6 m^3), but the scouring quantity extend to 47,000 t. This indicates that scouring after low tide and high tide is more effective than higher high water with the same drain volume.

Therefore, the tide features should be exploited as far as possible, which means open the gate in the low tide and high tide ebb period to drain water make the scouring effect getting best. In addition, draining after low tide and high tide can still resist the sediment tracing in the next high tide, which benefit to the gate deposition reduction.

2.5   Top tide sedimentation reduction

The experiment in Sheyang River Sluice demonstrates that: the sedimentation amount is 49,100 t when shutting the gate before spring tide, nevertheless, draining 5.86×10^6 m^3 water before spring tide can reduce the sedimentation amount to 34,500 t, and if the draining water is 11.3×10^6 m^3, the sedimentation amount will reduce to 2,880 t.

For the long port channel, draining water no matter in the non-flood season or spring flood season, shutting the gate when the water level between upstream and downstream is flush(that is, gate hole flow velocity is zero), then exploiting fresh water jack the tidewater into the port way to scouring silts. On the other hand, for the short port channel, draining water before rising tide especially before the spring tide, filling the port way under the gate with fresh water to withstand the sediment carried by the peak tide. This way is benefit to the tidewater reduction, thus reducing the siltation.

2.6   Scouring according to the wind power and direction

Most of the Jiangsu coastal sluices are distributed along the Yellow Sea coast and the port toward northeast. The east wind tide rises faster and higher as a result of wind power, which also leads to a slower ebb tide and a higher low tide. On the contrary, the west wind tide is slower and lower, while ebb tide is faster and the low tide is lower. So making the most of the much lower end up west wind tide to scouring silts is very effective.

3   Operate the gate group on schedule

Jiangsu coastal areas have a prosperous and dense water system, and parts of the sea-entering rivers are concentrated in a certain area. With the sluices construction, those sluices form a restrict relationship in water coming or draining, so the gate group formed. Each single gate of the gate group works the same way just as mentioned before.

The gate group scouring should be in line with the principle of "centralized scouring, external water scouring in turns, high water scouring, clearing the top silts in the flood season and draining continuous to scour." Making the most feasible operating schedule combined with the actual working condition to achieve the best scouring effect.

Centralized scouring—completing the regimen among the gate group and selecting the seriously silted port to scour on the prior. External water scouring in turn—using external water to scour the silted port in turn. High water scouring—using bigger water level difference to scour at an appropriate time. Clear the top silts in the flood season—drain water to scour the top silts before the spring tide. Drain continuous to scour—using drained water to scour continuously when in the flood season.

3.1   The scheme of upstream with the same drainage line, downstream merged into the sea gate group

The schedule of such gate group should take the design parameters and using frequency of each gate into account, and scour the main channel in priority and then the sub channel in turn. The Yantong four check gate is a typical representative.

The Yangdong four check gate include Wuzhang River Sluice, Liutang River Sluice, Longgou River Sluice and Yize River Sluice, all of these sluices had been constructed in 1969 successively, and the corresponding downstream tributaries all enter into Guan River. At last, the downstream scouring has been changed after the sluices construction.

In order to maintain the dynamic balance of the four check gate scouring, it is essential to coordinate draining water and scouring. In the meantime, according to the demand of the lockage channel of Wuzhang River, it is important to take the riverbed balance at the first place to prevent the channel from over silted. The Yandong Hydraulic Engineering Office and Hohai University exploit a one-dimensional flow and sediment mathematical model to numerical simulate and forecast the scouring variation. Furthermore, basing on this establish a reasonable gate open and close operation scheme, just shown in Tab. 1.

Tab. 1　Four check gate operation scheme

Check gate	Wuzhang River Sluice	Liutang River Sluice	Longgou Sluice	Yize River Sluice
Switching-off flow(m^3/s)	300	200	250	250
Switching-off time	Ebb tide (about 7 h)	Ebb tide (about 7 h)	Ebb tide (about 7 h)	Ebb tide (about 7 h)
Switching-off cycle	Closing after 2 d	Closing after 4 d	Closing after 1 ~ 2 d	Closing after 6 ~ 10 d

3.2　The scheme of upstream with the same drainage line, downstream into the sea respectively gate group

About those gates group, the flood should have the priority and the navigation should be taken into consideration at the same time. On behalf of those gates group are the four sluices in the Lixia River.

The Sheyang River, Xinyang Port, Doulong Port and Huangsha Port are collectively referred to as the four sluices in the Lixia River. From 1956 to 1972, four large drainage tidal gates known as Sheyang River Sluice, Xinyang Port Sluice, Doulong Port Sluice and Huangsha Port Sluice had been built successively in the estuary. Among those four gates, the distance between each other is around 10 ~ 30 km, and the Tongyu River is through away from the gate upstream of 40 ~ 60 km to form a relatively loose gate group.

Because each gate upstream of the four sluices is river network, so the actual operation controlling is major based on the siltation degrees and water conditions. On the water sufficient condition, the gate should be controlled to scour; on the wanter insufficient condition, the water should be concentrated to scour the seriously silted channel. While in the non-flood season, the remaining water, which have been controlled by normal upstream, should use to drain in turn and make the concentration scouring and rotation scouring planned. Through this way not only increase the ebb flow velocity, saving water and also avoid the drainage jacking in the Sheyang River Sluice and Huangsha Port Sluice what is the con-port into the sea and one port with more than one sluice channel, thus improving the scouring effect.

3.3　The scheme of upstream with the different drainage lines, downstream merged into the sea gate group

This kind of gate group should clear their respective silts at first place, and the whole port channel should use the higher water level gate in the upstream to scour in priority. Represented by Huangsha Port—Liming River—Yunmian River Sluice, Xinyang Port—Xichao River Sluice and Doulong Port—Dafeng River Sluice, ect. The following use the Huangsha Port—Liming River—Yunmian River Sluice is an example to explain.

The Huangshan Port is one of the main river in the Lixia River area with 865 km^2 drainage area, and the gate warning water level is 1.3 m, the length of the under gate channel is 14.07 km. The Yunmian River and the Limin River are converge respectively from the gate 1.0 km and 1.44 km. The Yunmian River Sluice located on the left of the Huangsha Port Sluice with 405 km^2

drainage area, and the gate warning water level is 1. 2 m, the length of the under gate channel is 910m; While the Limin River Sluice is on the right with 620 km^2 drainage area, and the gate warning water level is 1. 00m, the length of the under gate channel is 1,190 m. Comparing to the Yunmian River and Limin River, the Huangsha Port has a higer water level and larger water yield , and the mutual restraint and influence is also much larger.

These three gates downstream channel are converged in the 1km. The Huangsha Port Sluice upstream water level is higher than the other two gates and the water yield is larger too. In addition, the Huangsha Port Sluice opening flow jack is heavier than the others, so the management of these three gates needs coordination. Except the flood season, an appropriate time should be arranged to scour in turn through the big water level difference. The other two gates should be shut as far as possible when the Huangsha Port Sluice is opened to store water for scouring. Meanwhile, those two gates should not scour in order to raise the water level difference and save water. If all three gates are required to scour, the opening time should be staggered to increase the water level difference of the first opening gate, which can improve the scouring effect.

4 Conclusions

Now the gate controlling mainly depends on the water level difference adjustment and other natural conditions such as tide or wind power and direction. All the tidal gates should be adapt to the local conditions, and scour according to its characteristics.

If more than gates are required to scour, it should be taken project's importance, deposition, difficult degree of the scouring method and effectiveness into consideration to distribute water rationally and adopt concentrated scouring, continuous scouring, rotation scouring and high water scouring.

References

Huang Jianwei, Zhang Jinshan. Regulation Techniques of Sediment Siltation Downstream Tidal Barriers in China [J]. Journal of Sediment Research, 2004(3): 46 – 51.

Liu Tao, Sun Hongbin, Tang Hongwu, et al. Analysis of Current Situation of Sediment Deposition in Downstream Area of Southern Liuduo Gate [J]. Advances in Science and Technology of Water Resources, 2008, 28(1): 54 – 57, 61.

Nanjing Hydraulic Research Institute. Measures of the Sedimentation and Reduction Downstream Tidal Barriers after Setting up the Floodgate [R]. 1997.

Xin Wenjie, et al. The Problem and Solution of the Sediment Siltation Downstream Tidal Barriers after setting up the floodgate [R]. 2006.

Min Fengyang, Wang Yaping. Study on the sediment siltation downstream tidal barriers in the North of Jiangsu Province [J]. Marine Sciences, 2008 (12): 97 – 91.

Jiangsu Flood and Drought Control Office. Analysis Report of Scouring and Silting on the Coast of Xialihe Area and Research of Improving Water Environment [R]. 2008.

Yandong Hydraulic Engineering Office. Study on Channels Scouring and Silting in Downstream Area of Yandongsijiezhi Gate [R]. 1994.

Flocculation Model of Cohesive Sediment Considering Surface Morphology

Huang Lei

The State Key Laboratory of Hydro-science and Engineering,
Department of Hydraulic Engineering, Tsinghua Univ ersity, Beijing, 100084, China

Abstract: Flocculation frequently occurs in estuary and coastal zone, and is one of the most important issues of sediment dynamics. Based on Population Balance Equations (PBE), this paper simulates flocculation process of cohesive sediment and flocs size distribution under equilibrium state considering the impact of surface morphology. Results show that this model can accurately describe the flocculation process, and complex morphology can promote flocculation which results in larger flocs. This paper expands the study of flocculation phenomenon from microscopic scale, and prepares for further study on the impact of surface charge and other factors.
Key words: sediment particle, flocculation model, surface morphology

1 Introduction

Water movement in natural environment often carries sediment particles of various properties. Sediment transport is one of the most important processes in hydraulic engineering to further predict the morphological evolution of river, estuary and coastal zone. The transport of cohesive sediment is a function of the effective settling velocity of the sediment, which in turn is affected largely by flocculation phenomenon. So study of flocculation is of great significance for sediment dynamics.

Flocculation is a process of contact and adhesion whereby the particles of dispersion form larger-size clusters. Flocculation is a nonlinear and complex phenomenon because several processes such as adsorption, desorption, aggregation and breakup take place simultaneously. And complex factors such as surface morphology, surface charge and surface potential also make it difficult to give a clear interpretation to flocculation.

However, the flocculation process has been studied by many researchers. Smoluchowski (1917) was the first who used population balance equations (PBEs) to model the rate of change of particle numbers due to collision, i. e. floc size distribution, after which PBEs are widely employed to model flocculation process. Smoluchowski made a number of assumptions in order to simplify calculation, such as ignoring the effect of fragmentation, assuming all collisions lead to adhesion and considering flocs as spherical particles. Subsequently, PBEs have been gradually improved, and could give a more accurate description of the flocculation process. Apart from aggregation kinetic, destruction kinetic is also included in the population balance equation, which eliminates the computational growth to a maximum size class. The classical DLVO theory is widely employed to predict the stability of suspensions. Runkana et al. have applied the DLVO theory to compute the collision efficiency factor and incorporated it into the PBEs. Also, some researchers employ the collision efficiency as a floating parameter for fitting experimental data. Anyway, not all collisions will result in aggregation. Flocs formed by flocculation usually do not have a spherical shape but rather a more irregular structure. So a fixed fractal dimension is introduced to PBEs to describe statistically self-similar floc geometry. And Maggi et al. improve these models using variable fractal dimension, which produces more realistic results.

Evidently, PBEs have been greatly improved to model flocculation. While the effect of the geometric property of individual flocs has been considered with the concept of fractal dimension, the impact of surface morphology of particles on flocculation has usually been neglected. One important aspect of understanding flocculation processes is the small-scale interaction. The only attractive interaction between two porous flocs which need to be considered is that between the two nearest primary particles since the other particles in the two flocs are separated by too large a distance for the interaction forces to be effective. Snoswell et al. also concludes that the surface asperities dominate the particle interactions, hence the stability of suspensions. So the main objective of this paper is to propose a modified flocculation model considering the effect of surface morphology.

2 Flocculation model

2.1 Population balance equations

Population balance equations are employed to model the evolution of flocs size distribution (FSD), including both aggregation and fragmentation. The rate of change of floc concentration in an integral form is given by:

$$\frac{\partial n(v, t)}{\partial t} = -\int_0^\infty \alpha(v, u)\beta(v, u)n(v, t)n(u, t)\mathrm{d}u + \frac{1}{2}\int_0^\infty \alpha(v-u, u)\beta(v-u, u)n \cdot$$

$$(v-u,t)n(u,t)\mathrm{d}u - S(u)n(v, t) + \int_0^\infty S(u)\gamma(v, u)n(u, t)\mathrm{d}u \tag{1}$$

where, n is the number density of flocs; v and u denote floc volume or mass; t is flocculation time; α is collision efficiency implying the probability of attachment due to collision, which is usually considered as fitting parameter for simplicity; β is collision frequency, which will be explicitly interpreted in next section; $S(u)$ is breakup frequency expressing the rate at which flocs break up due to turbulence shear rate; γ is breakup distribution function expressing how daughter flocs are distributed in different sizes after breakup.

The first two terms on the right hand side accounts for aggregation, i. e. the first term implies the loss of flocs of size v due to their collisions with flocs of all sizes and the second term represents the formation of flocs of size v due to collisions between flocs of smaller sizes. The last two terms accounts for fragmentation, in which the third term implies the loss of flocs of size v due to their fragmentation and the last term represents the generation of flocs of size v due to breakup of larger flocs. The fundamental principle of PBEs is the conservation of mass.

Two kinds of breakup frequency are widely used to model fragmentation, i. e. the exponential kernel and the power law kernel. The power law kernel is applied in this study, which is an empirical model based on experimental observations proposed by Pandya and Spielman. That is:

$$S(u) = b_1 G^{b_2} \cdot r_i \tag{2}$$

where, b_1 and b_2 are fitting parameters; r_i is the radius of floc, which is calculated by the concept of fractal dimension as follows:

$$r_i = r_p \cdot i^{1/d_f} \tag{3}$$

where, r_p is the radius of primary particle; i expresses the number of primary particles within a floc; d_f is the fractal dimension.

Eq. (2) and Eq. (3) show that the greater the shear rate and the larger the size of floc, the floc is more easily breakup.

Breakup distribution function expresses how daughter flocs are distributed. The most common breakup distribution functions are binary breakup, ternary breakup and normal breakup. Binary breakup and ternary breakup means two and three daughter flocs are produced after fragmentation respectively. Meanwhile, normal breakup represents a normal distribution of daughter flocs is produced. Zhang et al. concludes that no considerable discrepancy exists on the simulation results of the steady-state size distribution among three breakup distribution functions. So the easiest binary breakup is applied.

2.2 Estimate of collision frequency

In general, collisions occur due to Brownian motion, differential sedimentation and turbulent shear, and the collision frequency should be expressed as a linear superposition of the contribution of the three effects. Considering a dispersion of spherical particles, Levich discusses the probability of collision between two dispersed particles using the theory of diffusion. Levich supposes a sphere of radius R (interaction radius) surrounding a particle and any other particle entering the sphere will collide with the first particle.

However, cohesive sediment has complex surface morphology, not a spherical particle. Fig. 1 is an image of sediment in The Three Gorges Reservoir obtained by scanning electron microscope (SEM). It can be found that sediment is angular, and both convex and concave regions exist on the surface, which is much different from spherical particle. So modification to Levich's derivation should be carried out. Here, let us consider the effect of Brownian motion and derive the probability of collision between natural sediment.

Fig. 1　SEM image of sediment particle

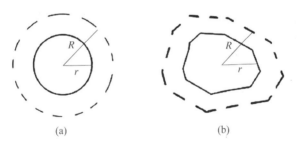

(a)   (b)

Fig. 2　Schematic diagrams of interaction radius

Based on Levich's derivation, the concentration n of particles around a spherical particle satisfies the diffusion equation:

$$\frac{\partial n}{\partial t} = D \frac{1}{r^2} \frac{\partial}{\partial r}(r^2 \frac{\partial n}{\partial t}) \tag{4}$$

Combining the boundary and initial conditions, the concentration profile of i_{th} particles around j_{th} particles is written as:

$$n_i(r, t) = n_i(t)[1 - \frac{R_{ij}}{r}] + \frac{2R_{ij}}{\sqrt{\pi}r} \int_0^{\frac{r-R_{ij}}{2\sqrt{D_{ij}t}}} \exp(-\xi^2) d\xi \tag{5}$$

where, $n_i(t)$ corresponding to the concentration of i_{th} particles in bulk water $n_i(t) = n_i(\infty, t)$; D_{ij}

is the average Brownian diffusion coefficient $D_{ij}(D_i + D_j)/2$; we take $R_{ij} = 2\sqrt{r_1 r_2}$ for easily introducing the effect of surface morphology.

Hence the total number of particles entering the interaction radius, i. e. the total number of collisions per unit time equals:

$$M = \beta(i,j)n_i(t) = \iint j\,dS = \iint D_{ij}\left(\frac{\partial n}{\partial r}\right)_{r=R_{ij}} dS \approx \iint \frac{D_{ij}n_i(t)}{R_{ij}}dS \tag{6}$$

For spherical particle, R is constant, so integration in equation (6) can be directly calculated by multiplying $4\pi R^2$. However, R is variable and independent at different position of the surface of natural sediment, as shown in Fig. 2, so numerical integration need to be conducted for the whole surface. That is:

$$\beta(i,j) \approx \int \frac{D_{ij}}{R_{ij}}dS = \iint \left(\frac{D_{ij}}{R_{ij}}\right)R_{ij}d\varphi d\theta = D_{ij}\iint 2\sqrt{r_1 r_2}\cos\varphi d\varphi d\theta \tag{7}$$

where, θ and φ represent the longitude and latitude respectively.

Fourier series are generally used to approximate a complicated periodic function. Fang et al. has applied Fourier series to reconstruct the boundary contours of sediment with the following equation:

$$r(\theta) = \bar{r} + \bar{r}\sum_{n=1}^{\infty}\left[a_n\cos(n\omega\theta) + b_n\sin(n\omega\theta)\right] \tag{8}$$

where, $\bar{r}$ is the average radius; a_n and b_n are Fourier coefficients, which reflect surface morphology.

If $\{a_n\} = \{b_n\} = 0$, the profile is a sphere. Substitute Eq. (8) into Eq. (7), and let $r'(\theta) = \sum_{n=1}^{\infty}\left[a_n\cos(n\omega\theta) + b_n\sin(n\omega\theta)\right]$, the collision frequency can be written as:

$$\beta(i,j) = 4D_{i,j}\int\sqrt{\overline{r_1 r_2} + \overline{r_1}r_2' + r_1'\overline{r_2} + r_1'r_2'}\,d\theta \tag{9}$$

For simplification, averaging process is carried out, so the first order terms of r' can be omitted. That is:

$$\beta(i,j) = 4D_{i,j}\int\sqrt{\overline{r_1 r_2} + r_1'r_2'}\,d\theta = 4D_{ij}T_n\sqrt{r_1 r_2} \tag{10}$$

where, $T_n = \int\sqrt{1 + \left\{\sum_{n=1}^{\infty}\left[a_n\cos(n\omega\theta) + b_n\sin(n\omega\theta)\right]\right\}^2}\,d\theta$, is a key morphological factor representing the impact of surface morphology on flocculation, which can be calculated with a numerical method. In order to correspond to the results of other researchers, here we replace $2\sqrt{r_1 r_2}$ with $(\bar{r_1} + \bar{r_1})$, so

$$\beta(i,j) = 2D_{ij}T_n(\bar{r_1} + \bar{r_1}) \tag{11}$$

Eq. (11) shows the collision frequency due to Brownian motion considering the impact of surface morphology. Expressions due to differential settling and turbulent shear can be similarly obtained.

2.3 Discretization and calculation of PBEs

Eq. (1) is the integral form of the rate of change of floc concentration. Numerical discretization is needed to calculate the evolution of flocs size distribution. Here the size of flocs is classified into different sections with a scale factor of 2, i. e. $V_i = 2V_{i-1}$, where V_i is the characteristic volume relate to floc mass. Calculation is carried out using 30 sections in this study. Hounslow and Kusters proposed the discretization procedure for aggregation and fragmentation, respectively, and the integrated result is as follows:

$$\frac{dN_i}{dt} = \sum_{j=1}^{i-2}2^{j-i+1}\alpha\beta_{i-1,j}N_{i-1}N_j + \frac{1}{2}\alpha\beta_{i-1,i-1}N_{i-1}^2 - N_i\sum_{j=1}^{i-1}2^{j-i}\alpha\beta_{i,j}N_j - N_i\sum_{j=1}^{i_{max}}\alpha\beta_{i,j}N_j -$$
$$S_iN_i + \sum_{j=1}^{i_{max}}\gamma_{i,j}S_jN_j \tag{12}$$

Submitting Eq. (2), Eq. (3) and Eq. (11) into Eq. (12), a set of ordinary differential equations is obtained, and the evolution of flocs size distribution can be calculated with the function of ODE45 in MATLAB.

3　Model calibration and validation

Fang et al. has made statistical analysis of 470 sediment particles and obtained two sets of Fourier coefficients as Tab. 1.

Tab. 1　Statistical values of Fourier coefficients for the sediment of Guanting Reservoir

Group	a_1	a_2	a_3	a_4	a_5	a_6	a_7	a_8	a_9	a_{10}
1#	−0.004	0.007	0.014	0.038	0.035	0.006	−0.022	−0.03	−0.016	0.001
2#	0.032	0.015	0.065	0.11	0.091	0.027	0.033	0.038	0.027	0.01

Group	b_1	b_2	b_3	b_4	b_5	b_6	b_7	b_8	b_9	b_10
1#	0.025	0.008	0.02	0.008	−0.028	−0.038	−0.026	0.001	0.019	0.019
2#	0.063	0.018	0.065	0.046	0.043	0.051	0.037	0.018	0.046	0.048

Here we only consider the first 10 coefficients, because studies found that it can effectively simulate sediment particle profiles when $n \geqslant 9$. Fig. 3 shows the re-constructed profiles with above coefficients by MATLAB, which are profiles of statistical concepts.

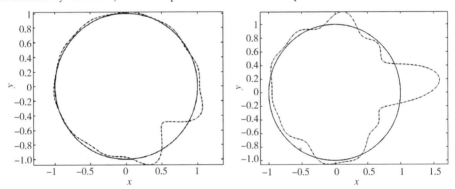

Fig. 3　Reconstructed boundary contours of sediment particles

Fig. 4 shows an example of calculation. A monodisperse system is assumed with the radius of primary particle $r_p = 4$ μm. Other parameters are as follows: $G = 100\ s^{-1}$, $d_f = 2.2$, $= 1$, $b_1 = 1$, $b_2 = 1.5$. Both Brownian motion and turbulent shear are considered, and the second group of Fourier coefficients are used for simulation.

Fig. 4 shows the flocs size distribution at the time of 24 s, 480 s and 960 s. At preliminary stage, small particles collide and adhere to generate large flocs, resulting in a quickly reduce of the number of small particles and a gradually increase of large flocs. With the growth of flocs, the effects of turbulent shear to the breakup of flocs increase obviously, and the size of flocs reaches equilibrium finally ($T = 960$ s). Turbulent shear accelerates the growth of flocs at the beginning; nevertheless, it inhibits the growth of flocs gradually.

Fig. 5 shows the change of average floc size with time. The curves of case 1 and case 2 represents the simulation results using Fourier coefficients of group 1 and group 2 respectively. And the curve of spherical particle is the result without considering the impact of morphology. As shown in Fig. 5, three curves have the same trend, i. e. growing first and then reach equilibrium at the time of about 800 s, consistent with Fig. 4. In addition, a greater average floc size can be reached after considering the impact of morphology, and the more complex of the morphology, the greater of the average floc size. The equilibrium floc sizes of three cases are 0.29 mm, 0.30 mm and 0.35 mm, respectively.

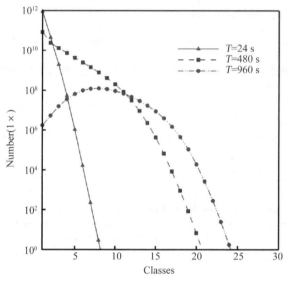

Fig. 4 Flocs size distribution at time of 24 s, 480 s and 960 s

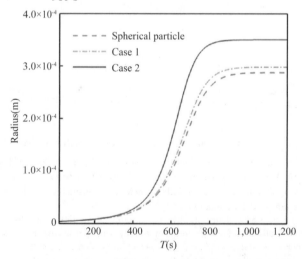

Fig. 5 Evolution of average floc size

4 Conclusions

This paper observes the surface morphology of sediment particles with scanning electron microscopy, and analyses the boundary contours using Fourier analysis, after which a sediment particle of statistical sense is obtained. Based on the characterization of sediment particles, the collision frequency considering the impact of surface morphology is derived. Combining the population balance equations, the evolution of flocs size distribution and average floc size are simulated, with a focus on the impact of surface morphology. Results show that this model can accurately describe the flocculation process, and complex morphology can promote flocculation which results in larger flocs. This paper expands the study of flocculation phenomenon from microscopic scale, and prepares for further study on the impact of surface charge and other factors.

References

Flesch J C, Spicer P T, Pratsinis S E. Laminar and Turbulent Shear-induced Flocculation of Fractal Aggregates [J]. AIChE Journal, 1999(45): 1114-1124.

Kramer T A, Clark M M. Incorporation of Aggregate Breakup in the Simulation of Orthokinetic Coagulation [J]. Journal of Colloid and Interface Science, 1999(216): 116-126.

Runkana V, Somasundaran P, Kapur P C. Mathematical Modeling of Polymer-induced Flocculation by Charge Neutralization [J]. Journal of Colloid and Interface Science, 2004(270): 347-358.

Winterwerp J C. A Simple Model for Turbulence Induced Flocculation of Cohesive Sediment [J]. Journal of Hydraulic Research, 1998(36): 309-326.

Maggi F, Mietta F, Winterwerp J C. Effect of Variable Fractal Dimension on the Floc Size Distribution of Suspended Cohesive Sediment [J]. Journal of Hydrology, 2007(343): 43-55.

Son M, Hsu T J. Flocculation Model of Cohesive Sediment Using Variable Fractal Dimension [J]. Environmental Fluid Mechanics, 2008(8): 55-71.

Kusters K A, Wijers J G, Thoenes D. Aggregation Kinetics of Small Particles in Agitated Vessels [J]. Chemical Engineering Science, 1997(52): 107-121.

Snoswell D, Duan J M, Fornasiero D, et al. Colloid Stability of Synthetic Titania and the Influence of Surface Roughness [J]. Journal of Colloid and Interface Science, 2005(286): 526-535.

Somasundaran P, Runkana V. Modeling Flocculation of Colloidal Mineral Suspensions Using Population Balances [J]. International Journal of Mineral Processing, 2003(72): 33-55.

Pandya J D, Spielman L A. Floc Breakage in Agitated suspensions: Effect of Agitation Rate [J]. Chemical Engineering Science, 1983(38): 1983-1992.

Spicer P T, Pratsinis S E. Coagulation and Fragmentation: Universal Steady-state Particle-size Distribution [J]. AIChE Journal, 1996(42): 1612-1620.

Zhang J J, Li X Y. Modeling Particle-size Distribution Dynamics in a Flocculation System [J]. AIChE Journal, 2003(49): 1870-1882.

Levich V G. Physiochemical Hydrodynamics [M]. Englewood Cliffs, EUA: Prentice-Hall, 1962.

Kusaka Y, Fukasawa T, Adachi Y. Cluster-cluster Aggregation Simulation in a Concentrated Suspension [J]. Journal of Colloid and Interface Science, 2011(363): 34-41.

Fang H W, Chen M H, Chen Z H. Surface Characteristics and Model of Environmental Sediment [M]. Beijing: Science Press, 2009.

Hounslow M J, Ryall R L, Marshall V R. A Discretized Population Balance for Nucleation, Growth and Aggregation [J]. AIChE Journal, 1988(34): 1821-1831.

Kusters K A, Pratsinis S E, Thoma S G, et al. Ultrasonic Fragmentation of Agglomerate Powders [J]. Chemical Engineering Science, 1993(48): 4119-4127.

Fang H W, Chen M H, Chen Z H. Analysis of Polluted Sediment Surface Pore Tension and Adsorption Characteristics [J]. Science in China (Series G), 2008, 51(8): 1022-1028.

Zhao H M, Fang H W, Chen M H. Floc Architecture of Bioflocculation Sediment by ESEM and CLSM [J]. Scanning, 2011(33): 1-9

Flood Characteristics of the TTs of the Yellow River in Inner Mongolia and Its Impact on Bed Evolution of both TTs and Main Stream

Wang Ping, *Hou Suzhen*, *Chen Hongwei* and *Chang Wenhua*

Yellow River Institute of Hydraulic Research, Zhengzhou, 450003, China

Abstract: Mainly based on the field data of Xiliugou gully, the flood characteristics of the Ten Tributaries (TTs) in Inner Mongolia reaches of the Yellow River (YR) are analyzed. The impacts of the flood on bed evolution of both TTs and main stream of the Yellow River are also explored. The results shows that the flood of ten tributaries features steep rise and steep drop, short duration, large flood peak, high sediment concentration and large sediment discharge, synchronization for the change of flow and sediment concentration. A separate flood presents a good flow-sediment concentration relation, different flood has different sediment concentration under the same discharge, with maximum around 1,500 kg/m^3. The high sediment concentration flood of the TTs has strong sediment transport capacity, when sediment concentration is more than 300 kg/m^3, the flow intensity will not increase any more, even tending to decrease. In addition, the flood has a distinct characteristic of silting floodplain and scouring channel, when the peak flow is more than 1,000 m^3/s, the riverbed can be scoured distinctly. The flood of the TTs may result in heavy siltation in the main stream of the Yellow River to form sand bar and raised water level.

Key words: TTs, flood characteristics, sediment transport capacity, scouring and depositing, sand bar

1 Overview

On the south bank of Inner Mongolia reaches, the Ten Tributaries (TTs) (Mongolian, i. e. torrential gullies) including Xiliugou gully are seasonal sediment load tributaries, which during the rainstorm season, are prone to form flood with high peak flow and sediment concentration, imposing great harm on the lower reaches and main streams of the Yellow River, mainly behaving in three aspects: Firstly, it may wash off the villages, houses, roads and other facilities in downstream, flood farmlands and grasslands and cause the deaths of people and livestock; Secondly, it may form a sand bar to block the Yellow River, resulting in difficulty for the water taking of Baotou Steel Corporation, thus resulting in production halts and huge economic losses, meanwhile, it may raise water level of the Yellow River, affecting the safety of flood control; Thirdly, it may aggravate channel siltation of the Yellow River, affecting flood and ice discharge. Zhi Junfeng and Shi Mingli (2002), Yang Zhenye (1984), Wu Sheng and Yu Linghong (2001) have made analysis of the course which the flood from Xiliugou gully blocks the main stream of the Yellow River, its harm and conditions for siltation and blocking. Liu Tao, Zhang Shifeng and Liu Suxia (2007) has made an analysis of the characteristic of sediment yield and transport of Xiliugou in the flood produced by cloudburst, and established a relation between flow and sediment produced by rainstorm. Yuan Jinliang (1995), from a long-time scale, has made an analysis of characteristic of runoff and sediment discharge of the TTs. In view of the short of field data and other factors, the existing researches scarcely mention the sediment transport capacity of TTs' flood and its characteristics of erosion and deposition.

There are a large number of research results on the high-sedimentation concentration floods which are of ubiquity in the middle and lower reaches of the Yellow River. For example, Wan Zhaohui and Shen Shoubai (1978) in his research on sediment transport capacity of high-sediment flood in the middle and lower reaches of the Yellow River reveals that high-sediment flood has strong sediment transport capacity, when sediment exceeding 200 kg/m³, may transport the sediment smoothly under weak flow intensity. Zhang Shengli and Meng Qinmei (1980) studies the characteristic of the high-sediment flood of Huangfuchuan river, proves it is Bingham fluid and points out that when bulk density of turbid water reaches a certain value or above, the critical velocity of erosion and deposition balance may depress with the increase of the density, which means that the energy demanded for the transportation of sediment will not increase with the increase of sediment concentration. There is difference for the high-sediment floods between the middle and lower reaches and the upper reaches of the Yellow River, however, these results, especially the research approach, are critical reference for the research on the erosion and deposition behavior of the TTs.

This paper, focusing on Xiliugou gully, uses the field flood data to have a research of the high-sediment flood characteristics and its impact on both TTs and the YR main stream's bed evolution.

2 Study area

The TTs runs from south toward north before entering into the Yellow River, for details of locations see Fig. 1. TTs have a length ranging from 65 km to 110 km, covering a total area of 11,000 km², with average channel gradient ranging from 2.67‰ to 5.25‰. The upstream region is the feldspathic sandstone hilly and gully area, suffering from serious soil erosion, is the main sediment source of the TTs. The ground surface is covered with sandstorm residual soil, made of rough grains of which the grains greater than 0.05 mm account for 60% of the total. In the middle it is the Kubuqi Desert. During monsoon, a large number of sand are blown by the wind to stack in the channel, thus be an important sand source. The downstream of the TTs is the fan-shaped alluvium, with relative even topography.

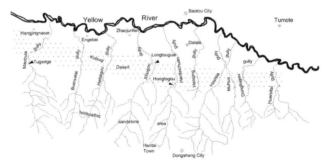

Fig. 1 Sketch of the TTs Basins in Inner Mongolia

Of the TTs, Xiliugou, Maobula and Hantaichuan each establish a hydrological station, they are respectively Longtouguai, Tugerige and Hongtagou. Xiliugou gully is 106.5 km long, featuring upper steep but lower gradual, as shown in Fig. 2. The feldspathic sandstone area is 43 km above from the estuary, with stream gradient more than 5 / 1,000. From Longtonguai (31 km from the estuary) to the area (43 km from the estuary) is the wind-blown sand accumulation area running through Kubuqi Desert, where the stream gradient decreases gradually, down to 2/1,000 when reaching Longtouguai, corresponding to it, the section of channel becomes narrow and deep. Below Longtouguai is the alluvial plain, with even topography, gradient gradually depressing down to 1/1,000, corresponding to it, the section of channel becomes gradually wide. Giving priority to rainstorm producing flow, as a result, the residual soil of sandstorm, gravel-stone in hilly and gully area in the upstream, as well as the wind-blown sand cumulated in the channel of middle reaches offer ample sand for the rainstorm to produce sediment. The large stream gradient causes strong sediment transport capacity and fast flow and sediment yield. Thus the flood usually contains high sediment, even as high as hundreds or thousands kilogram per cubic meter.

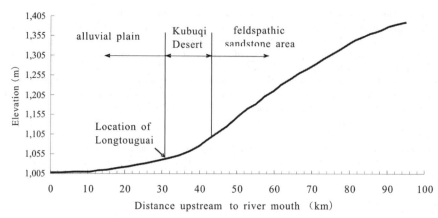

Fig. 2　River profile of the Xiliugou gully

3　Flood characteristics

3.1　Characteristics of high sediment concentration flood

The flood of the TTs is formed mainly by rainstorm. Due to sparse vegetation, poor ability of water and soil conservation, short channel, steep gradient and fast flow and sediment yield, the flood has the following characteristics:

3.1.1　Steep rise and steep drop, short duration

According to the statistic of 20 events of floods of Longtouguai Station from 1961 to 1989 (Tab. 1): The time difference from the beginning of rise to the flood peak ranges from 0.1 h to 4.9 h, averagely 1.4 h, flooding duration is 5 h to 35 h, averagely 18 h. Some floods rise especially quickly, for example, on July 31, 1961, Longtouguai Station at 17:59 has a fractional flow only of 11.6 m^3/ s, but at 18:00 soaring to 2,360 m^3/ s, increasing by 2,349 m^3/ s in one minute, at 18:12, the flow reaches its peak value of 3,180 m^3/ s, only taking 13 min from the beginning of rise to the flood peak (Fig. 3). Other tributaries have the same characteristics. For example, the flood at Tugerige station of Maobula (Fig. 4), on August 5, 1967, only takes 38 min from the beginning of rise (21:40, 0.04 m^3/s) to the appearance of flood peak (22:18, 5,600 m^3/ s), flooding duration is 15 h.

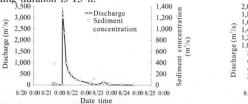

Fig. 3　Folld occurring at Xiliugou Longtouguai Station on August 21,1961

Fig. 4　Folld occuirring at Maobulang Guanchangjin Station on August 6,1967

3.1.2　High sediment concentration, large peak flow and large sediment discharge

The flood of the TTs has high sediment concentration, even a small flood (several hundreds of cubic meters per second) may have a sediment concentration up to hundreds or thousands kilogram per cubic meter (Tab. 1), e. g. floods occurring on August 31 to September 1, 1961, August 11 to

13, 1979, July 1, 1981 and September 9, 1988. Some typical floods have a peak flow up to thousands cubic meters per second, with sediment concentration of more than 1,000 kg/m³, large flood volume and large sediment discgarge. For example, floods occurring in Longtouguai Station on August 21, 1961, August 13, 1966 and July 21, 1989, of which, the flood occurring on July 21, 1989 (Fig. 5) had a peak flow of 6,940 m³/s, maximum sediment concentration up to 1,240 kg/m³, flood volume of 0.727,5 × 10⁸ m³, sediment discharge of 0.401,6 × 10⁸ t, all refreshing historic record. The flood occurring at Tugerige station at Maobula on July 21, 1989 (Fig. 6) had a peak flow up to 5,600 m³/s, maximum sediment concentration up to 1,500 kg/m³, flood volume 6.11 × 10⁶ m³, sediment load 6.69 × 10⁹ t, all refreshing historic record.

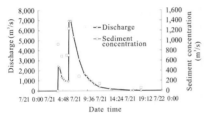 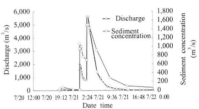

Fig. 5 Flood occurring at Xiliugou Longtouguai Station on July 21, 1989

Fig. 6 Flood Occurring at Maobulang on July 21, 1989

3.1.3 Basically synchronization for flow and sediment process

During flooding, sediment concentration increases or decreases with the flow, as shown in Fig. 3 to Fig. 6. The flood peak and the sediment peak appear mostly at the same time. According to the statistic of 20 floods in Tab. 1, there are 13 floods for which the flood peak and sediment peak appear with synchronization, the remaining 7 floods go inconsistently, of which 6 floods see time difference limited within 1 h, 1 flood sees time difference limited within 3 h.

3.1.4 Water and sediment quantity of floods have a high proportion compared to the whole year

For example, a single flood at Xiliugou has a flood volume accounting for 9% ~ 85% of yearly runoff, averagely 29%, sediment discgarge accounting for 9% ~ 94%, averagely 56%, taking a larger proportion. In some floods which have an outsize flood volume and sediment discharge, such a proportion will be larger, for example, Xiliugou floods occurring on "1961.8.21", "1966.8.13", "1976.8.2" and "1989.7.21" respectively, with flood volume and sediment load accounting for 63% and 89%, 54% and 94%, 51% and 81%, 85% and 85% of the whole year respectively.

3.2 Water-sediment relationship of high sediment concentration flood

If plotting different flood discharge and sediment concentration on a figure, it is easily found that sediment concentration tends to increase with the increase of flow, however, the sediment concentrations under identical discharge have a large difference (Fig. 7), for example, under discharge 50 m³/s, sediment concentration ranges from 30 kg/m³ to 600 kg/m³, with amplitude of 20 times, under discharge 100 m³/s, sediment concentration ranges from 50 kg/m³ to 800 kg/m³, with amplitude of 16 times. The cause relates both to the rainstorm distribution and the sediment production condition of underlaying surface, and to the different scouring topsoil volume at early stage and late stage of precipitation (Chan Ning, 1989). The field data of loess hilly and gully area in the middle Yellow River shows that owning to the impact of boundary condition of channel and sediment supply capacity of river basins, when flow increases to a certain value, the sediment concentration will not increase with the increase of flow, for example, at loess gully area, the small basin has a maximum sediment concentration of 1,000 kg/m³ (Fei Xiangjun and Shao Xuejun, 2004), while Huangfuchuan River, Gushanchuan River, Kuye River and so on, the tributaries in the middle Yellow River, have a maximum sediment concentration ranging from 1,400 kg/m³ to 1,600 kg/m³ (Chan Ning, 1989). As viewed from Fig. 7, when the flow reaches 300 m³/s, the sediment concentration may reach 1,500 kg/m³, thereafter will not increase with the increase of flow, namely that the floods of the TTs have a maximum sediment concentration of 1,500 kg/m³.

Tab. 1　Separate flood at Xiliugou Longtouguai Station and yearly water-sediment eigenvalue

Year	Date (month, day)	Duration (h)	Flood peak (m³/s)	Maximum sediment concentration (kg/m³)	Flood volume (×10⁶m³)	Sediment discharge (×10⁶t)	Annual runoff (m×10⁶m³)	Annual Sediment discgarge (×10⁶t)	Proportion to the whole year(%)	
									Flood volume	Sediment discharge
1961	7.30	29.9	1,330	447	5.56	2.95	93.07	33.17	6	9
1961	8.21~22	28.0	3,180	1,200	58.42	29.68	93.07	33.17	63	89
1966	8.13~14	22.0	3,660	1,380	22.46	16.51	41.52	17.55	54	94
1971	8.31~9.1	12.3	602	1,420	3.56	2.17	19.96	2.44	18	89
1973	7.9~10	14.3	640	563	6.77	1.92	44.30	13.13	15	15
1973	7.17~18	10.8	3,620	1,550	13.72	10.65	44.30	13.13	31	81
1976	7.28~7.29	20.0	604	194	7.61	0.86	78.21	8.98	10	10
1976	8.2~8.3	35.5	1,330	371	39.66	7.28	78.21	8.98	51	81
1978	8.12~8.13	10.9	722	404	11.02	2.33	45.88	6.38	24	37
1978	8.30~31	19.0	618	342	13.45	2.92	45.88	6.38	29	46
1979	7.26~27	21.0	342	775	6.55	1.11	39.86	4.54	16	24
1979	8.11~13	12.5	701	1,150	7.36	2.91	39.86	4.54	18	64
1981	7.1	6.7	884	1,370	3.96	2.43	26.58	4.94	15	49
1981	7.26~7.27	14.5	312	955	7.11	2.09	26.58	4.94	27	42
1982	9.16~9.17	14.0	449	1,320	5.80	2.78	20.79	3.18	28	87
1984	8.9~8.10	21.7	660	651	9.24	3.24	25.02	4.36	37	74
1985	8.24~8.25	24.4	547	294	6.93	0.98	20.83	1.58	33	62
1988	7.20~7.21	18.0	609	631	1.69	0.52	18.73	3.85	9	14
1988	9.9	5.1	531	1,290	2.85	2.46	18.73	3.85	15	64
1989	7.21	14.1	6940	1,240	72.75	40.16	85.62	47.49	85	85

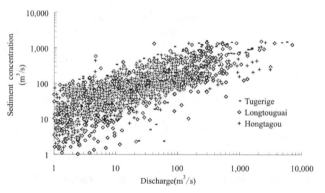

Fig. 7　Relationship between flood discharge and sediment concentration

As for a single flood, it still has a good flow-sediment concentration relationship. It is obvious that the sediment concentration increases with the flow, the difference of sediment concentration under identical flow is small (Fig. 8). The water-sediment relationship of high sediment concentration flood also finds expression in the relation between peak flow and flood volume and sediment discharge (Fig. 9). Obviously, the more the peak flow is, the larger the flood volume and sediment discharge are, it is especially distinct when peak flow is greater than 1,000 m³/s.

3.3　Flow characteristic of high sediment concentration flood

　　The ground surface particle in the feldspathic sandstone area is coarse, resulting in the coarse sediment carried by the flood. Sampling the sediment in suspension of Xiliugou on July 31, 2010 (the discharge and sediment concentration at Longtouguai station only being 31 m³/s and 25

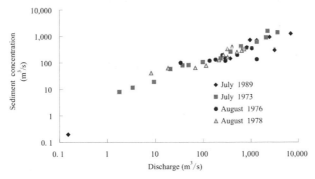

Fig. 8 Flow and sediment concentration relation of typical floods of Xiliugou

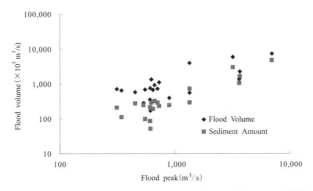

Fig. 9 Relationship between flood volume and sediment discarge of Xiliugou Gully

kg/m³) for measurement, the results show that median particle diameter is around 0. 05 mm (Fig. 10), distinctly greater than the median particle diameter at Bayangaole station located at the main stream of the Yellow River. For the reason both the discharge and sediment concentration are very small, such median particle diameter is slight fine for a big flood. Compared to the median particle diameter of suspended sediment under different flow and sediment concentration of flood of Huangpuchuan River, it is observed that when flow and sediment concentration are small, the median particle diameter is around 0. 033 mm, when reaching 2,070 m³/s and 1,150 kg/m³ respectively, median particle diameter may get to 0. 33 mm. Since both Huangpuchuan and the TTs originate from feldspathic sandstone area, it is believed that their suspended sediment grains have the similarity, thereby we can extrapolate that the sediment grain size of the TTs is very coarse. When the fine sediment concentration in water flow reaches to a certain value, it may change the fluid property of flow, turning from Newton fluid to Bingham body, in this way, the cohesive fine sediment in flow forms some structure to support the suspension of sediment grain not to silt such that hyper-concentration flow has a large sediment transport capacity (Chan Ning, 1989). The sediment of the TTs flood is coarse, but its sediment concentration is very high, fine sediment concentration can reach a certain value such that hyper-concentration flow of the TTs has the characteristic of non-Newton fluid. In a view of relation between sediment transport factor $U^3/(gh)$ and sediment concentration of Xiuliugou flood (Fig. 11), $U^3/(gh)$ increases with the increase of sediment concentration, when sediment concentration exceeding 300 kg/m³, it tends to decrease with the increase of sediment concentration, which conforms to the relationship of hyper-concentration flow sediment concentration and $U^3/(gh\omega)$ in the tributaries in middle Yellow River (Wan Zhaohui and Shen Shoubai, 1978). It shows that when sediment concentration of Xiliugou flood reaches a certain value, the required flow intensity will not increase any more, even tending to decrease.

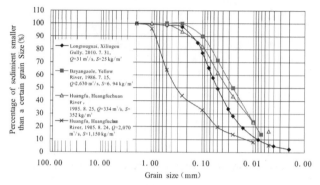

Fig. 10   Grain size composition of suspended load for different rivers

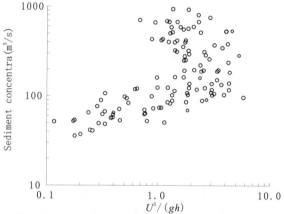

Fig. 11   Relationship between sediment concentration and sediment transport capacity factor $U^3/(gh)$ for Xiliugou flood

3.4   Erosion and deposition characteristics of flood

When a high sediment concentration flood results in an intensive scouring over the riverbed, it may carry a large number of sediment and run toward downstream. On the one hand, the sediment transport capacity of the flow gradually becomes saturation, even super saturation, the scouring of riverbed decreases gradually; on the other hand, bed slope gradually flattens out, when the flow runs over floodplain, the floodplain may suffer from siltation, while the main channel will be not silted or even be scoured. Thus the stream channel, especially the deep channel becomes narrow and deeper. For example, the flood occurring on September 9, 1988 has a peak flow of 531 m³/s, with maximum sediment concentration of 1,290 kg/m³, after the flooding, the river width where the deep channel locates narrows from 132 m to 84 m, channel depth increases from 0.55 m to 1.21 m, the average deposition thickness of floodplain is 0.42 m. For details of the change of channel morphology see Fig. 12.

In view of the relationship between channel scouring and the discharge during flooding (Fig. 13), with the increase of peak flow, the scouring of Longtouguai cross-section distinctly tends to increase, when peak flow exceeding 1,000 m³/s, riverbed may be scoured strongly.

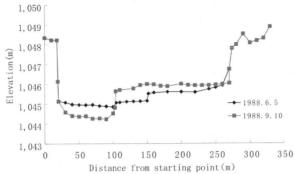

Fig. 12   Xiliugou Longtouguai cross-section before and after the flood occurring on September 9, 1988

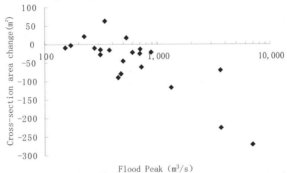

Fig. 13   Relationship between Longtouguai peak flow and riverbed scouring

4   Impact on channel of main stream

As the gradient of river reaches where the TTs enters into the Yellow River is only around 1/10,000, its sediment transport capacity is very low, as a result, the large number of sediment carried by the flood from the TTs usually silt up rapidly at the conjunction between main stream and tributaries of the Yellow River to form a sand bar, even blocking the Yellow River at a serious situation. Fig. 14 shows the sectional change of Zhaojunfen at the main stream of the Yellow River, about 500 m away from the upstream of conjunction reaches during the flood occurring in Xiliugou in August 1966 (Longtouguai station, peak flow of 3,660 m³/s, maximum sediment concentration of 1,380 kg/m³), it is observed that in several hours, such channel of the main stream is blocked and silted up, the floodplain also becomes higher, with maximum deposition thickness up to 7 m. The siltation not only expands horizontally, but has a large covering scope longitudinally along the river bed, thus forming the so-called sandbar, for example, in the high-sediment flood occurring in Xiliugou on July 21, 1989 which blocks the river, a sand bar as high as 2 m to 4m is formed, having a length of 600 m to 1,000 m, extending a 7 km scope from upstream to downstream.

During the formation of sand bar, the upstream water level of confluence area becomes higher. Fig. 15 shows the change of water level and discharge of the main stream during the Xiliugou Flood occurring on July 21, 1989, it is observed that after the flood enters into the Yellow River, due to the jacking of flood and sediment accumulation, the water level of the main stream rises rapidly and flow drops quickly, after the flood subsidence, the water level of the main stream still remains at a high level due to the existence of sand bar. Tab. 2 gives the statistics of several historical typical blocking floods, raising value of water level and scouring time of sand bar. In a major blocking event, e. g. , "1961. 8. 21", "1966. 8. 13"and "1989. 7. 21", all the raised water levels of the main stream exceed 2 m, with maximum up to 2. 42 m. Obviously, the raising degree relates to sediment discharge of flood, namely relating to the size of sand bar. For the reason riverbed is silted up to a great extent, sediment is too coarse to be scoured, and the small flow in Inner Mongolia reaches, slow gradient, the course for the scouring and disappearance of the sand bar is of slowness. In Tab. 2, all the scouring process usually take 4 d to 25 d. The sand bar formed by the flood occurring on July 21, 1989 experienced a longest scouring, up to 25 d, for details see Fig. 16.

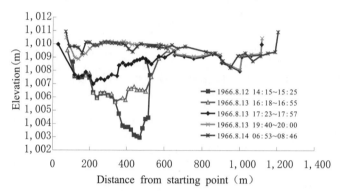

Fig. 14   Changes of Zhaojunfen cross-section during the Xiliugou flood in August 1966

Tab. 2   Siltation and blocking characteristic value in typical blocking years

Flooding date (y. m. d)	Characteristics of Xiliugou water and sediment				Water level Amplitude of Zhaojun fen (m)	Number of Days to restore normal water level (d)
	Flood peak (m³/s)	Maximum sediment concentration (kg/m³)	Flood volume (×10⁶m³)	Sediment discharge (×10⁶t)		
1961. 8. 21	3,180	1,200	58. 42	29. 68	2. 42	12
1966. 8. 13	3,660	1,380	22. 46	16. 51	2. 38	20
1976. 8. 2	1,330	383	39. 66	7. 28	1. 98	8
1984. 8. 9	660	651	9. 24	3. 24	0. 58	4
1989. 7. 21	6,940	1,240	72. 75	40. 16	2. 18	25

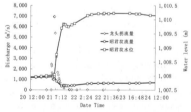

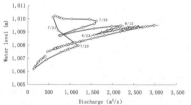

Fig. 15   Water level and flow of Zhaojunfen during "1989. 7. 21"flood

Fig. 16   Relationship between water level and flow of Zhaojunfen before and after the "1989. 7. 21"flood

5   Conclusions

(1) The flood of the TTs features steep rise and steep drop, short duration, large peak flow, high sediment concentration and large sediment load, synchronization for the change of flow and sediment concentration.

(2) A separate flood presents a good flow-sediment concentration relation; different flood has different sediment concentration under identical flow, with maximum around 1,500 kg/m³,

(3) The sediment concentration flood of the TTs has strong sediment transport capacity, when sediment concentration is more than 300 kg/m³, the required flow intensity will not increase any more, even tending to decrease.

(4) The flood has a distinct characteristic of silting floodplains and scouring channel, when peak flow is more than 1,000 m³/s, riverbed can be scoured distinctly. The flood may result in heavy siltation in main stream of the YR to form sand bar and raise water level.

Acknowledgements

This study was funded by the National Twelfth Five-Year Scientific and Technological Support Plan Subject (No. 2012BAB02B03) and the Basic Research Fund for Central Public-interest Scientific Institution(HKY – JBYW –2010 – 10).

References

Zhi Junfeng, Shi Mingli. Analysis of Sediment Blocking Yellow River During Ten Tributaries Floods Occurring in "1989.7.21" [C]// Research of Water and Sediment Change of Yellow River. Zhengzhou: Yellow River Conservancy Press, 2002.

Yang Zhenye. Analysis of Sedimentation Blocking Yellow River at Zhaojunfen reaches of Inner Mongolia in 1961,1966 [J]. Yellow River, 1984 (6) :15 – 19.

Wu Sheng, Yu Linghong. Harm of Xiliugou Flood over Baotou Steel Corporation and Countermeasures[J]. Science & Technology of Baotou Steel Corporation, 2001,27 (Suppl.): 159 – 161.

Liu Tao, Zhang Shifeng, Liu Suxia. Preliminary Study on the Relationship between Sediment Production and Rransport of Ten Tributaries Storm Floods[J]. Water Resources and Water Engineering, 2007, 18 (3):18 – 21.

Yuan Jinliang. Hydrology and Sediment Characteristics of Erdos Plateau[J]. Inner Mongolia Water Resources, 1995 (1):39 – 42.

Wan Zhaohui, Shen Shoubai. High – concentration Sediment Transport Phenomena Occurring in Main Streams and Tributaries of the Yellow River[R]// Research Report of the Yellow River Sediment [R]. Zhengzhou: Yellow River Institute of , 1978.

Zhang Shengli, Meng Qinmei. Preliminary Analysis of Huangfuchuan High – sediment Flow Characteristics [J]. Yellow River, 1980 (3):44 – 49.

Chan Ning. Hyperconcentrated flow motion [M]. Beijing: Tsinghua University Press, 1989.

Fei Xiangjun, Shao Xuejun. Calculation Method of Sediment Transport Capacity of Trenches at Sediment Source Zone[J]. Journal of Sediment Research, 2004 (2) :2 – 8.

Study on Runoff and Sediment Process Variation in the Lower Yellow River

Li Wenwen[1], *Fu Xudong*[1], *Wu Wenqiang*[2] and *Wu Baosheng*[1]

1. State Key Laboratory of Hydroscience and Engineering, Tsinghua University,
Beijing, 100084, China
2. China Institute of Water Resources and Hydropower Research, Beijing, 100084, China

Abstract: There is the temporal and spatial variability in river hydrology sequence. The trends change in the overall distribution of river hydrology sample which has significantly different in both before and after. The grasp of the trends of runoff and sediment and the cause of the their changes is the key to reasonably use and develop water resources and regulate runoff and sediment in lower Yellow River. The runoff and suspended sediment load were trend detection and mutation detection by the Mann – Kendall trend test and Pettit change point detection method at five hydrological stations in the lower Yellow River. The results show that first change point of the annual runoff and suspended sediment load appeared in 1985 in the lower Yellow River. Variation caused by climate change and the construction of Longyangxia Dam. Second change points were not exactly at the same year in the annual runoff and suspended sediment load. Second change point of the annual suspended sediment load appeared in 1996 in the lower Yellow River. The change year of annual suspended sediment load was later than the annual runoff load. The construction and operation of Sanmenxia, Liujiaxia, Xiaolangdi reservoir is a major factor of the second change points.
Key words: lower Yellow River, runoff and suspended sediment load, trend test, change detection

1 Introduction

The changes of runoff and sediment load caused by the hydrological cycle and the natural conditions of dramatic change and watershed human activities. The impact of dramatic changes of the hydrological cycle and the natural conditions and the watershed of human activities often leads to changes in Watershed Runoff and sediment, and that may mutate runoff sequence and show the phase characteristics. These reflect the significant changes of statistical characteristics of value. The Yellow River is the largest river with maximum sediment concentration in the world. It has a characteristic with less water and more sand and water and sediment coming from different area. Yellow River Basin was affected by human activities severely. Since 1950, the runoff and sediment load show decreasing trend in the lower Yellow River. From 1950 to 2000, the annual runoff load decreased from 4.97×10^{10} m^3 to 2.30×10^{10} m^3 with the proportion of 46%. The annual suspended sediment load also has the same trend as annual runoff. From 1950 to 2000, the suspended sediment load decreased from 1.819×10^9 t to 0.64×10^8 t with the proportion of 3.5%. In order to understand the law and the change mechanism of runoff and sediment load in the lower Yellow River, one of the study key steps is to understand the eigenvalue of the trends and characterization point of mutations of water and sediment sequence

There has been carried out a lot of variation of characteristics study of runoff and sediment load in the middle reaches of the Yellow River. Gao Peng(2010) analyzed the trends and change point of runoff and sediment transport of the past 60 years in the middle Yellow River, and discussed the driving factors. They found that water mutations occurred in 1985 and sediment change occurred significantly in 1981 in the middle Yellow River. Mu Xingming (2007) analyzed the change of runoff and sediment load of the 1952 ~ 2000 year period from Hekou to Longgmen with change point analysis, the duration curve method and the cumulative curve. The results show that runoff and sediment discharge were significantly reduced trend from 1971 and 1979. Yao Wenyi (2009) studied the trends of runoff load under climate change in the Yellow River. The results show that the Yellow River source region will likely continued warming trend in the next few years, and the magnitude of changes has smaller in precipitation, which is likely to deteriorate the shortage trend of water resources. Hu Chunhong(2003) studied the variation of runoff and sediment load and the reason of Yellow River estuary. They proposed management direction of Yellow River estuary.

Summary of previous studies, there have many study about mutation analysis of runoff and sediment load in the middle reaches of the Yellow River and the research results are nearly the same. It was affected by human activities seriously and has complicated change of runoff and sediment load in the lower Yellow River. There have few studies about mutation analysis of runoff and sediment load in the lower Yellow River. Therefore, runoff and sediment trends and characteristics of analysis are needed in the lower Yellow River. This paper test the trend of runoff and sediment load by Mann – Kendall nonparametric method and detect the change point by Pettitt non – parametric method at five hydrological station include Huayuankou, Gaocun, Sunkou, Aishan and Lijin station in the lower Yellow River. We look forward to find the variation point of runoff and sediment load in the lower Yellow River.

2 Data and methods

2.1 The selection of typical section and data

In order to study the variation of the law of runoff and suspended sediment load in the lower Yellow River, according to the different characteristics of the river, we select five typical hydrometric stations include Huayuankou, Gaocun, Sunkou, Aishan and Lijin station to be study. The data of Sunkou hydrological station was from 1964 to 2007, and that of the rest hydrological stations were all from 1951 to 2007.

2.2 Methods

The statistical test method is divided into parametric statistical test methods and non-parametric statistical test methods. The advantage of non-parametric tests is not to need to presuppose the sample to comply with a certain distribution, and do not affect by the interference of a few outliers. Two methods of non-parametric tests (Mann – Kendall test method, the Pettitt test method) were used in this paper to test the trend and the chang point seperately.

2.2.1 Trend analysis of Mann – Kendall method

The characteristics of Mann – Kendall(man H B, 1945; kendall M G,1975) test method is not to need assumption that the distribution characteristics of data, with a wide detection range and the less human interference. It was used to analyze trend of the data about runoff, precipitation, sediment, temperature and other hydrometeorological trends. Test statistic Z of Mann – Kendall is calculated as

$$Z = \begin{cases} \dfrac{S-1}{\sqrt{\mathrm{var}(S)}} & S > 0 \\ 0 & S = 0 \\ \dfrac{S+1}{\sqrt{\mathrm{var}(S)}} & S < 0 \end{cases} \qquad (1)$$

among

$$S = \sum_{k=1}^{N-1} \sum_{j=k+1}^{N} \mathrm{sgn}(x_j - x_k) \qquad (2)$$

$$\text{sgn}(x_j - x_k) = \begin{cases} 1 & x_j - x_k > 0 \\ 0 & x_j - x_k = 0 \\ -1 & x_j - x_k < 0 \end{cases} \qquad (3)$$

where, x_j and x_k represents the observed at j and k year, and $k > j$, respectively N represents the number of samples.

Using the above test statistic Z for sequence analysis, in fact, the test complied with the following assumptions. Null hypothesis H_0: no significant change in trend sequence. When $|Z| > Z_{(1-\alpha/2)}$, we reject the null hypothesis H_0, which means there are significant changes in trends. $Z_{(1-\alpha/2)}$ is the statistics value with the standard normal distribution value is $1 - \alpha/2$, when significant level is α. When the Z is positive, it indicates that the sequence has increasing trend. On the contrary, When the Z is negative value, it indicates that the sequence has downward trend.

2.2.2 Change point detection of Pettit method

Change point detection of hydrological sequence is one of the statistical methods that hydrological sequence response to climate change and human activities. Change point detection of Pettit method is based on mutations of a sequence of non-parametric testing. It was able to identify the sequence distribution of mutations and explicitly change the time. This method is based on Mann – Whitney statistic $U_{t,N}$ to test the two samples ($x_1, \cdots, x_t$ and $x_{t+1}, \cdots, x_N$) which have the same distribution. For the continuous sequence, $U_{t,N}$ is calculated as

$$U_{t,N} = U_{t-1,N} + \sum_{j=1}^{N} \text{sgn}(x_t - x_j) \qquad t = 2, \cdots, N \qquad (4)$$

and

$$\text{sgn}(x_t - x_j) = \begin{cases} 1 & x_t - x_j > 0 \\ 0 & x_t - x_j = 0 \\ -1 & x_t - x_j < 0 \end{cases} \qquad (5)$$

The number of test statistic calculated of the first sample sequence value is more than the second sample sequence value. A change point does not exist in the zero hypothesis of Pettit method. Statistics K_N represent the most significant change point t at the maximum $|U_{t,N}|$; its formula and associated probability (P) test of significance formula are as follows:

$$K_N = \max_{1 \leq t \leq N} |U_{t,N}| \qquad (6)$$
$$P \cong 2\exp[-6(K_N)^2 / (N^3 + N^2)] \qquad (7)$$

Due to the complexity of climate change and human activities, the number of change points of a hydrological sequence may be more than one. As Pettit examine method can only identify one possible change point, the method should be used for many times. In this paper, all the change points can be identified by these following steps: ①first change point can be identified using Pettit Examine Method; ②according to the first change point, the original sequence can be divided into two, the two samples can be calculated using Pettit examine method respectively. If there is no change point in the sequence, all possible hydrological change points have been identified. If there also is change point in the sequence, repeat the above steps.

3 Results

3.1 The trend analysis of the runoff and sediment load

The annual runoff and suspended sediment load were tested by Mann – Kendall method. The results are shown in Tab. 1. It can be seen from the Tab. 1 that statistics of annual runoff and suspended sediment load $|Z|$ are all greater than 2.58 at five stations in the lower Yellow River. That is statistics Z are all greater than 0.01 confidence level, and the statistics Z are all negative. It is shown that the annual runoff and suspended sediment load are all decreased trend at all station over time in the lower Yellow River.

Tab. 1　the trend test results of the annual runoff and suspended sediment load in the lower Yellow River

	otation	Huayuankou	Gaocun	Sunkou	Aishan	Lijin
Mann – Kendall method	Z-test statistics	annual runoff load	-5.369	-5.631	-4.825	-5.714	-6.14
		annual suspended sediment load	-5.273	-6.099	-5.947	-5.672	-5.948
	significance level α	0.01	0.01	0.01	0.01	0.01

3.2　Change point analysis of runoff and sediment load

The first change point of runoff and sediment discharge sequence is calculated using Pettitt examine method in Huayuankou, Gaocun, Sunkou, Aishan and Lijin stations. According to the first change point, hydrological sequence is divided into two samples, and the two samples can be calculated using Pettitt examine method respectively, until there is no change point. The Pettitt's test results of runoff and sediment discharge sequence in Huayuankou and Aishan stations is shown in Fig. 1. The change points in Huayuankou, Gaocun, Sunkou, Aishan and Lijin stations of runoff and sediment discharge sequence are shown in Tab. 2. Tab. 2 shows that the transition point of runoff is in 1985, the variable point of sediment load in Huayuankou station is in 1981, and the other stations are in 1985 with the significance level of 0.001. This is consistent with the past analysis results(Gao Peng, et al. , 2010).

The second change points are tested after the first change points are tested. It can be seen from Tab. 2, the change points are in 1968, from 1951 to 1985, in Gaocun, Aishan and Lijin stations, with a significant level of 0.1, while the Huayuankou and Sunkou stations are not significant. In the years 1986 ~ 2007, the change points are in 1994, with a significant level of 0.1. But Aishan and Lijin station are not significant.

The change points of runoff in Huanyuankou, Gaocun, Sunkou, Aishan and Lijin stations are different from that of sediment. In the years of 1951 ~ 1985 (1981), to meet the significant level of 0.1, there are no change points in 5 stations. But in the yerar of 1986 (1982) ~2007, the change points are in 1996 of five stations with the significant level of 0.1.

Therefore, the first change points in the lower reaches of Yellow River of runoff and sediment sequence are in 1985, but the other change points of runoff and sediment sequence are different. The change point of sediment is later than that of runoff and the mutations regularity of sediment load series is better.

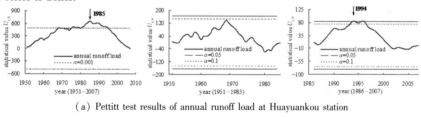

(a) Pettitt test results of annual runoff load at Huayuankou station

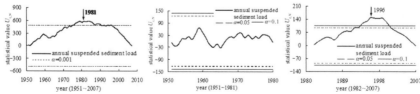

(b) Pettitt test results of annual sediment load at Huayuankou station

Fig. 1　Pettitt test results of annual runoff and sediment load at Huanyuankou and Aishan station

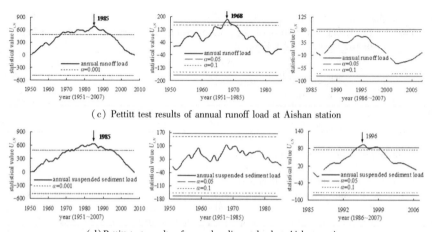

(c) Pettitt test results of annual runoff load at Aishan station

(d) Pettitt test results of annual sediment load at Aishan station

Continued to Fig. 1

Tab. 2   Pettitt test results of annual runoff and sediment load at five stations

	Station	Huayuankou		Gaocun		Sunkou		Aishan		Lijin	
Annual Runoff load pettitt test	First Transition point	1985 (1951~2007)		1985 (1951~2007)		1985 (1964~2007)		1985 (1951~2007)		1985 (1951~2007)	
	$\alpha = 0.001$	*		*		*		*		*	
	Second Transition point	1968 (1951~1985)	1994 (1986~2007)	1968 (1951~1985)	1994 (1986~2007)	1968 (1964~1985)	1995 (1986~2007)	1968 (1951~1985)	1994 (1986~2007)	1968 (1951~1985)	1994 (1986~2007)
	$\alpha = 0.1$	not significant	*	*	*	not significant	*	*	not significant	*	not significant
Annual sediment load Pettitt test	First transition point	1981 (1951~2007)		1985 (1951~2007)		1985 (1964~2007)		1985 (1951~2007)		1985 (1951~2007)	
	$\alpha = 0.001$	*		*		*		*		*	
	Second Transition point	1959 (1951~1981)	1996 (1982~2007)	1959 (1951~1985)	1996 (1986~2007)	1971 (1964~1985)	1996 (1986~2007)	1968 (1951~1985)	1996 (1986~2007)	1968 (1951~1985)	1996 (1986~2007)
	$\alpha = 0.1$	not significant	*	not significant	*	not significant	*	not significant	*	not significant	*

Note: * indicates significant.

4   Hydrological mutations cause analysis

According to the above study results, there has the obvious transition point of annual runoff and sediment load. In theory, the reasons of transition point are climate change and human activities. In addition to climatic factors, the Yellow River is affected by modern human activities, dry tributary water conservancy construction, the basin water resources development and utilization of water resources in a variety of soil and water conservation activities. These has a greater impact on water and sediment process in the middle and lower reach of the Yellow River. Tab. 2 gives the theoretical transition point in each site. This paper gives the reasons for runoff and sediment load at five stations from the climate change and human activities.

4. 1　Impacts of climate change

There are the same trends between climate change and runoff and sediment load in the lower Yellow River. The temperature increased in recent years and precipitation decreased in the Yellow River basin. After studying the climate in China and the northern hemisphere many scholars concluded that there had happened three temperatures mutation in the 1920s, the mid – 1960 s and early 1980 s(Wei Fengying,et al. ,1995；Tang Maocang,et al. ,1998). Gao Peng(2010) had test the mutation detection with a significance level of 0. 05 in the Lower Yellow River, and found that transition point was appeared in 1993. From 1960 to 2007, the rainfall of most area in Yellow River shows a downward trend, and significant variation occurs mainly area south of latitude 38 degrees in middle and lower reaches of the Yellow River area south of latitude 38 degrees. Variation occurs mainly in the 1963 ~ 1998 year period, and variation of time was earlier in the Lower Yellow River than the upper and middle reaches of the Yellow River(Liu Qiang,et al. ,2008). Zhao Fangfang (Zhao Fangfang,et al. ,2007) had test the mutation detection about the annual average temperature and precipitation, and they found that transition point of the annual average temperature and precipitation were all appeared in mid-1980s. In this paper the first transition point of runoff and sediment load was appeared in 1985 at five stations in the lower Yellow River. It have consistently with previous studies. Therefore, climate change of the entire basin has an important influence on hydrological variability in the downstream.

4. 2　Impact of human activities

The impact of human activities in the Yellow River basin is very intense. Including population growth, industrial and agricultural development, urban construction and other human activity that altered the underlying surface runoff and sediment yield in the conditions. Dams, water and soil conservation measures and irrigation changed the hydrological cycle and water spatial and temporal distribution.

4. 2. 1　First transition point Cause Analysis

From the overall situation, the first change point of the annual runoff and suspended sediment load appear in 1985 at five stations in the lower Yellow River. There are two main reasons：

(1) Soil and water conservation activities are the main influencing factors for variable point. It were began to control the soil erosion in the 1950s(Xiao Pengqing, et al. ,2008) in the Loess Plateau. Since 1960s, soil and water conservation began to change from disorder treatment to the overall planning and comprehensive management in the Loess Plateau. Sluicing dam, machine repair terraces and aerial seeding grass has made significant progress. At the end of 1970s, soil erosion on sandy coarse sediment producing areas, and noted that the sediment is mainly produced in the valley, and determined the important measures of reducing the sediment with strenthening the construction of dam and reservoir in the middle reach of Yellow River. So far, all measures are the early stage of implementation of water conservation measures. 1980 ~ 1985 is a comprehensive management stage based on a small watershed. This stage is a significant role with water conservation measures play.

(2) The main stream of large – scale water conservancy project construction and operation is another major factor for transition point produced. From 1961, many dams have been built in main stream in the upper reaches of the Yellow River, such as Yanguoxia (It started in september 1958 and been completed in 1970) , Qingtongxia (It was built in 1958, and began filling with water in 1967, and all generating units installed in 1978) , Liujiaxia (It started in september 1958 and been completed in 1974) , Longyangxia (It was closure in 1979, and began impoundment in 1986) and so on and other irrigations measures. The runoff and sediment load in main stream were began to substantial adjustments in the Yellow River. After Longyangxia reservoir was put into operation, the runoff load of the upper Yellow River decreased 54%. First transition point was appearing in 1985 in a variety of factors.

4.2.2 Second transition point cause analysis

The various water conservancy project construction and operation lead to the second variable points, but we found that the second variable points are different between annual runoff load and annual sediment load. Sanmenxia reservoir was completed in September 1960, and the reservoir area had a large number deposition with operation mode of storing water and retaining sediment. It entered another operation stages of flood detention and silt sluicing and in april 1962. It was the first alterations for the implementation Sanmenxia hub from 1965 to 1968. The reservoir area below Tongguan has been changed from deposition into erosion. At the same time Liujiaxia reservoir was put into operation in October 1968, and the distribution of the runoff load was changed in Xiaobeiganliu of the Yellow River. The runoff load of flood season decreased. So the annual runoff load has the second transition point that appeard in 1986 at Gaocun, Aishan and Lijin in the Lower Yellow River. The main project of Xiaolangdi project started in september 1994, and the river was closure on october 1997. Lijiaxia reservoir started its water impoundment on december 1996. With the combined action the transition point of the runoff load was appeared in 1994, and the transition point of the sediment load was appeared in 1996 in the lower Yellow River.

5 Conclusions

In this paper, the runoff and suspended sediment load are trend detection and mutation detection by the Mann – Kendall trend test and Pettit change point detection method at five hydrological stations include Huayuankou, Gaocun, Sunkou, Aishan and Lijin stations in the lower Yellow River. By analyzing the results we get the following understanding.

(1) The first transition point of the runoff and sediment load were all appeared in 1985 with a significance level of 0.05 in the lower Yellow River. This is consistent with previous findings of the upper and middle reaches of Yellow River. The main variation reasons are climate change and Longyangxia and other water conservancy construction and operation.

(2) The main reason of second transition point is the result that various water conservancy project constructed and operated in the main stream. This study found that the second transition point were different between annual runoff load and annual sediment load in the lower Yellow River. Because Sanmenxia and Liujiaxia reservoir were built and operated, the second transition point of the annual load was appeared in 1968 in the lower Yellow River. With the combined action of Xiaolangdi water control project starts and Lijiaxia reservoir impoundment the second transition point of the runoff load was appeared in 1994, and the second transition point of the sediment load was appeared in 1996 in the lower Yellow River.

Acknowledgements

This study was sponsored by the National Natural Science Foundation of China (grant no. 51109116) and was supported by the Key Project of Chinese National Programs for Fundamental Research and Development (973 program) (grant no. 2011CB409901).

References

Gao Peng, Zhang Xunchang, Mu Xingmin, et al. Trend and Change – point Analyses of Streamflow and Sediment Discharge in Yellow River Mainstream During 1950 ~ 2005 [J]. Hydrological Sciences Journal, 2010, 55(2): 275 – 285.

Mu Xingmin, Ba Sangchille, Zhang Lu, et al. Impact of Soil Conservation Measures on Runoff and Sediment in Hekou—Longmen Region of the Yellow River [J]. Journal of Sediment Research, 2007 (2): 36 – 41.

YaoWenyi, Xu Zongxue, Wang Yunzhang. Analysis of Runoff Variation in Yellow River Basin on the Background of Climate Change [J]. Meteorological and Environmental Sciences, 2009, 32 (2): 1 – 6.

Hu Chunhong, Cao Wenhong. Variation, Regulation and Control of Flow and Sediment in the Yellow River Estuary II: Regulation Countermeasures [J]. Journal of Sediment Research, 2003(5): 9 – 14.

Mann H B. Non – parametric Tests Against Trend[J]. Econometric, 1945, 13(3): 245 – 259.

Kendall M G. Rank Correlation Measures[M]. Charles Griffin, London, UK, 1975.

Pettitt A N. A Non – parametric Approach to the Change – point Problem[J]. Applied Statistics, 1979, 28(2): 126 – 135.

Pettitt A N. Some Results on Estimating a Change – point Using Nonparametric Type Statistics[J]. Journal of Statistical Computation and Simulation, 1980, 11: 261 – 272.

Wei Fengying, Cao Hongxing. Detection of Abrupt Changes and Trend Prediction of the Air Temperature in China, the Northern Hemisphere and the Globe [J]. Scientia Atmospherica Sinica, 1995, 19(2): 140 – 148.

Tang Maocang, Bai Chongyuan, Feng Song, et al. Climate Abrupt Changein the Qinghai – xizang Plateau in Recent Century and Its Relation to Astronomical Factors [J]. Plateau Meteorology, 1998, 17(3): 250 – 257.

Gao Peng. Streamflow and Sediment Discharge Change Trend and Its Response to Human Activities in the Middle Reaches of the Yellow River [D]. Institute of Soil and Water Conservation, CAS & MWR. Xian, 2010.

Liu Qiang, Yang Zhifeng, Cui Baoshan. Spatial and Temporal Variability of Annual Precipitation During 1961 – 2006 in Yellow River Basin[J]. Journal of Hydrology, 2008, 361: 330 – 338.

Zhao Fangfang, Xu Zongxue, Huang Junxiong. Long – term Trend and Abrupt Change for Major Climate Variables in the Upper Yellow River Basin[J]. Acta Meteorologica Sinica, 2007, 21(2): 204 – 214.

Xiao Peiqing, Wang Wenshan. Study Strategy of Soil and Water Conservation in the Loess Plateau Based on Keeping Healthy Life of the Yellow River [C]// Soil and Water Conservation Development Strategy – Soil Erosion and Ecological Security of Comprehensive Scientific Investigation and Development Strategy of Soil and Water Conservation Symposium. Beijing, 2008.

Non – equilibrium Sediment Transport Simulation in Danjiangkou Reservior

Lin Yunfa, *Liao Changlu*, *Bi Yong*, *Yan Jianbo*, *Lu Danzhi*
and *Zhang Hongxia*

The Hanjiang River Bureau of Hydrology and Water Resources Survey,
The Yangtze River Water Conservancy Commission, Danjiangkou, 442700, China

Abstract: During 1970 to 1980, 28 cross sections were measured and observed along 91. 75 km to 117. 11 km within Danjiangkou Reservior to study the non – equilibrium sediment transport. In this study, a two – dimensional numerical model was used to simulate the non – equilibrium sediment transport in reservoir. The simulation results were analyzed and compared with the measurements.
Key words: Danjiangkou Reservoir, non – equilibrium, sediment transport, 2D numerical, model

1   The layout of measured cross sections

In the Hanjiang River reach of Dajiangkou(DJK) reservoir, 28 cross sections from 91. 95 km to 117. 11 km far from the dam (see Fig. 1), where the sediment transport capacity changes abruptly with significant sediment deposition and distribution changes, were selected to observe the trend of the sediment transport in the reservoir. The Hanku 35 – 1 (Youfanggou) is the inlet cross section, while the Hanku 24 – 1 (Shendinghe) is the outlet cross section. The averaged distance between cross sections is about 1 km to 1. 5 km.

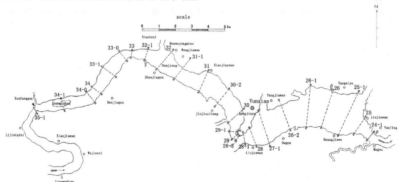

Fig. 1   The sections layout diagram of Danjiangkou Reservoir

2   Non – equilibrium sediment measurements

2. 1   Measurements

In each year from July to October or May to October, at both inlet and outlet cross sections, the time series of water surface elevation, discharge, sediment concentration, transport rate, and bed samples were measured; while for all the cross sections in the whole reach, cross section profiles and bed changes were measured.

2.2 Methods

(1) Water surface elevation: In normal season, it was measured twice at 08:00 and 20:00 in one day. However, in flood season, more frequent measurements were taken.

(2) Discharge: both flow and sediment discharges were measured using the velocity meter (three – points – of – 7 – lines or five – points – of – 7 – lines method). In normal season, discharge was measured once every 3 ~ 5 d, while in flood season, it was measured 1 ~ 3 times daily.

(3) Sediment Concentration: For clear water, no measurement was carried out; otherwise, 1 ~ 4 times measurements were taken daily. In 1970 ~ 1971, six – points – of – one – line method was used; and, in 1972 ~ 1973, nine to ten – points – of – 2 to 3 – lines method was applied.

(4) Transport rate: Usually three – points – of – 7 – lines method was used to measure the transport rate according to the changes of sediment discharge. And, sediment concentration was often measured simultaneously.

(5) Sediment sorting: For suspended load, it was measured according to sediment discharge. As for the bed material sorting, only 1 ~ 2 times measurements were performed for one month.

3 Sediment computation

The computation method of sediment deposition in reservoir can be derived from the observation of the sediment transport in reservoir. In general, the transport of the suspended load in reservoir and river channel is a 3D unsteady process. But it can be approximated by using a series of steady processes. If only the cross – section – averaged concentration or bed change is considered, this process can be further simplified using one – dimensional (1D) method.

(1) Equation of Suspended load concentration.

$$S_{i \cdot j} = S_{i \cdot j}^{*} + (S_{i \cdot j-1} - S_{i \cdot j-1}^{*}) \sum_{\ell=1}^{m_\ell} P_{\ell \cdot i \cdot j} u_{\ell \cdot i \cdot j} + S_{i \cdot j-1}^{*} \sum_{\ell=1}^{m_1} P_{\ell \cdot i \cdot j-1} \beta_{\ell \cdot i \cdot j} -$$

$$S_{i \cdot j}^{*} \sum_{\ell=1}^{m_\ell} P_{\ell \cdot i \cdot j} \beta_{\ell \cdot i \cdot j} \tag{1}$$

where:

$$u_{\ell \cdot i \cdot j} = E^{-\alpha \frac{\omega_\ell (B_{i \cdot j-1} + B_{i \cdot j-1}) \Delta x_j}{Q_{i \cdot j-1} + Q_{i \cdot j}}} \tag{2}$$

$$\beta_{\ell \cdot i \cdot j} = \frac{Q_{i \cdot j-1} + Q_{i \cdot j}}{\alpha \omega_\ell (B_{i \cdot j-1} + B_{i \cdot j}) \Delta x_j} (1 - u_{\ell \cdot i \cdot j}) \tag{3}$$

$$S_{i \cdot j}^{*} = k_0 \frac{Q_{i \cdot j}^{2.76} B_{i \cdot j}^{0.92}}{A_{i \cdot j}^{3.68} \omega_{i \cdot j}^{0.92}} \tag{4}$$

$$\omega_{i \cdot j}^{0.92} = \sum_{\ell=1}^{m_1} P_{\ell \cdot i \cdot j} \omega_\ell^{0.92} \tag{5}$$

In Eq. (1) to Eq. (5), S is the suspended load concentration; S^{*} is the sediment transport capacity; P_ℓ is the fraction of the ith class of sediment; ω_ℓ is the settling velocity for ith class of sediment; α is the adaption length of suspended load, also called the recover coefficient, 0.25 for reservoir deposition, 0.5 for lake, and 1 for reservoir erosion.

For reservoir downstream where erosion is dominant, for large sediment particles, α is related to the sediment size, denoted by α_ℓ, and evaluated by the ratio of the near – bed equilibrium concentration and the depth – averaged concentration. The coefficient k_0 (kg. m³) in sediment capacity formula includes the effects of non – uniform distribution of the velocity. In reservoir, it is 0.03 for deposition area and 0.02 for banks with erosions.

As can be seen in Eq. (1), The non – equilibrium sediment concentration is contributed by three parts, namely, the local sediment transport capacity, the excess sediment ($S_{i \cdot j-1} - S_{i \cdot j-1}^{*}$) transported to downstream with a distance of Δx_j, and the correction due to the changes of the sediment transport capacity along channel. Each part cannot be ignored arbitrarily.

(2)Equation for suspended load composition.

There are two cases to describe the changes of the suspended load composition. For deposition,

$$P_{\ell \cdot i \cdot j} = P_{\ell \cdot i \cdot j-1}(1 - \lambda_{i \cdot j})\left(\frac{\omega_{\ell}}{\omega_{\text{中}i \cdot j}}\right)^{\theta -1} \quad \ell = 1,2,\cdots,m_1 \tag{6}$$

where, $P_{\ell \cdot i \cdot j}$ is the fraction of ℓth sediment size class; θ represents the effects of the non – uniform distribution of suspended load along cross section, in river channel or banded areas, it is 0.75 for lake, and 0.5 for other area; λ is the percentage of deposition;

$$\lambda_{i \cdot j} = \frac{S_{i \cdot j-1}Q_{i \cdot j-1} - S_{i \cdot j}Q_{i \cdot j}}{S_{i \cdot j-1}Q_{i \cdot j-1}} \tag{7}$$

and, $\omega_{\text{中}}$ is the averaged settling velocity when the sediment transport rate changes from $S_{i \cdot j-1}Q_{i \cdot j-1}$ to $S_{i \cdot j}Q_{i \cdot j}$(also called the effective settling velocity) and can be determined by:

$$\sum_{\ell=1}^{m_{\ell}} P_{\ell \cdot i \cdot j-1} = \sum P_{\ell \cdot i \cdot j-1}(1 - \lambda_{i \cdot j})\left(\frac{\omega_{\ell}}{\omega_{\text{中}i \cdot j}}\right)^{\theta -1} \tag{8}$$

From Eq. (6), for coarse sediment, $\omega_{\ell} > \omega_{\text{中}i \cdot j}$, so $\left(\frac{\omega_{\ell}}{\omega_{\text{中}i \cdot j}}\right)^{\theta} - 1 > 0$, $(1 - \lambda_{i \cdot j})\left(\frac{\omega_{\ell}}{\omega_{\text{中}i \cdot j}}\right)^{\theta -1}$

< 1, $P_{\ell \cdot i \cdot j} < P_{\ell \cdot i \cdot j-1}$. As for fine sediment, $\omega_{\ell} < \omega_{\text{中}i \cdot j}$, then $\left(\frac{\omega_{\ell}}{\omega_{\text{中}i \cdot j}}\right)^{\theta} - 1 < 0$, $(1 -$

$\lambda_{i \cdot j})\left(\frac{\omega_{\ell}}{\omega_{\text{中}i \cdot j}}\right)^{\theta -1} > 1$, $P_{\ell \cdot i \cdot j} > P_{\ell \cdot i \cdot j-1}$. Eq. (6) describes a sediment refinement process: due to deposition, the percentage of the coarse sediment is reduced while the fine sediment increases. With $\lambda_{i \cdot j}$ increasing, this refinement process makes suspended load even finer. Fig. 2 illustrates this process.

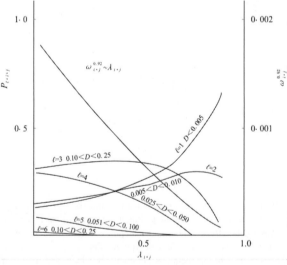

Fig. 2

4　Comparisons

In this study, Eq. (1) to Eq. (8) has been integrated into a 2D numerical model, which was used to simulate the study reach (28 cross sections from 91.95 km to 117.11 km far from the dam). Tab. 1 and Tab. 2 list the comparisons of the measurements and the simulated results during

53 d in 1970. As can be seen, except for the time section 2 when the lateral discharge was ignored, the simulation agreed well with the measurements. The error for the suspended sediment composition at the outlet is less than 2%.

Fig. 3 shows the comparisons of suspended load composition from 1970 to 1980 at the outlet cross section(hanku 24 – 1). It has shown that good agreements between the measurements and the simulations were observed.

Tab. 1   Sediment concentration comparison

Time No.	Days	Inlet cross section (35 – 1) (kg/m³)	Outlet cross section (24 – 1) Measured	Outlet cross section (24 – 1) Simulation	Notes
1	16	0.066	0.028	0.025	
2	1	3.520	0.034 *	0.824	
3	19	0.469	0.048	0.133	
4	3	1.190	0.297	0.490	* Lateral discharge is ignored
5	1	4.560	2.290	1.980	
6	3	2.830	1.590	1.720	
7	4	0.971	0.628	0.521	
8	6	0.112	0.014	0.051	
Total	53	1.540	0.759	0.794	

Tab. 2   Size grading of the comparison

Cross section	Days	Concentration kg/m³	Percentage λ	<0.01	0.01 ~ 0.025	0.025 ~ 0.05	0.05 ~ 0.1	0.1 ~ 0.25	0.25 ~ 0.5
35 – 1	53	1.540		19.8	20.9	24.8	26.6	6.1	1.8
24 – 1 (Measured)	53	0.759		34.5	28.7	22.9	13.5	0.4	
24 – 1 (Simulation)	53	0.794	0.49	34.5	29.0	23.1	12.3	1.1	

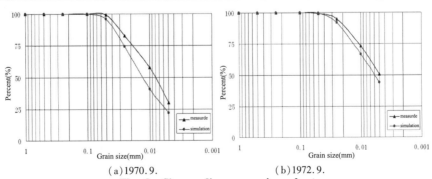

(a)1970.9.   (b)1972.9.

Fig. 3   Size grading comparison chart

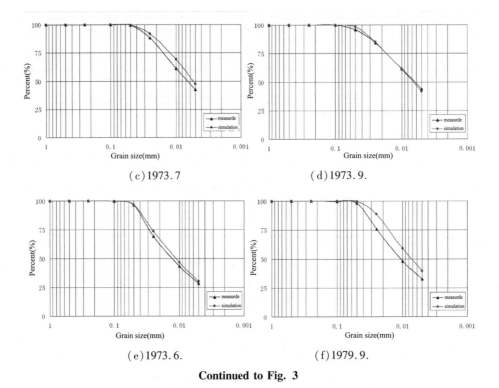

(c) 1973.7

(d) 1973.9.

(e) 1973.6.

(f) 1979.9.

Continued to Fig. 3

5 Conclusions

In this study, the simplified equations for the non − equilibrium sediment concentration and composition along cross section were derived based on the observation of the sediment transport in reservoir. These equations were applied in a 2D numerical model to simulate the non − equilibrium sediment transport in DJK reservoir. They were validated and proved by the comparisons with the measurements. It has shown that they can be used to analyze the long term sediment transport within DJK reservoir.

References

Han Q W. Deposition in Reservoir[M]. Beijing: Science Press, 2003.

Zhang H X. Lin Y F, Zhou X Y, et al. Study of Channel Evolution of Danjiangkou Hydraulic Scheme, 2003.

Experimental Studies on Sediment Transport in Danjiangkou Reservoir, 1983.

Study on Lower Yellow River Water – Sediment Regulation

Sun Qilong

Yellow River Conservancy Technical Institute, Kaifeng, 475003, China

Abstract: the Yellow River is believed as the mother of the whole Chinese nation, who is nurturing thousands of Chinese people. However, the Yellow River has become the aboveground river because of sediment deposition and the flooding has brought severe disasters to the people on both sides. Since 2001, in order to control the Yellow River, the Yellow River water – sediment regulation made the main channel in downstream to be scoured completely and had generally improved its over – current capacity, sediment transport capacity, sediment discharged from which reservoir prolongs its life, increased wetland area and improved the estuarine ecosystem. All these work were the first complete use of many reservoir infrastructures and modern technology productions on the basis of the previous the Yellow River's management experience. We should continue to promote and develop the water – sediment regulation, carry out the lower Yellow River management in an all – round way, more sand – blocked engineering construction in the Loess Plateau and Western water diversion from the source water measures to solve the "sediment" and "hanging river" problem of the Yellow River, to ensure the safety of the Yellow River every year and realize the continuous flow of it, so as to maintain the healthy life of the river.

Key words: water – sediment regulation, effect, enlightenment, countermeasures and thinking

The Yellow River is believed as the mother of the whole Chinese nation, who is nurturing thousands of Chinese people. It nourishes both river sides with its "milk", and brings so huge influence to human civilization. The middle reaches of the river flows through the Loess Plateau Region and has become the aboveground river because of sediment deposition. In order to control it, the Yellow River Conservancy Commission sticks to regulate the water and sediment since 2001, they carried out 3 times experiment and the later six practices of operating water and sediment regulations. They have got significant effect in the maintenance of healthy life of the Yellow River. The water – sediment regulation not only makes the main channel in the downstream to be scoured and had generally improved its over – current capacity, but also reduces the deposition in the reservoir area and improve the ecological environment.

1   The lower Yellow River Water – Sediment Regulation efficiency

1.1   With the main channel in the downstream scoured, its over – current capacity, and sediment transport capacity has improved generally

The Loess Plateau has severe soil erosion and large amount of sediment was brought to the downstream, more and more sand deposited in the main channel, leading the riverbed more and more high, gradually become "the suspended river". Being easy to migrate and deposit is the principal character of the downstream, the possibility of rush broken and migrate is higher. The total amount of sand took out of the reservoir is 219×10^6 t with nine times of water – sediment regulations. The total amount of sediment transported into the sea is 575×10^6 t and there are 356×10^6 t of sand rushing the main channel in downstream among it. In 2001, before the water – sediment regulation, the main channel in downstream minimum flow capacity is $1,800 \text{ m}^3/\text{s}$, it has been raised to $4,000 \text{ m}^3/\text{s}$ till 2011. The main channel has been lowered by 1.5 m, the minimum ability of flood discharge has improved more than double times, the ability to discharge flood and sediment have generally improved.

1.2 Discharging sediment prolongs reservoir's life

The middle Yellow River takes Longyang Gorge, Liujia Gorge, DaLiushu, Qikou, Guxian, Sanmenxia and Xiaolangdi seven backbone projects as its principal parts, undertaking the important function of "cutting sand, fencing sand and desilting". If the sediment in the reservoir cannot discharged or discharged very little, the life of the reservoir will be shorten with sediment is accumulated in the reservoir area. It has been 30 years since Luhun Gu Xian Reservoir was constructed and the Xiaolangdi and Wanjiazhai reservoir have been completed nearly 10 years, and these reservoirs are facing the problem of deposition increasing. From carrying out water storage in October 1999 to April 2006, there were about 700×10^6 m^3 of deposition and fine sediment covering 40 km long in the front of the Xiaolangdi dam, The sedimentation of the reservoir was about 2.166×10^9 m^3. How to discharge the reservoir sediment normally and to reduce sediment deposition has been a practical problem for the designers and the managers. Nine times of water – sediment regulations have proved that it can lower the reservoir water level to reduce sediment deposition by the way of reservoir operation or united operation by several reservoirs. This method not only enhances the reservoir flexibility and control ability, but also explores a new effective way of prolonging its life.

1.3 Wetland area are increasing, and the ecological environment is improved

The unreasonable water utilization and environmental destruction by human being is one of the main reasons for drying up in the Yellow River. Last century, the Yellow River had dried up for many times and caused the estuary wetland atrophy. The nine times of the water – sediment regulations guarantee the continuous flow of the Yellow River in the new century. Now, due to the water – sediment regulation with abundant water of the Yellow River, estuary ecological environment have been also changed in recent years. In 2009, for example, freshwater was filled to the 150,000 mu estuary wetland. The core area of the Yellow River wetland increased by 34,880 hm^2. The water area of estuary increased by 2,920 hm^2. Up to now, the former JianTan turn into the white bird paradiss with 200,000 mu of manual forages, 100,000 mu fast – growing trees, 130,000 acres of manual tamarix chinensis and 500,000 mu of manual reeds, in which the vast expanse of the natural reed looks more and more green and thick. Underground water level raised by 0.15 m to prevent the invasion of sea water and to reduce salinization. Groups of water birds jump between aquatic plants. Fresh water wetland area get restorative growth, and the ecological environment get constantly improving.

2 The enlightenment of the lower Yellow River water – sediment regulation

The lower Yellow River water – sediment regulation has gained rich innovation of science and technology achievements and obvious economic, social and ecological benefits, and opened up a new way of the Yellow River management. All these work was carried by complete use of many infrastructures of reservoir and modern technology productions for the first time on the basis of previous the Yellow River's management experience. It gives us some enlightenment as follows.

2.1 Retaining, discharging, warping, regulation, and dredging is the core principle to manage the Yellow River

For the present, the principal contradiction of the Yellow River is flooding, and the main problem causing flooding is too much sand. The relationship of water and sand is not harmonious. These factors lead to the river "take more, discharge more, and deposit more" "easy to migrate and deposit". The higher the embankment built, the higher the "hang river" increase. In 1997, State Development Planning Commission and Ministry of Water Resources approved the Yellow

River Management and Planning Development Programs which put forward the basic thought of "the upper holding up, the lower draining and both banks distributing", and dealing sand with "retaining, discharging, warping, regulation, and dredging". To realize that the sand retaining, discharging, warping, regulation in the Yellow River, a series of water conservancy project construction and the implementation of the water – sediment regulation were carried out. The regulation is the key part. The main task of the reservoir water – sediment regulation is to lower water level to realize the flood discharge, to transport sand in the flood period as far as possible, to reduce the downstream channel deposition and increase bed – forming and sediment transport function in a wet year flood period. From 2001, since the water – sediment regulation was put into effect, it singles out and tackles the principal contradiction to control the flood. For many years, the reservoir combined the sand regulation with clearing sands using converging flow closely to make flow carrying sand manually, in order to rush reservoir silt and river silt into the sea. It is the essential condition to lower the channel of the Yellow River downstream.

2.2　The lower Yellow River water – sediment regulation sublimes the Yellow River management theory and enriches the Yellow River management strategy

There are many the Yellow River management strategies in history and the Yellow River management theories appearing constantly. Someone of them have been quietly shelved. Someone have been realized partly and some of them were difficult to practice. And at the moment of the implementation of the water – sediment regulation, it is sublimating the Yellow River management theory and enriching the Yellow River management strategy. It is standing on the shoulders of giants and aims to great ambition. The water – sediment regulation borrows from the theory and experiences of predecessors "attack sand with water", and accurately use the modern new high – tech to explore a new way of river management. It is a great innovation of river management engineering in the world . The success experiences of the Yellow River water – sediment regulation also told us that only improving the ability to adjust and the management level can keep the Yellow River safe and stable. Of course, it must go through a long – term complex arduous process to rule the Yellow River. Therefore, we must make full use of modern technology and condition to overcome the difficulties with modern facilities and equipment. In this way we can learn the Yellow River operation rule and achieve the leap from "inevitable" to the "free".

2.3　The key point to control the Yellow River is to coordinate the relationship between man and nature

Because of historical reasons, in the lower Yellow River area, the population has more than 1.8×10^6 t now. The Yellow Rive residents constructed production dyke in 1950s. People competed for land with water, for water with zoology and the contradiction among them are increasing seriously. The Yellow River's normal flood diversion areas are being impacted by human beings. The river channel sedimentation caused floodplains, "the river channel is higher than benchland, and the benchland is above land level which is back of the river", the situation of "secondary Hanging River" becomes more and more serious.

We should coordinate relationship between human and nature well, put the ecological water on the agenda, implement development strategy in the water use, and consider the needs of the ecosystem. At the same time of carrying out the water – sediment regulation, we should prevent the destruction from water resources exhaust to the ecological environment, to build the harmonious relationship between human and nature.

3　Plan on water and sediment regulation

Chinese people have been expecting the efficient management of the Yellow River for thousands of years. So the success of water and sediment regulation of the Yellow River brings us a number of thoughts below.

3.1 Upholding and developing water and sediment regulation

On the harnessing of the Yellow River, some experts put forward that we should make efforts to make the lower Yellow River change its waterway. However, neither the floods nor the sediment can be controlled accordingly because the channel's modeling and maintaining are very challengeable. Moreover, the new channel is possible to be another suspended the Yellow River in 20 years. Compared with that assumption, the project on water and sediment regulation has the characteristic of low costs, high speed and high efficiency. Up to now, there are no comparable ways of dealing with this problem. For the current methods, such as warping and dredging, are only auxiliary treatments. Thus, the upholding and developing of this project will be achieving remarkable success.

3.2 Promote harnessing of the lower Yellow River to maintain the harmony and balance of human and nature

With the increase of basin population and the rapid development of economic society, the lower Yellow River has a close relationship with the lives and properties of 1×10^9 people in Huanghuai region, (especially, the 1.895×10^8 people in benchland area) and the development of economic society. With the increasing pressure of the Yellow River and growingly strained relations between human and water, water and land. The ecosystem of the Yellow River presents overall an deterioration trend. Moreover, the shortage of water resource and the water pollution's aggravating have exerted an increasingly negative influences on economic society. It is hard to implement the migration of 1.895×10^8 people in beach areas and it costs greatly. Because there are still conflicts between human and nature, we should maximize the benefit according to the abilities and technic of human. At the same time, the needs of nature should be taken into account to maintain the harmony and balance of human and nature and to realize the human sustainable development. In order to keep the positive interaction and development between human and nature, we should strive to recycle floods, sediment, river channel, and levee, carry out standardized levee construction, widen the river, strengthen the levee, reinforce the harnessing of dangerous exposed foundation stones of riverside projects caused by water – sediment regulation and the declination of riverbed, dredge river channel and harness the secondary suspended river and estuary.

3.3 Carrying out erosion control in the Loess Plateau, especially sandy and coarse sandy areas

The amount of 65 percent of water in the Yellow River is from the areas before Lanzhou, and 90% of the sediment is from the middle reaches of the Loess Plateau. The Loess Plateau area is wide and large. The intensity of water erosion and wind erosion is very serious. The average annual amount of mud and sand putting into the Yellow River from the Loess Plateau of the Yellow River Basin is up to 1.6×10^9 t which is the sum of the sediment load of the Nile, the Mississippi, Amazon and Yangtze, and is equivalent to the average annual loss of 1 cm arable layer of 2×10^6 acreages and is $200 \sim 400$ times faster than the formation of the soil. It reduces the yellow earth's productive capacity greatly, and even causes the loss of productive capacity. The coarse sediment washed down from the Loess Plateau is the main factor of deposition in the lower reaches of the Yellow River. So carrying out erosion control in the Loess Plateau, especially in sandy and coarse sandy areas is a fundamental measure to reduce the sediment into the Yellow River. Using biological engineering in the middle reaches of the Yellow River region to stop water and sediment in the mountains and slopes, and using hydraulic Engineering to stop water and sediment in the ditches, can provant the sediment from dropping into the river, curb the deterioration of ecological environment, and prevent soil erosion. This is a long – term and arduous task. The soil erosion in the Loess Plateau should be controlled according to the principle of "fine after rough". The governance of 18,800 km² of sandy and coarse sandy areas, particularly the concentrated source area of coarse sediment is especially prominent. Adhereing to the co – ordination of the whole river, and taking the upper, middle and lower reaches into account are necessary. In addition, with the further development of economy and society, water consumption continues to increase. In the case of a comprehensive water – saving, water shortage will reach 11×10^9 m^3 in the upper and middle reaches of the Yellow River, so the implementation of the western lines projects can solve water shortage. With the launching of the western lines projects and construction of East midline and north – south project, it will solve the drought and water shortage in Northwest China, and will optimize the water resources between east and west of China. This is a major strategic project to promote the management and development of the Yellow River.

4 Conclusions

In one word, the Yellow River water – sediment regulation is with the modern technical support, by using engineering facilities and scheduling methods fully , the reservoir's sand and riverbed's sedimentation are brought into the sea by the water. With reducing the sedimentation of reservoir area and improving the flooding capability of waterway, the healthy life of the Yellow River has been maintained.

References

Zhai Haohui, et al. Practice and Innovation as to Theory and Methods of Harnessing Yellow River [N]. Economic Daily, 2010 – 03 – 10(15).

Guo Xiaomei, et al. Analysis of the Influence of Yellow River Seventh Water and Sediment Regulation on the Section of Heze [J]. Jonural of Economic Technolony, 2007 (35): 302 – 302.

Su Qiyun, et al. Disturbance Process and Effect Analysis about the Third Water and Sediment Regulation Experiment of the Low Reach of Yellow River[C]// Papers of the Second Youth Forum of Science and Technology of Chinese Hydraulic Engineering Society, 2005.

Qi Pu, et al. Discussion and Harnessing Strategy on Water and Sediment Regulation as to Stabilizing Main Channel of Yellow River[J]. 2006(11): 7 – 10.

Study on Flood and Sediment Joint Operation of Dadu River Downstream Cascade Reservoirs

Tao Chunhua[1], *Yang Zhongwei*[1], *He Yubing*[1], *Ma Guangwen*[2], *Lu Liyu*[3] and *Zhong Qingxiang*[1]

1. Dadu River Hydropower Development Company Limited, Chengdu, 610041, China
2. College of Water Resource and Hydropower, Sichuan University, Chengdu, 610065, China
3. Architecture and Civil Engineering College, Xihua University, Chengdu, 610065, China

Abstract: There is large amount of high hardness sand in the water flow in Dadu River. The problem of sediment deposition is very prominent in the two early reservoirs, Gongzui Reservoir and Tongjiezi Reservoir. In order to decrease sediment deposition effectively and extend the service life of them, the sediment characteristics of Dadu River cascade hydropower stations are analysed, as well as changes of sediment deposition in Gongzui Reservoir and Tongjiezi Reservoir after Pubugou Reservoir puts into operation. Flood forecast and pre – discharge operation theory are studied for utilizing the discarding water to improve sediment – flushing effect, and flood and sediment joint dispatching schemes of cascade reservoirs are proposed. In order to evaluate effects of joint dispatching schemes, SBED extended one – dimensional flushing sediment mathematical model of deposited reservoir is built, and the mathematical model is used for simulative computation of coming 20 – years sediment – flushing effect. After that, the effects of different schemes are analyzed comparatively. Three conclusions can be drawn from the studying. Firstly, the sediment in Gongzui Reservoir and Tongjiezi Reservoir is obviously reduced after Pubugou reservoir puts into operation. Secondly, flood forecast, pre – discharge operation theory and creating sand – flushing conditions can obviously improve sediment – flushing effects and avoid wasting power generation water when the water levels and outflows of Pubugou Reservoir, Gongzui Reservoir and Tongjiezi Reservoir are rationally controlled. Finally, the larger the inflows and the lower the water level of Gongzui Reservoir and Tongjiezi Reservoirr is, the more obvious the sediment – flushing effects are. According to situations of power plant at the time, such as rules of flood controlling and operation requirements of flood discharge facilities, it is recommended that the inflow of Gongzui Reservoir is controlled at about 4,500 m^3/s, and timely lower the water levels of Gongzui Reservoir and Tongjiezi Reservoirr to their dead water levels, when carrying out joint dispatching schemes.

Key words: Dadu River, cascade reservoir, flood and sediment joint dispatching, flood forecast and pre – discharge operation for sediment – flushing

1 Introduction

Dadu River is located in the midwest regions of Sichuan Province in China, and originated in Qinghai Province. It is the largest tributary of the Minjiang River with total length 1,062 km and basin area 77,400 km^2 excluding Qingyijiang River. Water losses and soil erosion is not serious in the upper reaches of Dadu River, annual average modulus of sediment runoff is 164 t / ($km^2 \cdot a$). Soil erosion is serious in the middle reaches with modulus 16,704 t/ ($km^2 \cdot a$). Because of good condition of ground integrity and vegetation, water losses and soil erosion is not serious in the lower reaches.

In the 1970s and 1990s, Gongzui and Tongjiezi hydropower station were put into operation separately, both located in the downstream of the river. Another two stations, Pubugou and Shenxigou hydropower station in the middle reaches began to operate in 2009 and 2010. Pubugou Reservoir is the only multi seasonal regulating reservoir, and its sand blocking capacity can effectively block most of the sand from upstream and midstream. Before Pubugou Reservoir puts into operation, the problem of sediment deposition of Gongzui Reservoir and Tongjiezi Reservoir is becoming more and more serious for the large amount of sand. As of 2009, the total sediment deposition of Gongzui Reservoirs is $2.501,8 \times 10^{8}$ m³, accounting for 66.95% of the total capacity. 18.31% of the adjustment capacity is lost. The total sediment deposition of Tongjiezi Reservoir is $1.321,8 \times 10^{8}$ m³, about 62.44% of the total capacity, and 8.97% of the regulation capacity is lost. To some extent, the serious sediment deposition of reservoirs is affecting the safety of the reservoirs and generator units operation.

After Pubugou Reservoir impoundment, bed load and most of the suspended sediment above over the dam were stored in the reservoir. Because Shenxigou is a runoff reservoir, we can ignore its role of sand blocking. Therefore, the sediment of Gongzui Reservoir is composed of the sediment discharged from the Pubugou Reservoir and the sand from Pubugou − Gongzui section.

The ration of total capacity and annual sediment of Pubugou Reservoir is 191, reservoir service life will be long only when sedimentation is considered. According to the simulation results, after 50 years' operation, the total deposition volume of Pubugou Reservoir will be 1.11×10^{9} m³, the blocking ratio of suspended sediment will be 86.5%, and annual average sediment concentration of outflow water 112 g/m³. Meanwhile, the deposition will cause 5.95×10^{8} m³ loss of adjustment capacity with loss ratio of 14.3%. Remaining adjustment capacity will be 3.24×10^{9} m³. After 100 years' operation from now, the total deposit volume of Pubugou Reservoir will be 1.86×10^{9} m³, the blocking ratio of suspended sediment will be 85.9%, and annual average sediment concentration of outflow water 120 g/m³. Meanwhile, the deposition will cause the adjustment capacity loss of 9.60×10^{8} m³, or 25.0%. Then the remaining adjustment capacity will be 2.88×10^{9} m³, As shown in Tab. 1.

Tab. 1   The calculation result of Pubugou Reservoir sediment deposition

Running time length (a)	Deposit volume ($\times 10^{9}$ m³)	Deposit volume of adjustment capacity ($\times 10^{6}$ m³)	Loss ratio of adjustment capacity (%)	Blocking ratio of suspended sediment (%)	Annual average sediment concentration of outflow water (g/m³)
20	0.43	254	5.28	86.8	110
50	1.11	595	14.3	86.5	112
100	1.86	960	25.0	85.9	120

In the first 100 years of Pubugou Reservoir operation, blocking ratio of suspended sediment is within the range of 85.9% ~ 86.8%, with minor change. Particle size of suspended sediment in the outflow is 0.025 mm. In natural condition, the annual average sediment − transportation of suspended sediment is 3.17×10^{7} t at the dam site of Pubguou hydropower station, and 3.75×10^{7} t at the dam site of Gongzui hydropower station. 100 years later, considering the blocking sediment ratio of Pubugou Reservoir to be 85%, the annual average sediment − transportation of suspended sediment in outflow will be 4.76×10^{6} t at the dam site of Pubguou hydropower station. Taking sediment from Pubugou − Gongzui section into consideration, 5.80×10^{6} t, the annual average sediment − transportation of suspended sediment is 1.06×10^{8} t at the dam site of Gongzui hydropower station.

2   Study on flood and sediment joint dispatching schemes

2.1   General joint dispatching schemes

Since Gongzui hydropower station and Tongjiezi hydropower station has put into operation, erosion and deposition have reached a balanced state. Their dead storages have basically lost because of the sediment deposition, more and more sediment goes into hydroturbine units, and it is threatening the normal operation of the hydropower stations. In order to slow the sediment deposition of Gongzui Reservoir and Tongjiezi Reservoir, the current scheme is lowering the water level of the two reservoirs to reduce deposition when the inflow increasing. Pubugou Reservoir lies in the upper reaches of Gongzui Reservoir and Tongjiezi Reservoir. It was put into operation in 2009. Because it has regulation and storage capacity and with function of sand blocking, the amount of sediment entering Gongzui Reservoir and Tongjiezi Reservoir is substantially reduced. A chance is provided for moderately raising the water level of these two reservoirs. If the flood and sediment joint dispatching scheme is suitable, not only may it be avoided that raised operation water level aggravates the deposition, but also the flushing effect can be improved and part of the capacity of these two reservoirs be recovered.

2.1.1 Timing of sand sluicing joint dispatching

Sand sluicing joint dispatching is an effective way of creating suitable conditions to reduce the sediment deposition.

The water level requirements of Pubugou Reservoir: when the water level of Pubugou Reservoir is up to 841 m (the flood controlling limited level) or other water level limit, the abandoned water can be used to improve the sand sluicing effect of Gongzui Reservoir and Tongjiezi Reservoir. What's more, sand sluicing joint dispatching can avoid wasting the power generation water by forecast and pre-discharge, for the water which discharged can be supplemented by impounding flood.

The water level requirements of Pubugou reservoir, Gongzui Reservoir and Tongjiezi Reservoir: when the floods is coming and the inflow is approximately equal to maximum power flow, abandoning water of Pubugou Reservoir in advance can be used for flushing the sediment of Gongzui Reservoir and Tongjiezi Reservoir in the lower reaches. In order to ensure the security of the Pubugou dam, faceplate rock-fill dam, the water level decline should be less than 1 m every day.

In order to achieve the expected flushing effect, sand sluicing joint operation should be about 5 times every year, and each time should be about 12 h.

2.1.2 Flow of joint sand sluicing

According to flood controlling requirements of these power plants and running requirement of their flood discharge facilities, it is recommended to control the inflow of Gongzui Reservoir Q_{need} in the range of 4,000 m^3/s to 4,500 m^3/s to achieve good flushing effect. During the actual operation process, the inflow of Gongzui Reservoir can be adjusted within this range based on flood forecast.

The outflow of Pubugou Reservoir Q_{out} is equal to the inflow of Gongzui Reservoir Q_{need} minus the interval flow $Q_{int erval}$ between Pubugou Reservoir and Gongzui Reservoir. The bigger the interval flow is, such as 600 m^3/s, the less the abandoned water will be for joint sand sluicing.

2.1.3 The time length of joint sand sluicing

$$t = \frac{\sum_{i=0}^{n}(Q_i - Q_{power}) \times T}{Q_{need} - Q_{int erval}}$$

where, Q_i is the forecasting flood flow of Pubugou Reservoir in the i period; Q_{power} is the maximum power flow of Pubugou hydropower station; T is the length of the period; t is the time length for joint sand sluicing, It is better to control the time length to about 12 h each time for joint sand sluicing.

2.1.4 Operation requirements of Gongzui Reservoir and Tongjiezi Reservoir

In order to improve the sediment flushing effect, the water levels of Gongzui Reservoir and Tongjiezi Reservoir should be dropped to their dead water levels which are 520 m and 469 m seperately, and the sand-sluicing gates should be opened before the joint flushing flow reaches the dams.

2.2 Simulation schemes of water and sediment joint dispatching

In order to compare the effects of different water and sediment joint dispatching schemes, we educe the following schemes, referring to Tab. 2, to simulative calculate and analyze the flushing effect.

Tab. 2   The schemes of water and sediment joint dispatching

| Scheme | Pubugou reservoir | | Gongzui Reservoir | | Tongjiezi reservoir | |
	Inflow (m^3/s)	Water level (m)	Flushing flow (m^3/s)	Water level (m)	Flushing flow (m^3/s)	Water level (m)
Scheme 1	≥2,200	840.3 ~ 841	4,000	520	4,000	469
Scheme 2	≥2,200	840.1 ~ 841	4,500	520	4,500	469

3   Effect analysis of water and sediment joint dispatching schemes

In this paper, SBED extended one – dimensional flushing sediment mathematical model of deposited reservoir is built to calculate sand sluicing effect. Period, reaches of the river and sediment grain size are considered during the simulation.

3.1   Sand sluicing effect of different schemes of Gongzui Reservoir

After carrying out flood and sediment joint dispatching between Pubugou Reservoir and Gongzui Reservoir, annual maximum flushing amount increases by 1.06×10^6 m^3 when flushing flow is 4,000 m^3/s, while annual maximum flushing sediment amount increases by 2.47×10^6 m^3 when flushing flow is 4,500 m^3/s. Flushing rate improves 7.4% and 17.2% respectively. According to a series of water and sediment simulative calculations, both existing schemes and joint operation schemes of flushing flow 4,000 m^3/s or 4,500 m^3/s, sediment will be accumulatively flushed in the first 15 years of Pubugou reservoir operation, after that there will be no accumulative flushing any more. Sediment flushing effects of Gongzui Reservoir for different schemes are shown in Tab. 3 and Fig. 1

Tab. 3   Gongzui Reservoir sediment accumulative flushing amount of different schemes

Year		1	2	3	4	5	6	7	8	9	10
Accumulative flushing amount	Existing scheme	−527	−599	−682	−704	−787	−1,011	−1,024	−1,059	−1,054	−1,097
	Flushing flow 4,000 m^3/s	−545	−669	−820	−828	−964	−1,125	−1,155	−1,235	−1,212	−1,294
	Flushing flow 4,500 m^3/s	−567	−695	−865	−871	−1,008	−1,154	−1,182	−1,278	−1,262	−1,364

Year		11	12	13	14	15	16	17	18	19	20
Accumulative flushing amount	Existing scheme	−1,268	−1,270	−1,279	−1,258	−1,284	−1,428	−1,409	−1,406	−1,384	−1,403
	Flushing flow 4,000 m^3/s	−1,412	−1,409	−1,443	−1,416	−1,465	−1,524	−1,511	−1,534	−1,501	−1,521
	Flushing flow 4,500 m^3/s	−1,471	−1,475	−1,529	−1,491	−1,554	−1,660	−1,628	−1,655	−1,607	−1,675

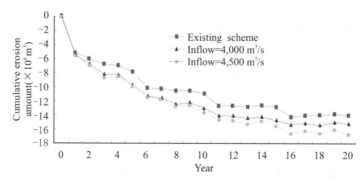

Fig. 1   Gongzui Reservoir sediment accumulative flushing effects of different schemes

3. 2   Sediment flushing effect of different schemes of Tongjiezi Reservoir

After carrying out flood and sediment joint dispatching schemes, annual maximum flushing amount of Tongjiezi Reservoir increases by 0.51×10^6 m³ when flushing flow is $4,000$ m³/s, while annual maximum flushing sediment amount increases by 1.03×10^8 m³ when flushing flow is $4,500$ m³/s. Flushing rate improves 4.8% and 9.7% respectively. According to a series of water and sediment simulative calculations, Tongjiezi Reservoir flushing sediment amount of water and sediment joint dispatching schemes is a little less than that of the existing scheme in the early stage, because in the above two joint dispatching schemes, sediment amount of Gongzui Reservoir increases significantly . As time goes on, the accumulative flushing amount of these two joint dispatching schemes is more than that of existing scheme, thus the effect of joint dispatching is shown gradually. After first 15 years, the accumulative flushing amount of joint dispatching schemes would be basically stable, Tongjiezi Reservoir would not be back silting anymore. After first 20 years, accumulative flushing amount differences between the two joint dispatching schemes and existing scheme are 3.32×10^6 m³ and 3.72×10^6 m³ respectively. Sediment flushing effects of different schemes for Tongjiezi Reservoir are shown in Tab. 4 and Fig. 2.

Tab. 4   Tongjiezi Reservoir sediment accumulative flushing amount of different schemes

	Year	1	2	3	4	5	6	7	8	9	10
Accumulative flushing amount	Existing scheme	−987	−1,062	−1,039	−1,043	−973	−1,028	−990	−914	−887	−806
	Flushing flow 4,000 m³/s	−945	−977	−969	−1,003	−999	−1,113	−1,052	−971	−969	−929
	Flushing flow 4,500 m³/s	−975	−986	−1,009	−1,033	−1,036	−1,165	−1,107	−1,054	−1,047	−989

	Year	11	12	13	14	15	16	17	18	19	20
Accumulative flushing amount	Existing scheme	−843	−786	−723	−716	−651	−658	−631	−610	−639	−595
	Flushing flow 4,000 m³/s	−1,004	−954	−943	−958	−937	−967	−926	−925	−945	−927
	Flushing flow 4,500 m³/s	−1,027	−983	−968	−976	−937	−978	−953	−966	−990	−967

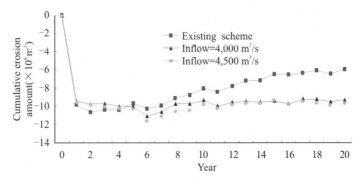

Fig. 2　Tongjiezi Reservoir sediment accumulative flushing effects of different schemes

　　The effect of Gongzui Reservoir and Tongjiezi Reservoir flood and sediment joint dispatching schemes can be briefly summed up as below.

　　From the increased proportion of sediment flushing, joint dispatching scheme with flushing flow of $4,500, m^3/s$ is superior to that of $4,000$ m^3/s. On the other hand, if the flushing flow is $4,500$ m^3/s, there would be enough water to open the bottom outlets of Gongzui dam and Tongjiezi dam, without affecting power generation. So it is preferred to carry out the joint dispatching scheme with flushing flow $4,500$ m^3/s, and lower water levels of Gongzui Reservoir and Tongjiezi Reservoir to near their dead water levels in time

　　From the results of simulation, the sediment flushing effect of Gongzui Reservoir will not be obvious if we carry out the joint dispatching scheme, because it does not obviously increase flushing times compared with the existing schemes, and sediment would be less and less. However, the effect of Tongjiezi Reservoir joint dispatching schemes would be significant once the sediment deposition phenomenon happens again in the current reservoir running mode.

4　Summaries

　　In this paper, based on analysis of sediment deposition of Gongzui Reservoir and Tongjiezi Reservoir downstream of Dadu River Basin, changes of sediment flowing into the two reservoirs were analysed, and flood and sediment joint dispatching schemes of all the reservoirs downstream the river, including Pubugou Reservoir, were studied. Effect of each scheme was analysed using reservoir sediment simulation mathematical model. Both the method and the results have clear guiding significance for flood and sediment joint dispatching.

References

Jin Jian, Ma Guangwen, Lu Jinbo. Study on Power Generation Plan and Water – sediment Joint Dispatching Plan of Reservoirs in the Lower Dadu River Reach[J]. Journal of Hydroelectric Engineer,2011,30(6): 210 –214, 236.

Xiong Min, Ma Wenqiong. Sedimentation Situation Analysis of Gongjiui Reservoir[J]. Sichuan Water Power,2008,27(S1) : 82 –86.

He Xianpei. Sediment Control Effect and Impact on the Downstream Station of Pubugou Reservoir [J]. Sichuan Water Power,2008,27(4) : 91 –94.

Han Qiwei, He Mingmin. Mathematical Modeling of Reservoir Sedimentation and Fluvial Process [J]. Journal of Sediment Research,1987(3) : 16 –28.

Xie Jianheng. River Modeling [M]. Beijing: China WaterPower Press,1992.

Experiment on Suspended Sediment Transport Rate in Weak Dynamic Cohesive-Sediment-Laden Flow

Xu Wensheng[1] , *Zhou Wei*[2] , *Chen Li*[3] and *Liu Lin*[4]

1. Division of Soil and Water Conservation, Changjiang River Scientific Research Institute, Wuhan, 430010, China
2. Changjiang Survey Planning Design and Research Co. , Ltd. , Wuhan, 430010, China
3. State Key Laboratory of Water Resources and Hydropower Engineering Science, Wuhan University, Wuhan, 430072, China
4. Changjiang Waterway Institute of Planning & Design, Wuhan, 430000, China

Abstract: As a basic representation of flow movement in the Yellow River estuary, the weak dynamic cohesive-sediment-laden flow plays an important role in the evolution of this area. By means of indoor flume experiment, the basic movement characteristics of weak dynamic cohesive-sediment-laden flow were first summarized, and then the (Suspended Sediment Transport Rate) SSTR impacted by hydrodynamic conditions, sediment supply, and water quality conditions was subsequently explored in this paper. It is concluded from the results that variation of the SSTR differs largely for different conditions. When hydrodynamic conditions become weak the SSTR decreases, of which the decreasing range depends on sediment supplies. The range is smaller in lower sediment supply, and larger in higher sediment supply, with the fact that when sediment supply becomes higher and exceeds a critical value, the ratio of SSTR for different hydrodynamic conditions will remain approximately invariable. As sediment supply increases the SSTR increases as well, of which the increasing range is dependent upon hydrodynamic conditions. The weaker the hydrodynamic conditions, the smaller the range. However, when hydrodynamic conditions get stronger and exceed a critical value, the ratio of SSTR for different sediment supplies will roughly not exhibit variation. It is also found that a lamination will occur in sediment-laden flows induced by both high sediment concentration and weak hydrodynamic conditions appearing at the same time, of which the heavier the ion concentration, the lower the SSTR, however, when the ion concentration becomes higher and exceeds a critical value the SSTR will nearly not vary with it.

Key words: weak dynamic flow, cohesive sediment, suspended sediment transport rate, flow condition, sediment supply, water quality condition

1 Introduction

As the second largest river in China and the river with highest sediment concentration in the world, the Yellow River is famous for her annually more than 1×10^9 t of sediment discharge into oceans around the world (Wang et al. , 2010; Liu et al. , 2011). Abundantly transported to the estuary by flows, sediment causes considerable impacts to the evolution of this area (Wang et al. , 2007). It is observed that on one hand, the flow in estuary is mainly presented as weak dynamic (Xu et al. , 2009). On the other hand, the grain sizes of sediment transported by flow become finer and finer along the way from upstream to downstream and finally get the finest at the estuary where sediment principally shows cohesive and is considerably affected by electrolyte ions plentifully contained in seawater (Wang et al. , 2010; Danchuk et al. , 2011). Therefore, in order to further investigate the evolution of estuary like the Yellow River mouth it is essential and of great advantage to explore the sediment transport rules in weak dynamic cohesive-sediment-laden flow.

Lots of achievements about the transport of cohesive sediment have been accomplished. By analyzing plenty of experimental data, Le (1983) revealed the settling properties of cohesive sediment in static water. Huang (1989) studied the effects of deposit concentration on the threshold shear stress of cohesive sediment in terms of the principle of flocculation electrochemistry. With Lagrangian model, the random flocculation characteristics of cohesive sediment were analyzed

(Maggi et al. , 2007, 2008). Based on field data, Giardino et al. (2009) established a 2-D hydrodynamic model for the transport of cohesive sediment in the IJzer estuary, Belgium. Wang et al. (2010) performed a numerical study on the flocculation settling of cohesive sediment influenced by salinity, sediment concentration and hydrate and so forth. Du et al. (2010) numerically simulated the cohesive sediment transport in Hangzhou Bay with a 3-D model. In addition, the scour rate for consolidated cohesive sediment was experimentally discussed (Tan et al. , 2010). However, due to the complexity of cohesive sediment transport and lack of systematically understanding of the transport law of cohesive sediment in weak dynamic flow areas like estuary, port, lake, and wetlands and so on, it is found that the previous achievements are obviously not sufficient for practical application.

The writer, as well as his collaborators in the same research team, has ever investigated the vertical distribution characteristics of flow velocity and sediment concentration in weak dynamic cohesive sediment-laden flow with and without lamination occurring (Wang et al. , 2006, 2008; Xu et al. , 2009, 2010). Since the SSTR is impacted by both flow turbulence shear stress and sediment gravity settling force, which is considerably similar to the results of particle sedimentation in hydrostatic water (Le et al. , 1983; Wang et al. , 2010), the SSTR influenced by hydrodynamic conditions, sediment supply, and water quality conditions in weak dynamic cohesive sediment-laden flow was necessarily further analyzed and explored based on the previous achievements by indoor flume experiment. The results will help to provide a theoretical support for revealing the evolution laws of weak dynamic flow areas like the Yellow River estuary.

In addition, it is necessarily pointed out that the weak dynamic flow here is generally referred as the flow status that often take place in the areas like estuary, port, lake, wetland and so forth, however, without specific definition.

2 Experiment design in brief

2.1 Methods and parameters

All experiments were carried out in a slip glass flume with a volume of 0.2 m wide, 0.25 m deep and 3.5 m long, whose bottom slope could be artificially changed with a range from 0 to 1%, as shown in Fig. 1. In the whole flume experimental system an energy dissipation measure and thermometer was fixed at the entrance, a velocity detector of MicroADV at the middle, and a water level measuring needle at the tailgate. There was no sediment on the flume bottom before experiment. For experiment the muddy water with determined concentration of sediment and $CaCl_2$ was first obtained by adding sediment and solid $CaCl_2$ into the clear water in reservoir. Then the muddy water was adequately stirred and mixed by a pump to generate sediment-laden flow which would cycle in the flume experimental system. When cycling the flow passed through electromagnetic flow meter, dish valve and water pipe in turn, and then entered the glass flume where the SSTR was tested, and finally went back to the reservoir again at the drive of a pump. The cross-section for measurement of flow velocity and sediment concentration was located at the middle of the flume, and the distribution of measurement dots on section was manually determined. Flow velocity was measured by a MicroADV detector, sediment concentration by a siphon based on a drying-and-weighing approach, water level at the tailgate of flume by a measuring needle, and water quality through potentiometric titration according to the "Monitoring and Analysis Method Guide for Water and Waste Water (middle volume)" (Wei et al. , 1997).

The experiment content includes three aspects: ① revealing the effects of hydrodynamic conditions on SSTR under the invariable sediment supply and water quality conditions; ② discussing the impacts of sediment supply on SSTR in the same hydrodynamic and water quality conditions; and ③ exploring the influences of water quality on SSTR when hydrodynamic and sediment supply conditions retain unalterable.

The sediment used here is from the Yellow River Basin with fine grain sizes, whose median diameter is around 3.5 μm as shown in Fig. 2. And the variation ranges of parameters concerned for flume experiment controls are as shown in Tab. 1.

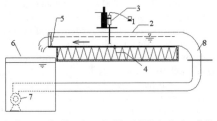

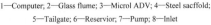

1—Computer; 2—Glass flume; 3—Microl ADV; 4—Steel sacffold;
5—Tailgate; 6—Reservior; 7—Pump; 8—Inlet

Fig. 1   System for flume experiment

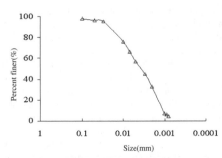

**Fig. 2   Particle size distribution curve
for experiment sediment**

Tab. 1   Variation ranges of controlling parameters in flume experiment

Discharge (m^3/h)	Water depth(cm)	Slope J	Reynolds NO. Re	Sediment supplies (kg/m^3)	Temperature($^\circ$C)	pH value
1~6	16	5‰	>500	0.9~5.0	10~25	7.5~8.5

For the sake of easily discussion in the following, parameters mainly considered in this experiment are classified and marked as following: ① q_1 (=1.0 m^3/h), q_2 (=2.0 m^3/h), q_3 (= 3.0 m^3/h), q_4 (= 4.0 m^3/h) and q_6 (= 6.0 m^3/h) represents hydrodynamic conditions respectively; ② S_1 (=0.9 kg/m^3), S_3 (=2.5 kg/m^3) and S_4 (=5.0 kg/m^3) stands for sediment supplies respectively; and ③ C_0, C_1, C_3 and C_4 refers to water quality conditions respectively. All parameters of water quality conditions are presented in Tab. 2.

Tab. 2   Variation ranges of parameters for water quality conditions

No. of water quality conditions	C_0	C_1	C_3	C_4
Concentration of Ca^{2+} (mg L)	13.3~14.0	43.9~47.7	66.8~72.4	91.9~98.7
Total hardness(mg/L)	15.2~16.1	53.3~58.3	76.2~82.3	101.1~106.7
Concentration of Cl^-(mg/L)	0	20.8~22.4	59.6~62.6	92.0~93.5
Concentration of HCO_3^-(mg /L)	42.1~44.4	142.1~146.8	142.5~148.7	137.8~144.6
pH value	7.6~7.8	8.1~8.4	8.1~8.4	7.8~8.3
Total salinity(mg/L)	82~91	271~288	359~374	416~433

Note: For water qualities in the flume, C_0 is referred to as distilled water, C_1 tap water, and C_3, C_4 solution of $CaCl_2$ by adding 40 g and 80 g $CaCl_2$ into tap water respectively.

2.2   Determination of suspended sediment transport rate

The SSTR is obtained by the product of flow velocity and sediment concentration. For a given result of vertical distribution of suspended sediment concentration (S_y) and flow velocity (u_y), the SSTR for per unit area at height of y from bed in per unit time can be presented as $u_y S_y$, where S_y and u_y stand for suspended sediment concentration and flow velocity at height of y above bed, respectively, and can be directly measured during experiment. As a result of vertical integration of $u_y S_y$, the SSTR per unit width is subsequently acquired. Then by summing up all the SSTR along the width, we can obtain a SSTR for the whole cross-section. The SSTR (in kg/h) here is referred to as the total suspended sediment discharge for whole cross-section and unit time, which is severely controlled by vertical distribution of both suspended sediment concentration and flow velocity.

3 Primary characteristics of weak dynamic cohesive sediment-laden flow

Because of the close dependence of SSTR on hydrodynamic conditions, the primary rules and characteristics of weak dynamic cohesive-sediment-laden flow movement were first summarized in terms of the previous achievements and the obtained experimental results in present paper. Two typical vertical profiles of flow velocity in weak dynamic cohesive-sediment-laden flow, as well as corresponding sediment concentration, are shown in Fig. 3, respectively.

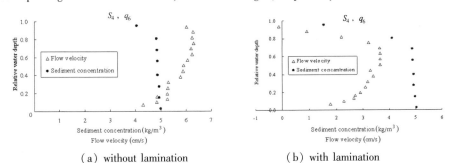

(a) without lamination   (b) with lamination

Fig. 3 Vertical profiles of flow velocity and sediment concentration

It can be found that the movement of weak dynamic cohesive-sediment-laden flow exhibits two different forms, viz. with and without lamination. When there is no lamination occurrence the vertical distribution of flow velocity and sediment concentration is approximately uniform, which is similar to that in general sediment-laden flow. Nevertheless, when lamination comes up the vertical distribution profile presents a curve like a transversal parabola with a catastrophe point, which is similar to that in density current (Yao et al., 1996). In addition, it is observed that in weaker dynamic flow with higher sediment concentration the lamination has more possibility to appear (Wang et al., 2008), and once this phenomenon occurs the water quality will bring considerably obvious influences on it (Xu et al., 2010).

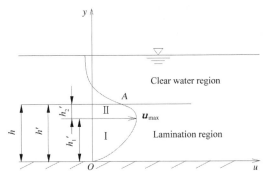

**Fig. 4 Vertical distribution of regions
for lamination**

As per experimental results and the primary movement characteristics of sediment-laden flow with lamination, the flow for a vertically longitudinal section can be divided into two regions (Xu et al., 2009), namely clear water region and lamination region (as shown in Fig. 4).

The longitudinal line where point A is located stands for the interface for both regions. Moreover, according to the location of maximum velocity, the lamination region can be further divided into two sub-regions, viz. (I) near-bottom-region and (II) turbulence-mixed-region.

4 Results and discussions

4.1 Impacts of flow conditions

Fig. 5 shows the comparison of variation of SSTR with hydrodynamic conditions from q_6 to q_1 for different sediment supplies with water quality of C_1. When hydrodynamic conditions become weaker from q_6 to q_1 in the same sediment supply, the SSTR is seen to gradually decrease. This is in respect that under the invariable conditions of water quality and sediment supply, when hydrodynamic conditions become weaker resulted from the reduction of flow discharge, the sediment settlement, however, will gradually become relatively more obvious on the contrary, which subsequently leads to falls of sediment carrying capacity of flow and the corresponding decline of the SSTR.

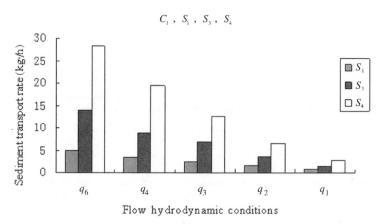

Fig. 5   Variation of suspended sediment transport rate with flow conditions for different sediment supplies

Nevertheless, the reduction range of SSTR differs for different sediment supplies, which presents as variation of the ratio of SSTR for different flow discharges with sediment supplies. In sediment supply of S_1, as the hydrodynamic conditions become weaker, the ratio of SSTR corresponding to different flow discharges from q_6 to q_1 is calculated to be 6.79 : 4.57 : 3.28 : 2.14 : 1, compared to 10.82 : 6.86 : 5.39 : 2.78 : 1 in S_3 and 10.45 : 7.22 : 4.68 : 2.41 : 1 in S_4, which shows that as the hydrodynamic conditions become weaker the SSTR is observed to decrease with relatively a small variation range in low sediment supplies, but a large range in heavy supplies. It is also found that when sediment supplies exceed a critical value the ratio of SSTR for different hydrodynamic conditions hardly registers variation with sediment supplies, namely the ratio depends upon only the hydrodynamic conditions but not sediment supplies.

This is due to that in low sediment supplies with a determined water quality condition the sediment sources which the flow can carry are limited, resulting in that the sediment carrying in strong hydrodynamic flow is unsaturated and the carrying amount approximates to that in weak hydrodynamic flow. Therefore, hydrodynamic conditions have an unconspicuous effect on SSTR. However, in heavy sediment supplies there are sufficient sediment sources for flow carrying, which leads to that both of the strong and weak flow always show a saturated status of sediment carrying. As a result, the SSTR is mainly dependent on hydrodynamic conditions. Therefore, there is a threshold value for sediment supplies. Once this critical value is exceeded, all the flow in experiments will come to a saturated sediment carrying status, in which the SSTR for different hydrodynamic conditions is approximate to the corresponding sediment carrying capacity and is almost independent upon sediment supplies.

With respect to other water quality conditions, the SSTR is found to exhibit similar variation with hydrodynamic conditions to the discussed above.

4.2 Impacts of sediment supplies

Variation of the SSTR with sediment supplies for different hydrodynamic conditions (q_1, q_3, q_6) in water quality of C_1 is shown in Fig. 6. It is observed that the SSTR increases with sediment supplies for any hydrodynamic condition, resulting from that when water quality and hydrodynamic condition are predetermined, as sediment supplies continually become higher, the sediment carrying amount of flow will arrive saturated or even super-saturated triggered by the occurrence of lamination.

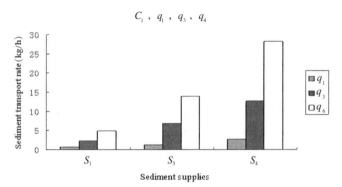

$$C_1, q_1, q_3, q_4$$

Fig. 6　Variation of suspended sediment transport rate with sediment supply in different flow conditions

Nevertheless, it can be seen that the increase amplitude of SSTR differs for different hydrodynamic conditions, that is, the ratio of SSTR corresponding to different sediment supplies varies with the hydrodynamic conditions. In hydrodynamic conditions of q_1, when sediment supplies become heavier, the ratio of SSTR for different sediment supplies from S_1 to S_4 is 1:1.81:3.81, compared to 1:2.97:5.43 in q_3 and 1:2.88:5.85 in q_6, which presents us that as sediment supplies become higher the SSTR is observed to increase with relatively a small variation range in weak dynamic flow conditions, but with a large range in the opposite flow conditions. It is also found that when hydrodynamic condition exceeds a critical value the ratio of SSTR for different sediment supplies rarely varies with it, viz. compared to sediment supplies the hydrodynamic conditions have less influences on SSTR at this time and the ratio chiefly depends only on sediment supplies but not the hydrodynamic conditions.

This is resulted from that in weak dynamic flow with predetermined water quality the capacity of sediment carrying of flow is relatively low and accordingly a low sediment supply can make the flow easily get saturated. As a result, dependent on sediment carrying capacity of flow the SSTR is unconspicuously affected by sediment supplies, and as sediment supplies become higher the increase range of SSTR is not obvious in this situation. Whereas in strong dynamic flow the capacity of sediment carrying is also high, resulting in that the flow does not get saturated until the sediment supplies become sufficiently heavy. Moreover, when lamination occurs, the flow will probably arrive at a supersaturated situation. Therefore, when sediment supplies become higher, the increase range of SSTR will become larger as well. But it is required to point out that when hydrodynamic condition exceeds a threshold value the sediment carrying capacity will become so heavy that the flow is hard to get saturated. Then the ratio of SSTR for different sediment supplies can be approximate to the ratio of the corresponding quantity of sediment supplies, and consequently the ration tends to be nearly invariable.

With respect to other water quality conditions, the SSTR is observed to register similar variation with sediment supplies as demonstrated above.

4.3 Impacts of water quality conditions

Fig. 7 presents the variation of SSTR with water quality in different hydrodynamic conditions of q_3, q_2 and q_1, respectively, with the same sediment supply of S_4. It is observed that in q_3 hydrodynamic condition, the ratio of SSTR for C_0, C_1, C_3 and C_4 is 0.96: 0.92: 0.99: 1, which indicates that in this situation the SSTR is weakly dependent on water quality condition (Fig. 7 (a)). In hydrodynamic condition of q_2, the ratio of SSTR for C_0, C_1, C_3 and C_4 is calculated to be 2.02: 1.11: 0.98: 1 (Fig. 7(b)), compared to 1.80: 1.16: 0.94: 1 in q_1 hydrodynamic condition (Fig. 7(c)), which shows us that for hydrodynamic conditions of q_2 and q_1, the SSTR abruptly decreases by about 35% ~45% as water quality condition varies from C_0 to C_1, compared to unconspicuous variation with water quality from C_1, C_3 to C_4.

For the above meaningful observations, some theoretical analysis has been explored. In the same sediment supply, when flow discharge is large and hydrodynamic condition is strong, the movement of sediment particles is then mainly dependent upon the flow turbulence shear stress, of which the vertical distribution of flow velocity and sediment concentration is relatively uniform and there occurs no lamination (Fig. 3(a)). However, if the sediment-laden flow is obviously affected by water quality conditions, it is necessary that there is lamination appearing in flow (Wang et al., 2008; Xu et al., 2009, 2010). So the SSTR for different water quality conditions is basically equal to each other in this situation. However, when flow discharge is small and hydrodynamic condition is weak, the vertical distribution profile of flow velocity and sediment concentration is observed to presents a curve like a transversal parabola, and the lamination comes up in flow (Fig. 3(b)). When lamination appears the movement of sediment particles in flow primarily depends on the gravity force of grains, the water quality shows an obvious influence on the sediment-laden flow, and the transport of sediment mostly happens in the near-bottom-region. In terms of the achievements by the writer (Xu et al., 2010), in lamination flow, increase of the ion concentration will lead to aggravation of the lamination and decrease of the thickness of lamination region. Whereas, once the ion concentration exceeds a critical value, the intensity of lamination will approximately not vary with it any more. As a result, as the water quality varies from C_0 to C_1, the SSTR presents an abrupt decrease, compared to no distinct difference between the SSTR corresponding to water quality conditions of C_1, C_3 and C_4.

5 Conclusions

As a basic representation of flow movement in the Yellow River estuary, the weak dynamic cohesive-sediment-laden flow plays a significant part in the evolution of this area. So it is certainly helpful to investigate it for further study on the evolution of the Yellow River estuary. By means of indoor flume experiment, the basic movement characteristics of weak dynamic cohesive-sediment-laden flow were first summarized, and then the SSTR was explored. The conclusions were obtained as followings:

(1) As the hydrodynamic conditions become weak the SSTR decreases, of which the decreasing amplitude depends on sediment supplies. The amplitude is smaller in lower sediment supplies, and larger on the contrary. There is critical value for sediment supplies. When it is exceeded, the ratio of SSTR for different flow conditions tends to be invariable.

(2) As sediment supplies increase the SSTR increases as well. Nevertheless, it is observed that the increase range varies with the hydrodynamic conditions. The weaker the hydrodynamic condition, the smaller the increasing range. But once the hydrodynamic condition exceeds a critical value, the ratio of SSTR for different sediment supplies will roughly incline not to exhibit variation.

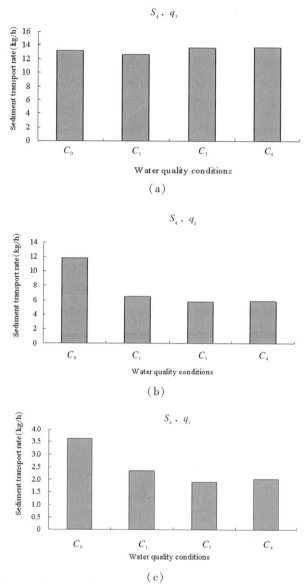

Fig. 7　Variation of suspended sediment transport rate with water quality conditions

（3）It is found that when lamination occurs, water quality condition plays an obvious role in SSTR. The increase of the ion concentration leads to a decrease of the SSTR. Whereas, it is also concluded that once the ion concentration is beyond a critical value, the SSTR will approximately not vary with it.

Because of the complexity of the hydrodynamic conditions in estuary, the movement of sediment here is not only affected by runoff, but also by tidal current, wave and coastal current （Wang et al. , 2010; Danchuk, 2011）, however, only the influences of discharge are concerned here. In addition, while water body of the estuary areas contains lots of ions which have considerable and complicated impacts on sediment movement, only the affects of Ca^{2+} are chosen to be explored in this paper. Thereby, as the development of the estuary resources extensively goes on, it obviously becomes urgent to further investigate the movement of sediment-laden flow synthetically impacted by complex hydrodynamic conditions and multiplicate ions in the future.

Acknowledgements

This work is financially sponsored by the National Natural Science Foundation of China (No. 10672125 & 10932012).

References

Danchuk, Samantha, Willson C S. Influence of Seasonal Variability of Lower Mississippi River Discharge, Temperature, Suspended Sediments, and Salinity on Oil-mineral Aggregate Formation [J]. Water Environment Research, 2011, 83(7): 579 –587.

Du P J, Ding P X, Hu K L. Simulation of Three – dimensional Cohesive Sediment Transport in Hangzhou Bay, China [J]. Acta Oceanologica Sinica, 2010, 29(2): 98 –106.

Giardino A, Ibrahim E, Adam S, et al. Hydrodynamics and Cohesive Sediment Transport in the IJzer Estuary, Belgium: Case Study [J]. Journal of Waterway Port Coastal and Ocean Engineering – ASCE, 2009, 135(4): 176 –184.

Huang J W. An Experimental Study of the Scouring and Setting Properties of Cohesive Sediment [J]. The Ocean Engineering, 1989, 7(1): 61 –70.

Le P J. Experimental Study on the Setting Properties of Cohesive Sediment in Static Water [J]. Journal of Sediment Research, 1983(2): 74 –78.

Liu F, Chen S L, Peng J, et al. Multi – scale Variability of Flow Discharge and Sediment Load of Yellow River to Sea and its Impacts on the Estuary during the Past 60 Years [J]. Acta Geographica Sinica, 2011, 66(3): 313 –323.

Maggi F, Mietta F, Winterwerp J C. Effect of Variable Fractal Dimension on the Floc Size Distribution of Suspended Cohesive Sediment [J]. Journal of Hydrology, 2007, 343(1 –2): 43 –55.

Maggi F. Stochastic Flocculation of Cohesive Sediment: Analysis of Floc Mobility within the Floc Size Spectrum [J]. Water Resources Research, 2008, 41(1): W01433.

Tan G M, Jiang L, Shu C W, et al. Experimental Study of Scour Rate in Consolidated Cohesive Sediment [J]. Journal of Hydrodynamics, 2010, 22(1): 51 –57.

Wang H J, Bi N S, Saito Y, et al. Recent Changes in Sediment Delivery by the Huanghe (Yellow River) to thesea: Causes and Environmental implications in its Estuary [J]. Journal of Hydrology, 2010, 391(3 –4): 302 –31.

Wang J S, Chen L, Liu L, et al. Experimental Study of Feature of the cohesive sediment lamination movement [J]. Advances in Water Science, 2008, 19(1): 13 –18.

Wang J S, Chen L, Wang Z G, et al. Study on the Fine Sediment Settling Velocity Formula with the Parameter of Ca^{2+} Concentration [J]. Advances in Water Science, 2006, 17(1): 1 –6.

Wang K R, Ru Y Y, Chen X T, et al. Discussion on the Dynamic Equilibrium Problem of the Delta Coastline of the Yellow River Estuary [J]. Journal of Sediment Research, 2007(6): 66 –70.

Wang L, Li J C, Zhou J F. Numerical Study of Flocculation Settling of Cohesive Sediment [J]. Acta Physica Sinica, 2010, 59(5): 3315 –3323.

Wang X H, Wang H J. Tidal Straining Effect on the Suspended Sediment Transport in the Huanghe (Yellow River) Estuary, China [J]. Ocean Dynamics, 2010, 60(5): 1273 –1283.

Wei F S, Qi W Q, Hua X, et al. Monitoring and Analysis Method Guide for Water and Waste Water (Middle Volume) [M]. Beijing: China Environmental Science Press, 1997: 389 –395.

Xu W S, Chen L, Liu L, et al. Experimental Study on Vertical Distribution of Flow Velocity in Weak Dynamical Flow with Cohesive Sediment [J]. Journal of Sediment Research, 2009 (6): 37 –42.

Xu W S, Chen L, Liu L, et al. Experiment on Vertical Distribution Characteristics of Flow Velocity in Cohesive Sediment Lamination Movement [J]. Engineering Journal of Wuhan University, 2010, 43(3): 315 –319.

Yao P, Wang X K. Study on the Plunging Laws of Turbidity Density Current [J]. Journal of Hydraulic Engineering, 1996(8): 77 – 83.

Study on Evolvement of High Beach and Deep Channel of Reservoir in the Sediment – laden Rivers

Li Kunpeng, *Zhang Junhua*, *Ma Huaibao* and *Wang Ting*

Yellow River Institute of Hydraulic Research, YRCC Key Laboratory of Yellow River
Sediment Research, MWR, Zhengzhou, 450003, China

Abstract: According to datas of Sanmenxia Reservoir and physical model test of Xiaolangdi Reservoir during later sediment retaining period, the paper study evolvement of high beach and deep channel of reservoir in the sediment – laden rivers. The results show that: ①the evolution of high beach and deep channel is as follows, firstly, beach and channel increase synchronously, and then scour happens in channel. Channel scour and scour amount is crucial to form high beach and deep channel. ②It is found that tributaries were equivalent to the horizontal extension of mainstream riverbed. The deposition process of tributary is closely related to many factors of natural topographic condition, deposition morphology of mainstream at the confluence area of mainstream and tributary and so on. The cross – section of tributary kept uplift horizontally, and sand bar has formed in some tributary estuary. ③Rainfall erosion formed in flood period is beneficial to forming hyper – concentrated flow, to a certain extent; it can recover channel storage capacity, and play an important role on utilizing long – term comprehensive benefits. Studying evolvement of high beach and deep channel can supply some technical supports for regulating operational mode of Xiaolangdi Reservoir during later sediment retaining period and prolonging service life.

Key words: high beach and deep channel, rainfall erosion, sediment – laden river, Sanmenxia Rteservoir, Xiaolangdi Reservoir

1　Introduction

By the end of 2010, and the amount of sediment in Xiaolangdi Reservoir reached $2.822,5 \times 10^9 \mathrm{m}^3$, and exceeded the cut off value of initial sediment retaining period and later sediment retaining period, that is 2.1×10^9 to 2.2×10^9 m^3 (referring to the cone deposition and the deposition surface elevation in front of the dam is 205 m), but the deposition surface elevation in front of the dam is only about 189 m. According to terrain of 2011, the reservoir storage capacity below 205 is only $0.208 \times 10^9 \mathrm{m}^3$. This means that the operation of Xiaolangdi Reservoir is about to turn into later operation of sediment retaining.

The most obvious difference of the early period and the later period of sediment retaining in Xiaolangdi Reservoir is that the reservoir will carry man – made precipitation washout at right occasion, recover channel storage capacity and form high beach and deep channel. After operation mode turning into the mode of storing clear water and discharging muddy water, under the condition of ensuring the safety of flood control in flood season, water and sediment regulation can be carried out so as the reservoir channel storage capacity can keep equilibrium of scour and deposition, an reservoir will play a role on reducing deposition of the Yellow River in long term.

2　Formation of high beach and deep channel in Sanmenxia Reservoir

After the construction of the Sanmenxia Reservoir, influenced by reservoir operation and water and sediment condition from 1960 to 1973, the morphology of high beach and deep channel formed in the reach below Tongguan. That is to say, because of inadequacy of the reservoir spillway and continuous high sediment concentrated flood from Tongguan station, deposition happened in beach and channel of reach bellow Tongguan, and by the October of 1964, the beach reached the highest deposition elevation and uplifted 2.4 ~ 5.6 m than that in October of 1963, and the high beach formed. Channel happened only in the reach near dam (dam to HH22) because of rainfall erosion, and channel in other reach was as high as beach. With the reconstruction of project in 1964 and beneficial water and sediment condition from Tongguan station, six times intense scour had occurred downstream Tongguan section, and channel was scoured continuously. Compared with October of 1964, the channel elevation between dam and HH22 decreased 7.3 ~ 7.7 m in October of 1973, and the channel elevation between HH31 and HH41 decreased 0.2 ~ 3.9 m. Since October of 1964, the beach elevation changed little, the elevation difference increased, and high beach and deep channel bellow Tongguan formed.

2.1 Scouring time and inflow and outflow water and sediment

Six times intense scour had occurred downstream tongguan section from March of 1962 to October of 1973. Namely, from Oct. 9, 1964 to Mar. 18, 1965 (the first time), from Oct. 5, 1966 to May 13, 1967 (the second time), from Oct. 8, 1968 to Oct. 8, 1969 (the third time), from Jun. 1, 1970 to Oct. 12, 1970 (the fourth time), from Jun. 25, 1972 to Jul. 31, 1972 (the fifth time), and from Jul. 1, 1973 to Aug. 19, 1973 (the sixth time).

According to the actual measurement of sediment discharge and flow, it is counted that the total water content and sediment of six scour time into the Tongguan Reservoir are 100.643×10^9 m^3 and 4.02×10^9 t. The total drainage and sediment in Sanmenxia Reservoir are 103.721×10^9 m^3 and 5.209×10^9 t. The ratio of total water and sand outbound accounted for inbound are 103.1% and 129.6% respectively. on October 9, 1964 to 1965 on March 18th, the ratio of desilting in reservoir is run up to as high as 261.2% ; the total sediment of reservoir is 1.189×10^9 m^3 t in six scour time (See Tab. 1).

Tab. 1 The changes of water and sediment according to entering – flowing out of reservoir in six scour times

Period of time	The total duration (d)	Tongguan		Sanmenxia		Out of proportion accounted into the library	
		Water content ($\times 10^9$ m^3)	Sand content ($\times 10^9$ t)	Water content ($\times 10^9$ m^3)	Sand content ($\times 10^9$ t)	Water content (%)	Sand content (%)
The first	161	20.697	0.250	21.283	0.653	102.8	261.2
The second	221	20.591	0.272	22.186	0.517	107.7	190.1
The third	366	33.306	1.342	34.334	1.538	103.1	114.6
The fourth	134	17.187	1.625	16.890	1.797	98.3	110.6
The fifth	38	4.482	0.234	4.669	0.317	104.2	135.5
The sixth	51	4.380	0.297	4.359	0.387	99.5	130.3
Combination meter	971	100.643	4.020	103.721	5.209	103.1	129.6

Hydrologic and sediment is the main part of reservoir sediment, and the deformation of riverbed power. The flow characteristic values in six scour time of Tongguan and Sanmenxia station at all levels are shown in Tab. 2. The analysis of Tab. 2 shows that six scouring time flow is greater than 3,000 m^3/s, into the sediment amount to 1.413×10^9 t, which first, second and fifth session sand storage amount is small, slightly scoured under the Tongguan Reservoir; in the sixth time the volume of inflow including water and sediment are small, that is flood detention and occurring deposition under tongguan reservoir; the third, fourth time into the reservoir water and sediment are very big, because in the midst of the reservoir after modification for the first time and second time before the renovation, although for the first time after modification of reservoir discharge capacity increases, but due to insufficient discharge sediment delivery facilities, reservoir sedimentation caused by backwater of still very serious, Tongguan Reservoir sedimentation under these two period up to 4.22×10^9 t. The rest of the flow level have been varying degrees of scour. The discharge water is during $2,000 \sim 3,000$ m^3/s in Sanmenxia Reservoir, which expending water at least of desilting, On average every one hundred million tons of sediment water consumption of 1.239×10^9 m^3. But when the discharge water is less than 1,000 m^3/s, the expending water at most of desilting, on average every one hundred million tons of sediment water consumption of 2.69×10^9 m^3.

Tab. 2　The flow characteristic values in six scour times

Station name	Period of time	Maximum flood peak flow (m^3/s)	Date of occurrence (Year, month and day)	Flow rate>3,000 m^3/s			Flow rate>2,000 m^3/s			Flow rate>1,000 m^3/s		
				Diachronic (d)	Water content ($\times10^9$ m^3)	Sand content ($\times10^9$ t)	Diachronic (d)	Water content ($\times10^9$ m^3)	Sand content ($\times10^9$ t)	Diachronic (d)	Water content ($\times10^9$ m^3)	Sand content ($\times10^9$ t)
Tong-guan	The first	5,210	1964.10.18	24	7.935	0.117	14	2.925	0.042	25	3.460	0.041
	The second	3,310	1966.10.05	10	2.736	0.056	19	3.566	0.128	68	7.741	0.095
	The third	5,680	1969.07.28	14	4.798	0.318	16	3.405	0.296	115	12.690	0.503
	The fourth	8,420	1970.08.03	13	4.419	0.819	20	4.020	0.325	53	6.175	0.360
	The fifth	8,600	1972.07.21	1	0.418	0.071	1	0.180	0.004	27	3.292	0.152
	The sixth	4,840	1973.07.19	1	0.285	0.032	2	0.393	0.040	15	2.067	0.156
	Combination meter			63	20.591	1.413	72	14.489	0.835	303	35.425	1.307
San-men-xia	The first	4,350	1964.10.19	25	8.497	0.123	12	2.554	0.151	27	3.730	0.181
	The second	3,430	1966.10.05	12	3.354	0.101	21	4.377	0.146	94	11.168	0.200
	The third	5,080	1968.10.13	16	5.452	0.243	25	5.452	0.366	126	14.396	0.645
	The fourth	4,930	1970.08.30	9	3.112	0.472	23	4.865	0.644	48	5.854	0.469
	The fifth	5,000	1972.07.21	2	0.565	0.076	1	0.181	0.050	28	3.376	0.170
	The sixth	3,350	1973.07.19	0	0	0	4	0.836	0.117	17	2.115	0.169
	Combination meter			24	20.980	1.015	86	18.265	1.474	340	40.639	1.834

2.2 The distribution and longitudinal adjustment between scourring and silting

2.2.1 Scour and silting amount and distribution

According to the test statistics of reservoir section, it is scouring sediment 8.86×10^8 m^3 under Tongguan Reservoir during the six scour time (See Tab. 3). It is mainly scouring in the 12 – 36 channel segment silting in the Yellow River, which totally scouring 7.08×10^8 m^3, accounting for 79.9% of the total scouring.

Tab. 3 The different river scourring and silting under Tongguan Reservoir in six scour times

Period of time	Sediment($\times 10^6$ m^3)					
	From 01 to12	From 12 to22	From 22 to31	From 31 to36	From 36 to41	Combination meter From 01 to41
The first	– 36.27	– 121.33	– 93.80	– 74.55	1.82	– 324.13
The second	– 17.40	– 44.19	– 45.91	– 6.41	– 2.54	– 116.45
The third	– 31.28	– 73.60	– 59.84	– 9.99	9.02	– 165.69
The fourth	– 18.68	– 23.51	– 40.60	– 35.45	– 30.87	– 149.10
The fifth	– 25.28	– 14.97	– 6.36	– 15.73	0.01	– 62.35
The sixth	– 15.50	– 11.46	– 14.63	– 15.33	– 11.28	– 68.20

2.2.2 Longitudinal adjustment

From March 1962 to October 1964, due to changes in reservoir operational mode, the reservoir silting eased. But due to insufficient discharge facilities, encounter wet abundant sediment during flood season in 1964, the reservoir still happen serious flood and sand detention, mass deposition, forming high beach trough, the capacity of a large number of loss, fill form for cone by delta development, longitudinal gradient changes between 1.1‰ ~ 1.7‰. In January 1965 the Sanmenxia dam rebuilt starts for the first time, which was completed in 1966 and enlarged the discharge capacity, the reservoir using water level lowering, Tongguan under reservoir flushing. In December 1969, the dam began the second renovation, the discharge capacity of reservoir, in June 1970 to October 1973 Tongguan under reservoir flushing during 4.0×10^8 m^3, forming high beach along cross section.

2.3 The water level changes along the reservoir upstream to downstream

Sanmenxia Reservoir operation is reflected in the change of water level before dam. As the conservancy hub for the two big rebuilding and control using the water level before dam, wash adjust the tongguan under reservoir and the reservoir water level along also change accordingly. Six time scouring tongguan water level of reservoir area along the below analysis: longitudinal surface slope from 0.54 ‰ to 2.02 ‰ for the first time; The second time longitudinal surface slope from 1.42 ‰ to 1.83 ‰; The third period longitudinal surface slope increased from 1.4 ‰ to 2.17 ‰. The fourth time longitudinal surface slope increased from 2.2 ‰ to 2.45 ‰. The fifth time longitudinal surface slope increased from 2.27 ‰0 to 2.88 ‰. The sixth time longitudinal surface slope increased from 2.96 ‰ to 3.29 ‰. Due to lower control using the water level before dam (the average water level dropped to 300.3 m, even under 290 m), six longitudinal surface slope scouring time the general trend is increase gradually, that Tongguan under reservoir area along river alluvial sediment of scour, riverbed were adjusted accordingly, slope increases.

3 Simulation test of Xiaolangdi in later sediment retaining period

Forming high beach and deep channel in the late blocking sand of Xiaolangdi Reservoir, the main idea of test is "multi – year sediment regulation and man – made precipitation washout at right occasion. It is refers to the reservoir in the process of using, when there is a large flood process, it will be reduced the reservoir appropriately using water level, erosion of the warp, the recovery capacity of the main channel, in order to extend the life of the reservoir sand bar.

3.1 Test conditions

Simulation experiment is at a north suburb of Yellow River Institute of Hydraulic Research, on the base of Xiaolangdi Reservoir model test of topography in 2007 after the sin of the measured terrain on the 1990 ~ 1999 + 20 years from 1956 to 1965 in water and sediment series and the proposed reservoir regulating mode gradually shaped naturally. This series of xiaolangdi into the reservoir is an average of 24.91×10^9 m^3, an average annual sediment is 8.54×10^8 t, an average of sediment concentration is 34.29 kg/m^3, among them 7 ~ 9 months cover an average of 9.21×10^9 m^3 of water and sediment an average of 7.82×10^9 t, the average sediment concentration of 84.94 kg/m^3. The maximum annual water amount and the maximum annual sediment for 19 years, respectively is 47.32×10^9 m^3, 1.88×10^9 t, the minimum water 14.22×10^9 m^3 (2 years), minimum annual sediment of 2.39×10^8 t (20 years).

3.2 The test results

3.2.1 Precipitation washout

In the process of simulation test, in mid to late August at the 13th year, water level before dam gradually declined, until August 28, water level before dam reduced to 230 m for the first time. Reservoir water level dropped dramatically, top – down strong beach groove flushing, flushing 4. 269×10^8 t, the reservoir of sediment delivery rate is 122.5%. During September 11 to 21 in the 18 th year, the water level before dam from 235 m to 231 m gradually decline in September 17; On September 21, water level rise to 235 m, and to September 24 again declined to 230 m, precipitation erosion happened the second time in this year. In July 19 the reservoirs is in storage period, with the large sediment concentration (maximum sediment concentration of 350 kg/m^3), the topography is depositing, during the August 13 to 28, warehousing maximum flow rate of 6,561 m^3/s, maximum sediment concentration of 260 kg/m^3, the water level before dam from 248 m to 230 m, The whole reservoir are characterized by strong main channel scour broadening, individual sections show the full face flushing (HH13, HH17).

3.2.2 Sediment volume

In 20 years the water – sediment series of annual average incoming sediment 8.54×10^8 t, 5.57×10^8 t of annual average outbound sediment, an average of sediment delivery rate is 65.3%. The first to second years, the reservoir storage capacity is bigger, cover most of the heavy flow desilting. Then in state of reservoirs, reservoir, the streamwise tend to show different weight range of flow sediment flow. Reservoir used to the 13th years, it meet a larger flow rate when the water level dropped dramatically, the lowest to the minimum water level reservoir operation period, the reservoir from the accumulated sedimentation trend to scour, the sediment delivery rate of reservoir is up to 122.5%. the 19th years for the series water sediment were the largest for a year, from July to September water sediment of 22.54×10^9 m^3 respectively and 22.54×10^9 t, an average of 244.8% and 244.8% respectively in 20 years series. Larger flow rate in low water level before dam, forming a full reservoir channel incised sharply broadening status, the quantity is larger, the reservoir of fly up to 131.3%. 20 years series river alluvial 3.584×10^9 m^3, a tributary of deposition in 1.412×10^9 m^3, sedimentation volume of 4.996×10^9 m^3. River alluvial, relative to the capacity of the original 5.559×10^9 m^3, a tributary of deposition in 1.824×10^9 m^3, the reservoir silting total 7.383×10^9 m^3.

3.2.3 The deposition in form of main stream

In the process of using, the sediment deposition in the reservoir by water level change is uneven, the deposition morphology will constantly adjust accordingly. The first year is basically on the basis of original terrain synchronous lifting, longitudinal gradient changed little; the second year in control water level before dam, alluvial delta in the downstream of the process, delta vertex height is lower, the delta plane and longitudinal gradient is increased to 3. 9 ‰; 8 ~ 10 years of the main longitudinal gradient increased year by year, 2. 93 ‰ of which related to the conditions of coming water and sediment, the smaller flow with high sediment concentration, the section of the sediment deposition in reservoir area is larger, and often the reservoir storage capacity is relatively low, making the longitudinal gradient increased; 11 years cover coming water and sediment amount is larger, reservoir area along the scour deposition in the upper surface is reduced, the reservoir storage capacity big, picked up front silting dam surface, the reservoir profile to slow; 13 years back flushing precipitation, increased longitudinal gradient; after 14 years to 17 years, longitudinal gradient gradually decreased; 18 ~ 19 years of precipitation erosion, making the longitudinal gradient is increased significantly.

3.2.4　The deposition in form of tributary

In the process of shaping in high beach and deep channel, the tributary is equivalent to the horizontal extension of main stream, a tributary of the riverbed deposition process and terrain conditions and intersection of main stream – tributary river siltation forms and process closely related to factors such as, a tributary of cross – sectional silting form mostly parallel rise trend, part of a tributary of the obvious at the entrance of the bar. A tributary of evolution process is as follows:

(1) When the main channel of the main stream is close to the entrance of the tributary, the water in the tributaries is started to flow slowly to the main stream during the process of precipitation, and the tributary of longitudinal is pulling a small channel, as the small channel is gradually deep and broadening, and tributary flowing into the flow of main stream gradually increases, eventually a tributary of the main channel and the river main channel at the entrance of the elevation. Began after the impoundment of the reservoir, when the water level of river water is higher than at the entrance of the tributaries of a tributary of the main channel, river water flow along the tributaries of the main channel to tributaries upstream flow backward, tributary under water sedimentation, not easy to form the bar. When river water from tributaries entrance beaches, a tributary of the whole section at the entrance of the flow, in the form of water flow backward. As main stream water level rise gradually, with density flow at the entrance of the tributary stream flow backward, tributary beach trough will rise gradually deposition, in the case of a long time, finally beach trough deposition in the same height, easy to form a tributary of sand hom.

(2) Main stream channel of river from the entrance of the tributary, the reservoir water is not easy to flow to the main tributaries in the process of precipitation, tributary profile is not scouring. Reservoir began filling, after the river water level above the channel at the entrance of the tributary, a tributary of the whole section at the entrance of the flow, in the form of free flow backward; As main stream water level rise gradually, with density flow at the entrance of the tributary stream flow backward, will gradually fill up at the entrance of the tributaries, in the case of a long time, easy to form a tributary of sand hom.

(3) The using of reservoir with high water level for a long time, at the entrance of the tributaries is small, but inside is big, it is easy to form the sand hom. At the entrance of the tributary as big as inside or larger than inside, it is not easy to form the sand hom.

4　Evolution of high beach and deep channel

Simulation test of Sanmenxia and Xiaolangdi Reservoir in later sediment retaining period, the forming of high beach and deep channel is based on slot synchronization depositing in uplift and the main channel scouring process of falling again. In the process of main channel scour should not only take away in storage of sediment, but also need to scour the reservoir main channel the early deposition of sediment, to increase the beach trough, recover flat beach under the water level of the tank volume, so whether the main channel scour and scour and silting amount about whether can form deposition in high beach and deep channel shape is very important.

Built on the sand river reservoir in storage period for one year, the reservoir backwater deposition will occur. In sediment delivery period, reservoir discharge capacity is insufficient, the reservoir backwater deposition also will occur. Early reservoir with this due to backwater reservoir sedimentation is in whole section, that is to say, beach have sedimentation tank. After the flood, water level dropped, which creates a scour from top to bottom. With further water falling, even drain empty, then the reservoir not only has from top to bottom along the erosion, but also from bottom to top of back flushing. The joint action of two kinds of scouring, it is forming a channel on the surface of the reservoir sedimentation, As the channel scourring has decreased, it gradually formed the shape of high beach and deep channel.

5 Conclusions

According to datas of Sanmenxia Reservoir and physical model test of Xiaolangdi Reservoir during later sediment retaining period, The results show that: ①The evolution of high beach and deep channel is as follows, firstly, beach and channel increase synchronously , and then scour happens in channel. Channel scour and scour amount is crucial to form high beach and deep channel. ②In the process of forming high beach and deep channel, it is found that tributaries were equivalent to the horizontal extension of mainstream riverbed. The deposition process of tributary is closely related to many factors of natural topographic condition, deposition morphology of mainstream at the confluence area of mainstream and tributary and so on. The cross – section of tributary kept uplift horizontally, and sand bar has formed in some tributary estuary. ③Reservoir impoundment it will occur backwater sedimentation, when the desilting discharged capacity is insufficient, the reservoir backwater sedimentation, also occurs due to backwater of deposition takes place in the whole section. ④In the reservoir precipitation period, water level dropped, which creates a scour from top to bottom. With further water falling, even drain empty, then the reservoir not only has from top to bottom along the erosion, but also from bottom to top of back flushing. The joint action of two kinds of scouring, it is forming a channel on the surface of the reservoir sedimentation, As the channel scourring has decreased, it gradually formed the shape of high beach and deep channel. ⑤Rainfall erosion formed in flood period is beneficial to forming hyper – concentrated flow, to a certain extent, it can recover channel storage capacity, and play an important role on utilizing long – term comprehensive benefits.

Influences on Erosion and Sedimentation of Lower Weihe River Estuaries and Reaches by Rechanneling of North Luohe River Directly into Yellow River

Wang Puqing, *Jiang Naiqian* and *Wang Guodong*

Institute of Hydraulic Research of Yellow River, Zhengzhou, 450003, China

Abstract: The rechanneling of North Luohe River directly into Yellow River involves the elevation of Tongguan, flood – prevention works at the lower Weihe River and bed evolution problems of the confluence area of the three rivers and generates huge social and environmental influences. Confluence estuary of the rechanneled course into Yellow River relies on the Niumaowan project, and the included angle between its inflow angle and the main stream channel is larger and the Yellow River is changeable continuously, so the rechanneled North Luohe River will definitely generate certain influences over the pattern courses in the North Main Stream and the North Luohe River estuary. According to the test, when North Luohe River is hit by highly sediment concentrated flood, the reaches around the confluence estuary of the North Luohe River is easily to be silted and the Niumaowan project at lower reaches of the confluence estuary may easily off river, finally result in weakening of control over the reaches. But the rechanneling of North Luohe River has not changed the river patterns at the confluence area of the North Main Stream station fundamentally.

Key words: rechanneled course, Weihe River estuary, confluence estuary, current course, scour – and – fill change

1 Problem raising

Since the sediment concentration of North Luohe River flood is comparatively larger and its mean sediment concentration of accumulated year is 117 kg/m^3, much more than the 30 kg/m^3 of the Yellow River and 51.6 kg/m^3 of Weihe River, then when its highly sediment – concentrated flood inflows into the Weihe River and if the inflow of Weihehe River is little and the Yellow River floods jack or encroach the Weihehe River, the North Luohe River estuary is usually severely silted and the reaches above North Luohe River estuary and some reaches at lower Weihe River might even be deadly silted. For example, the Yellow River flood encroached Weihe River on August 24, 1967, when a highly sediment – concentrated flood whose peak discharge was 374 m^3/s and maximum sediment concentration was 850 kg/m^3 hit North Luohe River at the same time, then the estuary where the North Luohe River enters Weihe River silted and such siltation went upwards, finally resulted in a drying – up length of 8.8 km and Weihe River dyke breaching and the flood discharged downwards through the Erhua channel. Though the silted reaches have been excavated in 1968 and have been scoured by floods of that year, the recovery of the channel's flood discharge capacity is still very limited, and the Weihe River dyke was still breached in 1968. In 1994 and 1995, similar floods happened and the lower Weihe River course was also severely silted.

Rechanneling the North Luohe River directly into Yellow River seems can directly prevent the large sediment accumulation problem in the lower Weihe River estuaries and reaches caused by North Luohe River sediments and reduce the siltation in the Weihe River estuary resulted from the Yellow River flood encroaching Weihe River. In order to research and argue whether the rechanneling against North Luohe River directly into Yellow River can truly reduce the sediment siltation in the estuaries and reaches of Weihe River, this paper has employed the physical model test and its observation data to carry out researches.

2 Model introduction

The mode test is done based on the Sanmenxia Reservoir model built, and the model inlets are Weihe River Huaxian, North Luohe River Chaoyi, North Main Stream of Yellow River 45 section (the upper source) while the outlet is set on Sanmenxia Dyke. The model's horizontal scale $\lambda_l =$ 420 and its vertical scale $\lambda_h = 50$. The sediment concentration scale of the model $\lambda_s = 1.8$ while the scour and fill transfigured time scale $\lambda_{t2} = 57.5$.

2.1 Test boundary conditions

According to the test requirements and actual model site conditions, the model inlets have been jointly formed by Weihe River Huaxian, North Luohe River Chaoyi, North Main Stream of Yellow River 45 section (the upper source) while the outlet is set in Sanmenxia Dyke. The initial model terrain is formulated according to the sectional data after 2003 flood and the topographic map of the river courses. The operational mode of Sanmenxia Reservoir is: if the inflow volume in Tongguan exceeds 1,500 m³/s in flood season, then conduct ungated flow; if it is less than 1,500 m³/s, control and operate according to 305 m requirement while control and operate it if the highest water level is not larger than 318 m in non – flood season.

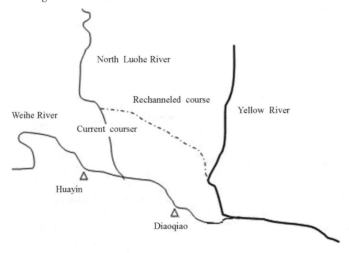

Fig. 1 Sketch map of North Luohe river rechanneling directly into Yellow River

The position where the North Luohe River directly joins Yellow River after rechanneling lies in the place 1.42 km upstream from the 4# section and enters into Yellow River at the 44# Dyke of Niumaowan project, the whole course is about 7.50 km long, see Fig. 1. The reach scale from Chaoyi to the rechanneled confluence estuary after rechanneling reduces to 1.71‰. The cofferdam distance of new channels is 1,200 m and the channel is designed as trapezoid section.

2.2 Water – Sediment situation

The inflowing water and sediment features of the North Main Stream is illustrated in Tab. 1, and its annual water volume at maximum is 2.401,9 × 10¹⁰ m³ (1994), while the aggregate water volume of the non – flood season in 1997 plus that of the July and August of 1967 is 2.387,5 × 10¹⁰ m³; the maximally annual sediment concentration is 7.36 × 10⁸ t (1994), while the largest flood discharge happed in July to August of 1967, and the maximum peak discharge on August 11 was 8,454 m^3/s and its corresponding sediment concentration was 10^6 kg/m^3. The maximum annually water volume at Weihe River Huaxian Station is 3.662×10^9 m^3 (1994) and its maximally annual sediment concentration is 4.12×10^8 t (1996), the maximum flood discharge happened at the end of July, 1996, and the maximally daily mean flow on July 29 was 3,426 m^3/s and its corresponding sediment concentration was 442 kg/m^3. The maximum flood at Zhuangtou Station of North Luohe River happened at the end of August and early September of 1994, its maximally daily mean flow on September 2 was 2,050 m^3/s and its corresponding sediment concentration was 560 kg/m^3.

Tab. 1 Water – sediment statistics for model inlet controlled tation

Station	Project	1994 Non – flood season	1994 Flood season	1995 Non – flood season	1995 Flood season	1996 Non – flood season	1996 Flood season	1997 Non – flood season	1967 Non – flood season
North main stream	Water content	129.5	110.7	128.4	99.36	117.34	100.6	84.1	154.65
	Sediment concentration	1.61	5.75	1.83	4.69	1.91	5.27	1.12	9.94
Zhuangtou	Water content	4.03	5.8	1.47	2.95	1.35	4.45	3.03	2.25
	Sediment concentration	0.04	2.17	0	0.34	0.01	0.81	0	0.76
Huaxian	Water content	19.8	16.8	10.94	11.42	8.76	22.9	17.6	20.58
	Sediment concentration	0.24	3.57	0.04	2.37	0.09	4.03	0.1	1.37

3 Scour – and – fill changes against lower Weihe River estuaries and reaches

The water level change is the major physical parameter reflecting the evolution of course erosion and sedimentation. In this paper, the information about Huayin and Diaoqiao Gauging Stations in Weihe River estuaries and reaches are employed to research the influences of rechanneling of North Luohe River on the lower estuaries and reaches, of which, Huayin Station is about 6.2 km upstream the North Luohe River estuary while Diaoqiao Station is 6.1 km downstream the current North Luohe River estuary. According to the test observed data, certain changes will occur compared with the current courses after rechanneling directly into Yellow River.

The flood happened on July 7 to 19, 1994 was the first flood in testing water and sediments of Weihe River, when the maximal daily mean flow at Huaxian was 1,300 m^3/s and the maximal daily sediment concentration was 600 kg/m^3. During this stage, a small flood whose maximal daily mean flow was 423 m^3/s and maximal daily sediment concentration was 688.7 kg/m^3 hit North Luohe River accordingly, while a small flood whose maximal daily mean flow was 3,191 m^3/s and maximal daily sediment concentration was 147 kg/m^3 hit North Main Stream.

Before and after the floods and under the current conditions, the water levels at Huayin Station and Diaoqiao Station have increased by 0.12 m and 0.16 m respectively, while the levels of Huayin and Diaoqiao after rechanneling increased by 0.08 m and 0.10 m respectively. It is obvious that the increase of water level in Huayin and Diaoqiao reduces after rechanneling. The reduction of water level in Huayin and Diaoqiao after rechanneling and the larger increase of water level in Diaoqiao indicate that the siltation degree in Weihe River estuaries and reaches reduces after rechanneling.

Catastrophic floods continuously hit North Main Stream Station in July to August of 1967, when Weihe River and North Luohe River were also hit by medium floods during this period. From

July 6 to July 20 of 1967, the maximally daily mean flow at Huaxian was 1,440 m³/s and its maximally daily sediment concentration was 53 kg/m³, while the corresponding daily mean flow of North Luohe River was 100 ~ 180 m³/s and its maximally daily mean sediment concentration was 389 kg/m³, generally about 10 kg/m³. Under the current conditions, the water levels at Huayin Station and Diaoqiao Station before and after Huaxian floods reduced by 0.18 m and 0.23 m respectively, while the water level of the same flux in Huayin Station and Diaoqiao Station after rechanneling reduced by 0.25 m and 0.30 m, which indicates that when the North Main Stream Station is continuously hit by floods and once the Weihe River is also hit by flood, the beds at Weihe River estuaries and reaches shall be scoured to a certain degree, and according to the test observed data, the scouring degree after rechanneling is greater than that under the current reach conditions.

In general, no matter the siltation problem at Weihe River estuaries and reaches is under unfavorable water and sediment conditions or scouring of reaches is under favorable water and sediment condition, the rechanneling of North Luohe River is beneficial to the siltation reduction or scouring against Weihe River estuaries and reaches compared with the current courses.

4 Conclusions

North Luohe River is a sediment – laden river and though its peak discharge is small, its sediment concentration is high and the floods hit it usually are those with less water but more sediments. According to the plentiful research achievements, the interplay factors between Noth Luohe River and Weihe River is complex and extensive. It is affected by the water – sediment conditions of the reach itself and is also closely affected by the combined water – sediment conditions of North Luohe River, Weihe River and North Main Stream. Meanwhile, the boundary conditions of reach including the overflow conditions, water pattern of the confluence area and the scouring condition of reaches are key factors resulting in reaches evolution. This research is only restricted to a specific topographic conditions and water – sediment conditions, and the influences of rechanneling of North Luohe River directly into Yellow River on Weihe River estuaries and reaches involve almost every party's interests and its social and environmental influences still needs further extensive and deep analysis and all – around argumentation.

Reference

Institute of Hydraulic Research of Yellow River. Overall Model Test Research on Diverting North Louhe River Directly into Yellow River (August 2006).

Study and Application on Flow – Sediment Mathematical Model of Dongzhuang Reservoir Area in Jinghe River [①]

Fu Jian, *Chen Cuixia*, *Li Qingguo* and *Wei Shitao*

Yellow River Engineering Consulting Co. ,Ltd. ,Zhengzhou,450003, China

Abstract: Hyperconcentrated flows have a significant influence on sediment carrying capacity and settling velocity of sediment particles, which leads to the sediment transport problem being more complex than common sediment – laden flows. Taking the influences above into account, packet sediment carrying capacity formula reflecting different states of riverbed erosion and deposition is established, and a mathematical model of reservoir sediment scour and deposition for silt calculation of hyperconcentrated flows is developed by applying correction method for settling velocity proposed by Zhang Hongwu to revising free settling velocity of a sediment particle and utilizing corresponding saturation recovery coefficient for different particle size based on the condition of erosion and deposition. The mathematical model is verified with measured data of Xiaolangdi Reservoir on the Yellow River and Bajiazui Reservoir on the Puhe River, the calculation of the reservoir siltation and deposition features agree well with measured data, and the process of sediment scour and deposition on the high sediment – laden river with reservoirs built is also well reflected. The mathematical model is adopted on the Dongzhuang Reservoir scheduling of water and sediment regulation for the calculation of reservoir sediment scour and deposition with using data of three series of flow and sediment. The results provide essential technical support for Jinghe river basin planning and reservoir planning and design.

Key words: reservoir sediment, mathematical model, verify, scour and deposition calculation, Dongzhuang Reservoir

1 Introduction

Flow – Sediment mathematical model on the reservoir is an important method for studying reservoir sedimentation problems and closely related to production practices such as river basin planning, project construction and management application, etc. As the fundamental theories of sediment transport is further researched and widely acknowledged, together with modern computational methods and computer technology rapidly developing, there has been a considerable development in the mathematical model for flow and sediment, and great progress has been made in computing model, numerical calculation method, preprocessing and postprocessing of calculation results, parameter selection and so on.

The Jinghe River has an extremely high flow sediment concentration. According to the observed data of Zhangjiashan hydrological station from July 1932 to June 2008, the annual average storage of water and sediment were 1.744×10^9 m^3 and 2.52×10^8 t respectively, the annual average sediment concentration was 145 kg/m^3, wherein the average sediment concentration in July and August were 317 kg/m^3 and 303 kg/m^3 respectively, and the measured instantaneous maximum sediment concentration reached 1,428 kg/m^3. There is a significant meaning in how to adjust the model to such hyperconcentrated flows for scour and deposition calculation in order to reasonable simulating the process of reservoir sediment erosion and deposition, which is also a hard task.

Taking into account the influences of flow sediment concentration on the sediment carrying capacity and settling velocity, the author establishes a formula of packet sediment carrying capacity combining high and low sediment concentration, and develops a mathematical model of Dongzhuang

① Fund Project: the Special Scientific Research of the Ministry of Water Resources for Public Welfare Industry (200901017)

Reservoir sediment scour and deposition. The mathematical model is verified with observed data of Xiaolangdi Reservoir on the Yellow River and Bajiazui Reservoir on the Puhe River, with which the process of sediment scour and deposition on the high sediment – laden river with reservoirs built is well reflected. The mathematical model is adopted on the Dongzhuang reservoir scheduling of water and sediment regulation for the calculation of reservoir sediment scour and deposition, and the achievements are utilized in reservoir planning and design.

2 Basic principles

2.1 Basic equations

The model is a one – dimensional steady suspended – load flow and sediment mathematical model, of which fundamental equations include flow continuity equation, flow movement equation, sediment continuity equation and riverbed deformation equation.

(1) Flow continuity equation

$$\frac{dQ}{dx} = 0 \tag{1}$$

(2) Flow movement equation

$$\frac{d}{dx}\left(\frac{Q^2}{A}\right) + gA\left(\frac{dZ}{dx} + J\right) = 0 \tag{2}$$

(3) Sediment continuity equation (for different particle size)

$$\frac{\partial}{\partial x}(QS_k) + \gamma\frac{\partial A_{dk}}{\partial t} = 0 \tag{3}$$

(4) Riverbed deformation equation

$$\gamma\frac{\partial Z_b}{\partial t} = \alpha\omega(S - S^*) \tag{4}$$

where, Q is the discharge; x the distance; g the acceleration of gravity; A the area of cross section; Z the water level; J the energy slope; k the grain size groups; S the sediment concentration; A_d the area of scour and deposition; t the time; γ the dry density of deposits; Z_b the thickness of scour and deposition; α the coefficient of saturation recovery; ω the sediment settling velocity; S^* the sediment carrying capacity.

2.2 Formula of sediment carrying capacity

Sediment carrying capacity is a comprehensive index reflecting the sediment carrying capacity of flow under the conditions of equilibrium of scour and deposition with certain incoming water and sediment, and a necessarily critical content of research on flow – sediment mathematical models. For a long time, many domestic and overseas experts and scholars in the field of engineering and academia have proposed a lot of semi – analytic, semi – empirical and experiential formulas of sediment carrying capacity from the theories, or depending on the data from different open – channel experiments and laboratories. At present, most of formulas of sediment carrying capacity at home and aboard are only suitable for low – sediment concentration flow, of which Zhang Ruijin's general formula has extensive application value. As for hyperconcentrated flows, the change of rheological behavior, fluid characteristic and sediment transportation because of the high sediment concentration and existence of fine particles results in the sediment transport problem being more complex than common sediment – laden flows. Among numerous formulas of sediment carrying capacity, though the processing procedure of Zhang Hongwu's formula is kind of empirical, the inclusiveness of calculation range is comparatively better and calculation results of full – loaded flow are more in accordance with practice. Therefore, the model employs Zhang Hongwu's formula of sediment carrying capacity.

$$S^* = 2.5\left[\frac{0.002,2 + S_v}{\kappa}\ln\left(\frac{h}{6D_{50}}\right)\right]^{0.62}\left(\frac{\gamma_m}{\gamma_s - \gamma_m}\frac{V^3}{gh\omega}\right)^{0.62} \tag{5}$$

where, D_{50} is the median diameter of bed material, mm; γ_s the bulk density of sand particles, 2,650 kg/m^3; γ_m the turbid water density; h the water depth, m; V the velocity, m/s; κ is the karman constant, $\kappa = 0.4 - 1.68\sqrt{S_v}(0.365 - S_v)$; S_v is the average sediment concentration in the inlet cross section under the calculation by volume ratio.

2.3 Gradation of sediment carrying capacity

Sediment with varied grain sizes has different effect on river scour and deposition. The gradation of sediment carrying capacity is usually obtained by multiplying size distribution of carrying capacity by total sediment carrying capacity. The computational method for gradation of sediment carrying capacity is put forward under condition of suspended silt transport non – equilibrium based on the exchange relationship between the size distribution of carrying capacity and bed material. The equation below is used for the calculation of gradation of sediment carrying capacity, which reflects the characteristics of size distribution of carrying capacity being siltation of most income sediment from upstream and scouring of most bed material.

$$S_k^* = \left\{ \frac{P_k \dfrac{S}{S + S^*} + P_{uk}\left(1 - \dfrac{S}{S + S^*}\right)}{\sum\limits_{k=1}^{nfs}\left[P_k \dfrac{S}{S + S^*} + P_{uk}\left(1 - \dfrac{S}{S + S^*}\right)\right]} \right\} S^* \tag{6}$$

where, S is the average sediment concentration in the upstream section; P_k is the size distribution of sediment from the upstream section; P_{uk} is the size distribution of surface bed material.

2.4 Settling velocity

Free settling velocity of single sediment particle is calculated according to formulae recommended by standard.

$$\omega_{0k} = \left(\begin{array}{ll} \dfrac{g}{1,800} \cdot \left(\dfrac{\rho_s - \rho}{\rho}\right) \cdot \dfrac{d_k^2}{v} & (d_k < 0.062 \text{ mm}) \\ (\lg S_a + 3.665)^2 + (\lg\varphi - 5.777)^2 = 39 & (0.062 \text{ mm} \leqslant d_k < 2 \text{ mm}) \end{array} \right. \tag{7}$$

where, φ is the discriminating number of particle size, $\varphi \dfrac{g^{\frac{1}{3}}\left(\dfrac{\rho_s - \rho}{\rho}\right)^{\frac{1}{3}} d_k}{10 v^{\frac{2}{3}}}$; S_a is the

discriminating number of settling velocity, $S_a = \dfrac{\omega_{0k}}{g^{\frac{1}{3}}\left(\dfrac{\rho_s - \rho}{\rho}\right)^{\frac{1}{3}} v^{\frac{1}{3}}}$; ρ_s is the sediment density; ρ is

the water density; d_k the representative diameter; v the kinematic viscosity coefficient of water.

The Jinghe River has a quite high flow sediment concentration, which has a considerable influence on the settling velocity of sediment particles, so modification should be done on free settling velocity of single sediment particle. Here correction method for settling velocity proposed by Zhang Hongwu is adopted.

$$\omega_{sk} = \omega_{0k}\left[\left(1 - \frac{S_\gamma}{2.25\sqrt{d_{50}}}\right)^{3.5}(1 - 1.25S_v)\right] \tag{8}$$

In the equation above, d_{50} is the median grain size of suspended sediment, mm.

The average settling velocity of mixed sand ω_s is gained by following equation:

$$\omega_s = \sum_{k=1}^{nfs} p_k \omega_{sk} \tag{9}$$

where, nfs is the total group number of particle sizes; p_k is the sediment weight percent of the k – th group of sediment particle size。

2.5 Coefficient of saturation recovery

When solving S, for different particle size corresponding α is utilized as follows:

$$\alpha_k = 0.001/\omega_k^{0.5} \tag{10}$$

After the state of scour and deposition is judged by trial calculation, the equation below is applied to updating the calculation results of coefficient of saturation recovery.

$$\alpha_k = \begin{cases} \alpha_*/\omega_k^{0.3} & S > S^* \\ \alpha_*/\omega_k^{0.7} & S < S^* \end{cases} \tag{11}$$

where, ω_k is the settling velocity of the k – th group of sediment particle size, m/s; α_* is a parameter calibrated based on observed data, which is often small in the inlet cross section and increasing as approaching the dam.

3 Numerical examples

3.1 Calculation example of Xiaolangdi Reservoir on the Yellow River

With the model employed, the calculation of sediment scour and deposition on the Xiaolangdi Reservoir is carried out depending on the measured incoming and outgoing process of water and sediment from November 1999 to October 2007.

3.1.1 Reservoir sedimentation

From November 1999 to October 2007, the observed deposition amount of Xiaolangdi Reservoir with method of sediment transport rate is 2.57×10^9 t, while the computational deposition amount calculated by mathematical model is 2.694×10^9 t, the calculation result of model is 1.24×10^8 t more than the observed amount, and the calculation error is about 5%. The process of reservoir sedimentation is shown as Fig. 1.

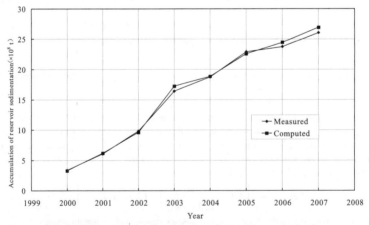

Fig. 1 Process of accumulation of reservoir sedimentation

3.1.2 Morphology of reservoir deposition

The comparison between longitudinal section of reservoir calculated by mathematical model and the measured one is shown as Fig. 2. As seen in the figure, the process of scour and deposition of Xiaolangdi reservoir simulated by the model agrees well with observed data.

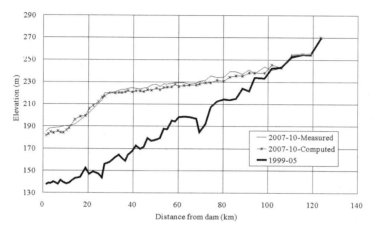

Fig. 2　Longitudinal profile of thalweg points in October 2007

3. 2　Verification of Bajiazui Reservoir

The incoming flow and sediment of Bajiazui Reservoir mainly come from the mainstream Puhe River and tributary Heihe River. Verification calculation is conducted on the basis of observed data from 1965 to 1970 with initial terrain of large section of Bajiazui Reservoir measured on May 1965.

3. 2. 1　Reservoir sedimentation

From May 1965 to May 1970, the deposition amount of Bajiazui Reservoir with method of section is 3.07×10^7 m^3, while the computational deposition amount calculated by mathematical model is 3.24×10^7 m^3, the calculated value is pretty close to the observed amount. The process of accumulation of reservoir sedimentation is shown as Fig. 3.

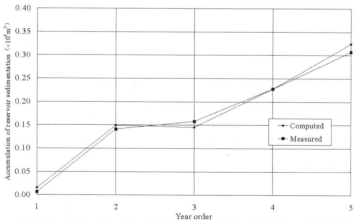

Fig. 3　Process of accumulation of Bajiazui Reservoir sedimentation
from May 1965 to May 1970

3. 2. 2　Morphology of reservoir deposition

The deposited longitudinal sections of mainstream Puhe River and tributary Heihe River of Bajiazui Reservoir computed by mathematical model are both quite close to the observed ones,

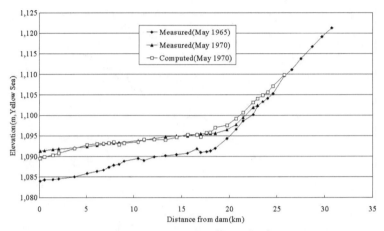

Fig. 4 **Longitudinal – section set graph of mainstream Puhe River of Bajiazui Reservoir**

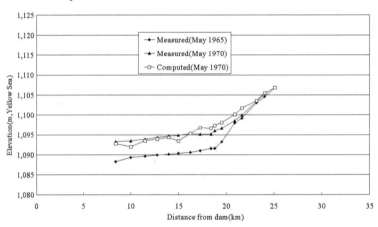

Fig. 5 **Longitudinal – section set graph of tributary Heihe River of Bajiazui Reservoir**

which are shown as Fig. 4 and Fig. 5.

4 Calculation of sediment scour and deposition of Dongzhuang Reservoir

4.1 Reservoir sedimentation

Three series of flow and sediment in the level of 2030 are used for the calculation of Reservoir sediment scour and deposition, which are the 1968 series of average flow and sediment, the 1961 series of relatively abundant flow and sediment at prophase, and the 1991 series of relatively poor flow and sediment at prophase respectively. The series length is all 50 years.

The computational result of sediment scour and deposition of Dongzhuang Reservoir is shown as Fig. 6. As seen in the calculation of reservoir sediment scour and deposition, the reservoir deposition in the 1961 series of relatively abundant flow and sediment in earlier stage is much faster, and the reservoir sediment storage periods of series 1968, 1961, 1991 are 27 years, 22 years, and 29 years respectively. After the sediment storage period of Dongzhuang Reservoir is over, the storage capacity for water and sediment regulation is employed as the function of water and sediment regulation during the main flood season, and the reservoir keeps the equilibrium of scour and deposition roughly for a long time. At the end of the calculation period of reservoir, the deposition amount of Dongzhuang Reservoir in three series of flow and sediment is all about 2.3×10^9 m^3.

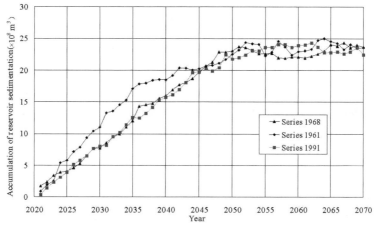

Fig. 6   Process of accumulation of Dongzhuang Reservoir sedimentation

4.2   Morphology reservoir deposition

The deposition morphology of Dongzhuang Reservoir in series 1968, 1961 and 1991 computed by mathematical model is shown as Fig. 7 to Fig. 9. As demonstrated in the figures, the changing process of deposition morphology of Dongzhuang Reservoir is almost same in different series of flow and sediment. In the early operation period, the reservoir is in the form of delta deposition, as the sediment scour and deposition is developing, the top of delta is gradually approaching towards the front of dam, when the peak arrives at the dam the deposition morphology of reservoir becomes cone shape, and then the sedimentation surface gradually uplifts afterwards; In the normal operation period, the reservoir sedimentation surface fluctuates between flood control level and dead water level.

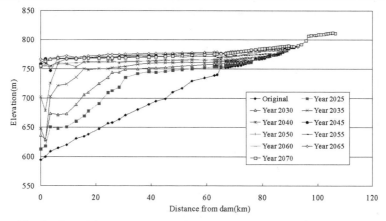

Fig. 7   Deposition morphology of Dongzhuang Reservoir(series 1968)

284

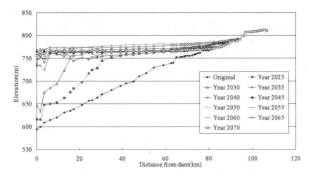

Fig. 8　Deposition morphology of Dongzhuang Reservoir(series 1961)

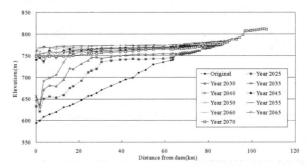

Fig. 9　Deposition morphology of Dongzhuang Reservoir(series 1991)

5　Conclusions

　　Taking into consideration the effects of hyperconcentrated flows on the sediment carrying capacity and settling velocity fully, packet sediment carrying capacity formula reflecting different states of riverbed erosion and deposition is established, and a flow – sediment mathematical model of Dongzhuang Reservoir on the Jinghe River is developed. The mathematical model is confirmed with observed data of Xiaolangdi Reservoir on the Yellow River and Bajiazui Reservoir on the Puhe River. The model is competent for simulating the calculation of scour and deposition of reservoir built on the high sediment – laden river, and the computational results conform to the practice. And the model is adopted in the calculation of sediment scour and deposition of Dongzhuang Reservoir and provides essential technical support for reservoir planning and design.

References

Zhang Ruijin, Xie Jianheng, et al. River Sediment Dynamics [M]. Beijing: China Water Power Press, 1998.

Yu Xin, An Cuihua, Guo Xuanying, et al. Study and Application of Sediment Hydrodynamic Mathematical Model for the Xiaolangdi Reservoir[J]. Yellow River, 2000 (8): 18 – 32.

Zhang Hongwu. Research on Similarity Law of Models for Flood in the Lower Yellow River[D]. Beijing: Tsinghua University, 1995.

Qian Yiying, Qu Shaojun, Cao Wenhong, et al. Mathematical Model for the Fluvial Processes in Yellow River[M]. Zhengzhou: Yellow River Conservancy Press, 1998.

Experiment Study on Effect of Grass on Runoff and Sediment Yield

Xiao Peiqing, *Yao Wenyi* and *Yang Er*

Institute of Yellow River Hydraulic Research, Key Laboratory of
Soil and Water Loss Process and Control on the Loess Plateau of the Ministry of
Water Resources, Zhengzhou,450003, China

Abstract: It is important to evaluate the impact of grass on soil erosion process so as to use them effectively on the Loess Plateau. Effect of grass on runoff and sediment yield was studied under rainfall intensities of 45 mm/h, 87 mm/h and 127 mm/h with 20° slope gradient using simulated rainfall experiment. The results showed that average runoff rates ranged from 39.7 L/min to 126.0 L/min for bare plots and 0.34 L/min to 6.22 L/min for grass plots, and the runoff rates from grass plots were much less than from bare plots. Average sediment yields varied from 3,636.7 g/min to 9,436.3 g/min for bare plots and from 26.7 g/min to 581.5 g/min for grass plots. Runoff and sediment yield increased with the increase of rainfall intensity. Compared with grass plots, runoff and sediment yield process on bare plots were closely related with surface condition, especially with rill development.

Key words: runoff, sediment yield, grass, simulated rainfall

1   Introduction

Soil erosion is one of the most severe eco – environmental problems in the world, particularly on the Loess Plateau of China. To improve the eco – environments in the region, the Chinese Government proposed the project of transforming the cultivated land to forest or grasslands in 1999. In the loess hilly area, the annual precipitation is only about 400 ~ 600 mm, which greatly limited the growth of arbor, and so, grass and grass coverage were taken as the main biologic measures to control soil loss as the project has been carried out (Li M., et al., 2009). Evaluation of grass and grass influences on soil erosion process can provide important information in soil and water conservation.

More researches have been done on runoff and sediment reducing benefit of grasses or trees in the past years (Zhu X M., 1960; Hou X.L. and Cao Q.Y., 1990; Tang K.L., et al., 1983; Wainwright J., et al., 2009). Zobeck and Onstad (1987) pointed plant measures increased soil roughness and decreased the ability of flow detach and transport soil. Chatterjea (1999) studied runoff and sediment generation on bare and grassplots under natural rainstorm, and concluded that the responses of the bare surfaces to incoming rainfall were more instantaneous and more significant than those of grassplots. Pan and Shangguan (2005) studied vegetative covers have the ability to reduce flow velocity, which reduces the erosive force Li, et al. (2007) studied runoff resistance of hill – gully slope with grass coverage and concluded grass coverage and slope have significant influence on runoff. Wu, et al., (2007) proposed that the flow regimes in grass slope belong to laminar flow and the capability of resisting soil erosion and sediment movement in grass slope is stronger than that in the bare slope based on the water scouring experiments. Moreover, the differences in vegetation types, soil properties, slope surfaces conditions, etc. in field experiments tend to have negative effects on the finding. Neave and Abrahams (2002) predicted water yields at the community level, the grass land sheds 150 per cent more water for a given storm event than the grass land based on simulation experiments in the Chihuahuan desert. However, on the loess plateau in China, Shen, et al. (2006) collected runoff and erosion data on filed plot in Yanhe Watershed and showed the amount of erosion on bare slope was 82 times greater than that on grass slope, and 150 times greater than that on grass slope. Although numerous studies (Moore T.R., et al., 1979; Braud I., et al., 2001) have mentioned vegetation impacts on soil erosion mainly concerning on reduction effects of grass on runoff and sediment compared that of bare slope, little is known about effects of grass on runoff and sediment yield process. So, it is difficult to use grass reasonably due to lack of sufficient reliable data to understand the soil erosion process.

The objective of this research is to investigate effects of grass on runoff and sediment under simulated rainfall experiments, which will present a theoretical guidance for the construction of grass on the loess plateau, and can offer basic data for the building of soil erosion mechanics model on vegetation – covered slopes.

2 Materials and methods

2.1 Experimental model design

The experiments were conducted at the Physical Model Yellow River Base of Yellow River Institute of Hydraulic Research, Zhengzhou city, Henan Province of China. Two experiment boxes were constructed with dimensions of 5 m (length) × 3 m (width) × 0.5 m (depth). One experiment box was planted for grass, and the other was control box of bare plots. Each box was divided into three experimental plots of 5 – m – long and 1 – m – wide by inserting the polyvinyl chloride (PVC) border into the ground. The experiment boxes were placed under simulator with side – nozzle. The height of side – nozzle is 7.5 m and the height of side – nozzle spewing was 1.5 m, which let all raindrops reach terminal velocity. Three different rainfall intensities were designed for 45 mm/h, 87 mm/h and 127 mm/h.

2.2 Grass measures

Alfalfa(Medicago sativa L.), a very commonly seen grazing grass on the loess plateau, which was used for the vegetative cover. The alfalfa grass was in good condition of 35 cm long and the grass root of 40 cm long averagely under the experiment condition. So Alfalfa has the ability of protecting soil, which was similar to the alfalfa grass on the field plot. The grass coverage degree was 65% or so.

2.3 Soil sample collection

The clayey loess collected from slope surface soil beside Mangshan Mountain nearby Zhengzhou city was used in this study. This soil had about 8.3% sand, 67.4% silt, and 24.3% clay. The tested soil median diameter was 0.012,17 mm. The soils collected from field were passed through a 10 mm sieve.

2.4 Soil box preparation

The depth of surface layer was approximately 40 cm and a 10 cm layer of sand at the bottom for better drainage. Soil pan box was packed in 5 cm layers to ensure uniform density. The surface layer simulated to the cultivation field and the soil buck density was about 1.10 g/cm^3. The bottom layer simulated to the furrow field and the soil buck density was about 1.35 g/cm^3. The soil moisture was 15% or so.

2.5 Experiment procedure

Once the boxes were prepared, a 10 min rainfall of 30 mm/h was applied prior 24 h to erosion run, and no any runoff occurred at soil box surface. Runoff samples from test boxes were collected in 2 – minutes interval, which was used to analyze soil erosion process. Each slope was divided into 5 sections from the top to bottom. On each section velocity was measured with a dye tracer in 5 – minutes interval and flow widths and flow depths measured by a steel ruler. The rainfall lasted for 60 min after runoff producing.

3　Results and discussions

3.1　Effects of grass on runoff

Under simulated rainfall, average runoff rates ranged from 39. 7 L/min to 126. 0 L/min for bare plots and from 0. 34 L/min to 6. 22 L/min for grass plots, and the runoff rate from grass plots were much less than from bare plots (Tab. 1). The greater runoff from bare plots may be due to the decrease of soil anti – erodibility causing the formation of erosion pits and rill development, which increased slope flow velocity and produced higher runoff. The discrepancy in runoff between bare plots and grass plots under the same rainfall mainly was determined by the surface condition. The differences in runoff between bare plots and grass plots varied from 95. 1% to 99. 1%. Rainfall intensity had significant influences on average runoff rate for each treatment. The runoff rate from both bare and grass plots increased with the increase of rainfall intensity. This is due to higher rainfall causing higher runoff velocity and lowered infiltration rate, and hence inducing increased runoff.

Tab. 1　Average runoff rate and sediment rate for bare and grass plots

Rainfall intensity (mm/h)	Average runoff rate (L/min)			Average sediment yield rate (g/min)		
	Bare plot	Grass plot	Grass plot reduction cf. bare plot (%)	Bare plot	Grass plot	Grass plot reduction cf. bare plot (%)
45	39.7	0.34	99.1	3,636.7	26.7	99.3
87	110.3	3.03	97.2	9,128.3	102.3	98.8
127	126.0	6.22	95.1	9,436.3	581.5	93.8

The runoff processes of bare plots and grass plots under different rainfall intensities were presented in Fig. 1 and Fig. 2. The runoff rate over time on grass plots basically kept steady under rainfall intensities of 45 mm/h and 87 mm/h and increased evenly under rainfall intensity of 127 mm/h, whereas runoff rate from bare plots increased rapidly and kept steady after rainfall duration of 20 min under rainfall intensity of 45 mm/h and showed fluctuant increase under rainfall intensity of 87 mm/h and 127 mm/h. One possible reason for the behavior is the grass slowed down the water flow and enhanced water infiltration as well as of soil surface morphology changed little under rainfall intensities of 45 mm/h and 87 mm/h.

Under rainfall intensity of 45 mm/h, the runoff from shrub plot kept a constant level of 0. 34 L/min and reached to a maximum of 0. 53 L/min, 48 min after rainfall started. Under rainfall intensity of 87 mm/h, the runoff kept relatively steady after rainfall duration of 44 min. However, runoff from grass plots did not reach to equilibrium under rainfall intensities of 127 mm/h. , this may be due to litters on grass slope can not detain flow and prevent flow runoff scouring effectively under higher rainfall intensity. The runoff process from bare plots showed a fluctuant increase and behaved similarly under rainfall intensity of 87 mm/h and 127 mm/h, the reason was rill formation and development at 20 min duration after rainfall started. The behavior indicated that surface condition and rainfall intensity had evident impacts on runoff. When rainfall intensity decreased to 45 mm/h, lower rainfall and even surface condition caused relatively less runoff on bare plot consequently.

3.2　Effects of grass on sediment yield

The differences in sediment yield from bare plots and grass plots were much more significant than that in runoff. Under the experiment conditions, the average sediment yields varied from

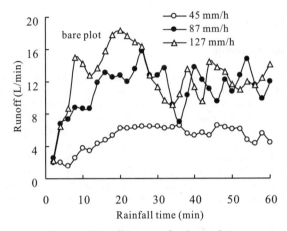

Fig. 1   Runoff process for bare plots

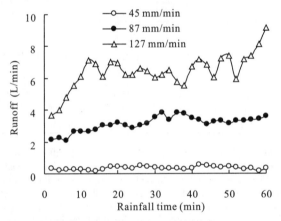

Fig. 2   Runoff process for grass plots

3,636. 7 g/min to 9,436. 3 g/min for bare plots and from 26. 7 g/min to 581. 5 g/min for grass plots and those from grass plots were less than those from bare plots. The percentage reduction of sediments from grass plots, compared with bare plots, ranged from 93. 8% to 99. 3% (Tab. 1).

The sediment yield processes of bare plots and grass plots under different rainfall intensities were presented in Fig. 3 and Fig. 4. Experiment data showed the sediment yield was closely related with surface condition, especially with rill development. Under rainfall intensity of 45 mm/h there was no rill erosion on bare and grass plots, so, sheet erosion was dominated and the sediment yield kept a relatively constant level. Under rainfall intensity of 87 mm/h for bare plot, the surface morphology on bare plot showed crescent - shaped scouring pits and there was rill erosion on slope after rainfall duration of 16 min. So, the sediment yield from bare plot increased sharply and kept a fluctuant trend affected by the advance of rill head and rill width and depth development. Whereas for grass plot, the sediment kept a constant trend over time and was less than that from bare plot. This may due to the grass has strong capacity of resisting the raindrop impacts and runoff scouring and the runoff scouring abilities were lower than the critical shear stress of the soil with grass roots. For rainfall intensity of 127 mm/h, the relatively greater availability of soil materials and the stronger detachment from raindrop splash caused more sediment yield at the beginning of rainfall and sediment increased dramatically once rill was formed after rainfall duration of 20 min on bare plot. For grass plot, the erosion pit came into being after 12 min causing a sharp increase in sediment yield and kept a fluctuant trend after that. So, the sediment yield of bare plot is much larger than that of grass plot under rainfall intensity of 127 mm/h.

By comparison of sediment yield for bare and grass plots, the results showed grass had stronger ability to resist flow detach and transport under experiment condition, but higher rainfall causing more runoff and sediment on grass plot must be paid more attention on the loose loess plateau. To control soil erosion well, this is important to adopt comprehensive harness of planting and engineering measures.

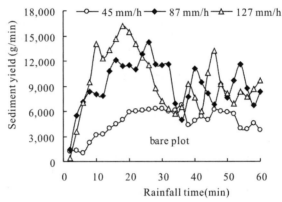

Fig. 3   Sediment yield process for bare plots

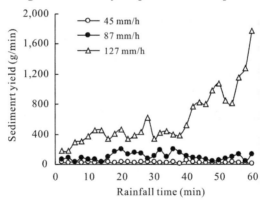

Fig. 4   Sediment yield process for grass plots

4   Conclusions

Laboratory experiments were conducted to compare runoff and sediment yield process on bare and grass plots under simulated rainfall of 45, 87 mm/h and 127 mm/h. The following conclusions were derived:

Under simulated rainfall, average runoff rates ranged from 39.7 L/min to 126.0 L/min for bare plots and from 0.34 L/min to 6.22 L/min for grass plots, and the runoff rate from grass plots were much less than from bare plots. The runoff rate over time on grass plots basically kept steady under rainfall intensities of 45 mm/h and 87 mm/h and increased evenly under rainfall intensity of 127 mm/h. The runoff rate from bare plots increased rapidly and kept steady after rainfall duration of 44 min under rainfall intensity of 45 mm/h and showed a fluctuant increase under rainfall intensity of 87 mm/h and 127 mm/h.

The differences in sediment yield from the bare and grass plots were much more significant than those in runoff. Average sediment yield varied from 3,636.7 g/min to 9,436.3 g/min for bare plots and from 26.7 g/min to 581.5 g/min for grass plots. The percentage reduction of sediment yield from grass plot, compared with bare plot, ranged from 93.8% to 99.3%. Under rainfall intensity of 45 mm/h, the sediment kept a relatively constant level for bare and grass plots. Under rainfall intensity of 90 mm/h and 127 mm/h for bare and grass plots, the sediment yield process were closely related with the changeable surface morphology, especially with rill development.

Acknowledgments

This study was funded by National Basic Research Program of China (2011CB403303), National Natural Science Foundation of China (41071191), Non – profit Industry Program of the Ministry of Water Resource (201201083) and YRIHR Foundation (HKY – JBYW – 2012 – 04).

References

Li M, Yao W Y, Ding W F, et al. Effect of Grass Coverage on Sediment Yield in the Hillslope – gully Side Erosion System. [J]. Journal Geography Science, 2009, 19(3): 321 – 330.

Zhu X. M. Effect of Vegetation on Soil Loss in Loessial Region[J]. Acta Pedagogical Sinica, 1960, 8(2):110 – 120.

Hou X L, Cao Q Y. Study on the Benefits of Plants to Reduce Sediment in the Loess of North Shaanxi[J]. Bulletin of Soil and Water Conservation, 1990 10(2):33 – 40.

Tang K L, Zheng S Q, Xi D Q, et al. Soil Loss and Treatment in Sloping Farmland in Xingzihe Watershed[J]. Bulletin of Soil and Water Conservation, 1983,3(5):43 – 48.

Wainwright J, Parsons A J, Abrahams A D. Plot – scale Studies of Vegetation, Overland Flow and Erosion Interactions: Case Studies From Arizona and New Mexico [J]. Hydrological Processes, 2000, 14(16):2921 – 2943.

Zobeck T M, Onstad C A. Tillage and Rainfall Effects on Random Roughness: A review[J]. Soil and Tillage Research, 1987, 9(1):1 – 20.

Chatterjea K. The Impact of Tropical Rainstorms on Sediment and Runoff Generation From Bare and Grass – Covered Surfaces: A Plot Study from Singapore [J]. Land Degradation and Development, 1999, 9(2):143 – 157.

Pan C Z, Shangguan Z P. Influence of Forage Grass on Hydrodynamic Characteristics of Slope Erosion[J]. Journal of Hydraulic Engineering, 2005, 36(3):371 – 377.

Li M, Yao W Y, Chen J N. et al. Experiment Study on Runoff Resistance of Hill – Gully Slope with Grass Coverage[J]. Journal of Hydraulic Engineering, 2007, 38(1):112 – 119.

Wu S F, Wu P T, Feng H, et al. Effects of Forage Grass on the Reduction of Runoff and Sediment and the Hydrodynamic Characteristic Mechanism of Slope Runoff in the Standard Slope Plot [J]. Journal of Beijing Forestry University, 2007, 29(3):99 – 104.

Neave M, Abrahams A D. Vegetation Influences on Water Yields From Frassland and Grassland Ecosystems in the Chihuahuan Desert[J]. Earth Surface Processes and Landforms, 2002, 27(9):1011 – 1020.

Shen Z Z, Liu P L, Xie Y S, et al. Study of Plot Soil Erosion Characteristic under Different Underlying Horizon[J]. Bulletin of Soil and Water Conservation, 2006, 26(3):6 – 9.

Moore T R, Thomasand D B, Barber R G. The Influence of Grass Cover on Runoff and Soil Erosion from Soils in the Machacos area, Kenya[J]. Tropical Agriculture, 1979, 56(4):339 – 344.

Braud I, VichA A I J, Zuluaga J, et al. Vegetation Influence on Runoff and Sediment Yield in the Andes region: Observation and Modeling[J]. Journal of Hydrology, 2001, 254(14):124 – 144.

Analysis of Dominant Factors for Density Current Sediment Discharge Efficiency of Xiaolangdi Reservoir in 2011

Jiang Siqi[1], *Ma Huaibao*[1], *Zhang Junhua*[1], *Wang Ting*[1] and *Wang Hongwei*[2]

1. Yellow River Institute of Hydraulic Research, YRCC, Key Laboratory of Yellow River Sediment Research, Zhengzhou, 450003, China

2. Yellow River Engineering Consulting Co. , Ltd. , Zhengzhou, 450003, China

Abstract: Based on measured data of water and sediment regulation in the process of shaping density current of Xiaolangdi Reservoir before flood season of 2011, this paper has analyzed impact on shaping density current forward of the Reservoir and sediment discharge of Sanmenxia Reservoir water storage capacity before flood season and discharge method, and pointed out that Sanmenxia Reservoir shall store water as much as possible before flood season and gradually enlarge water outflow before discharge in the process of regulation. On the basis of correlation between docking water level and delta vertex elevation, this paper has pointed out that Xiaolangdi Reservoir occurs retrogressive erosion and longitudinal erosion which can greatly increase silt concentration at plunging point of density current when docking water level is close to or is lower than the sedimentation delta vertex elevation of Xiaolangdi Reservoir and is combined with large flow discharging process of Sanmenxia Reservoir. It has combined with water and sediment process of Tongguan Section to make analysis and brought up that longer duration of more than 1,000 m³/s Tongguan water inflow makes higher water level of Sanmenxia Reservoir during large clean water discharging period of Sanmenxia Reservoir and particularly the emptying and sediment discharging period, Xiaolangdi Reservoir forms stronger follow-up power of density current and makes erosion or reduced siltation at the end of Xiaolangdi Reservoir under the condition of better water and sediment proportion. And it has reported current boundary condition of the reservoir and pointed out that siltation delta vertex of Xiaolangdi Reservoir is near to the dam, delta vertex gets more fine siltation before flood season and topography condition at the end of Xiaolangdi Reservoir is good for sediment discharge of density current.

Key words: Xiaolangdi Reservoir, reservoir operation, retrogressive erosion, density current

1 Introduction

Water and sediment regulation for shaping density current artificially before flood season of Yellow River is to make full use of water reserve upper water level limit of Wanjiazhai and Sanmenxia Reservoir, to erode siltation during non-flood season of Sanmenxia Reservoir and sediment deposition at the end of Xiaolangdi Reservoir through integrated regulation of Wanjiazhai, Sanmenxia and Xiaolangdi Reservoir under the condition of no flood in the middle stream, and accordingly to shape density current in Xiaolangdi Reservoir with sediment discharge out of the Reservoir, and finally to realize target of sediment discharge of Xiaolangdi Reservoir and siltation shape regulation at the end of the Reservoir (Fig. 1). To shape density current can change waste

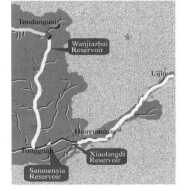

Fig. 1 Sketch Map of Reservoir Location

reservoir water into sediment discharge water and then make reservoir sediment discharge and siltation reservoir reduce, and finally make reservoir water flow into the sea with sediment outflow.

Based on integrated regulation of main stream reservoir, eight artificial density current shaping were made during water and sediment regulation period from 2004 to 2011, and sediment discharge was different from different water and sediment inflow, boundary condition and reservoir application method. Sediment inflow was 0.26×10^8 t, sediment outflow was 0.378×10^8 t and sediment discharge ratio was up to 145.4% during water and sediment regulation period of 2011.

2 Water and sediment regulation process of Xiaolangdi Reservoir before flood season of 2011

Water and sediment regulation started at 9:00 June 19, 2011, and Sanmenxia Reservoir began leakage greatly at 5:00 July 4, 2011 and got maximum water outflow 5,290 m³/s at 3:12 July 5, 2011, began sediment discharge at 6:00 July 5, 2011, and got maximum silt concentration outflow 329 kg/m³ at 11:00 July 5, 2011.

Water level of Xiaolangdi Reservoir was 248.45 m and water storage capacity was 4.36×10^9 m³ at 8:00 June 19, and water level of the Reservoir kept reduction after water and sediment regulation, water level was reduced to 215.39 m at 5:00 July 4, and maximum water outflow of Xiaolangdi Reservoir was 4,310 m³/s at 19:06 June 22. Sediment discharge began at 18:12 July 4 and silt concentration was 2.42 kg/m³, and maximum silt concentration was up to 311 kg/m³ at 22:00 July 4. Water and sediment regulation was finished at 8:00 July 8, and water level was 216.34 m and water storage capacity was 0.625×10^9 m³, which was 3.735×10^9 m³ before water and sediment regulation.

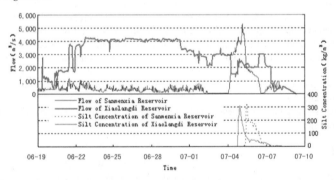

Fig. 2   Water and sediment Inflow and outflow process of Xiaolangdi Reservoir (Instantaneous)

According to water and sediment process of Xiaolangdi Reservoir, water and sediment regulation before flood season of Yellow River in 2011 was divided into two stages: the first stage was clean water discharge stage of Xiaolangdi Reservoir (water regulation period) from 9:00 June 19, 2011 to 5:00 July 4 i. e. 14.83 d, total flood volume was about 4.295×10^9 m³, sediment discharge volume was 0, and maximum flow was 4,310 m³/s at 19:06 June 22. The second stage was sediment discharge outflow stage of Xiaolangdi Reservoir (sediment regulation period) from 5:00 July 4, 2011 to 8:00 July 8 i. e. 4.13 d total flood volume was about 0.629×10^9 m³, sediment discharge volume was 0.378×10^8 t, persistent maximum flow was 3,000 m³/s from 10:36 to 20:00 July 6 and maximum silt concentration was 311kg/m³ at 22:00 July 4.

Before sediment discharge of Sanmenxia Reservoir (6:00 July 5), Xiaolangdi Reservoir was discharging sediment (18:12 July 4) and silt concentration was up to 311 kg/m³ (22:00 July 4), which indicated that flood peak resulted from emptying of Xiaolangdi Reservoir got strong erosion in Xiaolangdi Reservoir and eroded sediment formed into density current in Xiaolangdi Reservoir and discharged sediment out of the Reservoir.

During water and sediment regulation of 2011, sediment inflow of Xiaolangdi was 0.26×10^8 t and sediment outflow was 0.378×10^8 t. According to water and sediment process of Sanmenxia Reservoir, sediment regulation period of Xiaolangdi Reservoir can be divided into two stages with spread time into consideration, the first stage is to shape density current forward in delta vertex of Xiaolangdi Reservoir before sediment discharge of Sanmenxia Reservoir, and sediment outflow was 0.285×10^8 t in 2011, which took up 75.4% total sediment discharge volume of Xiaolangdi Reservoir during water and sediment regulation period; the second stage only got 0.093×10^8 t sediment outflow of density current resulted from sediment discharge of Sanmenxia Reservoir and sediment discharge ratio of Xiaolangdi Reservoir was 35.8% (Tab.1).

Tab. 1   Water and sediment inflow and outflow of Xiaolangdi Reservoir during water and sediment regulation period of 2011 (Instantaneous)

Statistical time	Water inflow ($\times 10^8$ m^3)	Sediment inflow ($\times 10^8$ t)	Water outflow ($\times 10^8$ m^3)	Sediment outflow ($\times 10^8$ m^3)	Sediment discharge ratio (%)
June 19 ~ July 8	10.235	0.260	49.524	0.378	145.4
5:00 July 4 ~ 8:00 July 8	5.785	0.260	6.286	0.378	145.4
18:12 July 4 ~ 19:12 July 5(First stage)	—	0	—	0.285	—
19:12 July 5 ~ 8:00 July 8 (Second stage)	—	0.260	—	0.093	35.8

3   Main influencing factors for reservoir sediment discharge

3.1   Operation of Sanmenxia Reservoir

According to analysis of water and sediment regulation before each flood season, sediment of density current of Xiaolangdi Reservoir mainly comes from erosion sediment of Xiaolangdi Reservoir delta section and erosion sediment of Sanmenxia Reservoir. The former water flow is mainly water reserve of Sanmenxia Reservoir, and the latter is mainly decided by incoming water condition of Tongguan.

According to different water flow condition, Sanmenxia Reservoir regulation can be divided into two stages including emptying period and sediment discharge period of Sanmenxia Reservoir (Fig. 3).

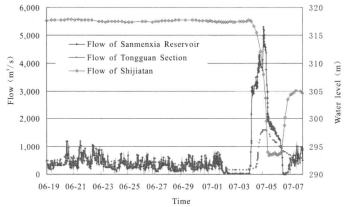

Fig. 3   Inflow, outflow and water Level of Sanmenxia Reservoir in 2011

（1）Emptying period of Sanmenxia Reservoir（5:00 July 4 ~ 10:00 July 5）

This stage mainly makes good use of water reserve of Sanmenxia Reservoir to shape large flow peak process and erode sediment of Xiaolangdi Reservoir delta section and then to cause density current under proper condition.

Sanmenxia Reservoir water level was reduced to 293.98 m during emptying period, and water storage capacity was reduced from 0.409×10^9 m^3 to 0.23×10^6 m^3, water outflow of Sanmenxia Reservoir was 3.825×10^8 m^3, taking up 37.4% total water outflow during water and sediment regulation period, and sediment outflow was 1.0 million tons.

（2）Sediment discharge period of Sanmenxia Reservoir

Sanmenxia Reservoir was basically under even and open channel flow state during sediment discharge period, and got erosion in Sanmenxia Reservoir. Silt concentration outflow was up to 329 kg/m^3 at 11:00 July 5, and outflow maintained at 1,500 m^3/s and formed high silt concentration water for a long duration to supply follow-up energy for continuous running of density current. Sanmenxia Reservoir water level maintained at 293.43 ~ 293.98 m due to sustainable water supply of Wanjiazhai reservoir, water outflow was 1.356×10^8 m^3, taking up 13.2% total water outflow during water and sediment regulation, sediment outflow was 0.248×10^8 t, taking up 95.5% total sediment outflow during water and sediment regulation period of Sanmenxia Reservoir.

It can be seen that more water storage capacity of Sanmenxia Reservoir makes greater peak flow, greater flow of Tongguan Section, longer duration and better sediment discharge efficiency.

3.2 Analysis of Xiaolangdi Reservoir application and boundary condition

Xiaolangdi Reservoir kept delta sedimentation state in 2011, delta vertex forwarded from 24.43 km between HH15 Section of 2010 and the dam to 18.75 km between HH12 Section of 2011 to the dam, delta vertex elevation was reduced to 214.34 m, delta vertex section (lower than HH37) gradient maintained 3‰, which was basically the same as 5.5‰ ~ 29.3‰ density current running reservoir bottom gradient. We can see from Fig.4, density current running distance was 12.9

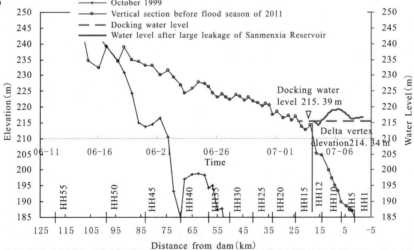

Fig. 4　Vertical section before flood season of 2011 and reservoir water Level during density current shaping period

Large leakage of Sanmenxia Reservoir was considered as the start of shaping density current before flood season, and corresponding water level was called as docking water level. Docking water level was 215.39 m in 2011 and higher than delta vertex elevation 214.34 m but far lower than balance water depth. Water level in retrogressive erosion stage was reduced to 214.28 m, and the section upper than plunging point occurred strong longitudinal erosion and retrogressive erosion.

3.3 Impact on fine inflow sediment concentration of Xiaolangdi reservoir

Density current is a kind of over saturation sediment transport, with running of density current, coarse sediment gets longitudinal sedimentation. Seen from density current sediment discharge data of Xiaolangdi Reservoir (Tab. 2), most sediment outflow are coarse sediment. Fine sediment content was up to over 83.94% in the years without retrogressive erosion in delta (2005 ~2007 and 2009); fine sediment content was still up to over 64.29% in the years with retrogressive erosion in delta like 2008, 2010 and 2011.

Seen from fine sediment inflow content statistics of Xiaolangdi Reservoir during density current shaping period before flood season, fine sediment took up 43.04% total sediment in 2006, and sediment discharge ratio was up to 30% even under poor density current shaping condition; while fine sediment took up 27.16% only in 2009, which was one reason for lower sediment discharge ratio in 2009. Fine sediment was 41.90% in 2011, and sediment discharge ratio was 145.38% in 2011, higher than 137.02% in 2010, as the maximum ratio in water and sediment regulation statistics. Thus, we can see that higher fine sediment inflow content can make higher sediment discharge ratio of density current during density current shaping period.

Tab. 2 Fine sediment outflow content of Xiaolangdi Reservoir

Year	Duration (Month-Day)	Sediment volume ($\times 10^8$ t)		Sediment Discharge Ratio (%)	Fine sediment inflow ($\times 10^8$ t)	Fine sediment outflow ($\times 10^8$ t)	Percent of fine sediment inflow in total sediment volume (5)	Percent of fine sediment outflow in total sediment volume (5)
		Sanmenxia	Xiaolangdi					
2004	07.07 ~07 – 14	0.385	0.055	14.29	0.133	0.047	34.55	86.13
2005	06.27 ~07 – 02	0.452	0.020	4.42	0.167	0.018	36.95	90.97
2006	06.25 ~06 – 29	0.230	0.069	30.00	0.099	0.059	43.04	85.40
2007	06.26 ~07 – 02	0.613	0.234	38.17	0.246	0.196	40.13	83.94
2008	06.27 ~07 – 03	0.741	0.458	61.81	0.239	0.361	32.25	78.78
2009	06.30 ~07 – 03	0.545	0.036	6.61	0.148	0.032	27.16	88.87
2010	07.04 ~07 – 07	0.408	0.559	137.02	0.141	0.359	34.52	64.29
2011	07.04 ~07 – 07	0.260	0.378	145.38	0.109	0.254	41.90	67.28

4 Conclusions

Based on the above analysis, this paper has brought up the main reasons for sediment discharge ratio during water and sediment regulation period before flood season of 2011 as follows:

(1) Water storage capacity before flood season and large clean water discharging process of Sanmenxia Reservoir have key impact on shaping density current forward and sediment discharge of the Reservoir. Sanmenxia Reservoir shall store water as much as possible before flood season and gradually enlarge water outflow before discharge. It is the mature experience that shall be kept on.

(2) Xiaolangdi Reservoir upper than the backward end occurs retrogressive erosion and longitudinal erosion which can greatly increase silt concentration at plunging point of density current when docking water level is close to or is lower than the sedimentation delta vertex elevation of Xiaolangdi Reservoir and is combined with large flow discharging process of Sanmenxia Reservoir. Xiaolangdi Reservoir sediment outflow that is led in large clean water discharging process during water and sediment regulation period of Sanmenxia Reservoir before flood season of 2011 takes up 75.4% total sediment discharge volume, which fully indicates that the regulation can improve sediment discharge efficiency of erosion-type density current.

(3) Water and sediment process of Tongguan Section takes great impact on shaping density current. Longer duration of more than 1,000 m³/s Tongguan water inflow makes higher water level of Sanmenxia Reservoir during large clean water discharging period of Sanmenxia Reservoir and

particularly the emptying and sediment discharging period, Xiaolangdi Reservoir forms stronger follow-up power of density current and makes erosion or reduced siltation at the end of Xiaolangdi Reservoir under the condition of better water and sediment proportion.

(4) Siltation delta vertex of Xiaolangdi Reservoir is near to the dam, delta vertex gets more fine siltation before flood season and topography condition at the end of Xiaolangdi Reservoir is good enough (more siltation and great gradient before flood season), all of these factors are good for increasing sediment discharge efficiency of reservoir density current.

Acknowledgements

The research was supported by public welfare industry research special fund of the Ministry of Water Resources (MWR) (No. 200901015, No. 200801024), the National Natural Science Foundation of China (NSFC) (NO. 51179072) and central-level nonprofit research institute fund (HKY – JBYW – 2009 – 6).

References

Li Shuxia, Zhang Junhua, Chen Shukui, et al. Studies on Key Technologies And Regulation Schemes of Density Current Forming in Xiaolangdi Reservoir [J]. Journal of Hydraulic Engineering, 2006(5): 567 – 572.

The Analysis and Calculation on Erosion – deposition and Effective Storage of Xiaolangdi Reservoir area [R]. Reconnaissance, Planning, Design and Research Institute of Yellow River Conservancy Commission. September, 1983.

River Channel Laboratory of Institute for Hydraulic Engineering of Shaanxi Province, Sediment Research Laboratory of the Department of Hydraulic Engineering of Tsing Hua University. Reservoir Sediment[M]. Beijing: Hydraulic and Electric Power Press, 1979: 125 – 128.

Zhang Qishun, Zhang Zhenqiu. The Calculations on Reservoir Scouring and Silting Patterns and Processes [J]. Sediment Research, 1982(1):1 – 13.

Han Qiwei. Reservoir Sedimentation[M]. Beijing: Science Press, 2003: 97 – 100.

Ma Huaibo, Zhang Junhua, Chen Shukui, et al. A Scheme Design of Sediment Discharging by Density current of Xiaolangdi Reservoir During Water and Sediment Regulation Before Flood Season in 2005 [C]// Proceedings of the 6th National Sediment Basic. Theory Research Academic Symposium . 2005: 1028 – 1034.

Dual State – parameter Estimation of One Dimensional Sediment Transport Using Ensemble Kalman Filter

Lai Ruixun[1,2] , *Yu Xin*[2] , *Yang Ming*[2] , *Zhang Fangxiu*[2] and *Zhang Xiaojing*[3]

1. Dept. of Hydraulic Engineering, State Key Laboratory of Hydroscience and Engineering, Tsinghua University, Beijing, 100084, China
2. Yellow River Institute of Hydraulic Research, Zhengzhou, 450046, China
3. North China University of Water Resources and Electric Power, Zhengzhou, 450046, China

Abstract: Using the algorithm of Ensemble Kalman Filter and the observations to identify the sediment transport capacity and, simultaneously, to improve the accuracy of predicted sediment concentration. The observations of water level, discharge and sediment were distributed by reasonable error to construct its state – space equations, and then the Ensemble Kalman algorithm were adopted to obtain the optimal value. At the same time, the sediment transport capacity was also considered as a state variable and its optimal value was identified in a method of inverse problem. This data assimilation algorithm was applied to the lower reaches of the Yellow River, the 2009' regulation of both flow and sediment, from Xiaolangdi to Lijin. The ability of the ensemble Kalman filter in improving the accuracy of one – dimensional sediment model was analyzed, and the rationality of sediment transport capacity was discussed. Studies have shown that: ① using the ensemble Kalman filter algorithm can effectively improve the predicted accuracy of water level, discharge and sediment concentration. ② Using the optimal variables and the inverse parameters as the initial value of the next time step can improve the predicted value further. ③The algorithm is recursive and therefore dose not require storage of all past information, so the computational efficiency can meets the practical engineering applications.

Key words: Ensemble Kalman Filter, sediment transport capacity, Yellow River, regulation of both flow and sediment, inverse problem

1 Introduction

Data assimilation is the optimal process of blending the observations and the numerical model. Kalman Filter (abbreviated as KF) is the optimal algorithm for linear dynamical system with Gauss distribution. For the solution of nonlinear problem, Maybeck developed the Extend Kalman Filter (abbreviated as EKF). Both KF and EKF are needed to define the error covariance explicitly. However, it is difficult to define the error covariance both in hydrodynamic and sediment transport models. To overcome these limitations, Evensen introduced the Ensemble Kalman Filter (abbreviated as EnKF) using Monte Carlo method to define the error covariance of nonlinear system. For recent decade, the implementation of EnKF for data assimilation in hydrodynamic model has been discussed to identify the roughness parameter.

Both in hydrodynamic and sediment transport model, the state variables (such as water level, discharge, sediment concentration) coupled with the parameters (such as sediment transport capability) determine the accuracy of prediction. This paper uses EnKF method to improve the accuracy of discharge and sediment transport. At the same time, the sediment transport capability was also considered as one of the state variable to identify its value using inversed numerical algorithm. The data assimilation method was applied in the lower reaches of the Yellow River Xiaolangdi to Lijin, during the 2009' regulation of both flow and sediment.

2 Principle of sequential data assimilation

2.1 The nonlinear description of sediment transport

The evolution of sediment transport coupling with state variables and parameters can be described as a process of nonlinear stochastic system:

$$X_{t+1} = F(X_t, U_{t+1}, \theta) + \omega_{t+1} \tag{1}$$

where, t is the time; X is the state variable; U is the forcing terms; θ is the parameters; f is the nonlinear operator driving the model from time t to time $t+1$ and ω is the model error including the uncertainty of variable and parameters. The observations can be written as

$$Y_{t+1} = H(X_{t+1}) + \varepsilon_{t+1} \tag{2}$$

where, Y is the observations; H is the operator mapping observation value to state variable and ε is the error of observation deriving from the measurement instruments and the observers.

2 Ensemble Kalman Filter (EnKF)

In classical KF, the error covariance needs to define in advance, making the method infeasible for models with nonlinear system. The real process of sediment transport, however, is so complicated that the numerical model is affected by the change of parameters and variables. So the identification of error covariance is difficult or even impossible. Associated with the classical KF, the Ensemble KF disturbs the boundary condition and the observations using the kind of Monte Carlo method to obtain the model error. The steps of En KF can be described as follows.

Step 1. Disturbing the initial value and parameters:

$$U_{t+1}^i = U_{t+1} + \zeta_{t+1}^i, \zeta_{t+1}^i \sim N(0, R_{t+1}), \quad i = 1, \cdots, n \tag{3}$$

where, t is the time; i is the number of ensembles; n is the total number of ensemble; U is the forcing term; ζ_{t+1}^i is the error of ith ensemble and $N(0, R_{t+1})$ denotes the normal distribution with a mean of 0 and covariance of R. The parameters of the system must be disturbed in a reasonable area

$$\theta_{t+1}^i = \theta_{t+1} + \zeta_{t+1}^i, \quad i = 1, \cdots, n \tag{4}$$

where, ζ is the reasonable area.

Step 2. Calculating the ensemble members of the state variables.

The ensemble of the state variable, with the number of n, can be yielded when driving the boundary conditions and forcing variables or initial conditions as well:

$$X_{t+1}^{i-} = f(X_t^{i+}, U_{t+1}^i, \theta) + \omega_{t+1}^i \tag{5}$$

the meaning of t, i, n are as the same meaning in Eq. (3), X_t^{i+} denotes the i th member's optimal value at time t, X_{t+1}^{i-} denotes the i th member's predict value at time $t+1$ and ω_{t+1}^i is the error of i th member.

Step 3. Disturbing the observations.

Similar to the initial conditions, the members of the observations can be described as

$$y_{t+1}^i = HX_{t+1} + \varepsilon_{t+1}^i, \varepsilon_{t+1}^i \sim N(0, S_{t+1}) \tag{6}$$

where, ε_{t+1}^i is the error of i th member, Y_{t+1}^i denotes the i th observation and $N(0, S_{t+1})$ denotes the normal distribution with a mean of 0 and covariance of S.

Step4. Calculating the error covariance of model.

The error covariance of model system can be described as

$$P_{t+1}^- \approx \frac{1}{n-1} \sum_{i=1}^n \left[(X_{t+1}^{i-} - \bar{X}_{t+1}^-)(X_{t+1}^{i-} - \bar{X}_{t+1}^-)^T \right] \tag{7}$$

$$\bar{X}_{t+1}^- = \frac{1}{n} \sum_{i=1}^n X_{t+1}^{i-} \tag{8}$$

where, $\bar{X}_{t+1}^-$ is the average value of all the members and P_{t+1}^- is the predict value of state variable.

Step 5. Calculating the optimal value:

$$X_{t+1}^+ = X_{t-1}^- + K_{t+1}(Y_{t+1}^i - HX_{t+1}^-) \tag{9}$$

$$K_{t+1} = P_{t+1}^- H^T (HP_{t+1}^- H^T + S_{t+1})^{-1} \tag{10}$$

where, X_{t+1}^+ is the optimal value and considered to be the initial of next time step; K_{t+1} is the Kalman grain weighting the value between observation and model; H is the observation operator.

3   Study area and equations

3.1   Study area

The study area locates in the lower Yellow River from Xiaolangdi to Lijin, about 65 km long, crossing Henan and Shandong provinces. Eight hydro stations are distributed along the river including Xiaolangdi, Huayuankou, Jiahetan, Gaocun, Sunkou, Aishan, Luokou and Lijin.

3.2   One – dimensional hydrodynamic equation

Using the Preissmann scheme, the state – space equation of water level and discharge can be described as

$$
\begin{pmatrix} Q_1 \\ Z_1 \\ Q_2 \\ \vdots \\ Z_{L-1} \\ Q_L \\ Z_L \end{pmatrix}^{n+1} = \begin{pmatrix} Q_1 \\ Z_1 \\ Q_2 \\ \vdots \\ Z_{L-1} \\ Q_L \\ Z_L \end{pmatrix}^{n} + \begin{pmatrix} 1 & & & & & \\ A_{11} & B_{11} & C_{11} & D_{11} & & \\ A_{21} & B_{21} & C_{21} & D_{21} & & \\ & \vdots & & \vdots & & \\ & & A_{1(L-1)} & B_{1(L-1)} & C_{1(L-1)} & D_{1(L-1)} \\ & & A_{2L} & B_{2L} & C_{2L} & D_{2L} \\ & & & & & 1 \end{pmatrix}^{-1}
$$

$$
\begin{pmatrix} f(Q,t) \\ E_{11} \\ E_{21} \\ \vdots \\ E_{1(L-1)} \\ E_{2L} \\ f(Z,t) \end{pmatrix} + \begin{pmatrix} \omega_1 \\ \omega_1 \\ \omega_2 \\ \vdots \\ \omega_{L-1} \\ \omega_L \\ \omega_L \end{pmatrix} \tag{11}
$$

and the observations of all the cross sections can be list as

$$
\begin{pmatrix} y_{Q1} \\ y_{Z1} \\ y_{Q2} \\ \vdots \\ y_{Z(L-1)} \\ y_{QL} \\ y_{ZL} \end{pmatrix}^{n+1} = \begin{pmatrix} 1 & & & & & & \\ & 1 & & & & & \\ & & 1 & & & & \\ & & & \vdots & & & \\ & & & & 1 & & \\ & & & & & 1 & \\ & & & & & & 1 \end{pmatrix} \begin{pmatrix} y_{Q1} \\ y_{Z1} \\ y_{Q2} \\ \vdots \\ y_{Z(L-1)} \\ y_{QL} \\ y_{ZL} \end{pmatrix}^{n} + \begin{pmatrix} v_1 \\ v_1 \\ v_2 \\ \vdots \\ v_{L-1} \\ v_L \\ v_L \end{pmatrix}^{n} \tag{12}
$$

The abbreviated form of Eq. (11) and Eq. (12) is:

$$
\begin{cases} \begin{pmatrix} Q_j \\ Z_j \end{pmatrix}^{n+1} = \Phi_{QZ}^{n+1/n} \begin{pmatrix} Q_j \\ Z_J \end{pmatrix}^{n} + U_{QZ}^{n}, \\ \begin{pmatrix} y_{Qj} \\ y_{Zj} \end{pmatrix}^{n+1} = H_{QZ}^{n+1} \begin{pmatrix} Q_j \\ Z_j \end{pmatrix}^{n+1} + v_{QZ}^{n+1} \end{cases} \tag{13}
$$

where, Q is the discharge; Z is the water level; A, B, C, D, E is the coefficients in discrete scheme; ω_{QZ}^n denotes the error of model system; v_{QZ}^{n+1} denotes the error of observation; y_{Qj} is the measured discharge; y_{Zj} is the measured water level; H is the observation operator and U is the control term.

3.3   One – dimensional sediment transport equation

The sand size of the Yellow River is very small and the bed – load accounts from 0. 31% to 0. 7%. So the sediment transport of the lower Yellow River can be described with suspended – load:

$$\frac{\partial(AS)}{\partial t} + \frac{\partial(QS)}{\partial x} + \alpha\omega B(S - S_*) = 0 \tag{14}$$

where, S is the average sediment concentration of cross section; S_* is the average sediment tansport capacity; α is the recovery coefficient; ω is the sediment fall velocity and B is the width of the river. The calculation of sediment concentration can be described as

$$S_j^{i+1} = \frac{\Delta t[\alpha\omega B S_*]_j^{i+1} + [AS]_j^i + \dfrac{\Delta t}{\Delta x_{j-1}}\left|(QS)_{j-1}^{i+1}\right|}{A_j^{i+1} + \Delta t[\alpha\omega B]_j^{i+1} + \dfrac{\Delta t}{\Delta x_{j-1}}|Q_j^{i+1}|} \tag{15}$$

where, i denotes the time and j denotes the cross section. The observation of sediment with error can be described as

$$S_{obs}^{i+1} = H^{i+1}S^{i+1} + v \tag{16}$$

where, S_{obs}^{i+1} is the observation; v is the error of sediment; H is the operator transferring from observation to numerical model.

The Zhang Ruijing equation was adopted to describe the sediment transport capacity:

$$S_* = k\left(\frac{v^3}{gR\omega}\right)^m \tag{17}$$

where, k is the coefficients; v is the average velocity; R is the hydraulic diameter; ω the particle fall velocity. The sediment carrying capacity can not be observed directly, but can be obtained by the numerical result:

$$S_{*,obs}^{i+1} = \frac{\dfrac{Q_j^{i+1}S_j^{i+1} - Q_{j-1}^{i+1}S_{j-1}^{i+1}}{\Delta x_{j-1}} + \dfrac{A_j^{i+1}S_j^{i+1} - A_j^i S_j^i}{\Delta t} + (\alpha\omega BS)_j^{i+1}}{(\alpha\omega BS)_j^{i+1}} \tag{18}$$

4   Process of data assimilation

Data assimilation scheme described above is applied to the lower Yellow River from Xiaolangdi to Lijin. The total assimilation time is 672 h ranging from June 15 th , 8 am. to July 13 th , 8 am, 2009. A dual state parameter estimation flowchart using EnKF is shown in Fig. 1 Totally, the process can be divided into 3 parts: ① calculating the optimal water level and discharge; ②calculating the optimal sediment using observation value; ③ estimating the sediment transport capacity using the inverse method and considered as the initial value of the next time step. EnKF using a Monte Carlo method to simulate the error covariance, so it is necessary to disturb the observation sediment concentration and parameters. The range or its initial value of each parameter is shown in Tab. 1.

The total number of the ensemble should be estimated at first, and then disturb the observations of water level and discharge. According to the regulation of river measurement, the stochastic error of the discharge is restricted to normal distribution and with confidence level of 95%. So the error of discharge is accounted for about 5% of its real observation. The measurement uncertainty of water level is less than 3 centimeters while the suspended – load is approximately 4. 2%. Then, the boundary condition is disturbed and the optimal value of water level and discharge is calculated once there has observations. Finally, the sediment transport capacity is calculated after acquiring the optimal discharge and sediment value.

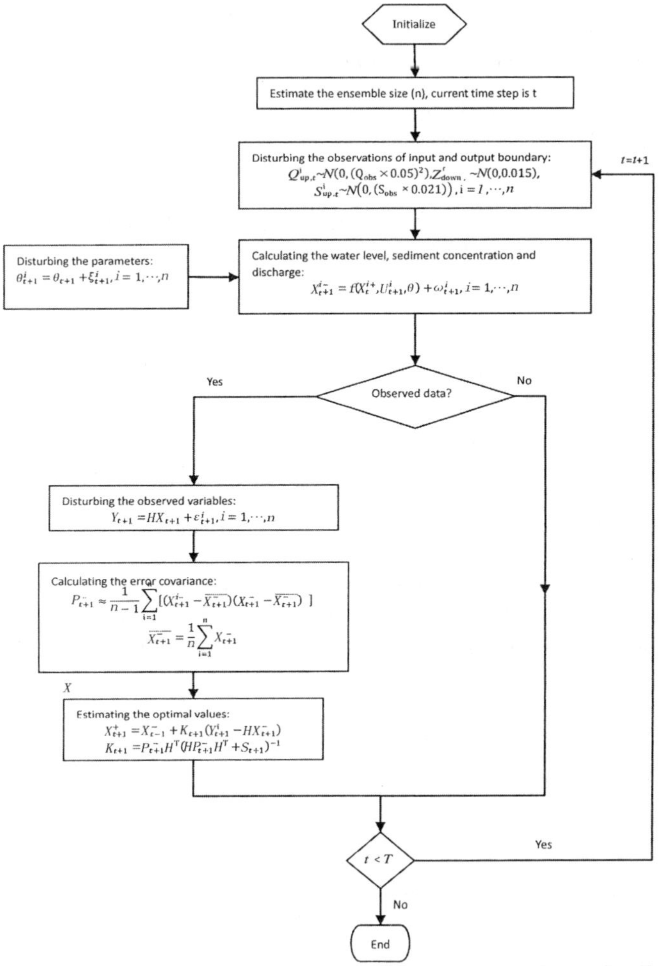

Fig. 1   Dual state – parameter estimation flowchart using the EnKF

Tab. 1   The estimation of uncertainty range associate with the parameters

Parameter	Description	Range	Initial value
Z	Measurement error of water level	$N(0,0.015)$	—
Q	Measurement error of discharge	$N(0,(Q_{obs}*0.05)^2)$	—
S	Measurement error of sediment concentration	$N(0,(S_{obs}*0.021))$	—
α	The range of recovery coefficient	—	0.01 when erosion while 0.005 when deposition
k	Coefficient k	—	0.5
M	Coefficient m	—	0.7

Note: N denote the normal distribution while the U denote the uniform distribution.

5 Results

5.1 Error analysis of discharge and sediment concentration

Discharge and sediment concentration are the key variables in the equation of sediment transport. To testify the accuracy of assimilation, the mean absolute error (abbreviated as MAE) and the root mean square error (abbreviated as RMSE) were calculated and is shown in Tab. 2.

Tab. 2　The MAE and RMSE of 6 hydro stations

	Huayuankou		Jiahetan		Gaocun		Sunkou		Aishan		Luokou	
	MAE	RMSE	MAE	RMSE	MAE	RMSE	MAE	RMSE	MAE	RMSE	MAE	RMSE
Discharge	39.3	32.3	44.9	38.7	35.9	27.5	46.0	35.3	35.7	32.6	37.9	31.3
Sediment concentration	0.25	0.24	0.3	0.01	0.6	0.4	0.8	0.7	0.7	0.5	1.1	0.99

As shown in Tab. 2, the total error of discharge among 6 stations ranges from 27.6 m^3/s to 46.0 m^3/s, and the error of sediment concentration is about to 1.0 m^3/s.

5.2 The reasonableness of calculated sediment transport capacity

Fig. 2 demonstrates the comparison between sediment concentration and its transport capacity using the technique of Ensemble KF. It is shown that the value of sediment transport capacity is of a little larger than sediment concentration. So the main channel is under the condition of erosion, and the result is consistent with the real data at the year of 2009.

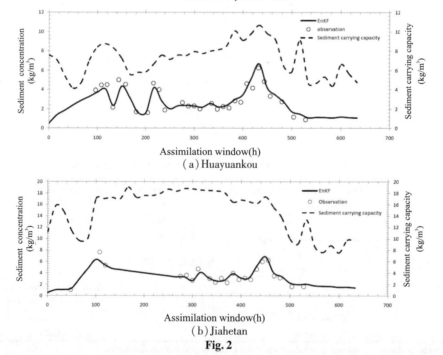

(a) Huayuankou

(b) Jiahetan

Fig. 2

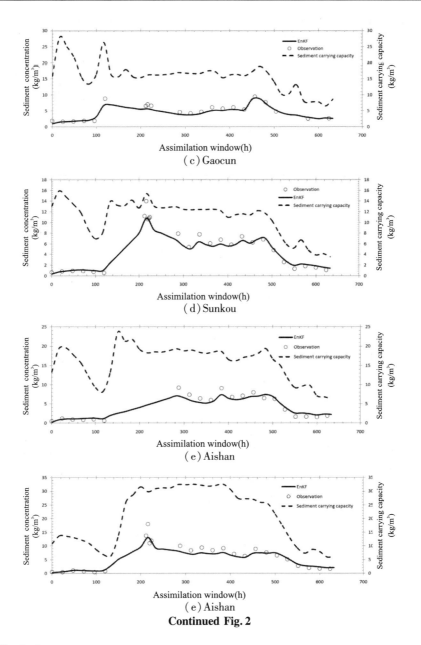

(c) Gaocun

(d) Sunkou

(e) Aishan

(e) Aishan

Continued Fig. 2

6 Conclusions

This paper constructed a data assimilation scheme to predict the water level, discharge and sediment concentration in the lower Yellow River during the 2009' regulation of both flow and sediment, using the method of ensemble Kalman filter and its inverse algorithm. Studies have shown that: ① using the ensemble Kalman filter algorithm can effectively improve the predicted accuracy of water level, discharge and sediment concentration; ② using the optimal variables and the inverse parameters as the initial value of the next time step can improve the predicted value further; ③ the algorithm is recursive and therefore dose not require storage of all past information, so the computational efficiency can meets the practical engineering applications.

Acknowledgments

This investigation was supported by the Central Foundation of Scientific Institute for Public Service, China, No. HKY – JBYW – 2012 – 5.

References

John M, Lewis, LakshmivAhan S, et al. Dynamic Data Assimilation a Least Suares Approach [M]. Cambridge: Cambridge University Press, 2006.

Kalmn R E. A New approach to Linear Filtering and Prediction Problems, Transactions of the ASME[J]. Journal of Basic Engineering, 1960; 82 (Series D): 35 – 45.

Maybeck P S. Stochastic Models, Estimation and Control [M]. Now York: Acdemic Press, 1979.

Evensen G. Sequential Data Assimilation with a Nonlinear Quasi – geostrophic Model Using Monte Carlo Methods to Forecast Error Statistics [J]. Journal of Geophysical Research , 1994, 99 (C5), 10143 – 10162.

Cunge J A. Practical Aspect of Computational River Hydraulics [M]. London: Pitman Publishing Program, 1980.

Long Y Q, Zhang Y F. Study on the Yellow River Sediment from the Viewpoint of Total Sediment [J]. Yellow River, 2002, 24(8). 28 – 29.

Han Q W. A Study on the Non – Equilibrium Transportation of Suspended Load [C]// Proc. , the First International Symposium on River Sedimentation, 1980.

Zhang R J. River Dynamics[M]. Beijing: China Industry Press, 1961.

Fang Hongwei, He Guojing, Lin Jinze, et al. 3D Numerical Investigation of Distorted Scale in Hydraulic Physical Model Experiments [J]. Journal of Coastal Research, Fall 2008, Special Issue 52.

Fang Hongwei, Han Dong, He Guojian, et al. Flood Management Selections for the Yangtze River MidStream after the Three Gorges Project Operation [J]. Journal of Hydrology, 2012, 432 – 433(11):1 – 11.

He Guojian, Fang Hongwei, Chen Minghong. Multidimensional Upwind Scheme of Diagonal Cartesian method for an Advection – diffusion Problem [J]. Computers & Fluids, 2009, 38 (5): 1003 – 1010.

He Guojian, Fang Hongwei, Bai Sen, et al. Application of a Three – dimensional Eutrophication Model for the Beijing Guanting Reservoir, China [J]. Ecological Modelling, 2011, 222(8): 1491 – 1501.

Han Dong, Fang Hongwei, Bai Jing, et al. A Coupled 1 – D and 2 – D Channel Network Mathematical Model Used for Flow Calculations in the Middle Reaches of the Yangtze River [J]. Journal of Hydrodynamics, 2011, 23(4)(Ser. B): 521 – 526.

Real – time Measuring the Yellow River Sediment Based on Turbidity Measurement

Liu Jinfu[1] , *Jiang Shuguo*[1] , *Yuan Xiaobin*[1] , *Liu Jimin*[1] and *Liu Yang*[2]

1. Jinan Yellow River Engineering Bureau, Jinan, 250032, China
2. Huawei Technology Co. , Ltd.

Abstract: The measurement of sediment concentration plays a significant role in the maintenance and regulation work of the Yellow River. However, existing sediment concentration measurement systems can get accurate value at expense of consuming long time and demanding manual operations, and cannot realize fast measurement. This essay presents a novel method that is able to realize real – time measurement of sediment concentration, and confirms the theory by experiments. Inspired by the wave – particle duality of light, the author concluded a relation between turbidity and an optical intensity pattern, which is generated by a light source projecting towards the Yellow River water samples in experiments. Then a function of sediment concentration and turbidity is established by analysing optical characteristics of suspended matters in water, such as reflection and refraction rate. Therefore, the value of sediment concentration can be obtained from the optical intensity pattern, and by implementing this method real – time measurement of sediment concentration can be realized. The essay also proposes a structure of measurement device and conducts experiments. The experiment proved that this method could be adopt when sediment concentration between $10 \sim 70$ kg/m^3.

Key words: sediment concentration, the Yellow River, turbidity, real – time measurement

Middle reaches of the Yellow River flows through the loess plateau, so that the water contains substantial silt. In the lower reaches, the Yellow River deposits a large amount of silt because the water slows down. Silting is the root cause of unstableness and complexity. Therefore, measuring the sediment concentration plays a significant role in conservancy of the Yellow River at lower reaches.

The system of measuring sediment concentration of the Yellow River has been established for a long time and become mature. There is also substantial academic research on real time measuring the sediment concentration. However, it has not been used in practice.

1 The lower reaches of the Yellow River characteristics

The Yellow River has the highest sediment concentration in the world. In the history, the Yellow River could transport up to 3.91×10^9 t (recorded in 1933) of sand, and the sediment concentration reached 920 kg/m^3 (recorded in 1977). Averagely, the annual transported sand is about 1.6×10^9 t and concentration is 35 kg/m^3.

The sand of the lower reaches of the Yellow River consists of three kinds mainly. Firstly, particles with diameter greater than 0.05 mm is called coarse sand and are the major part of the sand deposited in the lower reaches of the Yellow River. Secondly, the sand with diameter between 0.025 mm and 0.05 mm is call middle sand. Lastly, sand with diameter less than 0.025 mm is referred as silver sand. According to statistics, between September 1960 and June 1996, there was a total of 38.556×10^9 t of sand entering the lower reaches of the Yellow River, among which there was 8.907×10^9 t of coarse sand, 9.536×10^9 t of middle sand and 20.113×10^9 t of silver sand. The total deposited sand was 3.622×10^9 t, among which there was 2.935×10^9 t of coarse sand, 0.931×10^9 t of middle sand and 0.234×10^9 t of silver sand. According to this, silver sand is the major part of the floating sand in the lower reaches and mainly contributes to the sediment concentration.

2 Penetration of light

Light is electromagnetic radiation with very high frequency (430 ~ 790 THz) that is visible to human eyes (Visible light), and is made up of photons. Therefore, light is proved to have the characteristics of both wave and particles, which is referred as the wave particle duality.

A ray of light indicates the propagation direction of the light, and consists of a large number of photons. The energy carried by a photon might be different from others' in the same ray of light. When light propagates through a medium, atoms of the medium are affected by both the electric field component and the magnetic – field component. Meanwhile, a part of the energy of light is absorbed by the medium, and the light refracts.

3 Turbidity and its measurement

3.1 Turbidity definition

Turbidity is a term used to indicate how muddy the river water is. According to International Organization for Standardization (ISO), turbidity is defined as cloudiness or haziness of a fluid caused by individual particles (suspended solids). Turbidity is an important optical property of river water, and represents the opacity of river water caused by suspended solids. It is relevant to the size, shape and density of the suspended solids. Therefore, although turbidity is not directly linked to the density of the suspended solids, it is related to the quantity of the suspended solids.

There are plenty of unit to quantize the turbidity. The most common three are list below:

Formazin Turbidity Unit (FTU);

Nephelometric Turbidity Units (NTU);

Jackson Turbidity Unit (JTU).

Our national standard way (GB 570—85) of measuring turbidity is to measure the diatom mass of diatom solids suspension.

3.2 Methods of measuring turbidity

3.2.1 Scattering Measurement

Turbidity can be measured by measuring the intensity of scattered light through the object water, which is referred as Scattering Measurement. In China, this method is widely adopted. One typical way of using Scattering Measurement is to compare the sample water with a reference suspension, and to define the 1 mg of SiO_2 in the suspension as 1 unit of the turbidity. The measured value of the turbidity may vary when using different measuring methods. The turbidity cannot directly indicate the water contamination.

3.2.2 Measuring by Turbidimeters

Turbidimeter can also be used to measure the turbidity. The light from a turbidimeter goes through the sample water, and turbidmeter can measure the scattered light intensity in the direction of vertical to the incident light to determine the turbidity of the sample water. A turbidimeter can be used for outdoor measurement or laboratory use, and is also suitable for 24 h monitoring. A turbidimeter normally has a threshold, above which an alarm can be triggered.

3.2.3 Other method

Turbidity can also be obtained by measuring the intensity of light transmitted through the sample water with colorimeter and spectrometer. However, this method is neither commonly accepted nor meets the standard by American Public Health Association.

By measuring the light transmissivity can be affected by the various factors, such as colour absorption ratio and particle absorption factor. In addition, the turbidity measured based on scattering ratio can be quite different from the one based on transmissivity. In spite of that, the turbidity measured by colorimeter and spectrometer can also be used in some practical applications.

4　The experimental principle

4.1　Theoretical basis

This section introduces the methodology of real time monitoring the sediment concentration by measuring water turbidity.

The turbidity can be determined by illuminating the main stream of the Yellow River. Then the sediment concentration can be obtained based on the mapping between turbidity and sediment concentration. However, the turbidity is not only determined by the sediment concentration in the Yellow River. Different water samples of the Yellow River with the same sediment concentration have different turbidity, or vice versa. This is because of the proportion of different types of sand changes frequently. The sediment in the Yellow River mostly comes from loess, which contains more than 60 types of minerals. However, the particle size, mineral and chemical composition changes following a certain rule. As a result, the turbidity of the Yellow River is relatively stable. This provides practical support for the stable relation between sediment concentration and particle composition. Therefore, it is necessary to determine the function among turbidity, sand proportion and sediment concentration.

The most commonly adopted ways of measuring turbidity are using a spectrophotometer or turbidimeter. These are either transmissivity based or scattering based. The function between turbidity T and incident light intensity I_0 is as following:

$$T = \frac{2.303}{b}\log\frac{I_0}{I} = \frac{2.303}{b}D \qquad (1)$$

where, D is light intensity, namely the value read from spectrophotometer; b is light path; I is transmitted light intensity.

when measuring the particle size, the light intensity can be defined as

$$D = -CL(\log e)\sum_{r=0}^{r=st} K_r \kappa_r n_r d_r^2 \qquad (2)$$

Where, C is particle mass concentration; L is sedimentation slot length; K_r is light cancelation ratio; κ_r is particle shape determined constant (for sphere $k = \pi/4$); d_0, d_{st} are the largest and smallest particle diameter; n_r is particle number.

At the first stage of measuring, light intensity reaches max (DM), and

$$S = \sum_{r=0}^{r=\max} \alpha_{s,r} n_r d_r^2 \qquad (3)$$

$$W = \rho_s \sum_{r=0}^{r=\max} \alpha_{v,r} n_r d_r^2 \qquad (4)$$

where, S is Particle surface area; $\alpha_{s,r}$ is particle shape determined constant; W is particle mass; $\alpha_{v,r}$ is particle volume determined constant; ρ_s is particle density. And then we have:

$$DM = CL(\log e) K_m \kappa S_w \qquad (5)$$

where, K_m is average light cancelation ratio; S_w is mass to superficial area ratio.

$$S_w = \frac{\alpha_{sr,A}}{\rho_s} \sum \frac{X_r}{d_{r,A}} \qquad (6)$$

where, $\alpha_{sr,A}$ is particle shape, volume determined constant; X_r is $d_{r,A}$ particle proportion.

In Eq. (6), $\Sigma \dfrac{X_r}{d_{r,A}}$ is the particle diameter $d\theta\kappa\sigma$.

Combining Eq. (1) and Eq. (5) we have $K_m k \dfrac{L}{b} CS_w$ (7)

Eq. (7) and equation in Lambert – Beer law are similar, namely:

$$I_R \cdot \left(\frac{I_0}{I} \right) = \kappa ACL \tag{8}$$

where, A is the project area taken by 1 g particles; C is the particle density of suspension; L is light path; κ is constant related to light optical path.

Lu, Xu and Long pointed out the formula of light cancelation ratio:

$$K = 2.303 \frac{2\rho_n}{3L} d_{50} \frac{\log \dfrac{I_0}{I}}{C} \tag{9}$$

where, d_{50} is median value of particle diameter.

Eq. (9) can also be written as follow:

$$\log \frac{I_0}{I} = 0.65 \frac{KCL}{\rho_s} \cdot \frac{1}{d_{50}} \tag{10}$$

For a type of sediment sample:

$$\log \frac{I_0}{I} = \beta \cdot C \cdot \frac{6}{d_{50}} \tag{11}$$

Where $\dfrac{6}{d_{50}}$ can be defined as particle surface area S_0, so Eq. (11), Eq. (5) and Eq. (8) are similar.

According to our record, it is still not clear to use d_{50} represent S_0 or S_w. In most cases, d_{50} and S_0 have relatively stable relation, but in other cases, it is totally different.

Combining Eq. (5), Eq. (8) and Eq. (11), we have that turbidity is not only related to sediment concentration but also particle diameter and surface area.

From Eq. (1), Eq. (5) and Eq. (8), we have turbidity is proportional to the total particle surface area. However, according to the laboratory record, turbidity T, sediment concentration C_w and particle surface area S_0 is subject to the following equation:

$$T = a(s_0^m \cdot C_w)^b \tag{12}$$

where, n, a and b are empirical constant.

Eq. (12) can be rewritten as

$$T = 764(S_0^{0.70} \cdot C_w)^{1.024} \tag{13}$$

where, $r = 0.998,1$.

4.2 Implementation

In this laboratory, the turbidity is real – time measured and data is obtained real – time. The measurement position is the middle of the Yellow River course. Employing imaging equipment images of the illuminated sample water can be obtained and transferred to a computer for spectrum analysis. At last, the turbidity can be send to a base station, where the sediment concentration is calculated based on the turbidity.

The initial design of this laboratory is as follows: use industrial lighting slab as the light source, use high resolution industrial camera as imaging equipment, and use a computer as remote image processing platform, then transfer statistics to base station. The metal box should be waterproof sealed in case of water or sediment interference. The internal structure is shown in Fig. 1.

Experiment procedure:

(1) The experiment instrument is placed in the river water, and make sure water flow tunnel is parallel to the river flow.

(2) Connect base station and the experiment instrument so that base station remotely control the instrument.

(3) The base station control to capture images.

(4) The captured images are processed in the remote platform. And the analysis results are transferred back the base station.

Based on the received date, the turbidity can be obtained by the based station.

Captured image when sediment concentration 40 kg (in Fig. 2).

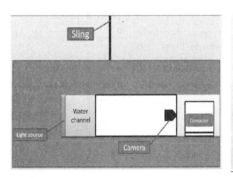

Fig. 1　　　　　　　　　　　　　　　　**Fig. 2**

Captured image when sediment concentration 50 kg (in Fig. 3).
Captured image when sediment concentration 60 kg (in Fig. 4).

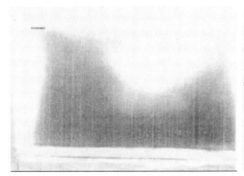

Fig. 3　　　　　　　　　　　　　　　　**Fig. 4**

According to the images above, in the water of the Yellow River, there is clear difference at different turbidity. Under certain condition, the sediment concentration can be determined by optical property of different turbidity.

5 Experiment conclusions

(1) When reduce other factors interference, we proved that sediment concentration can be real – time measured based on turbidity both theoretically and practically. When the sediment

concentration is between 10 kg/m^3 and 70 kg/m^3, turbidity images change clear.

(2) Experiment measuring should be conducted in the Yellow River water. The conditions of water keep changing, and turbidity is not only affected by sediment concentration and particle factors, but also by other conditions, which can cause measurement error. Therefore, in experiment design, the interference of other factors must be eliminated.

(3) Sediment concentration cannot be measured when sediment concentration is greater than 70 kg/m^3 according to experiment.

References

Alan T. Determination of Particle Size (third edition) [M]. Translated by Huapu La, Sandu Tong, Juan Shu. Beijing: China Building Industry Press, 1984.

Минц, д. М. шуберт, С. А. Гидрвликазернистыхматериалов. Нзд. МКХ РСХСР. 1955

三輪茂雄・日高重助である. Engineering Experiment Handbook [M]. Translated by Xie Shuxian, Yang Lun. Beijing: China Building Industry Press, 1987.

Lu Yongsheng, Xu Youren, Long Yuqian. The Extinction Method used in the Analysis of River Sediment Particle – river Sediment International Academic Conference (second volume). Beijing: Guanghua School Press, 1980.

Development of Depth-averaged Water and Sediment Routing Mathematical Model for the Lower Yellow River

Luo Qiushi[1,2] , *Liu Jixiang*[2] , *Wang Hongmei*[2] and *Cheng Ji*[2]

1. Postdoctoral Research Institute of Yellow River Engineering Consulting Co. Ltd. ,
Zhengzhou, 450003, China
2. Yellow River Engineering Consulting Co. , Ltd. , Zhengzhou, 450003, China

Abstract:Geometric properties of the lower Yellow River can be summarized as narrow channel, broaden floodplain and the flow velocity and direction in it may changes suddenly before and after overflowing the main channel. Grids generation of lower Yellow River was discussed and a hybrid grids arrangement which composed of quadrilateral grids in main channel and triangle grids in floodplain was suggested to adjust the characters of the lower Yellow River and control the grid number within a reasonable range to reduce the calculation work load. At the same time, a depth-averaged water and sediment routing mathematical model was built and validated with Yellow River water and sediment regulation test data. The validation results indicate that the model established here works excellently and the calculated results are in good agreement with the observed ones.

Key words:the lower Yellow River,water and sediment routing,mathematical model

1 Introduction

The lower Yellow River is the key reach which should be paid more attention during the river regulation process, because it is not only the area for flood mitigation, flood detention and sediment deposition but also the family of 1.89×10^6 people. The regulation of lower Yellow River involves several aspects, such as technical, economic and social problems, but the sediment is the root problem and the final goal. Prototype observation, physical model testing and mathematical model calculation are the three main tools for the research of lower Yellow River sediment problem, and mathematical model calculation is the most economic tool. But, until now lots of work has been done on prototype observation and physical model testing, but poor on mathematical model calculation. So research of the depth-averaged water and sediment routing mathematical model of lower Yellow River can provide vital important support for sediment research in lower Yellow River.

In recent years, great development has been made on sediment-laden flow mathematical model. The simulation technology by 1D mathematical model has been mature and the 2D and 3D mathematical model develops rapidly. The 1D mathematical model, such as HEC-6 model developed by the U. S. Army Corps of Engineers, SUSBED-2 developed by Yang Guolu and the river-net model developed by Li Yi tian have been widely used. The 2D and 3D models developed later than 1D model, but great success has also been achieved, especially in dealing with the complex river boundary and improving the calculation speed. And now, researchers tend to use unstructured grid to deal with the complex natural river boundary, for example: Tae developed a second-order upwind FVM on unstructured triangular grids using the HLL Riemann solver and applied it to the simulation of the malpasset dam break. Shi et al. constructed a depth-averaged flow-sediment model on unstructured grids and used it to the simulation of the riverbed deformation of the Zhangzhou Reach in the Tan River in China. Liu Shihe suggested an implicit scheme on the unstructured grid and applied it in Chenglingji reach. Hu Ningning suggested a hybrid girds arrangement which composed of quadrilateral grids in main channel and triangle grids in floodplain to adjust compound channel. As to improving the calculation speed, the first approach is to use the parallel calculation to improve the calculation efficiency , and the second approach is to use hybrid model, which consisted of 1D model or 2D model according to the characters of the river to improve the calculation speed with respect the accuracy.

Geometric properties of lower Yellow River can be summarized as narrow channel, broaden floodplain and the flow velocity and direction in it may changes suddenly before and after overflowing the main channel. Single grid cells can't take into account both grid arrangement and computational efficiency. And in this article, grids generation of lower Yellow River was discussed, a Depth-averaged water and sediment routing mathematical model for lower Yellow River was studied and the model was validated with Yellow River water and sediment regulation test data in year 2009.

2 Mathematical model and humerical method

2.1 Governing equations

Depth averaged mathematical model was employed to describe the sediment laden flow in nature river. The general form of governing equations in Cartesian Coordinate can be written as:

$$\frac{\partial}{\partial t}(H\rho\phi) + \frac{\partial}{\partial x}(H\rho u\phi) + \frac{\partial}{\partial y}(H\rho v\phi) = \frac{\partial}{\partial x}\left(H\Gamma_\phi \frac{\partial\phi}{\partial x}\right) + \frac{\partial}{\partial y}\left(H\Gamma_\phi \frac{\partial\phi}{\partial y}\right) + S_\phi \tag{1}$$

where, ϕ is universal variable; Γ_ϕ is universal diffusion coefficient; S_ϕ is the source term; ρ is water density; H is the flow depth.

The other variable are shown in Tab. 1. In Tab. 1, Z is the elevation of the river flow, u and v are the depth averaged velocity in the x and y directions respectively, n is the roughness coefficient, g is the gravity acceleration, V_T and V_{TS} are turbulent diffusion coefficient and sediment diffusion coefficient respectively, S_i and S_{*i} are the depth-averaged concentration and sediment-carrying capacity by weight for the ith group of suspended sediments respectively, ω_i is the settling velocity for the ith group of suspended sediments, α is the non-equilibrium adaptation coefficient of suspended load.

Tab. 1　Variable of the model

	ϕ	Γ_ϕ	S_ϕ
Continuity equation	1	0	0
Momentum equations of u	u	v_T	$-g\dfrac{n^2\sqrt{u^2+v^2}}{H^{\frac{1}{3}}}u - gH\dfrac{\partial z}{\partial x}$
Momentum equations of v	v	v_T	$g\dfrac{n^2\sqrt{u^2+v^2}}{H^{\frac{1}{3}}}v - gH\dfrac{\partial z}{\partial x}$
Transport equation of suspended sediments	S_i	v_s	$-\alpha\omega_i(S_i - S_{*i})$

(1) Equation for riverbed deformation.

$$\gamma' \frac{\partial Z_{0i}}{\partial t} = \alpha_i\omega_i(S_i - S_{*i}) + \left(\frac{\partial q_{bxi}}{\partial x} + \frac{\partial q_{byi}}{\partial y}\right) \tag{2}$$

where, g_{bxi}、g_{byi} is the bed load transport rate of unit width in x and y direction.

(2) Boundary conditions and the relative problems.

The boundary conditions are as follows: the distributions of the depth-averaged velocity of river flow and the depth-averaged concentration of the suspended sediment are given at upstream, at downstream the elevation of river flow is given, and no-slipping condition for the velocity of the flow and no-penetrating condition for the suspended sediments are employed at the side banks of the river.

The method employed here to calculate the non-uniform sediment-carrying capacity, bed load transport rate and the non-uniform sediment-carrying capacity was employed from literature 7 ~ 9 respectively.

2.2 Numerical method

A FVM was employed to discrete the governing equations of depth averaged mathematical model, and an implicit Semi-Implicit Method for Pressure Linked Equations (SIMPLE) scheme based upon collocated grids was proposed to settle the coupling relation ship between the pressure (or water level) and the discharge. The final discretization equations can be written as follows:

$$A_p \phi_p = \sum_{j=1}^{ED} A_{Ej} \phi_{Ej} + b_0 \tag{3}$$

where, A_p and A_{Ej} are coefficient; b_0 is source term, which can be seen in literature 7,12; ED is the edge number of the control volume.

The solving step of the discretization equations also can be seen in literature 7,12.

Compared with the previous methods, the numerical method suggested here has the following advantages: the model is more adaptable to complex terrain and it can be solved on whichever grid you selected, because the governing equations are discretized over the polygonal cells and the variations in the calculation program are saved with the unstructured storage format.

3 Grid generation and errain interpolation

The reach from HuaYuankou to Aishan has been employed for calculation. The calculation domain was generalized as in Fig. 1. According to the characters of the reach, a hybrid girds arrangement which composed of quadrilateral grids in main channel and triangle grids in floodplain was suggested to adjust the geometric properties of the lower Yellow River. The quadrilateral grids were generated by solving the differential equations, the triangle grids were generated by the modified Delaunay triangulation algorithm, and different grids were joined together by finding the common points at the boundary.

The river terrain was generated by parsing the GE (Google Earth) terrain information and modified based on the measured cross-section data in 2009. And in this article, a digital elevation model (DEM) of the lower Yellow River was built by the terrain data generated here and terrain interpolation was carried out on it to assign elevation value to grid points.

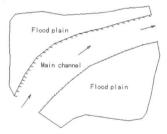
Fig. 1   Sketch of the lower Yellow River

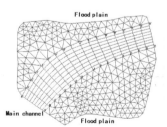
Fig. 2   Hybrid girds arrangement

4   Verification

The model was validated with the Yellow River water and sediment regulation test data in year

2009. The roughness coefficient is taken as 0.010 ~ 0.015 for main channel and 0.023 ~ 0.030 for floodplain.

4.1　Verification data

The model was validated with Yellow River water and sediment regulation test data from June 19 to July 4 in year 2009. Accroding to the information released by the Yellow River Conservancy Commission, the total amount of water released into the lower river is 4.5702 × 10⁹ m³. The measured discharge-time and sediment concentration-time curve of Huayuankou section was shown in Fig. 3.

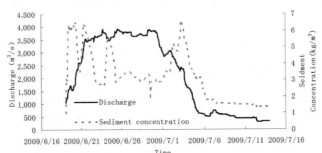

Fig. 3　Measured discharge-time and sediment concentration-time curve of Huayuankou section

4.2　Verification results

4.2.1　Results for discharge

The simulated discharge-time curve is shown in Fig. 4. The comparison between the simulated and observed discharge of Jiahetan section is shown in Fig. 5, which is in good agreement with each other. The comparison between the simulated and observed flood peak are shown in Tab. 2 with the relative deviation between −5.24% ~ 0.45%.

Tab. 2　Comparison between the simulated and observed flood peak

Station name	Huayuankou	Jiahetan	Gaocun	Sunkou	Aishan
Observed value(m³/s)	4,170	4,120	3,890	3,900	3,780
Simulated value(m³/s)	4,170	3,904	3,907	3,824	3,752
Deviation value(m³/s)	0	−216	17	−76	−28
Relative deviation value(%)	0	−5.24	0.45	−1.95	−0.73

4.2.2　Results for sediment concentration

The Comparison between the simulated and observed maximum sediment concentration are shown in Tab. 3 with the relative deviation between −20.33% ~ −15.96%.

Fig. 4   Simulated discharge – time curve

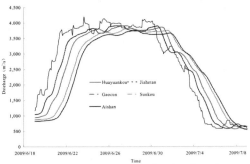

Fig. 5   Comparison between the simulated and observed discharge of Jiahetan section

Tab. 3   Comparison between the simulated and observed maximum sediment concentration

Station name	Huayuankou	Jiahetan	Gaocun	Sunkou	Aishan
Observed value(kg/m³)	6.25	7.65	9.41	11.20	11.30
Simulated value(kg/m³)	6.25	6.10	7.91	9.32	9.40
Deviation value(kg/m³)	0.00	− 1.56	− 1.50	− 1.88	− 1.90
Relative deviation value(%)	0.00	− 20.33	− 15.96	− 16.79	− 16.83

4.2.3   Results for water level

The Comparison between the simulated and observed maximum water level are shown in Tab. 4 with the relative deviation less than 6.17% .

Tab. 4   Comparison between the simulated and observed maximum water level

Station name	Huayuankou	Jiahetan	Gaocun	Sunkou	Aishan
Observed value(m)	93.44	73.87	62.28	48.46	41.42
Simulated value(m)	93.27	73.96	62.39	48.58	41.42
Deviation value(m)	− 0.17	0.09	0.11	0.12	0.00
Relative deviation value(%)	− 6.17	4.77	4.07	3.55	0.00

4.2.4   Results for bed deformation

The observed data shows that the river bed totally displays the erosion in the simulated period, and the observed amount of sediment erosion is $1\,611 \times 10^4$ t, yet the simulated amount of sediment erosion is $1,915 \times 10^4$ t, the deviation between the two is less than 304×10^4 t. The averaged erosion depth is 0. 042 m of observation and 0. 049 of simulation.

5 Conclusions

(1) Geometric properties of the lower Yellow River can be summarized as narrow channel, broaden floodplain and the flow velocity and direction in it may changes suddenly before and after overflowing the main channel. Single grid cells can't take into account both grid arrangement and computational efficiency. And in this article, a hybrid girds arrangement which composed of quadrilateral grids in main channel and triangle grids in floodplain was suggested to adjust the geometric properties of the lower Yellow River and control the grid number within a reasonable range to reduce the calculation work load.

(2) The model was validated with the Yellow River water and sediment regulation test data in year 2009. The validation results indicate that the model established here works excellently and the calculated results are in good agreement with the observed ones. The relative error is less than 5. 24% for peak discharge, 20. 33% for sediment concentration, 6. 17% for flood crest, 21% for sedimentation amount. The calculating result for each river section on the characteristics of scouring and deposition fits well with measured data.

References

Feld man A D. Hec Modeis for Water Resources System Simulation; Theory and Experience [J]. The Hydraulic Engineering Center. Davis California, 1981.

Yan Guolu, Wu Weiming. SUSBED-1 Movable Bed Modeling of Graded Sediments [J]. Journal of Hydraulic Engineering, 1994(4): 1 –9.

Li YiTian. A Junctions Group Method for Unsteady Flow in Multiply Connected Networks [J]. Journal of Hydraulic Engineering, 1997(3): 49 –57.

Sun Shaohua, Li Yitian, Cao Zhifang. Numerical Simulation of Sediment Transport in Unsteady Flow of River Network [J]. Advances in Water Science, 2004, 15(2): 166 –172.

Tae Hoon Yoon, F. ASCE, Seok-Koo Kang. Finite Volume Model for Two – dimensional Shallow Water Flows on Unstructured Grid[J]. Journal Hydraulic Engineering,2000(130):678 –688.

Shi Yong, Hu Siyi. A Finite Volume Method for Numerical Modeling of 2 – D Flow and Sediment Movements on Unstructured Grids[J]. Advances in Water Science, 2002, 13 (4): 409 –415.

Liu Shihe, Luo Qiushi, Mei Junya. Simulation of Sediment – laden Flow by Depth-averaged Model Based on Unstructured Collocated Grid [J]. Chinese Journal of Hydrodynamics, Ser. B, 2007, 19(4):515 –523.

Hu Ningning, Luo Qiushi, Liu Shihe. Numerical Simulation of Compound Channel by Depth Averaged Mathematical Model Based on Hybrid Grid [C]//The 4th National Academic Conference of Hydraulics and Water Information, 2009, 40(11):1381 –1385.

Yu Xin, Yang Ming, Wang Min. Parellel Calculation of Sediment Laden Flow in Lower Yellow River by 2D Model with MPI [J]. Yellow River, 2005, 27(3):49 –53.

Luo Qiushi, Liu Shihe, Liu Jixiang. Simulation of Sediment Laden Flow by Depth Averaged Mathematical Model on Muti-core Processor [J]. Chinese Journal of Hydrodynamic, Ser. A, 2011, 26(1):94 –99.

Zheng Guodong, Huang Dong, Zhao Mingdeng. Application of 1 – D Mathematical Model Coupled with 2 – D Model to Estuarine Engineering [J]. Journal of Hydraulic Engineering, 2004 (1).

Luo Qiushi. Simulation of Sediment – laden Flow by 2D/3D Mathematical Model Based on Unstructured Grid [D]. Wuhan: Wuhan University, 2007.

Design and Performance Analysis of the Low – disturbing Sampling Equipment in Deepwater Reservoir[①]

Yang Yong[1] ,*Zhang Qingxia*[2] ,*Chen Hao*[3] and *Zheng Jun*[4]

Yellow River Hydraulic Research Institute, Zhengzhou, 450003, China

Abstract: In order to obtain low disturbance sediment samples, this paper introduces how to develop the low – disturbing sampling equipment in deepwater reservoir of which inner diameter is 90 mm and length is 10 m. According to the conditions of the field, the designing inclndes guiding device, triggering device, sampling device, counter weight and so on. Taking the lever mechanism, which is based on support force controlling trigger principle, it can fast insert the sediment under own weight to achieve low – disturbing sampling. This paper also describes how to analyze the sampling process, mechanical parameters and sediment conditions, analyze the sampling performance and disturbance of the sediment in tube. By taking an example of calculation about disturbance in the Sanmenxia Reservoir. The results show that, when the spherical cavity expansion radius is 2.5 mm, the final expansion pressure of sediment is 65.58 kPa. In undisturbed elastic region, sampling disturbance makes excess pore water pressure increase 0.018 kPa. The disturbed proportion of sediment sample is 15% in the sampling tube diameter direction.

Key words: sediment, equipment development, disturbance analysis

1 Introduction

The Yellow River is the world recognized sandy river, resulting in serious silting of reservoir and bringing great difficulties to govern the Yellow River. The silting of reservoir shorten the project life, impact capacity of the project in flood control, irrigation and power generation, limit high efficient utilization of water resources. Treatment of reservoir sediment problem, firstly understand the sediment characteristics, origin, quantity and temporal and spatial variation, which requires reliable sediment data. However, the deepwater sediment sampling technology is immature in current reservoir, lack relevant source material. Therefore, mirror truth – preserving sampling techniques of the deep sea sediment, base on gravity piston principle, develop inner diameter 90 mm and length 10m of the low – disturbing sampling equipment in deepwater reservoir.

2 Overall structure

Low – disturbing sampling equipment in deepwater reservoir bases on gravity piston principle, mainly by guiding device, triggering device, sampling device and counter weight components, as shown in Fig. 1. When the sampler down to a position from a distance to the bottom, the hammer hits the bottom first, the lever loses balance, loosen the main cable, the low – disturbing sampling tube insert sediment under the action of own weight. Piston declines with the speed of winch cable, relative to the free fall sampling tube to rise. As the piston and liner tube have better sealing, the low part of the piston can form local vacuum, so the sample more smoothly enter into the liner tube.

---

① **Fund Project**: Public Benefit Research Foundation of MWR 201201085

2.1 Triggering device

The low – disturbing sampling equipment take the triggering device based on support force controlled, mainly including plywood, lever, main cable, cable of hammer, heavy hammer and the piston in the Fig. 1. Its principle is the triggering device in stress state at the beginning, when it meets certain support reaction force, trigger sampling action after force balance, plywood breaks away from the sampler, sampler rapidly inserts the sediment by own weight and samples; After the completion of sampling, take back the sampler by the main cable.

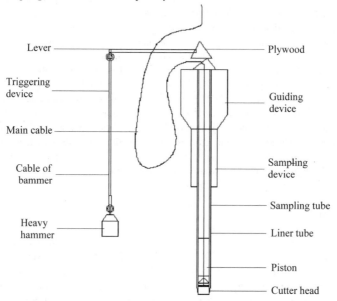

Fig. 1 The structure diagram of low – disturbing sampler

2.2 Low – disturbing sampling equipment

Low – disturbing sampling equipment is mainly composed of sampling tube, liner tube (in sampling tube) and cutter head. This part of the role is to sample the sediment, take out the liner tube from the sampling tube, get low – disturbing sediment samples, and realize low – disturbing sampling in the deepwater reservoir.

Designed the inner diameter 90 mm and length 10 m of sampling tube belongs to slender pole, use the stability theory of pole in the material mechanics to analyze the stability, need to choose hard steel tube of certain wall thickness, satisfy the request. The friction of liner tube will increase the difficulty of the samples into the sampling tubes, so should choose the materials of smooth wall like organic glass tube. In the soil mechanics, in order to gather the original samples, need to make reasonable design of cutter head in the area ratio, inner diameter ratio, outer diameter ratio, the shape and angle of the blade.

2.3 Accessories and sampling interface

The accessories of low – disturbing sampling equipment mainly include guiding device, transitional cone and counter weight, etc. Their roles are to reduce water resistance when the sampler free falls, make sampler vertically insert sediment as far as possible; Through increase counter weight in the guiding device, can further strengthen the own weight of sampler, sampler can insert smoothly the sediment and complete sampling operation.

3　Sampling performance analysis

The sampling tube for sediment sampling, often appear phenomenon of samples to be compacted in the tube, reduce the sampling rate of the sediment, make the sampling tubes into the soil like "pile", put all the sediment to the surrounding, form "pile effect" and produce adverse effect to sampling, as shown in Fig. 2.

Fig. 2　"Pile effect"

According to the knowledge of Coulomb Theory, the appeared condition of "pile effect" is: compressive stress of sample in cutter head space is greater than or equal to the ultimate strength σ of sample under the volume compression state, that is $p \geqslant \sigma$. When $p = \sigma$, the critical value of h is the solution of the following equation:

$$\sigma = (2/d)f\gamma h^2 \cdot \tan^2(45° + \varphi/2) + h(\gamma - \gamma_w) \tag{1}$$

The reasonable solution of Eq. (1) is:

$$h = \left[d(\gamma_w - \gamma) + \sqrt{d^2(\gamma_w - \gamma) + 8f\gamma\sigma d \cdot \tan^2(45° + \varphi/2)} \right] / \left[4f\gamma \cdot \tan^2(45° + \varphi/2) \right]$$

where, h is the critical value of filled height by sample in the sampling tube; d is the inner diameter of sampling tube; γ is the specific weight of sediment; φ is the internal friction angle of sediment; f is the friction coefficient between tube wall and samples; γ_w is the specific weight of water; σ is the ultimate strength of sample under the volume compression state.

By the analytical formula of h, h is related with d, σ, f and φ of the sediment sample. Therefore, in mechanical design, need to consider related factors. Taking the larger pipe diameter or smooth liner tube, can reduce the friction coefficient between sediment and wall of tube; Through designing reasonable shape, it is helpful for sampling tubes to insert sediment and reduce the friction between the samples and liner tube; Taking sampling tubes based on the principle of the piston, the piston produce suction to make samples and liner friction compensation, put off to appear "pile effect" and increase sampling length.

4 Disturbance analysis

At present, continuous static or dynamic injection theory of pile consists of pure shear theory, cavity expansion theory and finite element method in the soft clay. Pure shear theory essentially lies in the injection of pile tip as local soil shear destruction process, and use the ultimate equilibrium theory to solve ultimate penetration resistance of the pile tip, but it can't reflect the volume change phenomenon of plastic zone. The finite element method is used for complex soil constitutive relation and boundary conditions, for the difficult situation to solve analytical solution or approximate solution. In this article, take the theory of cavity expansion between the shear and compression mechanism, and analyze the disturbance of internal sample in sampling tubes.

Spherical cavity expansion theory is based on the Coulomb and Moore condition, solve the fundamental solution of the spherical cavity expansion in the infinite soil with internal friction angle and cohesive force. The process of low – disturbing sampling tube into the sediment can be regarded as the process of expansion n spherical holes, which have equivalent volume with sampling tube, in semi – infinite soil, as shown in Fig. 3. Calculation process is as follows:

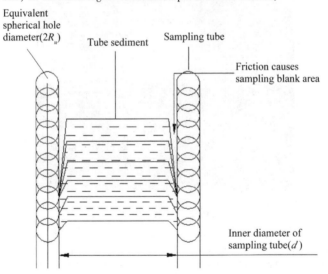

Fig. 3 The soil practical model of sampling pipe

(1) The relatively radius of plastic area.

$$\frac{R_P^3}{R_u^3} = \frac{E}{2(1+\mu)c} \tag{2}$$

where, R_p is the final expansion radius of plastic zone; R_u is the spherical hole final radius; E is the deformation modulus of sediment; μ is the Poisson's ratio of sediment; c is the cohesive force of sediment.

(2) The equation between the modulus of deformation and modulus of compression.

$$E = E_s(1 - \frac{2\mu^2}{1-\mu}) \tag{3}$$

where, E is the compression modulus of sediment; other symbols are the same with above.

(3) Final pressure.

$$P_u = \frac{3(1+\sin\varphi)}{3-\sin\varphi}(q+c\cdot ctg\varphi)(\frac{R_P}{R_u})^{\frac{4\sin\varphi}{1+\sin\varphi}} - c\cdot \cot\varphi \tag{4}$$

where, P_u is the final pressure of plastic zone; φ is the internal friction angle of sediment; q is the uniform distribution of pressure in the sediment; other symbols are the same with above.

(4) The excess pore water pressure of elastic zone.

$$\Delta u_R = 0.707B \cdot \frac{4c\cdot\cos\varphi}{3-\sin\varphi}(\frac{R_P}{R})^3 \tag{5}$$

where, Δu_R is the excess pore water pressure of elastic zone; B is the hole stress coefficient under confining pressure; R is the outer radius of sampling tube; other symbols are the same with above.

(5) Disturbance ratio of sample on diameter direction

$$\alpha = \frac{R_p}{d/2} \tag{6}$$

where, α is the disturbance ratio of sample on the sampling tube diameter direction; d is the inner diameter of sampling tube; other symbols are the same with above.

Because of no carry out the field test, based on the existing measured data of sediment in the Sanmenxia Reservoir in this paper, do not consider the friction between the sediment and sampling tube, analyze the disturbance of sediment sample on the direction of sampling tube diameter. According to the *soil mechanics*, Poisson's ratio is 0.36 for the silty clay. Taking the compression modulus E_s of the sediment is 690 kPa. Cohesion is 7.6 kPa and internal friction angle is 7.6°. Take 90mm as the inner diameter of sampling tube and $R_u = (100-90)/4 = 2.5$ as the diameter of the spherical hole expansion; When calculate excess pore water pressure, the pore water stress coefficient B is 1 for saturated soil. The results are as follows:

From Eq. (3) can be known, deformation modulus of sediment $E = 0.595$ kPa, $E_s = 410$ kPa. From Eq. (2) can be known, the final expansion radius of plastic zone $R_p = 2.7$ mm, $R_u = 6.77$ mm. From Eq (4) can be known, the final expansion pressure $P_u = 65.58$ kPa. From Eq. (5) can be known, the excess pore water pressure $\Delta u_R = 0.018$ kPa. From Eq. (6) can be known, disturbance ratio of sample on the sampling tube diameter direction $= 15\%$.

P_u, Δu_R and α are related with wall thickness of sampling tube, and properties of sediment. P_u affects the friction between sampling tube wall and sample, then affects the sample disturbance in vertical direction of sampling tube; The elastic region of sampling tube produces excess pore water pressure in the sampling process; Δu_R can react the effect of sampling to samples, directly reacts the α disturbance ratio of sample on the sampling tube diameter direction, these factors affect the sample disturbance within the sampling tube. If want to obtain the small plastic deformation zone and smaller excess pore water pressure, require the wall thickness as thin as possible when the sampler contact sediment.

5 Conclusions

(1) Based on gravity piston principle, design the low – disturbing sampling equipment in deepwater reservoir, it is able to collect 0 ~ 10 m sediment, and provide basic data for the further study to sediment of the Yellow River.

(2) Using the Coulomb theory obtain the formula about critical filling height of sediment in the sampling tubes, and analyze the parameters of "pile effect" caused. Take the smooth liner tube, reasonable shape of cutter head, larger inner diameter of the sampling tube to put off "pile effect", and increase sampling length.

(3) Using relevant knowledge of soil mechanics, calculate and analyze the disturbance and excess pore water pressure of the sediment sample in the sampling tube. The results show that, if the spherical cavity expansion radius is 2.5 mm, the final expansion pressure of sediment is 65.58 kPa by the sampling tube wall; Undisturbed elastic region, sampling disturbance makes excess pore

water pressure increase 0. 018 kPa; The disturbed proportion of sediment sample is 15% in the sampling tube diameter direction.

(4) The actual performance of low – disturbing sampling equipment in deepwater reservoir should be further validated in the field test.

References

Yang Taining, Bu Jiawu, Chen Hanzhong. Theoretic Discussion and Parameter Calculation of Subsea Sampler: the Fifth Introduction of Subsea Coring Technology [J]. Geological Science and Technology Information, 2001, 20(2):103-106.

Zhu Liang. Research on Truth-preserving Technique of Deep-sea Pressure Tight Sediment Sampler [D]. Hangzhou: Zhejiang University, 2005.

Qin Huawei. Research on Low-disturbing Sampling Theory and Truth-preserving Technique of Deep-sea Surface Layer Sediment[D]. Hangzhou: Zhejiang University, 2005.

Chen Zhongyi, Zhou Jingxing, Wang Hongjin. Soil Mechanics[M]. Beijing: Tsinghua University Press, 1994.

Analysis of the Characters of Recent Dynamic Variation of Runoff and Sediment in the Middle Yellow River Coarse Sediment Source Area

Gao Yajun[1], *He Xiaohui*[2], *Wang Zhiyong*[1], *Lin Yingpin*[1] and *Xu Jianhua*[1]

1. Hydrological Bureau of YRCC, Zhengzhou, 450004, China
2. School of Environment and Water Conservancy, Zhengzhou University
Zhengzhou, 450001, China

Abstract: The middle Yellow River coarse sediment source area is the most serious soil and water loss area of the Yellow River basin, to strengthen soil and water loss controlling of this area is one of the most direct and effective ways which reduced the sand into the Yellow River and the harm of siltation in the lower reaches. Mastering runoff and sediment of this area in time, is the premise of correctly analysing the variation of runoff and sediment, and is the basic data for formulating soil and water loss prevention measures and analyzing soil and water conservation benefit.

In this study, we take the middle Yellow River coarse sediment source area as object, the year of 1956 ~ 2004 is the basic period, 2005 ~ 2010 is the monitoring contrast period, areal precipitation is obtained by the way of drawing the isoline map of precipitation which based on collecting observed data of existing rain – gauge station, considering the difference of geomorphic type and river basin management degree in this area, runoff and sediment discharge use zone plan for calculation, the flood change reacts according to the annual maximum peak discharge of hydrologic station in this study area, and the management degree of soil and water conservation reacts by means of changing of soil and water conservation measures. The outcomes show that comparing with the basic period, precipitation in the monitoring period increased by 3.9%, runoff decreased by 48.2%, the total sediment, coarse sediment of grain – size larger than 0.05 mm and 0.1 mm are less than 85.9%, 90.8% and 93.0%, respectively. It is an estimation that there is no great flood in the area according to the annual maximum peak discharge of hydrologic station, precipitation shows a weak increasing trend, and runoff and sediment decrease remarkably, It relates to that in the monitoring period there is no happen to a wide – extent and great – intensity rainstorm, precipitation intensity is smooth relatively although total precipitation has a little change. In ultra – infiltration flow area, it goes against producing runoff and sediment. Additionally effective management of soil and water conservation has been condffcted in recent years, especially the implementation of "return farmland to forests or grassland" and banned for harness measures, the ecological environment in this area has improved evidently, appearing a sharp reduction of observed runoff and sediment in the monitoring period.

Key words: precipitation, runoff, sediment coarse sediment source area, dynamic variation

The middle Yellow River coarse sediment source area is located one of the three torrential rain source areas of the Yellow River Basin. Vulnerable ecological environment, serious soil and water loss, rich in energy resources and backward economy are the remarkable characteristics of this area. The centralized management of coarse sediment source area is one of the important measures to realize "riverbed not run – up" which is included in "Maintain the Yellow River healthy life". The study takes coarse sediment source area as research object, systematically analyzes dynamic variation characteristics of runoff and sediment in recent years, it provides the theoretical basis for the management and implementation of soil and water conservation in coarse sediment source area.

1 Study site

The middle Yellow River coarse sediment source area is that total sediment modulus is greater than 5, 000 t/(km² · a) and coarse sediment (grain – size larger than 0. 05 mm) modulus is greater than 1,300 t/(km² · a). It locates the reach from Hekouzhen to Longmen and above area of Jinghe River upstream Malian River, Beiluo River Liujia River, area of 7. 86 × 10⁶ km². The area accounts for only 11% of the total area of the middle and upper reaches of the Yellow River, however sediment accounts for 62. 8% of the total sediment, coarse sediment accounts for 72. 5% of the total coarse sediment. The area involves in Inner Mongolia, Shanxi, Shannxi, Ningxia and Gansu five provinces (areas) 45 counties (cities, flag). Coarse sediment source area belongs to arid and semiarid regions, few rainfalls, plenty of sediment in the river; it is the main Source area of Sediment deposition of downstream Yellow river. The study takes the year of 1956 ~ 2004 as the basic period, 2005 ~ 2010 is monitoring contrast period, analyzes dynamic variation characteristics of runoff and sediment of the area.

2 Dynamic variation characteristics of runoff and sediment of coarse sediment source area

2. 1 Precipitation variation

The average annual precipitation of 1956 ~ 2004 is 422. 0 mm in the middle Yellow River coarse sediment source area, the average precipitation of 2005 ~ 2010 monitoring period is 438. 3 mm, 3. 9% more than the average annual precipitation.

Precipitation year by year of 2005 ~ 2010 in coarse sediment source area is 371. 1 mm, 402. 6 mm, 551. 3 mm, 414. 0 mm, 463. 4 mm and 427. 2 mm, comparing with the average annual precipitation, precipitation in the year of 2005 is less than 12. 1% , 2006 is less than 4. 6% , 2007 and 2009 is more than 30. 6% and 9. 8% , respectively, 2008 and 2010 is mostly equal (in Fig. 1).

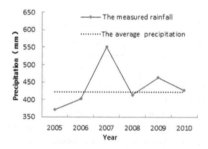

Fig. 1   Coarse Sand Area precipitation changes

2. 2 Runoff variation

The average annual runoff of 1956 ~ 2004 is 3. 03 × 10⁹ m³ in the middle Yellow River coarse sediment source area, the average runoff of monitoring period is 1. 57 × 10⁹ m³ , 48. 2% less than the average annual runoff.

Runoff year by year of 2005 ~ 2010 is 1. 35 × 10⁹ m³ ,1. 71 × 10⁹ m³ ,1. 77 × 10⁹ m³ ,1. 43 × 10⁹ m³ ,1. 56 × 10⁹ m³ and 1. 59 × 10⁹ m³ , comparing with the average annual one , runoff in the years is less than 55. 5% ,43. 3% ,41. 3% ,52. 7% ,48. 5% and 47. 5% , respectively.

In general, the measured runoff of 2005 ~ 2010 in the middle Yellow River coarse sediment source area is less than the average annual one obviously (in Fig. 2).

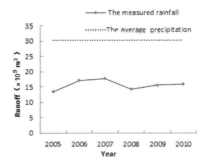

Fig. 2   Coarse Sand Area runoff changes

2.3   Sediment variation

The average annual total sediment of 1956 ~ 2004 is 7.88×10^8 t in the coarse sediment source area, the average total sediment of monitoring period is 1.11×10^8 t, 85.9% less than the average annual one. Total sediment year by year of 2005 ~ 2010 is 1.37×10^8 t, 1.77×10^8 t, 1.39×10^8 t, 0.55×10^8 t, 0.55×10^8 t and 1.03×10^8 t, comparing with the average annual one , total sediment in the years is less than 82.6% , 77.5% , 82.4% , 93.0% , 93.0% and86.9% , respectively (in Fig. 3).

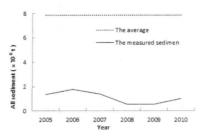

Fig. 3   Coarse Sand Area sediment changes(all sediment)

The average annual coarse sediment of grain – size larger than 0.05 mm of 1956 ~ 2004 is 2.53×10^8 t in the coarse sediment source area, the average coarse sediment of monitoring period is 0.23×10^8 t, 90.8% less than the average annual one. Coarse sediment of grain – size larger than 0.05mm year by year of 2005 ~ 2010 is $0.317,8 \times 10^8$ t, $0.481,6 \times 10^8$ t, $0.233,4 \times 10^8$ t, $0.092,6 \times 10^8$ t, $0.091,3 \times 10^8$ t and $0.180,6 \times 10^8$ t, comparing with the average annual one , is less than 87.4% , 80.9% , 90.8% , 96.3% , 96.4% and 92.9% , respectively(in Fig 4).

The average annual coarse sediment of grain – size larger than 0.1 mm of 1956 ~ 2004 is $0.868,9 \times 10^8$ t in the coarse sediment source area, the average coarse sediment of monitoring period is $0.061,1 \times 10^8$ t, 93% less than the average annual. Coarse sediment of grain – size larger than 0.1 mm year by year of 2005 ~ 2010 is $0.086,7 \times 10^8$ t, $0.166,4 \times 10^8$ t, $0.073,2 \times 10^8$ t, $0.011,7 \times 10^8$ t, $0.009,6 \times 10^8$ t and 0.019×10^8 t, comparing with the average annual one, is less than 90.0% , 80.8% , 91.6% , 98.7% , 98.9% and 97.8% respectively (in Fig. 5).

As runoff in the study area decreased, sediment is decreasing more obviously, and as the sediment of grain – size became coarser, reduction is more and more sharp.

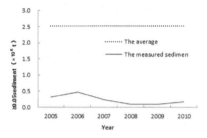

Fig. 4 Coarse Sand Area sediment changes($d \geqslant 0.05$mm)

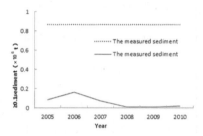

Fig. 5 Coarse Sand Area sediment changes($d \geqslant 0.1$ mm)

2.4 Flood variation

Seen from coarse sediment source area tributary hydrological station which lies on the river entrance, in monitoring period, Qingyang station in 2005 is a little more than the average annual, the changing extent is 1.6%, Gaojiachuan station and Baijiachuan station in 2006 is more than the average annual, the changing extent is 12.2% and 44.2% respectively. Yangjiapo station in 2007 is more than the average annual, the changing extent is 65.6%. Linjiaping station, Houdacheng station and Qingyang station in 2010 is more than the average annual, the changing extent is 138.1%, 31.2% and 29.2%, respectively. Other stations are less than the average annual. Seen from controlling area accounts for the coarse sediment source area, year of 2005, 2006, 2007 and 2010 is 14.6%, 19.0%, 0.4% and 18.1%, respectively (in Tab. 1). It shows that, in the monitoring period of 2005 ~ 2010, there is no big flood in coarse sediment source area, especially after 2008 and 2009, in all of the controlling station, not a annual maximum peak discharge is more than the average annual.

Tab. 1 Maximum peak flow of control stations in coarse sand area tributaries

River name	Station name	Peak flow(m^3/s)						
		Average for many years	2005	2006	2007	2008	2009	2010
Huangfuchuan	Huangfu	2,430	273	1,830	211	262	753	558
Kuyehe	Wenjiachuan	4,180	279	145	520	89.8	43.9	189
Tuweihe	Gaojiachuan	900	207	1,010	98.9	91	66.4	199
Jialuhe	Shenjiawan	771	22.8	346	217	139	16.8	4.61
Wudinghe	Baijiachuan	1,380	216	1,990	706	172	494	248
Xianchuanhe	Jiuxian	178	2.47	7.67	62.8	152	1.17	32

Continued Tab. 1

River name	Station name	Peak flow(m³/s)						
		Average for many years	2005	2006	2007	2008	2009	2010
Zhujiachuan	Qiaotou	376	42.8	34.4	319	307	72.7	5.71
Weifenhe	Xingxian	354	22	23.7	71.1	5.42	27.6	0.679
Gushanchuan	Gaoshiya	1,510	243	975	215	140	30.3	22.7
Qingliangsigou	Yangjiapo	288	152	117	477	6.44	54.6	149
Qiushuihe	Linjiaping	966	157	432	160	78.1	36.6	2,300
Sanchuanhe	Houdacheng	884	29.9	289	90	38.9	82.5	1,160
Quchanhe	Peigou	609	345	566	303	6.13	195	28.2
Xinshuihe	Daning	726	130	46.6	141	15.1	27.7	91.3
Qingjianhe	Yanchuan	1,500	270	410	253	44.7	131	127
Yanshui	Ganguyi	1,180	598	133	86.4	267	124	335
Beiluohe	Liujiahe	1,540	611	99.4	236	88.5	97.5	714
Puhe	Maojiahe	634	260	170	164	195	170	179
Malianhe	Qingyang	1,270	1,290	1,080	386	392	362	1,650

2.5 Changes in soil and water conservation measures

Seen from the changes in soil and water conservation measures of 1998 ~ 2007, terrace field had increased from 2,099 hm² to 2,766 hm², increased 31.8%, dam land had increased from 1,766 hm² to 3,588 hm², increased more than one double, forest land had increased from 51,375 hm² to 96,350 hm², increased 87.5%, grassland had increased from 11,383 hm² to 21,663 hm², increased 90.3%, area banned for harness had increased from 2 hm² to 1,679 hm², increased 1,677 hm², the above – mentioned show that, in Huangfuchun basin from monitoring period before to monitoring period, soil and water conservation measures in the basin had increased a lot, especially forest land and grassland had increased about 90%, the implementation of banned for harness measures had a great efforts, increases almost 8×10^4 times, ecological environment had improved obviously, in the monitoring period there was no rainstorm with wide extent and great intensity, led to observed silt and discharge of Huangfuchuan basin reduce sharply(in Tab. 2).

Tab. 2   Soil and Water Conservation measures changes

Year	Units	Terraced	Dam	Woodland	Grass	Closing Management
1998		2,099	1,766	51,375	11,383	2
2007	hm²	2,766	3,588	96,350	21,663	1,679
The amount of change		667	1,822	44,975	10,280	1,677
The rate of increase	%	31.8	103.2	87.5	90.3	78,374.8

3 Summaries

In the monitoring period, the annual precipitation of the middle Yellow River coarse sediment source area fluctuates near the average annual, but the observed runoff shows a sharp reduction, the observed sediment reduces more evidently, accompany with the sediment of grain – size becoming coarser, the reduction is more evidently, comparing with peak discharge of flood over the years, there does not appear rainstorm with wide extent and great intensity. This is because in the monitoring period there is no rainstorm with wide – extent and great – intensity in the area, additionally the implementation of "return farmland to forests or grassland" and banned for harness measures, the ecological environment improves evidently.

Studies on the Role of Guxian Reservoir in the Yellow River Water and Sediment Regulation and Control System①

Wan Zhanwei, *An Cuihua* and *Liao Xiaofang*

Yellow River Engineering Consulting Co. , Ltd. , Zhengzhou, 450003, China

Abstract: Guxian Reservoir is a vitally important component in the Yellow River Water and Sediment Regulation and control System because it has special geographical advantages for sediment retaining and Water and Sediment Regulation by Yellow River Water and Sediment Regulation and Control System. The combined operation plan of Guxian and Xiaolangdi was put forward according to the operation plan of Xiaolangdi Reservoir. Studies were carried out on the role of Guxian Reservoir in the Yellow River by observed data analysis and mathematical model calculation with the designeded water and sediment condition. Guxian Reservoir which operated with the existing reservoir projects will significantly optimize the water and sediment relationship, reduce the sediment dissipation in the lower yellow river, maintain the medium flood river channel with bankful discharge release capacity of $4,000 \text{ m}^3/\text{s}$, decrease Tongguan elevation by about 2 m, cut down the flood discharge and reduce the flood water level of Sanmenxia Reservoir. And at the same time, it can significantly improve the water supply conditions of the surrounding areas and promote regional economic and social development. Therefore, we should accelerate the construction of Guxian Reservoir to perfect the Yellow River Water and Sediment Regulation and Control System and keep the healthy life of Yellow River.

Key words: Guxian Reservoir, Water and Sediment Regulation and Control, medium flood river channel, Tongguan elevation

1 Survey of Guxian Reservoir

The proposal damsite of Guxian water control project locates in the downstream of the Yellow River north main stream. The damsite is 238. 4 km distant from the upstream damsite of Qikou and 10. 1 km and 74. 8 km distant from the downstream Hukou falls and Yu menkou steel bridgeat respectively. At the left bank of the damsite it is Ji County in Shanxi Province and at the left is Yichuan County of Shanxi Province. The damsite controls a drainage area of about $0.49 \times 10^6 \text{ km}^2$ which accounts for 65% of the total area. The average flux for many years at the damsite is $29.05 \times 10^9 \text{ m}^3$ which account for 76% of the total runoff of the Yellow River. The average sediment delivery for many years is 0.857×10^9 t which account for 64% of the total sediment delivery of the Yellow River. The duty of the development of the reservoir is mainly at flood prevention and sedimentation reduction, and at the same time gives consideration to power generation, water supply and irrigation etc.

2 The strategic role of Guxian Reservoir at the water and sediment control system of the Yellow River

The crux of the refractoration of the Yellow River is of the little water, much sediment and incongruous relationship between them. The basic way to solve this problem is to increase water, decrease sediment and adjust them. To adjust water and sediment is an important strategic measure

① **Fund Project**: the Special Scientific Research of the Ministry of Water Resources for Public Welfare Industry(200901017 ,201001012)

of the treatment and development of the Yellow River. It is an important way to seek the enduring of the Yellow River. But, because of the limited function of the reservoir at adjusting water and sediment, more reservoir projects and corresponding unproject measures which constitude a water – sediment system are needed. And only in this way can the measure play a better role.

The book "The Planning of Yellow River Basin" puts forward that the perfect Yellow River water – sediment control system should have the key projects such as Longyangxia, Liujiaxie, Heishanxia, Qikou, Guxian, Shanmenxia and Xiaolangdi at the main channel as the mainstay, the Haibowang, Wangjiazhai reservoir as the support, and combine with the controling reservoirs such as Luhun, Guxian, Hekoucun, Dongzhuan on the branch river. And the key projects at the upstream such as Longyangxia, Liujiaxia and Xiaolangdi constitute the subsystem of the water storage control. When the upstream subsystem coorperate with the downstream subsystem, the water – sediment relationship of the Yellow River with be better fitted so as to adapt the need of controlling and development of the Yellow River.

As one of the key project of water – sediment control system, Guxian water control project is not only with large storage capacity but also with an advantageous position. It control the main coming area of the Yellow River's flood and sediment, especially the coming area of coarse sediment. It is near to the Xiaolangdi water control project. And it has particular geographical advantage at sediment storage and adjusting water and sediment combined with Xiaolangdi Reservoir which can not only control the upstream water and sediment of the Yellow River affectively, but also provide flow dynamics. It serves as a connecting link at the overall arrangement in the Yellow River water – sediment control system, and has a very important strategical status.

3 Principles of joint control among reservoirs

Based on the recent project conditions, the action of water and sediment adjustment reach an obvious effect which putting the Xiaolangdi Reservoir first. But there also exist problems such as the difficulty for Xiaolangdi Reservoir to adjust water and sediment by itself and the lack of following dynamic power for artificial hypopycnal flow etc. In order to bring the combined role of reservoirs into full play, based on the theroy of water and sediment adjustment and achievements of the practices and combine the developing task of Guxian reservior, the principle of Guxian—Xiaolangdi reservoir joint controlling is put forward. See it as follows:

In the early days of Guxian Reservoir's sediment storage, Guxian and Xiaolangdi are combined for the sediment storage and water – sediment adjusting which is used to adjust the water and sediment's process when entering into the little north stem and the downstream channel and make full use of the hypopycnal flow for desilting application. When the water storage of the two reservoirs and the inflow from the channel fill the requirements of the water controlling of the downstream channel, the combined adjusting create a certain diachronic large discharge process, erode the downstream channel and keep the middle river channel.

In the late days of Guxian Reservoir's sediment storage, according to the downstream bankfull discharge of the Yellow River and the change of the storage of the Xiaolangdi Reservoir, store water at the approprite time or using the natural inflow to erode the downstream channel of the Yellow River and Xiaolangdi Reservoir, keep the flood drainage capacity of the middle river channel of the Yellow River downstream for a long time and keep the water – sediment storage of the Xiaolangdi Reservoir. When coming up with suitable water – sediment condition which can erode the sediment silting up at the Guxian reservoir and prolong the running life of the Guxian Reservoir.

At the normal operation period of Guxian Reservoir, on the premise of keeping the storage capacity for flood control of the two reservoirs, use the two reservoir's water and sediment adjusting storage capacity to jointly adjust water and sediment, and increase the opportunity of the flood water silting and erosion of the Yellow River downstream and the two reservoirs' region so as to play a long time role in water and sediment adjusting.

4 Study on Guxian Reservoir's function at the Yellow River's water – sediment control system

Guxian Reservoir's function at the Yellow River's water – sediment control system is mainly in two aspects. In order to fully prove its flood prevention and sedimentation reduction effect, combined with the achievement of the comprehensive plans for the Yellow River basins, considering the water and sediment reduction objectives of soil and water conservation measures in different target years, select 2020 as the target year 1968 ~ 1979 + 1987 ~ 1999 + 1962 ~ 1986 which is a fifty year system and 2050 as the taget year 1968 ~ 1979 + 1987 ~ 1997 which is a seventy two year system, use the method of "contract between yes or no" and mathematical model to contrast and analysis the calculation. The water – sediment charateristic parameter of Longmen and the four stations (Longmen, Huaxian, Hejin, and Zhuantouzhan) at the midstream is in Tab. 1.

Tab. 1   Table of water – sediment characteristic parameter of designed representative series

Hydrological station	Period	Water – quantity($\times 10^8$ m³)			Silt – quantity($\times 10^8$ t)		
		Flood season	Nonflood season	Whole year	Flood season	Nonflood season	Whole year
Longmen	2008 ~ 2020	107.44	120.20	227.64	6.67	0.84	7.51
	2020 ~ 2080	101.36	118.07	219.43	4.92	0.78	5.70
Four stations	2008 ~ 2020	146.62	143.54	290.17	10.48	1.00	11.48
	2020 ~ 2080	139.38	143.58	282.97	8.46	1.07	9.54

The planned time for Guxian Reservoir to be built up and go into force is in 2020. Before it goes into force, Xiaolangdi Reservoir runs by the designed way. After it go into force, Guxian Reservoir and Xiaolangdi Reservoir is jointly used. The initial boundary conditions of model calculation is 2008, the contrast time period is 2020 ~ 2080. The adopted water – sediment mathematical model is from Yellow River Engineering Consulting Co. , Ltd. (YRCC) and China Institute of Water Resourses and Hydropower Research(IWHR).

4.1   The function of coordinating the water – sediment relationship of Lower Yellow River

According to the designed water – sediment series and the selected reservoir's application way, and following an regulation of the midstream reservoir and an adjustment of scour and fill in the little north stem, the water and sediment quantity enter the downstream in different schemes (Xiaolangdi Station, Heishiguang Station, Wuzhi Station) can be refered to Tab. 2. The recent project proposal is that Xiaolangdi Reservoir's sediment storage capacity is fully filled in 2020, the average water quantity and sediment quantity entering the downstream is 28.92 $\times 10^9$ m³/a and 0.47 $\times 10^9$ t/a respectively, and the average sediment concentration is 16.2 kg/m³. During the time 2020 ~ 2080 when the sediment storage period is over, the average water quantity and sediment quantity entering the downstream is 28.85 $\times 10^9$ m³/a and 0.90 $\times 10^9$ t/a respectively, and the average sediment concentration is 31.3 kg/m³. Guxian Reservoir's scheme to come into effect in 2020 is that because the reservoir intercept the majority of sediment from the north stem, the sediment entering the downstream decrease obviously during the time 2020 ~ 2080, the average sediment quantity decrease by 0.2 $\times 10^9$ t when compared with the recent scheme, and the average sediment concentration decrease to 24.4 kg/m³.

Tab. 2 Table of characteristics of water and sediment quantity entering into the downstream when there have Guxian scheme or have not

Scheme	Period	Water quantity ($\times 10^8$ m^3)			Sediment quantity ($\times 10^8$ t)		
		Flood period	Nonflood period	Full year	Full year	Flood period	Nonflood period
Recent project scheme	2008 ~ 2020	136.42	152.74	289.17	4.67	0.01	4.67
	2020 ~ 2080	137.60	150.85	288.45	9.02	0.02	9.03
Guxian Reservoir scheme	2008 ~ 2020	136.42	152.74	289.17	4.67	0.01	4.67
	2020 ~ 2080	137.26	150.28	287.55	7.02	0.01	7.03

After Guxian Reservoir come into force, it not only make the downstream sediment concentration decrease obviously but also optimize the downstream flow process and coordinate the water – sediment relationship (refer to Tab. 3). Compared to the recent project scheme, when Guxian Reservoir come into force in 2020, the main flood period is benefit to the channel sedimentation reduction, the average number of days when the middle river channel keep four days' discharge of more than 4,000 m^3/s increase by 6.6 days per year, the rate of the discharge grade more than 2,600 m^3/s increase by 17.4%, and the increase of the probability when large discharge occurs and the rate of bringing sediment are more benefit to bring water's Sediment transport efficiency into full play.

Tab. 3 Statistical table of the average number of days per year when the discharge entering the downstream occur when there have Guxian scheme or have not

Scheme	Item	Discharge grade (m^3/s)				Main flood period
		≤800	800 ~ 2600	≥2600	≥4,000 in four consecutive days	
Recent project scheme	Days (d)	24.95	51.47	15.58	1.65	92.00
	Water quantity ($\times 10^9$ m^3)	1.11	5.87	5.19	0.78	12.17
	Sediment quantity ($\times 10^9$ t)	0.05	0.38	0.47	0.08	0.90
Guxian Reservior scheme	Days (d)	27.85	50.80	13.35	8.28	92.00
	Water quantity ($\times 10^9$ m^3)	1.37	5.82	4.69	3.16	11.87
	Sediment quantity ($\times 10^9$ t)	0.05	0.26	0.38	0.16	0.69

4.2 The function of slowing down the sedimentation of Yellow River's downstream channel

The water – sediment process of entering the downstream channel which is based on different schemes is calculated by the mathematical model from Yellow River Engineering consulting Co., Ltd. and China Institute of Water Resourse and Hydropower Research. The accumulative sedimentation process is shown in Fig. 1 (It is measured value in 2000 ~ 2008).

The recent project scheme: before 2020, on account of Xiaolangdi Reservoir's sediment retaining and water – sediment adjusting, the downstream channel basically take a state of continuous eroding, the accumulated maximum eroding quantity of the downstream channel

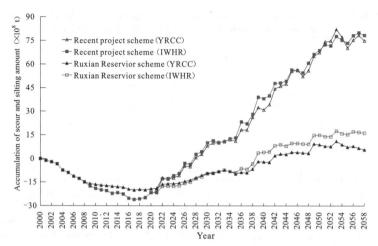

Fig. 1 Accumulative sedimentation process of the Yellow River's downstream when there is Guxian scheme or not

calculated by the two models is 2. 01 × 10⁹ t and 2. 61 × 10⁹ t respectively. After 2020, the sediment storage of Xiaolangdi Reservoir is fully filled, the reservoir's ability to adjust water and sediment become weaken, then the downstream channel gradually back – silting, the particular year when the caculated downstream channel' back – silting state go to the state when Xiaolangdi Reservoir have not built is 2028, during 2020 ~ 2080, the amount of the silt accumulation calculated by the two models is 14.90 × 10⁹ t/a and 14. 00 × 10⁹ t/a respectively, the average amount of silt accumulation is 0. 25 × 10⁹ t/a and 0. 23 × 10⁹ t/a respectively, the channel's quick back – silting will bring serious threaten to control safety of the channel.

When Guxian Reservoir go into force in 2020 and jointly used with Xiaolangdi Reservoir at silt retaining, the water – sediment relationship entering the downstream channel will obviously improved, and the serious siltation state will be relieved. During 2020 ~ 2080, the accumulated amount of sediment deposition of the downstream channel calculated by the two models is 5. 69 × 10⁹ t and 7. 92 × 10⁹ t which amount to that of the 37 years' and 26 year' amount respectively at the recent project condition. After Guxian Reservoir goes into force, the downstream channel will present a state of tiny silting, and the problems during the flood prevention is basically controled effectively.

4.3 Keep the function of mid – river channel of the Yellow River's downstream channel

The mid – river channel is not only the basical requirement for keeping the channel's function of desilting sediment transport, but also the important condition to ensure a long time stability of the downstream river regime. Research show that on the normal flow and sediment conditions, the proper mid – river channel scale of the Yellow River's downstream is with a flow capacity of about 4,000 m³/s.

On the recent project condition, after 2020, with the Yellow River's quick back – silting, Xiaolangdi Reservoir's mid – river channel whose sediment storing period become recovery begin to gradually atrophy. The time when the minimum bankfull discharge calculated and analysed by YRCC and IWHR decline to under 4,000 m³/s is 2026 and 2030 respectivily, after that, the mid – river channel is difficult to maintain. After Guxian reservoir brought into service, by means of combined use with the Xiaolangdi Reservoir, the situation that the discharge and silt transport ability of the Yellow River mid – river can be dramatically changed. YRCC's calculation shows that during the period of 50 years before 2070, the downstream mid – river channel's bankfull discharge can keep over 4,000 m^3/s; IWHR's calculation also shows that after Guxian Reservoir go into function between 2020 ~ 2058, downstream channel's bankfull discharge will increase by about 1,000 m^3/s comparing to the recent scheme.

Thus it can be seen that Guxian reservoir's function at keeping Xiaolangdi Reservoir's mid – river channel's flow capacity recuperate d at silt – retaining period is dramatical. Mid – river channel's long time keeping is not only benefit to the downstream's deflooding and sediment transport and strengthen the downstream channel's flood control capacity, but also make the downstream river regime become steady and reduce the risk of levee's bursting which is brought by the change of river regime.

4.4　The function of reduce Tongguan's height

The change of Tongguan's height is related to the water and sediment condition of Yellow River, Weihe River (include Beiluohe), the usage of Sanmenxia Reservoir and earlier channel siltation, and river regime condition etc. The related factors interlace and the regulation is complicated. When the other conditions are alike, the water – sediment condition is the main condition to affect the height of Tongguan. Improving Tongguan stream segment's water and sediment condition is an important way to reduce the height of Tongguan.

Guxian Reservoir's application in sediment storing and water – sediment adjusting will change the unfavorable water – sediment condition, bring about continuous eroding of the little north stem, thus make the height of Tongguan decline obviously. YRCC and IWHR's calculation indicate that after Guxian reservoir go into force, the little north stem' maximum accumulated amount of eroding reach 1.51×10^9 t and 1.28×10^9 t respectively; compared with that when the reservoir begin to go into force, the decreasing value of Tongguan height's maximum eroding is 2.57 m and 1.15 m respectively; because of Tongguan height's declining and the following declining of the downstream height of Weihe, it is very unfavorable to recover the flow ability of the mid – river channel of the downstream of Weihe River and improve the serious flood control stater of the downstream of Weihe River.

Compared with Guxian Reservoir's condition, the two model's calculation result indicate that when Guxian reservoir have been used for 60 years, the siltation of the little north stem can be reduced by 4.00×10^9 t and 2.34×10^9 t respectively, reduce the siltation rising value at Tongguan height by 2.27 m and 1.83 m respectively. During the 60 years after the reservoir go into force, it can keep the Weihe downstream flood control project reach or go over its criterion for a long time.

4.5　The function of reduce flood detention and sedimentation of Sanmenxia Reservoir

Guxian Reservoir basically controls the flood between Hekou and Longmen. The reservior's classify controlling to the heavy flood can reduce the peak flow of the flood which is a case of the Millennium from 38,500 m^3/s to 11,800 m^3/s, and reduce the peak flow of the flood which is a case of the centenary from 27,400 m^3/s to 10,300 m^3/s. It cut down the north stem's flood flow by objective, which is very important for reducing Sanmenxia Reservoir's flood detention water level and alleviating reservoir's bankfull storage loss caused by flood detention.

The results indicate that, for a case of the Millennium flood type 1933 years, can make the flood of Sanmenxia Reservoir water level lower 2.79 m, the capacity of the deposition of the reservoir reduced from 1.489×10^9 m^3 when no Guxian project to 7.43×10^8 m^3, the loss capacity of the deposition of the reservoir reduced 7.46×10^8 m^3; A case of the centenary can make the flood of Sanmenxia reservoir water level lower 4.40 m, the capacity of the deposition of the reservoir reduced from 8.57×10^8 m^3 to 1.82×10^8 m^3. The loss capacity of the deposition of the reservoir reduced 6.75×10^8 m^3. A case of 50 years once flood, the flood in the dam reservoir basicly not on the beach, it is detained and discharged in channel storage capacity.

4.6 The function of promoting economic and social development on river basin

Guxian reservoir has a greater regulation capacity, which can significantly improve the water supply conditions nearby, and joint runoff regulation with other engineering can alleviate the contradiction between supply and demand of water resources of the Yellow River. The results of runoff regulation (1956 ~ 2000 series), indicated that, after Guxian Reservoir has completed, compared with the plan of the present engineering, the total water flow in the basin of 2020 level are increase to 2.66 × 10⁸ m³, including the surface water supply of Shanxi and Shaanxi are increase to 1.34 × 10⁸ m³ and 4.79 × 10⁸ m³ respectively.

The shortage of river water in the basin at level years is significantly reduced each year, the reduction of low flow year which contain special year and successional year will be around 9.09×10^8 m³ to 3.465×10^9 m³, and 5.65×10^8 m³ to 1.817×10^9 m³ respectively.

Guxian water control project is located in border area between Hebei province and Shaanxi province and the grid of north China and northwest.

The construction of Guxian water control project can not only make the river water resources reutilization, but also has great regulation function, and can provide quality power load capacity and power; not only can make contributions to the regional economic development, but also can reduce the pollution of the environment, at the same time, it can create the conditions to send power transmission from west to east.

5 Conclusions

(1) Guxian water conservancy hub in water sediment regulation system is one of the important backbone projects. It is the key project to deal with the sediment in the Yellow River, coordinate water sediment relation, reduce the Tongguan elevation, and maintain the downstream channel of water flowing capacity. Its preceding strategic position is very important in the water sediment regulation system.

(2) Guxian Reservoir and Xiaolangdi Reservoir come into joint use of blocking sand and the water – sediment regulation. It can coordinate the lower river water sand relationship, reduce sediment in the lower Yellow River, and make the silting on downstream channel to be small amout in the long run. The capacity of Xiaolangdi Reservoir blocking the downstream channel sediment recovery de – siltation water flowing above 4,000 m³/ s could keep more than 50 years. And meanwhile, it can reduce the Tongguan elevation 2 m or so, greatly relieve the downstream of Weihe River flood control pressure; Guxian Reservoir hierarchically control flood, which objectively reduced the north main stream flood, and the Sanmenxia reservoir flood detention water level, and significantly promote economic and social development in the river basin.

(3) From the urgent requirement of flood prevention and sedimentation reduction in the downstream channel of the Yellow River, in order to give full play to joint control efficiency of the water sediment regulation system, Guxian hydraulic project must be completed as early as possible.

References

Yellow River Water and Sediment Regulation Theory and Practice [R]. Zhengzhou: Yellow River Conservancy Commission, 2009.

An Cuihua, Guo Xuanying, Yu Xin, et al. Hydrology and Sediment Mathematical Model and its Application of the Sandy Rivers' Reservoirs [J]. Yellow River, 2000, 22 (8):15 – 17.

Zhang Houjun, Liu Jixiang, Gao Guoming, et al. Research and Application of Hydrodynamic Mathematical Models of the Lower Yellow River [J]. Yellow River, 2000, 22 (8):21 – 22.

Hu Chunhong, Guo Qingchao. Sediment Mathematical Model of the Lower Yellow River and Study on the Dynamic Equilibrium Critical Threshold [J]. Science in China, Ser. E,

Technological Sciences,2004,34(Suppl. I): 133 – 143.

Hu Chunhong, Guo Qingcha, Chen Jianguo. Study on How to Shape and Maintain the Medium Flood River Channel in the Lower Yellow River [J]. Journal of Hydraulic Engineering, 2006 (4):381 – 388.

Zhang Yuanfeng, Jiang Naiqian, Hou Suzhen. Research on the Influencing Factors and the the Drop Height of Tongguan Elevation [J]. Journal of Sediment Research, 2005 (1): 40 – 45.

Zhenshui Sand Bar and Its Cause in Xiaolangdi Reservoir

Wang Ting, *Zhang Junhua*, *Chen Shukui* and *Li Kunpeng*

Yellow River Institute of Hydraulic Research, YRCC, Key Laboratory of
Yellow River Sediment Research, MWR, Zhengzhou, 450003, China

Abstract: The total original storage capacity of Xiaolangdi reservoir is 12.75×10^9 m^3, thereinto, the tributary storage capacity accounts for 41.3% of the total. Zhenshui is the biggest tributary of Xiaolangdi Reservoir, and its original storage capacity is 1.767×10^9 m^3 and accounts for 33.5% of the tributary. By October 2010, the amount of sediment in Zhenshui reached $0.169,1 \times 10^9$ m^3, and the sand bar was 7 m in height. The results show that it is influenced by operational mode of reservoir and terrain, and the formation of sand bar at Zhenshui estuary has certain inevitabilities. The sand bar can prevent water and sediment exchange between mainstream and tributary, and the internal storage capacity of tributary is not effectively used. Analyzing sand bar and its cause can supply some technical supports for prevention.

Key words: Xiaolangdi Reservoir, Zhenshui, sand bar, flowing backward, model test

1　General situation of Zhenshui

Xiaolangdi Reservoir is a comprehensive utilization project and its development mission was definitely oriented to flood control (including ice prevention) and sedimentation reduction, and balanced water supply, irrigation and hydropower generation. The highest operation water level is 275 m. The initial storage capacity is 12.75×10^9 m^3. The tributary storage capacity is 5.268×10^9 m^3 and accounts for 41.3% of the total. Zhenshui is the biggest tributary of Xiaolangdi Reservoir. It is in the right bank between mainstream cross section HH11 and HH12 and about 18 km apart from Xiaolangdi dam(Fig. 1). The original storage capacity of Zhenshui is 1.767×10^9 m^3 and accounts for 33.5% of the tributary. The backwater length at elevation 275 m is more than 20 km.

The Fig. 2 shows the designed deposition morphology of Zhenshui. It can be seen from the figure that the original slope of longitudinal profile is 56‰. The deposition morphology in sediment retaining period is as follows: the elevation of sand bar at tributary estuary is 257.1 m, and it is also equivalent to beach elevation at the confluence area of mainstream and tributary; the antislope of sand bar is 26‰; the elevation of tributary internal beach is 252.4 m; the height difference between estuary and internal of tributary is 4.7 m. The scouring pattern of tributary estuary in the normal operation period is as follows: the mainstream is in high beach and low channel; by many year scour of flood water from tributary, the sand bar gradually declines, and the final deposition morphology is corresponding to beach and channel of mainstream; the channel longitudinal profile connected by slope 16.8‰ and 6‰ is final profile scoured by tributary flood(Fig. 2). In the process of reservoir operation, the estuary channel is silted by mainstream flowing backward or is broken by tributary flood, and is in unstable state.

2　Deposition morphology of Zhenshui since Xiaolangdi Reservoir operation

Xiaolangdi Reservoir was put into use in May 2000, and the deposition morphology of mainstream was a cone. Because of sediment deposition, the morphology became a delta in October 2000. With the continuous deposition in reservoir, the elevation of delta continent surface uplifts

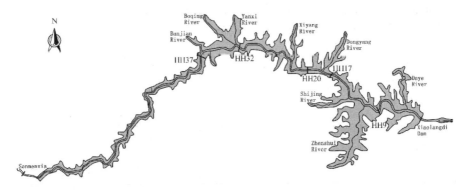

Fig. 1 Plan of Xiaolangdi Reservoir

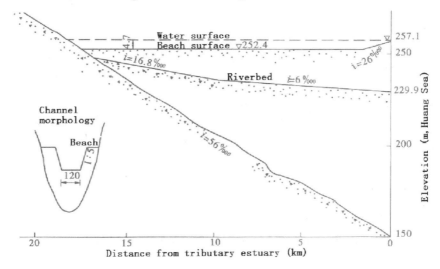

Fig. 2 Designed deposition morphology of Zhenshui

gradually, the delta vertex moves gradually downstream to the dam and correspondingly, and the plunging point of density current also moves downstream. By October 2010, the plunging point of density current had moved from cross section HH40 (69. 39 km to dam, in year of 2000) to HH11 ~ HH12, that is, the plunging point was at Zhenshui estuary.

Before flood season of 2010, Zhenshui was still in the downstream of mainstream delta vertex, the deposition form was density current flowing backward, and the deposition surface at estuary was flat and kept synchronous uplift with mainstream beach. In recent years of density current flowing backward, only because of terrain and sediment sorting, there was a certain antislope(Fig. 3(a)). The cross sections of tributary uplifted horizontally (Fig. 3(b) ~ (d)). With the continuous uplift of mainstream deposition surface, the deposition surface in tributary internal uplifts slowly, the height difference between estuary and internal increases year by year, and the antislope becomes more and more obviously. For example, in flood season of 2010, because delta vertex moved to Zhenshui estuary, the deposition surface of mainstream beach corresponding to Zhenshui uplifted rapidly, and the deposition surface at Zhenshui estuary also uplifted rapidly. By October 2010, the elevation of mainstream beach reached 215 m, and the average riverbed elevation at cross section ZS01 which is at estuary was 213. 6 m. However, the elevation at ZS04 which is 3. 5 km from ZS01 is less than 206.9 m, and at ZS05, which is 4.5 km from ZS01, it was even lower and only 206.7 m. Therefore, the apparent sand bar came into being at Zhenshui estuary and the height was about 7m.

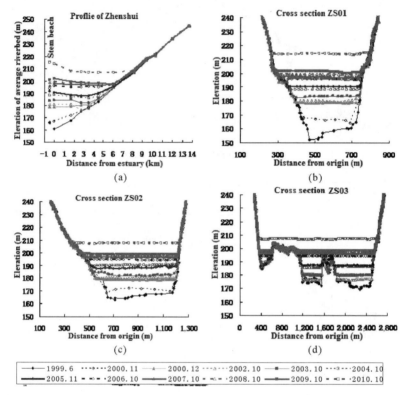

Fig. 3　Longitudinal profile and cross sections of Zhenshui

3　Results of model test

3.1　Introduction of test

In the research of operational mode of Xiaolangdi Reservoir for flood control and sedimentation reduction during later sediment retaining period, three groups of model test were conducted out. They were operational mode of gradually raising water level and trapping the coarse sand and discharging the fine sand (in short, mode one), operational mode of multi – year sediment regulation and man – made precipitation washout at right occasion (in short, mode two), and recommended scheme absorbing the advantages of the first two.

The scope simulated by model is from the section of Sanmenxia hydrologic station to Xiaolangdi dam. The length of prototype is 124.5 km and the elevation range is from 155 m to 290 m. The model includes all tributaries and main buildings. The horizontal and vertical scales of the model are 1:300 and 1:60 respectively. The initial boundary conditions of tests were all the terrains of October 2007. The incoming water – sediment condition of mode one and mode two were both 20 year series of less water and sediment, that is the series of 1990 ~ 1999 + 1956 ~ 1965 based on design level year of 2020. The incoming water – sediment condition of recommended scheme was 17 year series of more water and sediment, that is series of 1960 ~ 1976 based on design level year of 2020. The control conditions of outlet boundary were water level before dam calculated by mathematical model.

3.2 Results of model test

In the process of test, it is found that tributaries were equivalent to the horizontal extension of mainstream riverbed. When mainstream is in the downstream of delta vertex and the deposition form is density current, the deposition surface of mainstream riverbed uplifts horizontally, the deposition surface at tributary estuary corresponding to mainstream is flat, and only because of terrain and sediment sorting, there is a certain antislope. When mainstream is in the upstream of delta vertex, the obvious beach and channel is shaped in mainstream riverbed, and the sand bar of tributary is equivalent to mainstream beach. The deposition morphology of tributary changes constantly, and the antislope is gradually formed and becomes more and more obviously.

Fig. 4 ~ Fig. 6 show Zhenshui longitudinal profiles of mode one, mode two and the recommended scheme respectively. It can be obtained that: in the end of mode one, the elevation of deposition surface at estuary cross section ZS01 is 259.1 m, it is 213.6 m at internal cross section ZS04, and the height difference was 45.5 m. In the end of mode two, the elevation at ZS01 is 258.1 m, it is 219.7 m at ZS04, and the height difference was 38.4m. In the end of the recommended scheme, the elevation at ZS01 is 259.6 m, it is 233.1 m at ZS04, and the height difference was 26.5 m. We can see, no matter which test, the formation of sand bar at Zhenshui estuary was inevitable, and only the height was different.

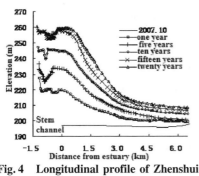

Fig. 4   Longitudinal profile of Zhenshui
(mode one)

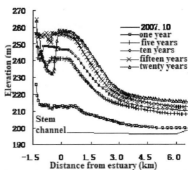

Fig. 5   Longitudinal profile of Zhenshui
(mode two)

4   Cause of sand bar in Zhenshui

The deposition process of tributary is closely related to many factors of natural topographic condition, deposition morphology of mainstream at the confluence area of mainstream and tributary, the process of incoming water and sediment and so on. Field observation data show that: there is little incoming water and sediment in Zhenshui, and basically there is no sediment. Therefore, nearly all the deposition comes from the flowing backward of mainstream. The sand bar is mainly caused by natural topographic condition. The cross section at estuary is very narrow and the valley width is only about 600 m, Fig. 7. The flow width of water and sediment laterally flowing into Zhenshui from mainstream is narrow, and it means that only a little sediment can flow into Zhenshui. The terrain in ZhenShui internal is open, such as 3 km apart from estuary, the valley width is more than 2,500 m, it means that the flow width in flowing backward direction increases suddenly, the flow velocity decreases rapidly, correspondingly, the carrying capacity decreases greatly, and a lot of sediment deposits. With the evolution of flow, the distance apart from estuary become longer and longer, the sediment carried by flow become less and less, however the deposition width become wide, therefore, the deposition thickness in the internal become thinner and thinner. Therefore, all these are the main causes that made little deposition thickness in Zhenshui internal. With the quickly uplift of deposition surface of mainstream riverbed, the deposition surface in tributary internal uplifts slowly, and the height difference between mainstream and tributary internal increases year by year.

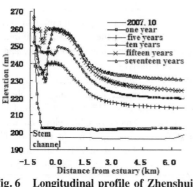

Fig. 6　Longitudinal profile of Zhenshui (recommended scheme)

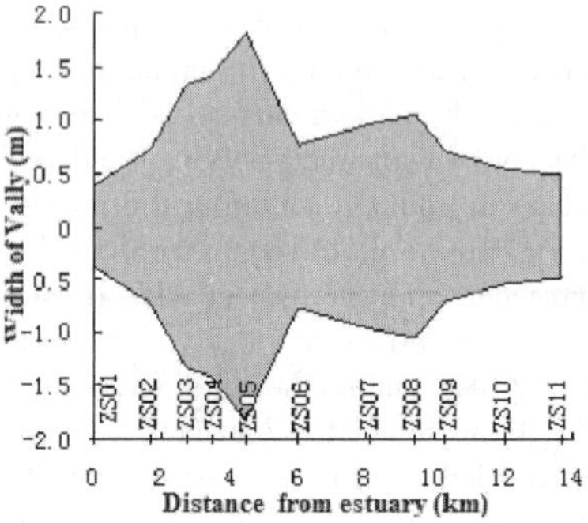

Fig. 7　Wally width of Zhenshui (275 m)

Influenced by operational mode of reservoir and terrain, the formation of sand bar at Zhenshui estuary has certain inevitability. The results show that the sand bar can prevent water – sediment exchange between mainstream and tributary. In non – flood season or period of flood control operation, reservoir water level is relatively high, and flood water overtops the sand bar and flows into tributary. When reservoir water level declines, if water level difference is not enough to influence the stability of the sand bar or the sand bar is not completely broken, then the isolated water body will come into being, the internal storage capacity of tributary will not be effectively used, and the efficiency of sedimentation reduction and even flood control will be affected. According to this phenomenon, it is necessary to carry out study on prevention of sand bar of Zhenshui so as comprehensive utilization measures of tributary are put forward.

5　Conclusions

By October 2010, the amount of sediment in Zhenshui reached $0.169,1 \times 10^9$ m^3, and the sand bar was 7 m. According to analysis, the sand bar can prevent water – sediment exchange between mainstream and tributary, and the internal storage capacity of tributary is not effectively used. In view of the particularity of Zhenshui terrain, it is necessary to carry out study on prevention of sand bar of Zhenshui so as comprehensive utilization measures of the tributary are put forward.

Acknowledgements

The research was supported by public welfare industry research special fund of the Ministry of Water Resources (MWR) (No. 200901015, No. 200801024), the National Natural Science Foundation of China(NSFC) (NO. 51179072) and central – level nonprofit research institute fund (HKY – JBYW – 2012 – 14).

References

Liu Xiushan, Li Jingzhong. Engineering Planning of the Planning and Design Series of Xiaolangdi Project of the Yellow River[M]. Zhengzhou: Yellow River Conservancy Press;2006 .

Wang Ting, Ma Huaibao. Characteristics of Water and Sediment and Scour and Silting Evolution of Xiaolangdi Reservoir in Early Sediment Retaining Period[R]. Zhengzhou: Yellow River Institute of Hydraulic Research, YRCC,2011 .

Zhang Junhua, Chen Shukui, Ma Huaibao, et al. Report on Model Test of the First Operational Mode of Xiaolangdi Reservoir for Flood Control and Sedimentation Reduction During Later Sediment Retaining period [R]. Zhengzhou: Yellow River Institute of Hydraulic Research, YRCC, 2010 .

Zhang Junhua, Chen Shukui, Ma Huaibao, et al. Report on Model Test of the Second Operational Mode of Xiaolangdi Reservoir for Flood Control and Sedimentation Reduction During Later Sediment Retaining period [R]. Zhengzhou: Yellow River Institute of Hydraulic Research, YRCC, 2010.

Zhang Junhua, Chen Shukui, Ma Huaibao, et al. Report on Model Test of Operational Mode of Xiaolangdi Reservoir for Flood Control and Sedimentation Reduction During Later Period of Sediment Retention [R]. Zhengzhou: Yellow River Institute of Hydraulic Research, YRCC, 2010 .

Analysis on Characteristics of Sediment Transport of Ning – Meng Channel of the Upper Yellow River

Zheng Yanshuang[1] , *Fan Jian*[2] and *Li Ping*[1]

1. Yellow River Institute of Hydraulic Research, Yellow River Sediment Key Laboratory
of the Ministry of Water Resources, Zhengzhou, 450003, China
2. Neimeng Hydrology and Water Resources Bureau, Baotou, 014030, China

Abstract: The paper analyzed the characteristics of sediment transport of Ning – Meng channel based on the observed hydrologic and sediment data of the upper Yellow River and by using analysis method of observed data. The outcomes show that the variation of sediment transport in the section above Qingtongxia of the channel is smaller and the section downstream Qingtongxia has the characteristics of more incoming sediment, more deposition and more desilting. The inflow of tributaries with higher sediment concentration witnesses deposition along the channel and the inflow of the main channel with bigger discharge and lower sediment concentration witnesses scouring along the channel. The scouring is affected by roughening of bed erosion and the increment of restoration of sediment concentration becomes smaller along with the increase of discharge. The research results can provide technical references for the harnessing of Ning – Meng channel on the upper Yellow River.

Key words: the upper Yellow River, Ning – Meng channel, incoming flow and sediment, characteristics of sediment transport

1 Foreword

The Ning – Meng channel on the upper reaches is a typical alluvial channel of the Yellow River. The neighboring area of the channel is an important energy base, grain production area and the centralized residential area of the minority nationality of the northwest China, having important economic and social positions. In recent years, the ice flood problem of the channel has become outstanding and several minors dyke breaches happened, causing life and property loss and bringing extreme great pressure to the related flood control departments. Therefore, clearly understand the characteristics of sediment transport of the channel are a premise for preparing appropriate countermeasures of management with a definite object. The paper exhaustively analyzed the characteristics of sediment transport in the different river sections of the channel based on the observed silt – discharge data of the upper river.

2 Characteristics of water and sediment variation of the upper Yellow River

The outstanding characteristic of incoming flow and sediment of the upper Yellow River is in different source area. The water mainly comes from the main river, 98% from the section above Lanzhou. The sediment mainly comes from tributaries and 10 major streams. Before the building of Liujiaxia Reservoir, the incoming sediment of Zuli and Qingshui rivers in the reach of Lanzhou – Qingtongxia occupied 51% of the total and high sediment concentration floods always happened. The major flood peaks of the main river rarely met with the sediment peaks of tributaries. Taking the maximum flood happened in the period of August 28 ~ October 6, 1981 (in Fig. 1) at Lanzhou station for an example, the sediment concentration of inflow was very low, but the sediment concentration of inflow of Zuli River in the period of July 12 ~ 18 was higher, the daily average maximum sediment concentration was 698 kg/m^3 and at this time the discharge of Lanzhou station was smaller. In addition, it can be clearly seen from the relation map of sediment concentration and discharge of Qingtongxia station in different years (in Fig. 2) that the discharge is not great but the variation of sediment concentration is greater when the inflow mainly comes from tributaries, the increase variation of sediment concentration along with discharge is not remarkable when discharge is greater and mainly comes from the main river, showing extremely not correlated silt – discharge relationship of the channel.

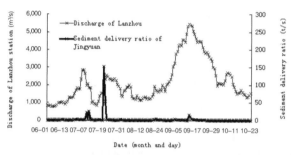

Fig. 1 Process of daily discharge and sediment delivery rate of Zuli River at Lanzhou station on the upper Yellow River

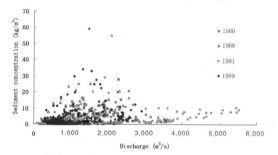

Fig. 2 Relationship between sediment concentration and discharge of Qingtongxia station in different years

3 Characteristics of sediment transport variation of Ning – Meng channel

3.1 Characteristics of temporal and spatial variation of sediment discharge

It can be see from the process of annual average sediment discharge of Tangnaihai – Toudaoguai section on the upper river in different periods (Fig. 3) that along with the operation of Yanguoxia (in 1961), Sanshenggong (in 1961), Qingtongxia (in 1967), Liujiaxia (in 1968), Bapanxia (in 1975) and Longyangxia (in 1986) on the upper river one after another, it has played the function of sediment control in a certain time. Comparing with the natural condition (in the period of 1952 ~ 1959), the quantity of sediment in each time interval has been reduced in varying degrees. After deducting debris – retaining of reservoirs, the sediment variation of each section above Qingtongxia is not very obvious and the sediment discharge at Qingtongxia station has reached the maximum value, showing the incoming water and sediment from the upper stream can be controlled by the station. It is mainly because the reach above Qingtongxia is a mountainous river and with greater channel gradient, in addition to the coarser sand and gravel composition of the bed. Viewing from the long river section, the variation trend of other river sections are basically the same except remarkable deposition in the sections of Qingtongxia, Liujiaxia and Longyangxia reservoir areas. The variations of sediment discharge of the section downstream Qingtongxia and the section between Qingtongxia and Shizuishan are greater. In the downstream section of Shizuishan, there is mainly Sanshenggong Reservoir. The scour and fill variation in other time intervals becomes smaller comparing with the section above Shizuishan except the period of 1960 ~ 1968.

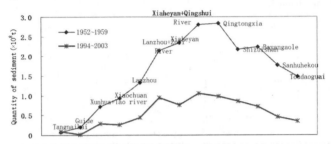

Fig. 3 Variations of annual average sediment of each station on the upper stream in different time periods

3.2 Characteristics of channel sediment transport

It can be see from the plotted relation between average sediment concentration and mean discharge in flood season of each station (Fig. 4) and the characteristic analysis of channel sediment transport that the distribution range of sediment concentration changed along with discharge of the section above Qingtongxia is wider, showing the sediment concentration of the section above Qingtongxia, no matter high or low, basically can be delivered and the channel deposition and the variation range of scour and fill is not great. The river section downstream Qingtongxia in Ning – Meng channel transits gradually to a plain river, especially the section of Sanhuhekou – Toudaoguai, the distribution range of sediment concentration narrows along with the variation of discharge, showing remarkable scour and fill regulation in the section downstream Qingtongxia, a certain discharge can carry a certain sediment and there will be deposition if it exceeds the carrying capacity, i. e. the channel will be deposited in high sediment concentration and the channel will be scoured in lower sediment concentration. The inflow of tributaries with higher sediment concentration forms deposition along the course and the channel would deposit more and drain more when there is more sediment incoming. The inflow of the main river with bigger discharge and lower sediment concentration scours along the river which is affected by roughening of bed scour and the increment sediment concentration restoration along the cause becomes smaller along with the increase of discharge (Fig. 5). It can be seen from Fig. 6 of relation between sediment delivery rate and discharge of Shizuishan station that the sediment delivery rate increases along with the increase of discharge when discharge is smaller than 3,000 m^3/s and the relation curve of sediment delivery rate and discharge becomes smooth or downward when discharge is greater than 3,000 m^3/s.

3.3 Cause of sediment transport variation in big discharge

The curve of sediment delivery rate becomes smooth when the inflow discharge is big in the section downstream Qingtongxia. The causes based on the preliminary analysis is that. ① the bed material becomes coarsening along with scouring and the supplied sediment is reduced; ② flooding causes the reduction of sediment transport. Fig. 7 is grading curve of bed material before and after the event of flood happened in autumn flood season in 1985 (from September 4 to October 16) at Shizuishan station. It can be seen from the Figure that the media particle diameter D50 before and

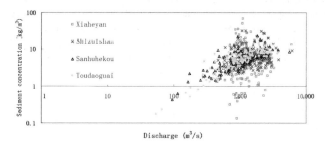

Fig. 4 **Relationship between average sediment concentration and mean discharge in flood peak period of flood season of each station on the upper stream**

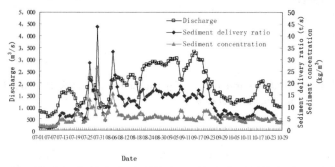

Fig. 5 **Hydrograph of daily discharge, sediment delivery rate and sediment concentration of Shizuishan station in 1989**

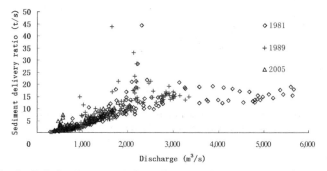

Fig. 6 **Relation between daily sediment delivery rate and discharge of Shizuishan station in different years**

after the event of flood happened at Shizuishan Station is 0.18 mm and 0.33 mm respectively and the bed material greater than 0.1 ~ 1 mm before and after the flood makes up 91% and 97% respectively, showing that the river bed becomes roughening obviously after the flood. Under the same flow conditions, the coarser the particle size, the greater the fall velocity and the weaker the flow carrying capacity will be. According to the principle of suspension work, the sediment concentration is in inverse proportion of fall velocity and the fall velocity of coarse sand is greater than that of fine sediment, resulting in lower capability of channel sediment transport. In addition, the reduction of peaks during a major flood in the section downstream Qingtongxia causes flooding, detention and deposition and decreases the capability of channel sediment transport as well. The hydraulic factor of sediment carrying (V^3/H) is the main important element that affects the capacity of channel sediment transpor. It can be seen from Fig. 8 that the hydraulic factor of sediment carrying increases along with the increase of discharge. It will not increase along with the increase of discharge when the discharge increases to about 3,000 m³/s. This is related to the reduction of mean velocity of cross section after flooding or increased roughness coefficient after roughening of bed scour.

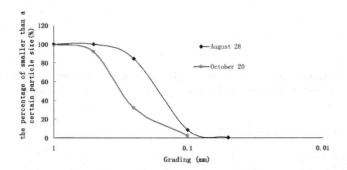

Fig. 7   Grading curve of bed material before and after 1985 flood of Shizuishan station

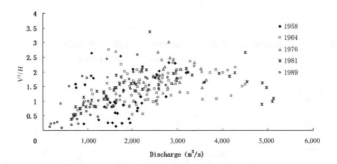

Fig. 8   Relationship between sediment transport factor and discharge of typical floods happened at Toudaoguai station

4   Conclusions

(1) The section above Qingtongxia of Ning – Meng channel is a mountainous river with greater gradient and smaller sediment transport variation. The sediment discharge of Qingtongxia station reaches to the maximum value and it basically can control the water and sediment coming from the upper stream.

(2) The section downstream Qingtongxia transits gradually to a plain river. The channel has the characteristics of more incoming sediment, more deposition and more discharge. The inflow of tributaries with high sediment concentration forms deposition along the river, the inflow of the main river with bigger discharge and lower sediment concentration scours along the river which is affected by roughening of bed scour and the increment sediment concentration restoration along the cause becomes smaller along with the increase of discharge.

Acknowledgements

This study was funded by National Basic Research Program of China (2011CB403305); Central – level public welfare scientific research institutes for basic research operating expenses of special funds(HKY – JBYW – 2012 – 18).

References

Zhao Wenlin. Yellow River Sediment [M]. Zhengzhou: Yellow River Conservancy Press, 1996.

Feng Guohua, Chao Bagen, Gao Ruizhong, et al. Study on Countermeasures for Ice Control of Inner Mongolia Reach of the Yellow River[J]. Hydrology, 2009(2):47 – 49.

Zheng Xiaohua, Zheng Yanshuang. Analysis on Scour and Fill Law and Affecting Factors of Ning – Meng Channel[R]. Zhengzhou: Yellow River Institute of Hydraulic Research, 2008.

Rao Suoiu, Huo Shiqing, Xue Jingguo, et al. Analysis on Characteristics of Silt Discharge Variation of the Upper and Middle Yellow River and Prospects for the Future Trend. Sediment Research [J]. 2001(2):74 – 77.

Zhang Xiaohua, Zheng Yanshuang, et al. Characteristics of Short – term Scour and Fill and Laws of Scour, Fill and Sediment Transport of Ning – Meng Channel on the upper Yellow River [J]. Transactions of Tianjin University, 2008(9):20 – 23.

Zeng Mao, Xiong Guisun, Dai Mingying. Characteristics of Incoming Water and Sediment of Ten Major Streams in Inner Mongolia and its Influence to Scour and Fill of Inner Mongolia Dection of the Yellow River[R]. Zhengzhou: Yellow River Institute of Hydraulic Research,2008.

Zhang Xiaohua, Zheng Yanshun, Shang Hongxia, et al. Study on Law of Scour and Fill and Characte Ristics of Sediment Transport of Ning – Meng Channe[J]. Yellow River,2008(11): 42 – 44.

Zeng Maolin, Xiong Guishu, Dai Mingying. Characteristics of Incoming Water and Sediment of Ten Major Streams in Inner Mongolia and its Influence to Scour and Fill of Inner Mongolia Cection of the Yellow River[R]. Zhengzhou: Yellow River Institute of Hydraulic Research, 2008.

Wang Ling, Dong Xuena, Li Xuemei. Data Compilation of Ice – run Regime of the Yellow River and its Characteristic Analysis the Upper and Middle Yellow River Part, Period from 1950 to 2005[R]. Zhengzhou: Yellow River Conservancy Commission, 2006.

Lu Binghui, Guo Decheng, Zhang Yatong, et al. Analysis on Characteristics of Ice Flood in Ning – Meng Reach of the Yellow River [R]. Hohehot Municipality: Inner Mongolia Water Resources, 2005.

Application and Discussion of Unidimensional Mathematical Model for Water Quality and Sediment in Lower Yellow River[①]

Li Lantao[1, 2], *Wu Zening*[1] and *Shao Xuan*[2]

1. School of Water Conservancy & Environment, Zhengzhou University,
Zhengzhou, 450002, China
2. Hydrology Bureau of YRCC, Zhengzhou, 450004, China

Abstract: It is aimed at the sandy characteristics in lower Yellow River, the physical process of sedimentation and resuspension has absorption and resolution role to contamination in water body. It is necessary to put the sediment of source leak as one main term. So this model separates sediment from source leak term, accedes sediment source leak term to unidimensional water quality mathematical simulation. The data gettting at the hydrology condition, is similar to runoff steadily, riverway well – proportioned and lab background. The data is coherent. The result of the calculation has little error compare with measurement data. It should take discrete role into account for complex hydrology condition riverway, check out the coefficients K_1, K_s. It should validate the role of sedimentation and resuspend to coefficients K_1, K_s.

Key word: silt, mathematic model, source leak term, unidimensional

1 Overview

Define certain water quality conditions corresponding to the number of water pollutants as water environmental capacity, it can use water environmental capacity to represent the number of water environment. The river is pollutants transfer carrier, pollutants transport with help of the water flow and sediment in the river. The size of flow which comes from Yellow River influences the river pollutants carry capacity, also influences river water environmental capacity and its self cleaning capacity. The sediment into the Yellow River cause certain pollution, at the same time, because the components contain a significant amount of clay mineral, inorganic gel and a certain number of organic gel, organic and inorganic compound gel, has the remarkable adsorption effect to the wide variety of pollutants.

The sediment has certain influence on the river water quality monitoring, mainly embodies in the following respects.

(1) The sediment itself as a great uncertainty the product of area pollution sources, also can desorb pollutants and might produce secondary pollution, it greatly increases the difficulty of water quality simulation.

(2) The randomness of water sand content variation and the complexity of pollutant interactions not only make river water quality model structure complex, but also make water quality model precision decline.

This text analyses that sediment concentration make great influence on river pollutant migration transformed, puts sediment factors in one – dimensional water quality mathematical model, puts ammonia nitrogen migration transformed water quality model into use at simple reach.

2 Basic theory

The pollutant migration transformed relate to many factors, is a very complicated process.

① Foundation item: Funded by the National Natural Science Foundation of China(51079131)

River geomorphology, rivers hydrological, hydraulics, river water temperature, pH, the initial concentration of the water pollution, sediment adsorption and analytical, sediment precipitation and again suspension are the main Influencing conditions.

The variation in the river includes all sorts of function make the water body pollutant in unit of time, is called Source Leak Term $\sum S_i$.

$$\sum S_i = -K_1 C + K_s \Delta S \tag{1}$$

where, $\sum S_i$ is the Source Leak Term of river contaminant, mg/($L \cdot d$); K_1 is Level 1 degradation coefficient of the contaminant, $1/d$; K_s is increase or decrease rate of pollutant concentration caused by sediment concentration changes, $1/d$; ΔS is sediment variation, mg/L.

Because dispersion coefficient E_d is much bigger than molecular diffusion coefficient E_m and turbulent diffusion coefficient E_t, the latter compared to the former often can be neglected.

$$E = E_m + E_t + E_d = E_d \tag{2}$$

Hence, river water quality migration and transfer basic equation form can be got.

$$\frac{\partial C}{\partial t} + u \frac{\partial C}{\partial x} = E \frac{\partial^2 C}{\partial x^2} + \sum S_i \tag{3}$$

where, $\sum S_i$ is a pollutant concentration in the river, mg/L; t is time, d; x is the flow of a river distance, km; u is the average flow velocity of river water, km/d; E is longitudinal dispersion coefficient of river flow, km^2/d; C is the pollutant concentration that the flow of water flow into the response unit.

For a more uniform sediment river, one dimension uniformity river water quality migration and transfer basic equations can be got as follow:

River one dimension water quality migration and transfer basic equations form:

$$\frac{\partial C}{\partial t} + u \frac{\partial C}{\partial x} = E \frac{\partial^2 C}{\partial x^2} - K_1 C + K_s \Delta S \tag{4}$$

For general from the tides inland rivers, diffusion and discrete function relative to move flow function often very small, this time, downstream concentration change caused by pollution discharge can be directly given:

$$C = C_0 \exp\left(-\frac{K_1 x}{u} \right) + \frac{K_s(S_1 - S_0)}{K_1} \tag{5}$$

where, C_0 is the upstream water pollutants concentration, mg/L; S_0 is the upstream water sediment concentration, mg/L; S_1 is the downstream water sediment concentration, mg/L.

3   Preliminary application verification

The reach of Yellow River from Gaocun to Luokou is a typical reach, which belongs to suspended river. There is no inflow branch and a little drain outlets in the reach, the river is simple, it provides a good model validation test condition.

The ammonia nitrogen degradation coefficient K_1 and sediment concentration changes of Lower Yellow River causes increase or decrease rate K_s of pollutant concentration, see as Tab. 1.

Tab. 1   The Yellow River downstream K_1, K_s value of every water period

Water period	K_1 (1/d)	K_s (1/d)
Wet period	0.26	0.000,405,8
Drought period	0.12	0.000,405,8

In order to validate the rationality and reliability of the model, the monitoring data of 2006 was used to validate the rationality and reliability of the model. Compared with simulation value, the result of ammonia nitrogen measured value is showed in Tab. 2.

Tab. 2 Reality measure value and simulation value contrast form of NH$_3$ - N

Upstream section measured value(mg/L)	Downstream section measured value(mg/L)	Simulation value (mg/L)	Relative error(%)
0. 24	0. 20	0. 21	5
0. 27	0. 22	0. 21	4. 5

In the Tab. 2, comparing the simulation value of ammonia nitrogen with measured value, the relative error is within 5%. The parameter of the model is reasonable. The model meet requirements and can be used to water quality forecast.

4 Conclusions

(1) In many river sediment, its factors should not be neglected, so it is necessary to separate sediment from source leak term, accede sediment source leak term to unidimensional water quality mathematical model.

(2) The datum are got at the condition that section hydrological conditions is similar, flow is relatively stable, river is smooth, laboratory monitoring background is same, so the datum are consistent and the result is good. Discrete function should be considered if the hydrological condition of the river is complex, the coefficient K_1 and K_s should be fit testing.

(3) The upstream coming water suddenly increases or decreases, sediment precipitation and resuspension function exert a tremendous influence on coefficient K_s, the coefficient K_s should be fit testing in conditions.

(4) The coefficient K_1 and K_s should be fit testing as the conditions permit , the function of sediment on water quality migration and transfer could be searched under complex conditions.

(5) The Lower Yellow River is simple, and the only c ammonia nitrogen is considered. The support should be given to search different pollution and reach, discuss pollution evolving laws under complex conditions and the the mechanism of the interaction between sediment and pollution, make science and technology support for Yellow River water resource and health life protection.

References

Wu Zening. Integrated Optimal Distribution of Water Quality and Quantity in a Region Based on Ecology Economy[D]. Nanjing HoHai University, 2004.

Luo Wensheng. The Water Environment Analysis and Prediction [M]. Wuhan: Wuhan University Press ,2000.

Bai Yuchuan, Gu Yuanyan, Xing Huanzheng, et al. Theory and Application of Mathematical Model for Water Flow Sediment and Quality [M]. Tianjin: Tianjin University Press ,2005.

Hu Guohua. Simulated Research on Water Quality for Such Heavy Sediment - containing Rivers like Yellow River [J]. Journal of Safety and Environment ,2004 ,4(4) :45 -48.

Sedimentation Disposal Methods in the Xiaolangdi Reservoir

Tong Yifeng[1], *Jiang Siqi*[2], *Wang Ting*[2] and *Chen Shukui*[2]

1. Press and Publication Center, YRCC
2. Yellow River Institute of Hydraulic Research, YRCC, Key Laboratory of Yellow River Sediment Research, MWR, Zhengzhou, 450003

Abstract: The main sedimentation disposal methods are analyzed and classified into six types, that is preventing soil erosion to reduce sediment inflow to reservoir, increasing sedimentation storage capacity to extend the service life of reservoirs, empting reservoir periodically for sediment washing and capacity restoring, dredging reservoir sediment with machinery, lowering water level during the flood period to scour sediment and reduce sedimentation, and sluicing reservoir sediment by density current. Take Xiaolangdi Reservoir as an example, the sedimentation disposal methods of lowering water level to scour sediment, and sluicing reservoir sediment by density current are studied. It is proposed that the main issues needed to study in reservoir sedimentation disposal includes theory of sediment movement and conveyance in reservoir area, the combination of artificial sedimentation disposal and reservoir operation, and the monitoring and observing technology in reservoir.

Key words: reservoir, sedimentation disposal, density current. sedimentation reduction

1 Introduction

From a global perspective, silt-laden rivers are roughly distributed over Central and Southwestern United States, Algeria and Egypt in North Africa, India and Pakistan in South Asia, and middle of China. Today's worldwide annual mean loss of reservoir storage capacity due to sedimentation is already higher than the increase of the capacity by construction of new reservoirs. The main purpose of dealing with reservoir sedimentation is to extend the service life of reservoirs.

2 Sedimentation disposal methods

2.1 Preventing soil erosion to reduce sediment inflow to reservoirs

To carry out soil and water conservation within basin to intercept or control sediment from the every source is an important measure. For reservoirs with small-area basins, this method has more notable effect. In the Yellow River basin, the check dams are effective engineering measures for soil and water conservation, which can not only intercept sediment and conserve soil and water, but also build up fields and increase grain yield.

2.2 Increasing sedimentation storage capacity to extend the service life of reservoirs

To extend the service life of reservoirs, the approach of reserving more so-called sedimentation storage capacity in planning is used to be adopted, thus the total storage capacity is increased. In fact, when the storage capacity increased, the sediment retaining ratio increased correspondingly. Under certain conditions, to increase storage capacity can not increase the service life proportionately. In addition, adding storage capacity by stages will involve very high spending, even 20 times of the original cost.

2.3   Emptying reservoir periodically for sedimentation washing and capacity restoring

Reservoir emptied periodically for washing has long been a practice for reservoir sedimentation reduction, which is mainly suitable for seasonal reservoirs. In China, emptying reservoir for washing has been adopted at many reservoirs and the achievements are significant. At the Heisonglin reservoir, with this method, 4.28×10^4 t of sediment was washed away in 1970, and 4.43×10^4 t in 1971. But, for navigable rivers, the economic losses for shipping and others in the washing period has to be considered.

2.4   Dredging reservoir sediment with machinery

Mechanical dredging is generally considered costly and can be carried out widely. Besides, the dredged sediment has to be transported away by powerful facilities. Yet, for small reservoirs, it might be one measure to deal with reservoir sediment. The dredging equipments are dredgers, suction pumps and suction vessels etc. In China, hydraulic suction machinery are commonly used, that is, to stir the sediment within the reservoir with hydraulic and mechanical power, then to discharge the highly sediment concentrated flood out of the reservoir through siphons by the help of water head difference between upstream and downstream of the dam.

2.5   Lowing water level during the flood period to scour sediment and reduce sedimentation

For most rivers, the sediment yield during the flood period accounts for a very large proportion, and most comes with the several big floods. Therefore, to lower water level and increase flow velocity within the reservoir during flood period can scour some or most sediment and reduce sedimentation.

2.6   Sluicing reservoir sediment by density current

For reservoirs built on sediment-laden rivers, density current is an important method for reservoir sedimentation reduction. When the density current is formed, the lower layer is high-density sediment flow, while the upper layer is clear water; at this time, ejecting the muddy water through the bottom outlet while keeping the clear water, which can achieve the purpose of both sediment scouring and water reserving. Averagely, the sediment delivery ratio with current density is 40% ~ 90% for one time and is 30% ~60% for several times.

For the above methods, the lowering water level to scouring sediment and sluicing reservoir sediment by density current are considered more economical and efficient.

3   Sediment disposal methods in the Xiaolangdi (XLD) Reservoir

3.1   General situation of the Xiaolangdi Reservoir

The XLD Reservoir is located in middle reach of the Yellow River and 40km far from Luoyang city. It is the last gorge reservoir and controls an area of 6.94×10^5 km^2, which accounts for 92.3% of the whole Yellow River basin. The XLD reservoir is also controls 90% runoff and almost all flow sediment of the Yellow River. It is a comprehensive complex, with main purpose for the lower Yellow River flood control (including ice), sedimentation reduction, and also providing service for water supply, irrigation, power generation, etc. According to design, the XLD Reservoir will trap about 10×10^9 t of sediment by using its storage capacity in 50 years, which can reduce 7.8×10^9 t of downstream sedimentation. Coupled with the effect of water and sediment regulation, the sedimentation disposal effect is as equivalent of 20 years without the lower Yellow River bed rising, which creates precious time for river harnessing.

At the initial operation period, the flow pattern in backwater area of the reservoir was density current and muddy water, the riverbed would gradually silt and rose. At the later operation period, multi-year water and sediment regulation, lowering water level for sediment flushing are the main methods for sediment disposal, which can help to achieve relative long and obviously flood control and sedimentation reduction benefit. Nowadays, the main sediment disposal approaches are density current outflow and lowering reservoir water level to scour sediment.

3.2 Sluicing reservoir sediment by density current

From September 1999 to October 2010, the amount of inflow sediment in XLD reservoir is $3.764,2 \times 10^9$ t, and the amount of outflow sediment is $0.702,9 \times 10^9$ t the deposition sediment is $3.061,3 \times 10^9$ t due to sand balance calculation, $2.822,5 \times 10^9$ m^3 in terms of cross-section method, which is about 0.26×10^9 m^3 per year. The total siltation accounts for 22.1% of the original storage capacity, among those, the main stream accounts for 79.3% and tributaries accounts for 20.7%. The sediment deposition status in XLD Reservoir is shown in Tab.1.

Tab. 1   The statistic of sediment deposition in the XLD Reservoir

Year	Sand balance calculation (×10⁸ t)			Cross-section method(×10⁸ m³)		
	Inflow	Outflow	Deposition	Main stream	Tributary	Total
2000	3.691	0.041	3.65	3.842	0.241	4.083
2001	2.949	0.221	2.728	2.549	0.422	2.971
2002	4.375	0.69	3.685	1.938	0.17	2.108
2003	7.564	1.18	6.384	4.623	0.262	4.885
2004	2.637	1.486	1.151	0.297	0.877	1.174
2005	4.148	0.449	3.699	2.603	0.308	2.911
2006	2.325	0.398	1.927	2.463	0.987	3.45
2007	3.125	0.705	2.42	1.439	0.848	2.287
2008	1.337	0.462	0.875	0.256	−0.015	0.241
2009	1.98	0.036	1.944	1.229	0.492	1.721
2010	3.511	1.361	2.15	1.156	1.238	2.394
Total	37.642	7.029	30.613	22.395	5.83	28.225

From 2002, 13 times of water and sediment regulation had been carried out before flood season on the Yellow River. From 2004, the water and sediment regulation began to be implemented on the basis of main stream reservoir group, that is, the Wanjiazhai, and Sanmenxia Reservoir was arranged to stored water firstly, then the three reservoirs were scheduled accurately to form artificial density current in Xiaolangdi reservoir, which can increase the amount of outflow sediment. The density current calculations equitation is as follows.

Free flow scouring section at the upstream of the reservoir delta:

$$G = \Psi \frac{Q^{1.6} J^{1.2}}{B^{0.6}} \times 10^3 \tag{1}$$

Free flow backwater section around the top of reservoir delta:

$$\eta = a \lg Z + b \tag{2}$$

The flow rule of dived density current:

$$S_j = S_i \sum_{i=1}^{n} P_{4,L_i} e^{\left(-\frac{\alpha \omega L}{q} \right)} \tag{3}$$

During the 13 times of water and sediment regulation experiments, the maximum sediment delivery rate is 145.4%. The relevant date is as Tab. 2.

Tab. 2　The data of density current during water and sediment regulation

Year	Period (m. d)	Hour (h)	Inflow (m³/s)	Average sediment concentration (kg/m³)	Sand(×10⁸t) Sanmenxia	Xiaolangdi	Sediment delivery rate (%)
2004	7.7 ~7.14	8	689.675	80.759	0.385	0.055	14.29
2005	6.27 ~7.2	6	776.917	112.238	0.452	0.02	4.42
2006	6.25 ~6.29	5	1,254.52	42.426	0.23	0.069	30.00
2007	6.26 ~7.2	7	1,568.71	64.582	0.613	0.234	38.17
	7.29 ~8.12	15	1,418	52.847	0.971	0.426	43.87
2008	6.27 ~7.3	6	1,324	92.56	0.741	0.458	61.81
2009	6.30 ~7.3	4	1,062.75	148.445	0.545	0.036	6.61
2010	7.4 ~7.7	4	1,655.5	73.1	0.418	0.553	132.30
	7.24 ~8.3	10	1,397	67.85	0.901	0.258	28.63
	8.11 ~8.21	11	1,626	70.65	1.092	0.508	46.52
2011	7.4 ~7.7	4	1,694.5	44.38	0.260	0.378	145.4

3.3　Lowering reservoir water level to scour sediment

At later period, the operation mode of XLD Reservoir will be turn from "storing water-sediment to regulate" to "lowering water level to scour sediment opportunely". The "lowering water level to scour sediment opportunely" means to decrease the reservoir water level appropriately while meeting big flood in order to scour the deposited sediment and restore the reservoir capacity and extend sediment storage time. The effect of "lowering water level to scour sediment opportunely" is decided by the boundary conditions of reservoir area (deposition patter and amount), inflow water and sediment (time and amount), reservoir regulation mode (reservoir water level varies range and rate). In order to study the rule and effect of "lowering water level to scour sediment opportunely", 4 series of physical model experiments were done by the Yellow River institute of hydraulic research. Based on the measured reservoir terrain in 2006 after flooding season, the initial model topography is gradually formed by the flow of 1978 ~ 1982 water and sediment series in conditions of the relevant reservoir regulation modes. When the reservoir sedimentation amount was 3.2×10^9 t, two lowering water level to scour sediment tests under different flood processes were tested, then the sediment scouring effect in conditions of same initial terrains and different water and sediment series were compared. When the reservoir sedimentation amount was 4.2×10^9 t, two lowering water level to scour sediment tests were done under different flood processes and different water level before dam, then the sediment scouring effect, outflow water and sediment process were compared. The test conditions are shown in Tab. 3.

Tab. 3 The test data of lowering water level to scour sediment

Siltation amount ($\times 10^8$ m^3)	Reservoir water level (m)	Period (d)	Test (scheme)	Inflow (m^3/s)		Inflow sediment concentration (kg/m^3)	
				Average	Range	Average	Range
32	210	16	1 (32/210/16)	2,962	1,240 ~ 4,660	103.22	43.0 ~ 189
	210	12	2 (32/210/12)	2,210	677 ~ 3,410	179.66	75.5 ~ 340
42	210	12	3 (42/210/12)	2,210	677 ~ 3,410	179.66	75.5 ~ 340
	220	12	4 (42/220/12)	2,210	677 ~ 3,410	179.66	75.5 ~ 340

The amount of scouring sediment is calculated by cross-section method and sediment balance method, respectively. The maximum scoured sediment are 0.521,7 × 10^9 m^3 by cross-section method, 0.613 × 10^9 t by sediment balance method, and the relevant maximum sediment delivery rate is 249%, shown as Tab. 4 ~ Tab. 5.

Tab. 4 The scoured sediment distribution along the reservoir area (cross-section method)

(Unit: $\times 10^8$ m^3)

Test series	After HH10	HH10 ~ HH18	HH18 ~ HH24	HH24 ~ HH31	HH31 ~ HH38	HH38 ~ HH50	Tributaries	Main stream	Total
1	−1.37	−1.47	−0.59	−0.50	−0.01	−0.01	−0.16	−3.96	−4.12
2	−0.89	−0.82	−0.46	−0.38	0.32	0.46	−0.09	−2.11	−2.20
3	−2.89	−1.97	−0.39	−0.00	0.11	0.33	−0.35	−4.82	−5.17
4	−1.46	−0.83	−0.21	−0.20	−0.14	0.14	−0.13	−2.70	−2.83

Tab. 5 The scoured sediment distribution along the reservoir area (sediment balance method)

Test series	Inflow			Outflow			Sediment delivery rate (%)	Scouring amount ($\times 10^8$ t)
	Water ($\times 10^8$ m^3)	Sediment ($\times 10^8$ t)	Average sediment concentration (kg/m^3)	Water ($\times 10^8$ m^3)	Sediment ($\times 10^8$ t)	Average sediment concentration (kg/m^3)		
1	40.945	4.226	103.22	48.672	9.157	188.13	216.7	−4.930
2	22.911	4.116	140.28	28.511	6.744	236.52	163.8	−2.628
3	22.911	4.116	179.65	31.962	10.248	320.62	249.0	−6.13
4	22.911	4.116	179.65	29.023	7.433	256.10	180.6	−3.32

3.4 Issues need to be studied

The sediment disposal technology of XLD Reservoir is complex and difficult for that it should be implemented in the conditions of high-water level, so the following three issues should be studied.

3.4.1 The study on theory of sediment movement and conveyance in the reservoir area

The study should include analyzing the reservoir siltation morphology and features; the deposition distribution of different grades sand in the reservoir area; the sand ejection rule in the Sanmenxia reservoir and inflow water and sediment features in XLD Reservoir; the distribution of flow field and rule of sediment transportation; the modeling and controlling technology of artificial density current.

3.4.2   The study on the combination of artificial sediment disposal and reservoir operation

The artificial sediment disposal should take full advantage of the natural forces, and combine with the reservoir operation, that is, the sediment disposal opportunity should be determined in terms of the incoming water sediment conditions and the reservoir operation status. The combination of sediment disposal and reservoir operation should be studied to improve the sediment sluicing rate.

3.4.3   The monitoring and observing technology in the reservoir area

The reservoir monitoring and observing should include the reservoir capacity, underwater topography, sedimentation patterns, density flow etc. The relevant monitoring and observing equipment should be developed.

4   Conclusions

The main sedimentation disposal methods are preventing soil erosion to reduce sediment inflow to the reservoir, increasing sedimentation storage capacity to extend the service life of reservoirs, empting reservoir periodically for sediment washing and capacity restoring, dredging reservoir sediment with machinery, lowering water level during the flood period to scour sediment and reduce sedimentation, and sluicing reservoir sediment by density current. Among these methods, the lowering water level to scour sediment and sluicing reservoir sediment by density current is considered more economical and efficient.

In the practice of XLD Reservoir sediment treatment, according to water and sediment regulation, the maximum sediment delivery rate is 145.4%, according to lowing water level to scouring sediment, the maximum scouring sediment are 0.517×10^{9} m^{3} by cross-section method, 0.613×10^{9} t by sediment balance method, and the relevant maximum sediment delivery rate is 249%.

Three issues should be studied for sediment disposal, these are the theory of sediment movement and conveyance in reservoir area, combination of artificial sediment disposal and reservoir operation, the monitoring and observing technology in the reservoir area.

References

Qian Ning. Hyperconcentrated Sediment Flow [M]. Beijing:Tsinghua University Press, 1989.
Han Qiwei. Reservoir Sedimentation [M]. Beijing:Science Press, 2003.
Qian Ning, Wan Zhaohui. Sediment Transport Mechanics[M]. Beijing:Science Press, 1982.
Jiang Enhui. Sediment-the Key Facter of Reservoir Life[J]. Water and China, 2011(10).

J. Application of Experiences and New Technologies of Water Resources Management

（ I ）

Application of Carrying Capacity & Pollution Load Allocation Simulation Model (CC + PLAS) in Implementation of Pollution Control Red Line in Water Functional Zones

Simon Spooner[1], *Lian Yu*[2], *Huang Jinhui*[2], *Wen Huina*[2] and *Cheng Wei*[2]

1. Atkins China, Beijing, 100022, China
2. Yellow River Water Resource Protection Bureau, Zhengzhou, 450004, China

Abstract: China's 2011 No. 1 policy document on Accelerating Water Conservancy Reform and Development emphasises the implementation of the strict water resources management policy of the "three red lines". Under the pollution load control red line (the 3rd line) concept each water function zone (WFZ) (control section) of a river must be managed so that it meets defined water quality targets by defined dates through control of discharges from cities and industry. A Carrying Capacity and Pollution Load Allocation Simulation model (CC + PLAS) has been developed to support implementation of this policy. The CC + PLAS model calculates the carrying capacity (amount of pollutions that can be added to each WFZ and still meet targets) and calculates the pollution load allocations (load of pollutants permitted from each discharge point). The current Chinese concept for calculating carrying capacity only operates under fixed, and rather unrealistic, scenarios. CC + PLAS integrates the basic Chinese approach with European concepts of river basin water quality modelling which take account of various factors impacting river water quality. For example, CC + PLAS can calculate the river carrying capacity under different flows and different seasonal conditions and use monitored data about actual river water quality to estimate the quantity and impact of all pollution sources and processes. It can then be used to forecast future water quality trends brought by social economic development as well as improvements from the application of various pollution reduction strategies with estimates of the likely water quality in comparison to WFZ targets. In this way it facilitates the river need oriented pollution management to help authorities to plan how to achieve their targets of compliance with WFZ targets by 2015, 2020 and 2030. The CC + PLAS model provides a scientific and convenient tool to water resources protection and river basin planning. In this paper pilot applications of the model in the middle reaches of the Yellow River are described.

Key words: Water Function Zone (WFZ), carrying capacity, pollution load control scenarios

1 Background of CC + PLAS model development

Currently water function zone (WFZ) management is the basis of water resources protection in China. The policy of Three Red Lines policy specifically emphasises the establishment of the WFZ pollution control red line, with calculation of WFZ carrying capacity and the strict control of pollution discharges. The basis of the pollution load allocation strategy in the integrated river basin planning on main rivers is the calculation of carrying capacity under the low flow conditions and the current estimated discharge load. As the existing calculated carrying capacity is restricted by factors including a fixed low flow scenario, it is difficult to carry out the pollution control planning efficiently and smoothly. Taking the middle reach from Longmen to Sanmenxia of the Yellow River Basin as a pilot area, the EU – China River Basin Management Programme (RBMP) comprehensively analysed the problems faced during the implementation of pollution load allocation under the existing water pollution control system and management mechanism in China. The RBMP introduced advanced European experience in water pollution management and integrated river basin planning to develop the Carrying Capacity plus Pollution Load Allocation Simulation model (CC + PLAS) for a Yellow River Project area. CC + PLAS can efficiently calculate carrying capacity and simulate the impact of pollution load allocation schemes under different scenarios on the actual water quality and likely compliance with WFZ targets. This gives decision makers an intuitive picture of the effect of the implementation of WFZ pollution control red line measures thereby supporting the formulation and direction of future pollution control strategies.

In China the "Carrying Capacity" of a river reach is a theoretical concept for calculation how much pollution can enter each river reach and WFZ water quality targets still be met at the boundaries. It is assumed that water enters at the standard of the upstream zone, then pollutants decay along the river allowing further pollution (as a flow and load) to be added part way along (which will also partly decay) such that by the time the flow reaches the downstream boundary the concentration exactly matches the target for that zone. The steps for this calculation are defined in Ministry of Water Resources (MWR) Guideline (GB 25173 – 2010). In the Yellow River the low flow scenario used to perform the calculation of carrying capacity is very conservative and is determined by ranking the 1957 to 1997 average annual flow data, selecting the 90th percentile low flow year and then selecting the lowest monthly average flow in that year. Typically this is would be close to a 98 th percentile low flow if calculated as a normal percentile of all monthly average flows in the data set, so represents an extreme low flow situation in which to comply with WFZ targets. Also this is done separately for each WFZ so each may use flows from different years and months. This makes the planning of realistic strategies of how to comply with targets very difficult.

The water quality in a river reach is influenced by a great number of factors with various pollution sources (point sources, non – point sources, unknown sources), bio – chemical reaction process, water abstraction process and interactions between upstream and downstream reaches within the river basin.

The current Chinese methods of calculating carrying capacity account for only a few of these processes ignoring non point sources, abstractions or the evidence that can be derived from actual monitoring data.

There is therefore a need to devise an improved approach to calculation of carrying capacity incorporating a more realistic definition of flow scenarios and consideration of the multiple pollution sources and then to use this for the calculation of pollution load allocations under likely future conditions in a way that accounts for actual flow and water quality conditions as indicated by observed flow and water quality monitoring data.

2 Concept of CC + PLAS model

The CC + PLAS model was developed by EU – China RBMP and YRCC WRPB for the Yellow River project area based on the SimBasinQ technology developed by Atkins of the UK. It introduces concepts of integrated river basin planning and calculation methods which have been used for decades in the UK and elsewhere in EU and then adapts these to the very different situation of China. It calculates the river carrying capacity under different flows, simulates and tests the real water quality impact of different pollution allocation scenarios (based on monitoring data), and forecasts the future water quality trend brought by social economic development as well as improvement in pollution reduction technology.

CC + PLAS is a general modelling system for use on any river. Under the RBMP a pilot CC + PLAS model was constructed covering the middle reaches of the Yellow River including parts of the Fen and Wei Rivers. Fig. 1 shows the area covered with a node at each WFZ boundary (where there is also often a water quality monitoring point) and reaches in between. There are also some additional nodes to take account of confluences in the river. Each node also has a catchment defined by the area of land draining to that point.

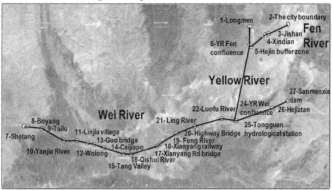

Fig. 1 Pilot area of CC + PLAS Model

For each class chemical targets are defined for COD and Ammonia, carrying capacity is separately calculated for each of these. CC + PLAS actually contains full modelling of the oxygen, nitrogen, phosphorous and pH systems and can be used to simulate more than 50 different key river water quality determinant. For the application in the yellow river only COD and Ammonia were used in detail.

CC + PLAS has 3 main modes of operation:

(1) The theoretical carrying capacity calculation mode. In which the concentration at the head of each WFZ reach is fixed exactly equal to the standard of the upstream reach and just the known point sources and decay processes are acting in the river. Basically each reach / WFZ is an isolated unit. The quality is reset to standard before passing to the next WFZ.

(2) The estimated river basin mode. In which the upstream water quality of each reach is set at the downstream water quality of the preceding reach (headwaters are set at an observed or estimated quality). This allows estimation of the impact and resulting water quality of different factors. This is a basic river basin simulation model with flow and quality passing along the river network based on estimated figures for loads.

(3) The calibrated model mode in which balancing flows and loads are calculated to exactly match to the observed water quality monitoring data. This is a calibrated river basin water quality model.

Each of these modes may be run with or without consideration of non point sources and abstractions, different decay coefficients, and different flow scenarios.

3 Functions of CC + PLAS model

The CC + PLAS model has three main functions: ① calculating WFZ carrying capacity and pollution load allocations; ② simulating river water quality considering all major river processes, calibrated to available data and assessing the likely compliance with WFZ targets under different pollution allocation scenarios; ③ forecasting the future water quality trend of the coming years and assisting analysis of the cost efficiency of river basin and investment planning.

3.1 Calculating carrying capacity and pollution load allocations

In different seasons of the year temperatures and velocity of flow and rates of self purification will vary and so the carrying capacity will vary. The first task of the study was therefore to understand how different factors affect the carrying capacity and how better and more realistic scenarios for the calculation of carrying capacity might be defined.

CC + PLAS can calculate carrying capacity based on a single flow scenario (e. g 90 percentile flow) or combinations of seasonal situations (e. g. a compound of capacity based on seasonal flows and temperatures).

3.1.1 Calculate carrying capacity at fixed flow scenario

The CC + PLAS model may be set up based on input data the same as the previous WRPB calculation methods using their definition of the 90 percentile low flow and the decay coefficient at 20 ℃. In this mode CC + PLAS not surprisingly comes to the same figures for the carrying capacity of each WFZ as the previous WRPB spreadsheets. Then in CC + PLAS we were able to incorporate flow:velocity relationships based on better data analysis which resulted in slightly different figures for CC with the flexibility to correctly calculate for any flow rather than use a fixed velocity for each reach .

In this very low flow scenario the standard method takes no account of diffuse (non – point) sources of pollution, but given that there would be very little surface runoff at this time this is not a serious omission. At higher flows, when there will be surface runoff to the river, non – point pollution will become significant.

3.1.2 Effect of temperature on carrying capacity at fixed flow

The carrying capacity is determined by the flow and the decay rate. The decay rate is very sensitive to temperature so in each season of the year the carrying capacity of the river will be very different. To represent this in a manageable way CC + PLAS was set up to calculate CC using 20 ℃ in summer, 7 ℃ in winter and the annual mean of 11 ℃ for Spring and Autumn.

In Fig. 2 the bar chart shows the calculation of carrying capacity for the WRPB 90 percentile flow for each WFZ for each of the 3 temperatures. This chart may be related to the schematic in Figure 1 as showing sequentially the network of the different branches running down the Yellow, Fen and Wei rivers on X axis and on Y axis the carrying capacity load for each temperature. The temperature makes a dramatic difference to the carrying capacity.

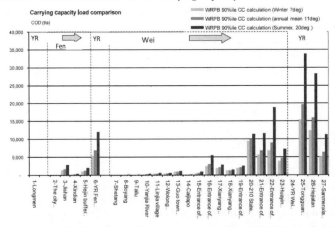

Fig. 2   Carrying capacity at different temperatures

3.1.3 Carrying capacity at different flows

Next multiple scenarios with different flows (90, 75 and 50 percentile flows) were tested to identify the effect on carrying capacity calculation results. The carrying capacity is calculated for

COD and for ammonia nitrogen, considering only the known point sources of pollution at the effective discharge point. This is as shown in Fig. 3. Note that in some WFZ there is less difference in CC for different flows, this is because of the flow velocity relationships, if the higher flow results in much higher velocity then there will be proportionately less increase in carrying capacity because there will be less time for decay.

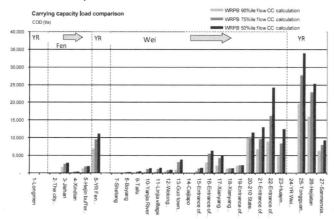

Fig. 3   Illustration of calculation of carrying capacity at 90, 75 and 50 percentile flows as defined by WRPB all at standard 20 ℃ temperature

3.1.4   Compound Carrying capacity for seasonal flows and temperatures

Through the RBMP research project it has been proposed by YRCC WRPB that an alternative flow scenario be defined based on a combination of the different seasonal flows and temperatures. This is 3 winter months of 90 percentile low flow (at 7 ℃) 3 months of summer flows as 50 percentile (20 ℃) and 6 months to represent Spring and Autumn at 75 percentile flow (11 ℃). Fig. 4 illustrates this calculation showing the annual carrying capacity for each of the seasonal scenarios and the compound carrying capacity combining these seasons. In this diagram each seasonal component is shown as though for a whole year of carrying capacity but the compound capacity is calculated as a weighted mean 1/4 Winter, 1/4 Summer and 1/2 the Spring and Autumn based capacities.

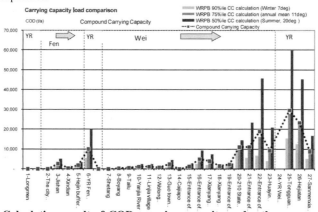

Fig. 4   Calculation result of COD carrying capacity under the compound flow
(with 3 winter months of 90 percentile low flow (at 7 ℃) 3 months of summer flows as 50 percentile (20 ℃) and 6 months to represent Spring and Autumn at 75 percentile flow (11 ℃))

Note that this compound carrying capacity works out to be typically slightly more than the 75% WRPB flow annual scenario. This compound carrying capacity may then be used to set the pollution load allocations rather than using the fixed 90 percentile 20 ℃ scenario.

In the scenarios of the 90 percentile and 75 percentile monthly mean low flows the precipitation will be small so the diffuse source pollution from surface runoff is likely to be much less; in the scenario of the mean monthly flow of 50% , the river flow is relatively large with the added surface runoff, so the diffuse source pollution should be taken into account. For future further definition of the Carrying Capacity consideration should be given to including estimates of the non – point diffuse flows in the calculation.

CC + PLAS calculates the carrying capacity for each WFZ for any given flow and decay rate or temperature for COD, Ammonia or many other pollutants. Therefore we can repeat this exercise using normal percentile calculations based on the annual average flows from 1996 to 2006. This gives generally higher flows than the WRPB method and so would result in higher carrying capacity. A further scenario to explore would be to use percentiles of just the relevant months from a long flow time series. The final decision on the best method for selection of flows for carrying capacity calculation should be derived after careful consideration and debate.

3.2　Simulating the impact of different pollution load allocations

The impact on the river of different flow scenarios for carrying capacity calculation may be looked at in two ways, firstly in the theoretical context of the carrying capacity calculation for each river reach, or in the context of the actual river water quality based on monitoring data and considering upstream and downstream interactions.

3.2.1　Application of CC + PLAS in theoretical carrying capacity calculation mode

CC + PLAS may be used to illustrate the impact of different pollution load allocations in the theoretical scenarios in which carrying capacity was calculated. This indicates if a pollution load allocation strategy will meet the WFZ objectives under the flow scenario in which carrying capacities were calculated.

To indicate how this may work Fig. 5 shows the concentrations of COD that would occur in the 90 percentile flow situation in the scenario with no discharges of pollution into each reach (i. e. this is just the self purification from the upstream to the downstream end of each WFZ) and also a scenario while discharges as reported in which the total load exceeds the possible self purification. If the pollution discharges were set at the carrying capacity then the concentration would equal the standards.

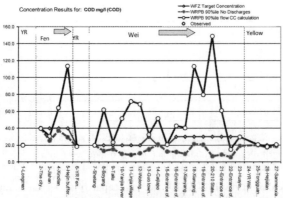

Fig. 5　Concentrations of COD under above scenarios

CC + PLAS incorporates the database of registered pollution discharge points along the river. The default method for pollution load allocation is to take the carrying capacity calculation as the allowable total discharge amount and allocate this around the individual discharges to meet this. If the actual discharge load is less that the carrying capacity, then the pollution load allocation is set at this lesser value. However if the actual pollution load is much higher than carrying capacity then for practical and economic reasons, it may not be immediately achievable so steps will be required to work towards this over time.

Some of the possible steps include:

(1) Apply industry standards. In China there are norms of water use and discharge concentrations set for each industry type, similar to Best Available Technology (BAT) standards in the EU. These limits can be applied to the database of pollution discharges to estimate a resultant "Industry Standard" pollution load and this may be compared to current loads and carrying capacity loads.

(2) Apply proportionate reductions. Further reductions may be applied to each individual discharger until the total load meets the carrying capacity. Different dischargers may be reduced by different amounts depending on the practical circumstances of each enterprise and of local government policy.

Ultimately the aim is to reduce the pollution allocations loads to be no more than the carrying capacity. CC + PLAS can help to organise data and rapidly perform the complex calculations for matching pollution loads to carrying capacity. Fig. 6 shows the predicted quality in the case of these different scenarios being applied.

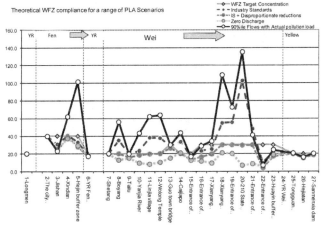

Fig. 6　WFZ theoretical compliance for COD concentration for different pollution load allocation scenarios at the WRPB 90 percentile flow at 20 ℃

However even setting discharges to carrying capacity will not guarantee that WFZ targets are met in all cases. Where there is a step change in water quality targets between upstream and downstream of a WFZ this may not be possible and also the final impact of the pollution discharges on the river is cumulative and combined with other non – point discharges and also unregistered discharges which may confound the efforts to control just the registered point sources.

The pollution load allocations could be managed on a seasonal basis, allowing more discharge during the warm wetter months than during the cold dry periods. This would require additional monitoring and supervision to enforce but may have less negative economic impact while achieving the general aims of the red line management system.

3. 2. 2　Pollution load allocation impact modelling – river basin water quality Model

CC + PLAS can be set into an alternative mode where the water flow and quality at the start of each reach is that of the upstream reach. Thus there is water quantity and quality continuity from one reach to the next to form a river basin quality model.

If this model is run for flow data that represent actual monitored situations then the flow and water quality data can be roughly matched to simulate the conditions of the river by adjusting the estimates of non – point inputs. These inputs are difficult to accurately calculate as the load reaching the river depends on the load generated and the proportion of that load that actually reaches the river. The source load may be estimated using data and a GIS to assign known sources to each model / WFZ catchment. This may include loads from rural and urban population, from agricultural runoff based on areas farmed and numbers of livestock, and from urban runoff. Also waste water treatment works may be included to indicate effective reduction in loads reaching the river. The "transport factor" of how much of each pollutant in each catchment actually gets to the river can be estimated by expert judgement and calibration to available data.

These factors can be adjusted for the various pollutants until, based on averaging of a few years of flow and water quality data in the river, the model's prediction of quality roughly matches the observed quality. This calibration process provides some insight into the likely proportions of load from the various sources.

Fig. 7 shows the results of this based on the annual average flows and quality for the 3 year period 2004 to 2006 for which data were available. The circles indicate the observed water quality in this period.

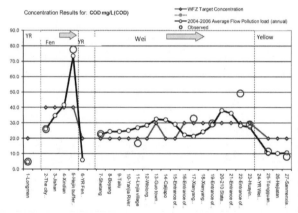

Fig. 7　Manually calibrated model

A process of revising the estimates of various pollution sources was used to manually calibrate the model to get a reasonable match between the model and the observed data. CC + PLAS has tools to assist this process by breaking down all of the pollution sources to their components and comparing impacts with monitoring data.

3. 2. 3　Pollution load allocation impact modelling – auto – calibrated river basin water quality model

CC + PLAS may be set to calculate the balancing flow and load between the estimated and the observed values and add this to the mass balance such that the model exactly matches the observed data.

These calibration flows represent the many unregistered (unknown) sources of pollution, sampling error and diffuse pollution, see Fig. 8.

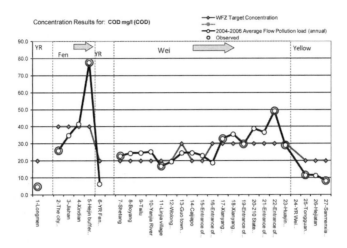

Fig. 8　Auto – Calibrated model

3. 2. 4　Applying CC + PLAS to simulate impact of pollution load allocation scenarios

Either the manually calibrated or the auto – calibrated model may be used to explore what would be the water quality under different pollution load allocation scenarios. Fig. 9 illustrates this for the manually calibrated model exploring the water quality impact at 75 percentile flows.

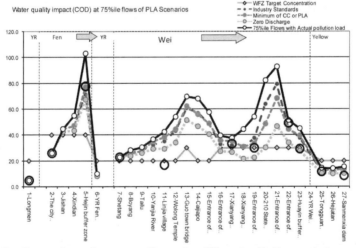

Fig. 9　Comparison of water quality under different pollution control scenarios
pollution control scenarios.

Fig. 10 then shows the simulated impact using the model calibrated against the 2004 to 2006 average data with the carrying capacity having been calculated for the seasonal compound flow and temperature scenario. This demonstrates that CC + PLAS is able to extrapolate from monitoring data to show the likely water quality compliance under different possible pollution load allocation scenarios.

This shows that while main course of the Yellow River is likely to comply with WFZ targets, the tributaries of the Fen and the Wei will struggle to meet WFZ targets even in the situation of zero discharges because of large amounts of non – point and unknown pollution sources. Under low flow conditions this situation will be more serious.

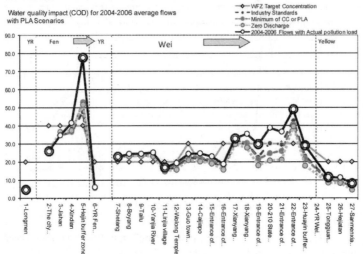

Fig. 10   Simulation of water quality impact of PLA scenarios in the case of a calibrated river basin water quality model

4   CC + PLAS as a river basin water quality planning tool

4.1   Forecasting future conditions

The CC + PLAS model incorporates the facility to allow for growth in population, industry and agricultural runoff, social economic development, improvement for pollution reduction technology, construction of wastewater treatment plant (WWTP) as well as other factors. This can aid planning for meeting targets for compliance levels by 2015, 2020 and 2030 etc.

4.2   River basin indexing and benefit cost

CC + PLAS incorporates a River Basin Pollution Indexing system (RBPI) which calculates the ratio between the actual (or predicted) water quality and the standard for each WFZ and then combines these to a weighted river basin index. This gives a quantitative measurement of the likely improvement that will be brought about by any pollution management strategy or scheme. It can also be used to quantify the impact of any single discharge on the overall quality of a river or whole basin. When compared against the cost of any given improvement this is an effective tool for cost benefit analysis and prioritisation of measures in a river basin water quality plan.

Such indexing can be used in the process of setting of key performance indicators for officials responsible for achieving river water quality improvements within the five year plan timetables

5   Conclusions

The main significance of the CC + PLAS model developed by the EU – China RBMP lies in the fact that it supplies a scientific and convenient tool for water resources protection and river basin planning, and enables practical operation and flexibility of pollution control schemes which promotes the implementation of the pollution control red line in water function zones. The model can also assist with active collaboration between water resources departments and environmental protection departments in terms of water resource protection. It can link the pollution control of industry and cities to the achievement of river water quality targets. The CC + PLAS model is a remarkable achievement of the EU – China RBMP after the five – year dedicated domestic promotion of the concept of integrated river basin management.

Acknowledgements

We gratefully acknowledge the significant contributions to CC + PLAS modelling work from the team members, from YRCC WRPB, Ms. Song Shixia, Mr. Zhang Shaofeng, Mr. Zhang Jianjun, Mr Wang Xingong, Ms. Yanli and Ms. Zhang Shikun, and from Atkins UK, Mr. Neil Upton and Mr. Cox Brain.

The CC + PLAS model development was undertaken as part of the EU – China RBMP which was funded by the EU.

Research on Multi-source Data Integration Method in the Process of Building the Yellow River Remote Sensing Analysis System Based on 3S

Zhao Shilai[1] and *Liu Guizhen*[2]

1. Yellow River Affair Bureau of Shandong, Jinan, 250011, China
2. Hydrology and Water Resources Bureau of the Yellow River in Shandong,
Jinan, 250011, China

Abstract: The data in the Yellow River remote sensing analysis system includes the paper topographic maps, bathymetric charts, estuaries map etc. geographic data, multi – temporal multi – sensor and different resolutions remote sensing data, the Yellow River runoff and sediment load , section of the Yellow River Estuary, water depth, estuarine bar etc. hydrological observation data. To meet the needs of the system, we need integrate these three types of data, to achieve data format conversion and normalization. Before the data integration, need formulate data standards for integration content such as multiple size, the projection, attribution data etc.. To be consistent with the local coordinate system and projection way of the system research area, this project adopts Dongying city local coordinate system, the central meridian longitude is east 118 °30'. In order to meet the feature requirements of the system, need to stratify the data, conduct symbolic processing for incoming data according to one of the benchmark documents of the Digital Yellow River, the digital map schema (SZHH04 – 2003) , and create an independent features symbols library. According to the integration standards, the integration process to the basis geographic data is: scan, registration, layered vectorization, attribution adding. Conduct Re – projection or geometric correction to remote sensing images, and pretreatment, and finally extract wetlands, rivers and coastline according to our needs. Hydrological data and river cross – section data should be edited in accordance with the format system needed. Data for scour analysis and kinetic studies which need consider about elevation changes, need to be normalized to data contains the 3D coordinates, using Surfer software. Coordinate system converted into a local coordinate system of Dongying City the system used, the last saved into shp format files. In order to get the data needed by the system, these data need to be edited several times collaboratively using Excel, Surfer, CASS, ArcMap and other software. In process of constructing Yellow River remote sensing analysis system based on 3S, according to the data requirements of the graphical information, space information and attributes information from system, these data are standardized and normalized integration. As it is found in long – term operation of this system, types of data obtained by this integration method can well satisfy the demand for this type of geographic information system.

Key words: data integration, data mining, format conversion, standardization

1 Data type analysis

The objective of the Yellow River remote sensing analysis system based on 3S is to establish a 3S space database integrated the Yellow River delta geography, hydrology, remote sensing, GPS etc. data, so as to realize visualization management and query of the modern Yellow River estuary 3S data, remote sensing data interpretation analysis , dynamics analysis of the estuary evolution process. It can be seen that, data system needed including basic geographic data, remote sensing data, the Yellow River runoff and sediment load and the Yellow River water depth observation data etc.

1.1 Basic geographic data

The basis geographic data including the Yellow River riverway topographic map scale of 10,000, the Yellow River estuary 25,000 water depth chart in estuarine bar area, 250,000 Yellow River estuary map and 100,000 Yellow River estuary 25,000 water depth chart in coastal area etc. data. For these data, the raw data is a variety of formats. Because of complex diverse acquisition means of these data, data with multiple time and space, multiple semantic, and the multi − scale expression, thus need data integration and data mining.

1.2 Remote sensing data

In order to meet the research of the riverway change, the coastline change, vegetation cover change etc. information, the system need to use a large number of multi − temporal remote sensing image. These images belong to different sensors, different resolution, span a long time, the earliest image got in 1976, the latest images got in 2005, covered the whole of the Yellow River delta area.

1.3 The Yellow River hydrological observation data

The Yellow River hydrological observation data collected in various hydrological station, river observation section and the bathymetric data near the estuary. Mainly include hydrological observation data, cross section observation data, bathymetric data and estuarine bar data etc. 4 types of data. The main storage means is excel spreadsheet or txt text format.

To meet the needs of the system, need integrate these three types of data, to achieve data format conversion and normalization. Data format conversion is to converse other formats data into the system data format through special data conversion program, and copy to the current system's database or file. This is the main way to integrate multiple format of current GIS software system.

2 The development of data integration standards

Since the data variety, data with multiple time and space, multiple semantic, and the multi − scale expression, before the data integration need formulate data standards for integration content such as multiple size, the projection, attribution data etc. .

2.1 The choice of a map projection

In the choice of system projection way, it should be consistent with the local coordinate system and projection way of the system research area. So it is easy to update and maintain the system information, increase reusability of the basic geographic information data, reduce the cost of repeated data collection. For this project, Dongying city local coordinate system is adopted, the central meridian longitude is east 118 °30′.

After determining projection way, there is geographic information basic framework used for data processing, for the topographic map scanning digitization, remote sensing image registration.

2.2 The topographic map stratification and symbolization

In order to meet the feature requirements of the system, need to stratify the data before the data processing. in accordance with the baseline document of the Digital Yellow River, one of the digital map type data storage symbolic creation of an independent coverings symbol library. The formulation of stratification standard should follow the national geographic information system stratification and the relevant literature provisions of "the Digital Yellow River". In accordance with the requirements of the system, conduct symbolic processing for incoming data. But the development environment do not have schema symbols library the project needed. According to one of the benchmark documents of the Digital Yellow River, the digital map schema (SZHH04 - 2003), create an independent features symbols library, symbols should cover independent symbols related to the Digital Yellow River, providing a convenient for the expansion of the system.

2.3 Standardization of database management

For adopting what management pattern to spatial data and attribution data, should design reasonable database, with E - R chart. The features attribution list system needed should be designed.

3 Data processing method

After development of data integration standards, can process data according to characteristics of different types of data. The core of the data processing is that different data sources should enter the GIS system. Specifically, is how to transfer, organize and manage original data. Here expound the key technology in the process of different data types.

3.1 Basic geographical data integration

The basic of geographic data is topographic maps in the existence of paper, which are in different scales. The basic workflows are as follows: scanning, registration, vectorization and adding attributes.

Registration should be done according to pre - established coordinate system and projection. Data vectorization should be done in hierarchical vector. Various types of GIS systems have their own special requirements, therefore, hierarchical requirements are widely different. Generally speaking, framework elements of geographic information are required. The other layers need to be classified according to user needs, which mainly divide into 10 layers to 30 layers. The modification of elements is the most tedious as well as the largest part of workload in data processing. There exist issues which need to be carefully modified, for example, the majority of linear features have a disconnection phenomenon, polygon features are not closed, annotation are in no property and so on. The amount of information is closely related to the graphic scale and purpose of the system, which requires well understanding of users needs. Overall, background or minor elements should be summarized integrated, while thematic elements to be detailed integrated. At the same time, graphical color should be reasonable matched. After process of images, you need to design attributes of tables and entry attribute information of all features shown in the images according to preliminary analysis of system requirements.

3.2 Remote sensing data integration

The use of remote sensing data can be divided into three aspects: firstly, as a raster map to navigate and query; secondly, the analysis of vegetation changes; thirdly, the extraction of the coastline and Yellow River shoreline. Because of different time, resolution and projection of original remote sensing images, it is necessary to re - project the images or geometric correct the image before processing. After correction, we should pre - process the image. The pre - processes of the image vary depending on the use of remote sensing images. In general, the processes of TM images include remote sensing image enhancement, histogram display and grayscale adjustment, linear transformation and nonlinear transformation, histogram equalization, normalization, matching, density segmentation, grayscale inversion, spatial enhancement, geometric correction of remote sensing images, radiometric correction of remote sensing images and histogram of the minimum value method.

After process of remote sensing image, we can extract wetlands, rivers and coastline in order to meet up with our needs.

3.3 Data integration of the Yellow River hydrological observations

Data file formats of the Yellow River Hydrological observations are generally as plain text or spreadsheet mode, and they can be said to vary widely. In order to find useful data, we need to analyze file – by – file before data integration.

Overall, processing target data of hydrological observation can be divided into two categories: one is dot pitch and elevation information without coordinates; the other is three – dimensional coordinates so as to graphic.

Process of hydrological data and river cross – section data is to extract given year data that system is in need from the water, sediment and hydrological data tables, and to edit data in accordance with format of specific system. The river cross – section data (Tab. 1) need to be extracted from the Yellow River cross – section data and twice – year – measured large cross – section data and edited into the Sectional drawing format so as to conform system. After adding corresponding IDs to sections, you should edit data into normalized format as Tab. 2 shown, then save as an excel file for system calls.

Tab. 1 The original recording format

1984 ~ 09 – 05	0	13. 91	10
1984 ~ 09 – 05	9	16. 49	10
1984 ~ 09 – 05	17	16. 51	10
1984 ~ 09 – 05	41	9. 51	10
1984 ~ 09 – 05	51	11. 5	10
1984 ~ 09 – 05	61	10. 96	10
1984 ~ 09 – 05	70	8. 67	10
1984 ~ 09 – 05	101	7. 48	10
1984 ~ 09 – 05	164	8. 83	10

Tab. 2 Processed data

1984 ~ 09 – 05	0	13. 91
1984 ~ 09 – 05	9	16. 49
1984 ~ 09 – 05	17	16. 51
1984 ~ 09 – 05	41	9. 51
1984 ~ 09 – 05	51	11. 50
1984 ~ 09 – 05	61	10. 96
1984 ~ 09 – 05	70	8. 67

After the end of data editing, you should draw all of the sections (Fig. 1), and compare data so as to check whether data is correct or not.

Fig. 1   sectional drawing based on normalized cross – section observational data

Data for scour analysis and kinetic studies which need consider about elevation changes on the space, you need to compile data in three – dimensional coordinates. Moreover, original data is in no regularity, which needs to be normalized. These data consist of table data of 36 fixed cross – section results, water regulation, sediment regulation, sand bar information and other data in the coastal area. Since using the 1954 Beijing coordinate system, these data should be processed in a uniform format, and be converted into a local coordinate system of Dongying City by software. After the coordinate transformation, you ought to edit data into a text file containing the x, y. Depth information in first line, then you can display and paint out three – dimensional coordinate data by adding "x, y data" in Arcmap, then output the data and save the temporary layer into shp file format , which can be used for system calculating. In this process, you are supposed to add projection information, add fields and edit the attribute information such as observational time, sounding line name and sounding line ID.

In order to obtain the data needed by the system, these observations need to be edited several times by collaboratively using Excel, Surfer, CASS, ArcMap and other software.

4   Conclusions

In process of constructing Yellow River remote sensing analysis system based on 3S(GPS,RS, GIS) , by analyzing different data obtained from basically geographic data, remote sensing image and hydrological observation data from multi – source, multi – temporal and multi – species, comprehensively using software such as CASS, Surfer, ArcView and so on according to the data requirements of the graphical information, space information and attributes information from system, these data are standardized and normalized integration. As it is found in long – term operation of this system, types of data obtained by this data integration method can well satisfy the demand for this type of geographic information system.

References

Zheng Chu. Graphics Data Processing of Digital Mapping [J]. Geospatial Information, 2007, 5 (2): 94 – 95.

A Decision Support System for Matching Irrigation Demand and Supply in the Near Real Time Environment

Mohsin Hafeez[1,2], *Kaleem Ullah*[3] and *Mahmood Khan*[2]

1. GHD Pty Ltd, 201 Charlotte Street, Brisbane QLD, 4000, Australia
2. School of Environmental Science, Charles Sturt University,
WaggaWagga, NSW 2678, Australia
3. NESPAK, Lahore, Pakistan

Abstract: This study deals with the application of a novel decision support system (DSS) for precise matching of irrigation demand and supply in a the near real time environment for the Coleambally Irrigation area (CIA), located in the southwest of New South Wales (NSW), Australia.

Forecasting of irrigation demand in the real time environment entails a complete understanding of the spatio – temporal variability of meteorological parameters and evapotranspiration (*ET*). For improved irrigation system management and operation, a holistic approach of integrating remote sensing derived *ET* from the SEBAL method with forecasted meteorological data and water use efficiency was used to forecast net irrigation demand. In order to capture the spatial variability, the CIA has been divided into 22 nodes based on direction of flow and connectivity. All hydrological data of inflow and outflow was estimated at all nodes of the CIA for the estimation of water use efficiencies. 10 Landsat 5 TM satellite images were used for mapping irrigated crops and estimation of actual *ET* for the summer cropping season of 2008 ~ 2009. This estimated actual *ET* and forecasted meteorological data was used for demand forecasting for 7 d. The results were compared with the data obtained for irrigation supplies. The methodology is very robust and cost effective for demand driven irrigation systems that have a good database, and daily demand can be forecast and updated by using remote sensing image analysis with minimum time input.

The outcome of irrigation demand forecasting on a daily basis for three summer months (December 2009 – February 2010) has been applied within the DSS by linking crop water demand with irrigation system management in the near real time environment. The developed model has been coupled with CICL's water ordering system and has been tested and refined for implementation across all spatial scales ranging from farm to irrigation system in the CIA. The developed DSS, featuring a web – based interface for near real time inquiry into the irrigation status from farm to system level, will enable irrigators to more closely match water application to crop water consumption within daily operational constraints in the CIA. This user friendly decision support model provides water managers with a deeper and more useful understanding about the irrigation demand and supply in the CIA.

1 Introduction

Water scarcity is rapidly becoming a vital issue for many countries in the developed and developing world, and could lead to a severe global water crisis (IWMI, 2009). It has been observed that in many parts of the world, water demand already exceeds supply, and a continuously rising population will result in water scarcity in many more areas of the world (Teixeira, 2008). Pressure to meet the growing water demands have resulted in greater competition among traditional water consumers namely agriculture, industry and cities for available water resources. Among all users, irrigated agriculture is the largest consumer, accounting for 70% of global water withdrawals; from this water, nearly 40% of the world's food is produced. Maintaining enough freshwater for agricultural production will become increasingly difficult in the near future due to climate change, growing water competition among various sectors and rapid population growth.

Water scarcity will force irrigated agriculture to produce more with less water in the future, which requires effective management of water.

Improving water management in irrigated areas and the assessment of irrigation performance are critical activities which require a complete understanding of all terms of the water balance at various spatial scales i. e. farm to basin levels (Khan and Hafeez, 2007). In such circumstances, water management in irrigation districts can be improved by analysing the irrigation water demand information (Pulido – Calvo et al. , 2007). Water demand information in irrigated areas is basic information for the development and implementation of successful tools for water resources management (Pulido – Calvo and Gutierrez – Estrad, 2009). For water demand estimation, it is important to understand how the irrigation system behaved in the past, what the current trends are, and what is expected in the future through accurate knowledge of various hydrological processes (ET, rainfall, runoff, seepage et al.) and different factors (land use changes) affecting these hydrological processes in space and time.

An appropriate and reliable irrigation demand forecasting tool based on a complete understanding of hydrological behaviours and novel remote sensing technologies can improve water management in an irrigation area. One of the main components of the water balance is evapotranspiration (ET), and its quantification is very important for matching irrigation demand and supply in any irrigation system, especially in arid and semiarid catchments where water shortage is a critical problem. ET is derived by solving an energy balance into various fluxes (net radiation, sensible heat, soil heat and latent heat). Conventional techniques to estimate ET employ only point measurements to estimate the components of the energy balance and can only be representative of local scales. Remote sensing can provide representative measurements of several relevant physical parameters at scales from a point to the whole system.

The potential of satellite image – based remote sensing for examining spatial patterns of regional ET has been investigated by a number of authors (Choudhury, 1989; Bastianssen 1995; Granger, 2000; Kustas and Norman, 2000; Menenti, 1993; and Su Z, 2002). These efforts have resulted in the development of remote sensing ET algorithms that are quite different in their spatial and temporal scales ranging from 30 m to 1,000 m and daily to monthly. Many methods for the estimation of actual ET have been developed by combining remote sensing data with ground based meteorological data for large areas.

This study deals with the application of a remote sensing based decision support system (DSS) for precise matching of irrigation demand and supply in a the near real time environment for the Coleambally Irrigation area (CIA), located in the southwest of New South Wales (NSW), Australia. This methodology for irrigation demand forecasting is based on spatial land use and land cover analysis, remote sensing derived actual ET (ET_a), crop coefficients and meteorological data at the node as well as the irrigation system level.

2　Study area

The Coleambally Irrigation Area (CIA) is located in the lower part of the Murrumbidgee River catchment in the southern part of the Murray Darling Basin (Fig. 1 (a)). The CIA contains approximately 79,000 hm^2 of intensive irrigation, 42,000 hm^2 of irrigation/dryland farms and 297,000 hm^2 of the Outfall District area, supplying water to 478 farms owned by 362 business units (CICL, 2007). Principal summer (November – April) crops grown include rice, soybeans, maize (corn), grapes, prunes, sunflowers and lucerne, while principal winter (May – October) crops include wheat, canola, sunflowers and lucerne. Pasture for grazing is grown in both seasons. Average rainfall is 396 mm per year and the average annual evaporation is 1,723 mm. The recent drought, a significant reduction in water allocations and concerns about climate change have all highlighted the need to manage water demand and supply more sustainably, especially in the CIA.

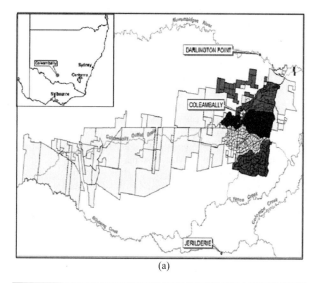

(a)

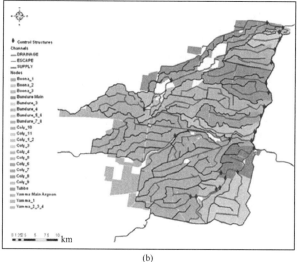

(b)

Fig. 1　(a) CICL operational area including CIA and
outfall district (CICL, 2009); and (b) spatial
distribution of nodes with supply and drainage
channels and selected control points

3　Material and methods

The CIA is a large irrigation area using a gravity flow system; water released from the dams takes seven days to reach the system, which makes the development of a robust methodology for water demand forecasting for the period of highest evaporative demand a necessity. High evaporative demand occurs in the months of December to February, when 75% of the area is planted to rice.

The irrigation system comprises of one main canal and a number of secondary and tertiary canals. As an initial step in water demand forecasting, spatial and temporal boundaries of the domain were specified and further divided into 22 nodes based on direction of flow and connectivity (Fig. 1 (b)). Due consideration was given to spatial distribution of rainfall, potential ET, soil texture and rooting depth of land cover.

The methodology for irrigation demand forecasting is based on spatial land use and land cover analysis, remote sensing derived actual ET (ET_a) and efficiencies, crop coefficients and meteorological data at the node and system level. To get information about the irrigated crops grown within the CIA and various nodes, land use and land cover classification was carried out using a simplified hybrid classification approach. This approach is based on a supervised classification algorithm in combination with false colour composite images, GIS based farm boundary data and water order data of each farmer from the SCADA system.

In order to get information about actual crop water consumption, the SEBAL model, based on the surface energy balance equation and developed by Bastiaanssen et al. (1998a & b), was applied to 10 Landsat 5 TM satellite images covering the period of October 2008 – March 2009. The pre – processing parameters required for the SEBAL model includes the Normalized Difference Vegetation Index (NDVI), surface emissivity, broadband surface albedo and surface temperature. The model was calibrated using the meteorological parameters measured from automatic weather stations at the selected farms. The daily ET_a was calculated from the instantaneous evaporative fraction, Λ, and the daily averaged net radiation, R_{n24}. The study provided ET_a maps on a monthly and seasonal basis for the entire study area.

For demand forecasting, irrigated crops in the CIA were clustered into rice and non – rice crops (others) in order to derive different hydrological and crop parameters. This developed methodology is not only based on analytical and empirical approaches, but also takes into account the physical processes as well as the spatial variability of different physical parameters influencing crop water requirement. A new and simplified methodology has been developed using remote sensing technology and field based estimates for forecasting irrigation demands for the CIA (Ullah, 2011). This methodology was applied in two steps; in the first step, efficiencies at the field scale as well as the system level were determined using remote sensing derived ET_a for different irrigated crops, and in the second step irrigation demand was forecast based on remote sensing derived crop coefficients and forecasted meteorological data. More details can be found in Ullah (2011) and Ullah&Hafeez (2011). The water use efficiencies adopted in this methodology are different from traditional efficiency terms used in irrigation system. The water use efficiencies defined here for both system and field or node levels were mainly used for the purpose of forecasting the demand for improvement of irrigation water management. These water use efficiencies were determined from local historical data of water use efficiencies, satellite based data of ET and actual applied water to the fields.

In this study, a DSS called Coleambally IRIS has been developed to represent a hydrological unit which has its own characteristics and requirements, and this DSS system will help irrigators to more precisely match water application to crop water consumption in a real time environment. The DSS helps farmers, managers and researchers to access real time hydrological data, and also helps to analyze the data for better water management. More detail on the Coleambally IRIS can be found in (Flugel, 2007) and Khan (2012). Three user categories, Farmer, Manager and Researcher are introduced in the Coleambally IRIS. Every user belongs to one of these categories. All user information including the user category is stored in PostgreSQL database tables. Based on the user login information (i. e. username and password) the DSS identifies the category of the user from the database and displays appropriate information. When a farmer accesses the web interface with his username and password (Fig. 2 (a)), he/she can only access information specific to his/her farm and other general information. Any information specific to another farm is not displayed due to strict privacy issues of each individual farmer. However, a manager or a researcher can access information for any farm, node or the whole irrigation area.

For example, after a farmer access the website, the DSS first shows the welcome page where the farmer can only see information, related to his farm only, such as area, perimeter, and soil type of the farm. The DSS also shows a map of the whole irrigation area with his farm map highlighted (Fig. 2(b)).

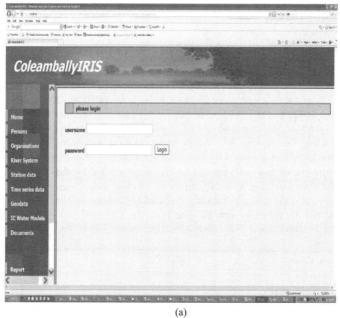

(a)

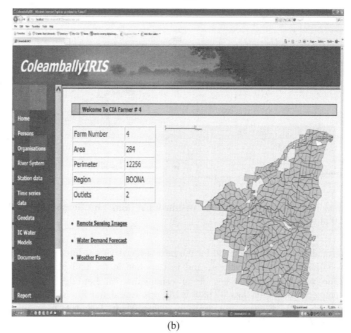

(b)

Fig. 2　(a) User login screen in the coleambally IRIS ; and
　　　　(b) Welcome page for a farmer on the coleambally IRIS

The DSS allows all users, including farmers, to access any general information. Due to higher access rights, managers and researchers can access information for any specific farm as well as the whole irrigation area. A farmer can only access relevant data, but he/she is unable to upload any data into the database.

4 Results and discussions

The irrigated area was mapped using the Landsat TM satellite image (spatial resolution of 30 m) of November 3, 2008 (Fig. 3(a)). Irrigated crops were classified into rice and other summer crops. Only 3.8% of the area was under all crops during summer 2008 ~ 2009, and rice was grown on an area of just 1,025 hm^2.

The seasonal actual evapotranpiration (ET_s) was estimated using 10 Landsat 5 TM images for 2008 ~ 2009, while the missing ET_a values were obtained by daily calculation of reference evapotranspiration (ET_o) as proposed by Tasumi et al. (2000). The output of SEBAL is a pixel based spatially distributed seasonal estimate of ET_a for the CIA as shown in Fig. 3(b). The seasonal actual ET values range from 20 mm to 1,705 mm with a mean value of 159 mm and standard deviation of 216 mm. To validate the remote sensing results, a comparison of ET_a produced by the SEBAL algorithm with ET data from the AWS and two eddy covariance systems installed in local rice and maize fields shows good agreement within the expected range (Ullah, 2011).

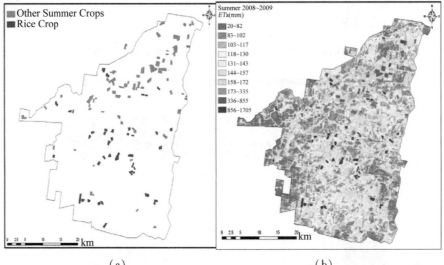

(a)   (b)

Fig. 3   (a) **Maps of irrigated crops within CIA; and** (b) **Seasonal actual ET of 2008 ~ 2009**

In this study, daily irrigation demand for the next seven days was forecast using remote sensing based water use efficiencies and K_c values, and was applied at the individual node as well as the system level. The water use efficiencies are a combination of field level water use efficiency (FWUE) and conveyance system loss. The FWUE was calculated for both rice crops and other crops. For rice fields, FWUE varies from 63% in YammaMain_Argoon to 81% in Tubbo and the average FWUE was 71%. Similarly for other crops, the highest and lowest water use efficiency was 81% for Coly 5 and 97% for Bundure 4 and the average efficiency was 89%. Overall, the average FWUE for rice fields was 71% and 89% for other crops at the system level.

The conveyance system loss based on the area under each node can provide reasonable estimation for water demand forecasting in systems that have fixed conveyance system loss regardless of the quantity of water that flows through system. For proportional distribution of conveyance system loss over the entire nodes, the average conveyance system loss of high evaporative demand months (i. e. December, January and February) was estimated using the average monthly value of these months. The conveyance system loss per day varies from 0. 79 mL for Coly 1_2 to 17. 30 mL for Yamma 2_3_4. This variation in daily conveyance system losses depends entirely on the total area of the node; the larger the area, the greater the daily conveyance system loss.

Similarly, the calibrated images of ET_a were used for the estimation K_{c_act} at the pixel level over each node for rice crop and other crops separately. The K_{c_act} values for rice vary from a minimum of 0. 86 in Coly 9 for the image of 3rd November 2008 to a maximum of 1. 06 for YammaMain_Argoon for the image from the 22nd January 2009. The average value derived for rice by Meyer et al. (1999) in this area ranges from 0. 9 at initial stage to 1. 1 at crop development and mid – season stage. The K_{c_act} values for other crops exhibited very high variability across the nodes and even within the nodes because of different crop combinations. It varies from 0. 08 in Bundure 7_8 for the 22nd January 2009 image to 1. 20 in Bundure 3 for the 31st January 2009 image. The lowest value is related to pasture which was completely finished by January in Bundure 7_8, while the highest value was found for lucerne in Bundure 3.

Results of irrigation demand forecasting indicate that the net irrigation demand forecast for fields at the node level (NIDFFN) shows a relatively high difference across the nodes. However net irrigation demand forecast for fields at the system level (NIDFFS) shows no significant difference and has good agreement with actual water being delivered to fields. For 2008 ~ 2009, NIDFF at the system level (NIDFFS) was 4,364 mL, 9,387 mL and 9,073 mL with demand supply ratio (DSR) of 0. 69, 1. 06 and 1. 20 for December 2008, January 2009 and February 2009 respectively. In terms of net irrigation demand forecast at system level NIDFS, making an allowance for conveyance system loss, results indicate that monthly forecasted irrigation water demand was higher for the selected months of 2008 ~ 2009 compared to actual water diverted into the system (Fig. 4). It was found that the forecasted demand was higher by 57%, 25% and 22% in December, January and February respectively.

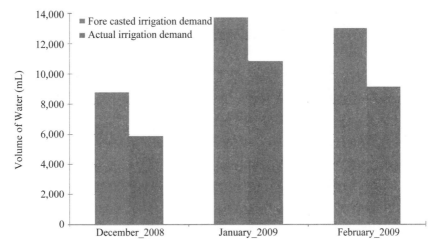

Fig. 4 Comparison of forecasted irrigation demand and actual diverted water in the high evaporative months of 2008 ~ 2009

The main reason for the higher values of net irrigation demand at the system level is due to the use of a fixed value of conveyance system loss to maintain system operation. The fixed value of conveyance system loss is proportionally significant as compared to total water diverted due to extremely limited irrigation supplies in this season. In addition to conveyance system loss, the lower forecasted demand is attributed to lower values of NIDFFN and NIDFFS, which was influenced by a lag time in updating K_{c_act}, inaccurate assessment of groundwater, change in water use efficiencies and remote sensing image analysis for ET_a estimation. However, overall the demand forecasts have reasonable agreement with actual water diverted to the system and can help in improving irrigation water management.

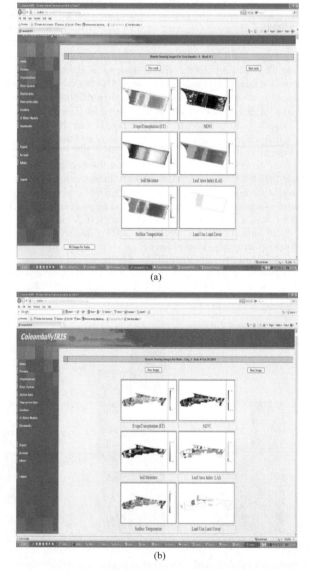

(a)

(b)

Fig. 5 (a) **Automatic Clipping and Masking of the Remote Sensing images for a farm within the CIA; and (b) Automatic Clipping and Masking of the Remote Sensing images for a node within the CIA**

Moreover, when a farmer wants to view the masked images of his farm, the DSS interface initially asks for a period. The DSS contains images of various parameters including vegetation health (e. g. Leaf Area Index (LAI) , and Normalized Difference Vegetation Index (NDVI) , Surface Temperature, Land use and land cover, actual crop water consumption and soil moisture. The farmer can choose a date such as 30 October 2009 using a drop down menu. The DSS then performs automatic clipping and masking of 30 October 2009 image of the whole irrigation area and shows only the masked image of the above mentioned parameters, which is only relevant to his/her farm Fig. 5(a). Additionally, the DSS allows the farmer to traverse through the images of different periods by using "prev" and "next" buttons and access all masked images of the previous or next available dates. This will be helpful for a farmer or manager to compare the changes happening in his/her farm over the times or crop stages. In the same way managers can view the information about all the farms, nodes and the irrigation area as shown below in Fig. 5(b) and Fig. 6.

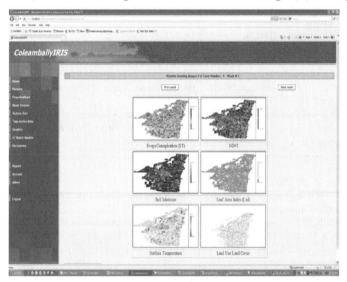

Fig. 6　Automatic Clipping and Masking of the Remote Sensing images for the CIA

5　Conclusions

The applied methodology is very simple and cost effective for a demand driven irrigation system that has a good database, and daily demand can be forecasted and updated by using remote sensing image analysis with minimum time input. This demand forecasting tool is based on a sound understanding of hydrological behaviour, novel remote sensing technology and forecasted meteorological data, and is useful for improved irrigation water management ranging from node to system levels. It may reduce the risks associated with over and under irrigation application by more accurately matching demand and supply. This methodology for demand forecasting will be tested for another cropping season in the near future.

In this paper we have presented a novel Decision Support System called Coleambally IRIS that has been developed for Researchers, Managers and Farmers in the CIA, NSW, Australia. The DSS has been developed by making necessary modifications to an existing DSS called AIDIS. This DSS is being presented to the farmers and managers of the CIA through a training program, as it will be practically used by the farmers and managers of CIA to refine water management decisions in the region. The DSS will help the CIA farmers and the managers in decision making by providing relevant information on the actual water use in a real time environment. Through the real life use of this DSS we expect to obtain useful feedbacks, which will be helpful for its further development.

References

Bastiaanssen W G M, Menenti M, Feddes R A, et al. A Remote Sensing Surface Energy Balance Algorithm for Land (SEBAL): Part 1. Formulation[J]. J. Hydrology, 1998, 212-213, 198-212.

Bastiaanssen W G M, Menenti M, Feddes R A, et al. The Surface Energy Balance Algorithm for Land (SEBAL): Part 2 Validation [J]. J. Hydrology, 1998, 212-213: 213-229.

Choudhury B. Estimating Evaporation and Carbon Assimilation Using Infrared Temperature Data: Vistas in Modeling[J]. Theory and Applications of Optical Remote Sensing: 1998, 628-690.

CICL. Annual Report, Coleambally Irrigation Cooperative Limited (CICL) [R]. Australia, 2007.

CICL. Annual Compliance Report, Coleambally Irrigation Cooperative Limited (CICL) [R]. Australia, 2009.

Flugel W A. The Adaptive Integrated Data Information System (AIDIS) for Global Water Research [J]. Water Resource Management, 2007, 21: 199-210.

Granger R Satellite-derived Estimates of Evapotranspiration in the Gediz Basin [J]. Journal of Hydrology ,2000, 229(1-2): 70-76.

IWMI. Water for A Food Secure World. International Water Management Institute (IWMI) Strategic Plan 2009-2013,2009.

Ullah M K. Ageoinformatics Approach for Spatial Water Accounting and Irrigation Demand Forecasting for a Gravity Irrigation System [D]. Charles Sturt University, 2011.

Kustas W, Norman J. A Two-source Energy Balance Approach Using Directional Radiometric Temperature Observations for Sparse Canopy Covered Surfaces [J]. Agronomy Journal, 2000, 92(5): 847.

Khan S, Hafeez M M. Spatial Mapping of Water Productivity in Irrigation System Using Geo-information Techniques[R]// International Congress on Modelling and Simulation. University of Canterbury, Christchurch, New Zealand, 2007.

Meyer W S, Smith D J, Shell G. Estimating Reference Evaporation and Crop Evapotranspiration from Weather Data and Crop Coefficients [R]. An Addendum to Research Project 84/162 Quantifying Components of Water Balance under Irrigated Crop. CSIRO Land and Water, Technical Report 34/98.

PulidoCalvo I P, Montesinosb P, Rolda'nb J, et al. Linear Regressions and Neural Approaches to Water Demand Forecasting in Irrigation Districts with Telemetry Systems [J]. Biosystem Engineering, 2007, 97: 283-293.

Pulido-Calvo I, Gutierrez-Estrada J C. Improved Irrigation Water Demand Forecasting Using Soft Computing Hybrid Model [J]. Biosystems Engineering, 2009, 102: 2;202-218.

Su Z. The Surface Energy Balance System (SEBS) for Estimation of Turbulent Heat Fluxes [J]. Hydrol. Earth Syst. Sci, 2002, 6: 85 – 99.

Tasumi M, Bastiaanssen WGM, Allen R G. Application of the SEBAL Methodology for Estimating Consumptive Use of Water and Stream Flow Depletion in the Bear River Basin of Idaho through Remote Sensing [R]. Appendix C: A step-by-step guide to running SEBAL. EOSDIS Project Final Report, The Raytheon Systems Company and the University of Idaho, USA, 2000.

Teixeira A H de C. Measurements and Modelling of Evapotranspiration to Assess Agricultural Water Productivity in Basins with Changing Land Use Patterns, A Case Study in the San Francisco River Basin, Brazil[D]. University of Delft, Netherlands, 2008

Ullah K, Hafeez M M. Real Time Irrigation Demand Forecasting Using Remote Sensing and Meteorological Data in Semi-arid Regions [J]. Remote Sensing and Ground-based Methods in Multi-Scale Hydrology International Association of Hydrological Sciences (IAHS) Red Book Series, Publ, 2011, 343: 157-162.

Principle and Verification of the Semi – scale Experimental Method for Soil Conservation

Xu Xiangzhou [1], *Zhang Hongwu* [2] and *Wang Guangqian*[3]

1. School of Civil & Hydraulic Engineering, Dalian University of Technology, Dalian, 116024, China
2. Department of Hydraulics and Hydropower Engineering, Tsinghua University, Beijing, 100084, China
3. Department of Soil and Water Conservation, Chung – Hsing University, Taichung, 40227, China

Abstract: Similarity plays an important role in the soil erosion experiments. This study presents a semi – scale method to simulate the processes of soil conservation by check dams in a small watershed on the Loess Plateau of China. During experiments, the model – prototype ratio of geomorphic variables was kept constant under each rainfall event. Consequently, experimental data are available for verification of soil erosion processes in the field and for predicting soil loss in a model watershed with check dams. Thus it can predict the amount of soil loss in a catchment. This study also mentions four criteria, similarities of watershed geometry, grain size and bare – land, Froude number (Fr) for rainfall event, and soil erosion in down – scaled models. The efficacy of the proposed method was confirmed using these criteria in two different downscaled model experiments. The B – Model, a large scale model, simulates watershed prototype. The two small scale models, D_a and D_b, have different erosion rates, but are the same size. These two models simulate hydraulic processes in the B – Model. When k_{DB}, the relative efficient of the D_a – Model to B – Model was 0.4 in these experiments, the amount of soil loss in the D_a – Model was very close to that in the B – Model, and the difference between two models was less than 50%, except the first rainfall event, in which field rainfall intensity (2.70 mm/min) in the D_a – Model was 63.9% more intense than the designed intensity (1.65 mm/min) resulting from human error. When k_{DB} was 0.4 in these experiments, the amount of soil loss in the D_b – Model was close to that in the B – Model, and the difference between two models was less than 55%, except for the second rainfall event (soil loss error was 81.2%). Moreover, the experimental results show that the "prototype" deposition depth was strong correlated with model deposition/erosion depths. Obviously, with a semi – scale physical model, experiments are available to verify and predict soil loss in a small watershed area with check dam system on the Loess Plateau, China.
Key words: loess plateau, check dam, semi – scale model, similarity condition, soil loss

1 Introduction

Under laboratory conditions, measurements are more accurate and many experiments can be conducted (Cerdà et al. , 2002). Thus, downscaled models are currently common in many different engineering fields, such as hydraulics and river engineering. The advantages of such models are well known (Zhang, 1994). However, few studies have simulated the process of soil loss using downscaled model experiments, because simulating hydrological, morphological, and geological conditions is extremely complex. Nevertheless, dimensional analysis links various observed phenomena of erosion and deposition into a unified process, and enables complete predictions in form changes to be anticipated when watershed treatment is altered (Strabler, 1958). Downscaled models are typically used to simulate physiognomy performance (Jin et al. ,

2003; Hancock et al. , 2003). Hancock and Willgoose (2004) investigated the effect of erosion on a back – filled and a capped earthen dam wall by constructing an experimental landscape simulator in the laboratory. Due to the design of the rainfall simulator, it is difficult to directly scale the rainfall runoff processes to the field. Consequently, no attempt has been made to match rate of gully development on the tailings dam to field – scale processes. Recently, downscaled model experiments on soil erosion in small watersheds of the Loess Plateau have been performed, and good progress has been made in similarity methodology, e. g. Shi et al. (1997a; 1997b), Jiang et al. (1994) and Yuan et al. (2000a; 2000b). These experiments were relatively successful; however, some aspects of the model design are still in dispute (Zhang et al. , 2000). The Chinese government has paid considerable attention to the theory underlying scale models for soil conservation on the Loess Plateau. The theory on model – based Loess Plateau has been proposed, which includes prototype, digital model and physical model (Li, 2001). National foundations have been given the task of developing the theory for designing a physical model, and especially the scale model experiments, that simulate soil erosion processes in small watersheds on the Loess Plateau.

2 Principle of the semi – scale method

2.1 Purpose and course

The proposed method predicts the effects of sediment retention by check dams based to rainfall, land cover and geological conditions in the prototype watershed before check dams are constructed. Suppose that a small watershed on the Loess Plateau must be managed, and rainfall – erosion data before check dam construction were available. To predict erosion while a check dam system is being constructed, we assume that other erosion factors (e. g. , rainfall, erosion material, and plant cover) are similar to those before the check dams were constructed.

In this study, four criteria are utilized to ensure that soil loss scale number remains constant. Firstly, for geometrical similarity, the initial dimensions of the landform, including the check dams, are scaled down according to the dimensions of the prototype watershed with the same proportion in the horizontal and vertical orientations. Secondly, for similarity of erosion form, simulated soil similar to that of the prototype is used, and model rainfall intensity exceeds the soil erosion threshold. Thirdly, rainfall duration was determined by Froude number (Fr) similarity. Finally, for similarity of rainfall – erosion, the relation between rainfall and erosion in the model experiment corresponds to that of the prototype. The proposed method provides the quantitative proportion of soil loss between the prototype and model, using similarity criteria. Nevertheless, it did not strictly meet the similar situation in the conventional scale model experiment. Thus, the proposed method is defined as the semi – scale physical model experimental method (SSPM).

The SSPM was tested using two downscaled experiments. A large scale model, the B – Model, at 1∶240 of the prototype landform, was used as the simulated prototype watershed. Two small scale models, D_a – Model and D_b – Model, with different erosion rates but the same geometrical size, each at 1∶960 of the prototype landform, were employed to simulate the hydraulic process in B – Model. Data from these three models are comparable because the experimental devices and observation measures in the small scale models are consistent with those in the B – Model. Moreover, an across – the – board comparison was possible as the soil erosion processes in the three models were all implemented in the laboratory using simulated rainfall.

2.2 Similarity requirements

The degree of similarity between the prototype and model is based on geometrical, kinematic and dynamic similarity. Geometrical similarity in a model is achieved when each geometrical length L_p in the prototype has a constant ratio corresponding to the model length L_m in the model. Subscripts p and m are a prototype parameter and model parameter, respectively. This ratio is called the length scale number λ_L of the model ($\lambda_L = L_p/L_m$). In the proposed downscaled model experiment, horizontal length scale is equal to the vertical length scale; therefore, the dam height scale number λ_H equals the length scale number. Likewise, due to geometrical similarity, the watershed area scale number λ_A could be calculated as λ_L^2, where subscript A is watershed area.

Dynamic similarity indicates that the corresponding forces in the prototype and model must have a constant ratio. For flows under gravity (e. g. , free surface flows) require geometrical similarity and equality of the Froude number (Fr) in model and prototype $\lambda_{Fr} = 1.0$. In a field observation or laboratory experiment, the gravity effect is greater than viscosity for turbulent flows caused by rainfall; consequently, the time scale number of rainfall duration, λ_t, is related to Fr similarity; that is,

$$\lambda_t = t_p / t_m = \lambda_L^{0.5} \tag{1}$$

where, t is rainfall duration, s.

The emphasis of the proposed method is that the ratio of geomorphologic evolvement ratio of the model corresponds to the prototype ratio, which remains constant after rainfall events, such that the prototype soil erosion processes can be measured based on model experimental results:

$$\lambda_{q_{sd}} = \lambda_{\overline{Y}} / \lambda_L = \frac{\overline{Y}_{p(1)}}{L_{p(1)}} \bigg/ \frac{\overline{Y}_{m(1)}}{L_{m(1)}} = \frac{\overline{Y}_{p(2)}}{L_{p(2)}} \bigg/ \frac{\overline{Y}_{m(2)}}{L_{m(2)}} \cdots = \frac{\overline{Y}_{p(i)}}{L_{p(i)}} \bigg/ \frac{\overline{Y}_{m(i)}}{L_{m(i)}} = k \tag{2}$$

where, i is the sequence of rainfall events; $\overline{Y}_{(i)}$ is the mean erosion/deposition depth of the landform during rainfall events, m; L is the length of the small watershed, m; Y/L is a dimensionless term, which stands for the geomorphologic evolvement ratio of a single rainfall event; and k is the ratio of erosion extent between the model and prototype. when $k = 1$, the erosion/deposition depth scale number, λ_Y, equals the length scale number, λ_L, illustrating that the erosion/deposition extent Y_p / L_p for a single simulated rainfall event in the prototype is equal to the product of the scaled model Y_m / L_m. However, when $k \neq 1$, a model is distorted because the erosion/deposition scale differs from the length scale of the physical soil erosion model. Thus, the length scale number, λ_L, in the scaled model can be obtained by multiplying the erosion/deposition depth scale number, λ_Y, by the ratio k.

Since simulated soil in the model experiment is similar to that in the prototype, the Loess Plateau, the soil density ρ_s (kg/m^3) in the prototype and model is similar. Thus, the soil loss scale number could be calculated as

$$\lambda_s = \frac{S_p}{S_m} = \frac{A_p \times \overline{Y}_p \times \gamma_{0p}}{A_m \times \overline{Y}_m \times \gamma_{0m}} = \lambda_L^3 \cdot k \tag{3}$$

when $k = 1$, the soil loss scale number, λ_S, equals the volume scale number, λ_L^3, which could estimate prototype soil loss volume based on the volume scale number in the model experiment. when $k < 1$, model erosion volume must be multiplied by the coefficient $k \cdot \lambda_L^3$ to estimate the amount of soil loss in the prototype, and the model landform varies more serious than the prototype landform under the same rainfall scale. when $k > 1$, the model erosion volume must be multiplied by a large coefficient to estimate the amount of soil erosion in the prototype. A comparison of sediment transport capacity and sedimentation similarity in river engineering indicates that the conditions of kinematic similarity and dynamic similarity are embodied in the scale of cumulative erosion volume for each rainfall event (Eq. (3)). Four measures are required to ensure that the ratio R is constant for the semi − scale model: ①the bare − land model with a corresponding erosion rate is employed in the experiment. ②the dimensions of the landform, including the check dams, are scaled according to the prototype watershed with the same proportions in the horizontal and vertical orientation. ③the soil is similar to that in the prototype. ④the initial water content in the soil layer before each simulated rainfall event remains constant.

In this study, if the ratio of the geomorphologic evolvement ratio in the small scale model to the corresponding ratio in the large scale model remains constant after rainfall events, then k_B / k_D is a constant; that is

$$K_{DB} = \frac{\overline{Y}_{B(1)}}{L_{B(1)}} \Big/ \frac{\overline{Y}_{D(1)}}{L_{D(1)}} = \frac{\overline{Y}_{B(2)}}{L_{B(2)}} \Big/ \frac{\overline{Y}_{D(2)}}{L_{D(2)}} \cdots = \frac{\overline{Y}_{B(i)}}{L_{B(i)}} \Big/ \frac{\overline{Y}_{D(i)}}{L_{D(i)}} = \frac{k_B}{k_D} \qquad (4)$$

where, k_{DB} is defined as the relative ratio of erosion extent between the small scale model and the large scale model; in short, relative ratio. In this study, two small scale models with $k_{DB} \approx 1$ and $k_{DB} \neq 1$ were used to verify the SSPM, respectively.

3　Validation of the SSPM

3.1　Methods and materials

　　The experiments were performed in the laboratory of the Yellow River Research Center, Tsinghua University, Beijing, China. Three experimental model landscapes, a large scale model (B – Model) and two small scale models (D_a – Model and D_b – Model) were developed. ①Each model was a small watershed that included elemental geomorphologic units (e. g. , gully, platform, and hillslope) with characteristics of small watersheds typical in the Loess Plateau. Each of the small scale models was a miniature version of the large scale model. ② Loess was used as erodible material in both models. ③The same instruments were used for measuring rainfall, soil loss and runoff in all models to ensure data comparability.

　　The prototype catchment of the rainfall experiments is abstracted according to the common erosion characteristics of the Loess Plateau. The conceptual watershed is also located in the *Loess Hill Ravine Region*, but it covers an area of 3. 32 km². The total length of the main gully is 3 km, the total height of 694 m and the gully bed is 960 m above from the outlet, with a "V" – shaped gradient of 2. 3%. Soil of the ground surface, the average annual runoff rate, and average annual soil loss rate in this small watershed are all correspondingly same to those in the Yangdaogou Catchment. The soil loss of the conceptual prototype catchment in a rainfall event, S_p is $6. 87 \times 10^8$ kg/a.

　　The landscape simulator for the B – Model consisted of a rainfall simulator suspended above a flume containing the large scale watershed model. The other two landscape simulators for the small scale models, the D_a – Model and D_b – Model, of which relative ratio R_{DB} was not equal to 1 and was close to 1, respectively, were used as to verify the SSFM. Experimental apparatus and techniques in this study have been used extensively to examine watershed behavior at a model scale, and could be used to generate model watersheds with many features that are similar to field watersheds (Xu et al. , 2006). During each rainfall event, the water – solid mixture was collected in a calibrated tank at the outlet of each plot. At 2 min or 1 min intervals, runoff samples were collected in 100 mL sampling bottles to determine the sediment concentration. Flow rates were synchronously measured using the label on the wall of the calibrated tank/bucket. After a rainfall event, the bed load deposited at the bottom of the tank was dried and weighed, and the weight of the suspended load was calculated by the concentration determined gravimetrically, and the volume was measured using the bucket gauge.

　　Several groups of experiments were conducted to determine model rain erosivity. Except that the initial landscape for the first rainfall event in a group of experiments was made by hand patting, the other subsequent landscapes were formed by the previous rainfall event. ①Two groups of experiments were conducted for the B – Model. Firstly, 7 rainfall events, generated by the SX2004 Sprayer – styled Rainfall Simulator with intensities of about 1. 60 mm/min and a duration of 20 min, were applied to the initial ground cover of the B – Model without check dams. Then recovered the landform to the initial state, and gave 10 rainfalls same to the above mentioned while 12 check dams were orderly constructed (as shown in Tab. 1). ②Two groups of experiments were conducted for the D_a – Model. Firstly, 6 rainfall events, generated by the SX2002 Net – styled Rainfall Simulator, had intensities of about 1. 65 mm/min and a duration of 10 min, were applied to the initial ground cover of the D_a – Model without check dams. The landform was then returned to the initial state. Then, 10 rainfall events, with the same properties as above mentioned, while 12 check dams were orderly constructed (Tab. 1). ③Two groups of experiments were performed for the D_b – Model. Firstly, 6 rainfall events, generated by the SX2004 Sprayer – styled Rainfall Simulator, had intensities of about 2. 45 mm/min and a duration of 10 min, were applied to the initial ground cover of the D_b – Model which was made of mixed soil and cultivated slopes, before check dams were constructed. The landform was then returned to the initial state, and 10 rainfall events, which had the same properties as above mentioned, while 12 check dams were orderly constructed (as shown in Tab. 2).

3.2   Results and discussion

3.2.1   Validation of soil loss scale number

Rainfall – erosion data for the natural watershed, especially those after constructing dams, are so scarce that the reliability of the SSPM had to be verified by comparing the experimental results for the D_a – Model and D_b – Model with those for the B – Model.

To validate the soil loss scale number as $k_{DB} \neq 1$. The D_a – Model was used to predict soil loss in the B – Model when check dams were constructed in an orderly fashion. Thus, k_{DB}, the relative ratio of the D_a – Model to B – Model was 0. 40 in these experiments. All soil losses in the model experiments were converted to prototype values by multiplying the soil loss scale number. The relative differences in soil loss for the corresponding rainfall events between the B – Model and D_a – Model, which represent the accuracy of the SSPM, were calculated using the following equation:

$$e = \frac{S_{PD} - S_{PB}}{S_{PD}} \times 100\% \tag{5}$$

where, e is the relative differences of soil losses; S_{PD} is the soil loss at the prototype scale that was transferred from the D_a – Model, $S_{PD} = S_D \times \lambda_{S_D}$; and S_{PB} is the soil loss at the prototype scale that was transferred from the B – Model, $S_{PB} = S_B \times \lambda_{S_B}$. Consequently, the amount of soil loss in the D_a – Model was very close to that in the B – Model, and the difference between two models was less than 50% (Tab. 1), except the first rainfall event, in which field rainfall intensity (2. 70 mm/min) in the D_a – Model was 63. 9% more intense than the designed intensity (1. 65 mm/min) resulting from human error.

To validate the soil loss scale number as $k_{DB} \approx 1$. The D_b – Model was used to predict soil loss in the B – Model when check dams were built in an orderly fashion. The value of k_{DB} was 0. 86 in these experiments. All soil losses in model experiments were converted to prototype values by multiplying the soil loss scale number. The relative difference of the soil losses in the D_b – Model caused by the difference of the rainfall intensities was counteracted using Eq. (5). Thus, the amount of soil loss in the D_b – Model was close to that in the B – Model, and the difference between two models was less than 55% (Tab. 2), except for the second rainfall event (soil loss error was 81. 2%). The absolute values of soil loss errors in the D_b – Model were larger than those in the D_a – Model, because increased error was caused when soil loss was observed due to the small soil loss during a run of rainfall event on the D_b – Model.

The precision in predicting the effect of intercepting sediment in the models was sufficiently good for model experiments on soil conservation. It could be concluded that the experimental method can be used to predict the volume of soil loss in the small watershed of Loess Plateau.

3.2.2   Qualitative Analysis of Erosion Depth

To compare erosion trends in the model gullies with that in the prototype, mean gully elevation in the D_a – Model and D_b – Model after each rainfall event was converted to that in the B – Model using the scale – modification method. Notably, if the B – Model is regarded as the simulated "prototype" and the D_a – Model as the downscaled model, then the length scale number is 4, and the soil erosion/deposition depth scale number is 1. 6 by Eq. (2); however, if the B – Model considered the simulated "prototype" and the D_b – Model as the downscaled model, then the length scale number is 4, and the soil erosion/deposition depth scale number is 3. 5 by Eq. (2).

The "SN5. 1" is a branch gully without check dams. The mean elevation of this gully bed after each rainfall event is shown in Fig. 1. Elevations in the experiments with the D_a – Model and D_b – Model were converted to corresponding values at the B – Model scale. The "prototype" deposition depth was strong correlated with model deposition/erosion depths. Thus, we conclude that the geomorphological evolvement trend of both scaled models was similar.

Conversely, according to regulations governing techniques for controlling erosion in gullies on the Loess Plateau (MWRC, 1996), the design deposition life of a medium sized check dam is 5 ~ 10 years, and that of a small sized check dam is only 5 years.

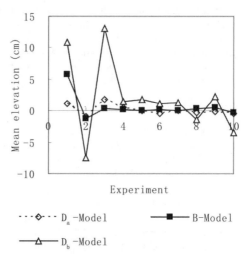

Fig. 1 **Erosion – deposition in gully SN5.1m where no dam was constructed. Depths in the D_a – Model and D_b – Model were converted to corresponding values in the B – Model scale**

Field data demonstrated that the deposition occurring on the dam – land is extremely rapid when hyper – concentrated flooding occurs. In this study, all check dams in the small scale model (the D_a – Model and D_b – Model), and almost all check dams except Dam 1, had the largest reservoir capacity in the B – Model and filled up after only one rainfall event. Even on dam – land of Dam 1, most deposition occurred during the first rainfall event, whereas the dam – land filled up after 2 rainfall events. Experimental results illustrate that deposition velocity on the dam – land simulated using model experiments was close to that in the prototype watershed on the Loess Plateau.

Nevertheless, additional laboratory experiments for natural watersheds on the Loess Plateau are required to further test the feasibility of the SSFM. This is a scientific challenge for future research.

4 Conclusions

The SSFM was employed to predict the effect of retaining sediment by check dams, which did not strictly meet the same conditions in classical scaled model experiments, but provided the quantitative proportion of soil loss between the prototype and models. When a check dam system is needed to retain sediment transport, it could be simulated using a downscaled model experiment. Data for the prototype watershed before dams are constructed should be utilized, including the rainfall, land cover and geological conditions of the prototype watershed. This study constructed the initial landform model in the laboratory, and applied simulated rainfall to the model. Simulated rainfall intensity was adjusted to conform to the ratio R. Finally, check dams were constructed according to the planned sequence and construction interval for the check dams for each simulated rainfall event. Soil loss accompanying each rainfall event was transferred to the prototype, which predicted the effect of retaining sediment by check dams.

To test the SSFM, two types of downscaled experiments were designed according to the prototype watershed. A large scale model experiment was used as the simulated prototype watershed, and two small scale model experiments with different erosion rates were utilized to verify the hydraulic processes in the B – Model. Soil losses in these models were all transferred to the values in the prototype scale according to the soil loss scale number, and were then compared to assess the precision of model experiments.

Data from the three models are comparable as experimental devices and observation measures

Tab. 1 Comparison of the soil loss between the D_a – Model and B – Model after the check dam system was constructed ($k_{DB} = 0.40$)

	Run designation	1	2	3	4	5	6	7	8	9	10
	Constructed dams	Dam 1	—	Dam 2	Dam 3, 4, 5, 6	Dam 7	Dam 8, 9, 10, 11, 12	—	—	—	—
B – Model	Observed soil loss, S_B (kg)	40.1	138.1	73	44.7	48.9	47.1	36.7	47.2	61.3	40.6
	Soil loss converted in the prototype scale, S_{pB} ($\times 10^6$ kg)	22.7	78	41.3	25.3	27.6	26.6	20.7	26.7	34.6	22.9
	Observed soil loss, S_D (kg)	5.04	6.72	4.48	1.51	1.4	2.11	1.65	2.73	1.64	1.55
	Soil loss free of rainfall difference, S_D * (kg)	*	6.06	3.90	1.48	1.51	2.28	1.79	2.77	1.73	1.50
D_a – Model	Soil loss converted in the prototype, S_{pD} ($\times 10^6$ kg)	*	91.9	60.2	21.5	21	31.6	24.7	39.6	24.3	22
	Observed rainfall intensity, I_D (mm/min)	2.7	1.71	1.74	1.64	1.56	1.56	1.56	1.61	1.58	1.65
	Soil loss error, e (%)	*	12.0	36.5	−15.5	−20.9	23.7	24.2	50.1	−28.0	−5.6

* human errors

Tab. 2 Comparison of the soil loss between D_b – Model and B – Model after the check dam system was constructed ($k_{DB} = 0.86$)

Run designation		1	2	3	4	5	6	7	8	9	10
Constructed dams		Dam 1	—	Dam 2	Dam 3, 4, 5, 6	Dam 7	Dam 8, 9, 10, 11, 12	—	—	—	—
B – Model	Observed soil loss, S_B (kg)	40.1	138.1	73	44.7	48.9	47.1	36.7	47.2	61.3	40.6
	Soil loss converted in the prototype scale, S_{pB} (10^6 kg)	22.7	78	41.3	25.3	27.6	26.6	20.7	26.7	34.6	22.9
	Observed soil loss, S_D (kg)	0.37	0.47	0.68	0.76	1.10	0.83	1.12	1.22	1.12	1.28
	Soil loss free of rainfall difference, S_D^* (kg)	0.45	0.47	0.69	0.80	1.16	1.10	1.02	1.14	0.87	1.12
D_a – Model	Soil loss converted in the prototype scale, S_{pD} (10^6 kg)	14.2	14.7	21.6	25	36.2	34.4	31.8	35.5	27.1	34.8
	Observed rainfall intensity, I_D (mm/min)	2.21	2.46	2.44	2.39	2.39	2.13	2.57	2.54	2.79	2.63
	Soil loss error, e (%)	-37.5	-81.2	-47.6	-0.9	31.2	29.4	53.6	33.1	-21.7	51.9

in the small scale models are consistent with those in the B – Model. An across – the – board comparison was available as soil erosion processes in the models were implemented in the laboratory using controlled rainfall.

The SSFM is presented for a check dam system on the Loess Plateau, China, in which the ratio of the model geomorphological variables to the corresponding prototype remains constant after rainfall events. Consequently, soil erosion processes in the prototype are reflected by model experimental results. Experimental data in the serial experiments verify that the SSFM can predict soil conservation by a check dam system in a small watershed on the Loess Plateau, China.

Acknowledgements

This study is supported by the National Science Foundation of China (51079016; 51179021), and the Open Research Fund Program of State key Laboratory of Hydroscience and Engineering (sklhse – 2008 – B – 01) for financially supporting this research. The first author wishes to acknowledge Professor Lei Tingwu, Professor Wang Xingkui, Professor Shao Xuejun, Professor Li YiTian, Mr. Liu Libin, Prof. Yu Cong, Prof. Laflen, and Dr. Samran Sombatpanit for their kind advices.

References

Cerdà A, Garcia – Fayos P. The Influence of Seed Size and Shape on Their Removal by Water Erosion, Catena, 2002(48): 293 – 301.

MWR (Ministry of Water Resource of P. R. China). Programming for Check Dams in the Loess Plateau [R]. 2003;47 – 48.

Hancock G R, Willgoose, et al. A Qualitative and Quantitative Evaluation of Experimental Model Watershed Evolution[J]. Hydrological Process, 2003(17): 2347 – 2363.

Hancock G R, Willgoose, et al. An Experimental and Computer Simulation Study of Erosion on a Mine Tailings Dam Wall[J]. Earth Surface Processes and Landforms, 2004(29): 457 – 475.

Li G Y. Construction of Physical Scale Model for the Yellow River[J]. Yellow River, 2001, 23 (12): 1 – 3.

Jiang D S, Zhou, et al. Simulated Experiment on Normal Integral Model of Water Regulating and Sediment Controlling for Small Watershed[J]. Journal of Soil and Water Conservation, 1994, 8 (2): 25 – 30.

Jin DS, Zhang, et al. Influence of Base Level Lowering on Sediment Yield and Drainage Network Development: an Experimental Analysis [J]. Geographical Research, 2003, 22 (5): 560 – 570.

Shi H, Tian, et al. Study on Relationship of Slope – gully Erosion in a Small Watershed by Simulation Experiment[J]. Journal of Soil Erosion and Soil and Water Conservation, 3(1): 30 – 33.

Shi H, Tian, et al. Study on Spatial Distribution of Erosion Yield in a Small Watershed by Simulation Experiment[J]. Research of Soil and Water Conservation, 1997, 4(2): 75 – 85.

Strabler A N. Dimensional Analysis Applied to Fluvially Eroded Landforms[J]. Bulletin of the Geological Society of America, 1958(69): 279 – 300.

Xu X Z, Zhang, et al. A Laboratory Study on the Relative Stability of the Check Dam System in the Loess Plateau, China[J]. Land Degradation & Development, 2006(17): 629 – 644.

Yuan J P, Lei, et al. Simulated Experimental Study on Normalized Integrated Model for Different Degrees of Erosion Control for Small Watersheds[J]. Transactions of the CSAE, 2000, 6(1): 22 – 25.

Yuan J P, Jiang, et al. Simulated Experiment on Normal Integral Model of Different Control Degrees for Small Watershed[J]. Journal of Natural Resources, 2000, 15(1): 91 – 96.

Zhang H W. The Study of the Law of Similarity for Models of Flood Flows of the Lower Reach of the Yellow River[D]. Beijing: Tsinghua University, 1994.

Zhang L P, Zhang. Research on Runoff Formation Distortion Coefficient in Soil Erosion Experiment of a Normal Model of Small Drainage[J]. Acta Pedologica Sinica, 37(4): 450 – 455.

Real Time Monitoring of Levees through Sensor Technology along the Yellow River

F. P. W. van den Berg[1] , *J. Heaton*[2] and *W. Moen*[3]

1. Deltares, Stieltjesweg 2, 2600BC Delft, The Netherlands
2. AGT International, Lintheschergasse 7, 8001 Zürich, Switzerland
3. AGT International, Kneuterdijk 18, 2514 EN Den Haag, The Netherlands

Abstract: This paper describes the successful application of a pilot with sensors in a groyne alongside the Yellow River near Dayulan. The aim of this innovative application of the sensors is to create a real time monitoring system. The used monitoring system was designed based on the knowledge gained in the Dutch IJkdijk project ("IJkdijk is the Dutch word for Calibration levee"). The deployed sensors are tri – axialy MEMS (Micro – Electrical – Mechanical Systems) and an optical glass fibre system in a geotextile. The management system, ReadyMIND, links the sensor output to a dike stability analysis model and provides a GIS based interface offering continuous status information on the dike stability. This system also provides sophisticated simulation capability, and operational support, offering benefits in areas of emergency response planning and crisis management .

Key words: sensor technology, dike stability, dike safety, flood protection, real time monitoring, operational support, emergency response planning, crisis management

1 Introduction

In the Netherlands several tests using sensor technology have been carried out at the so called IJkdijk. The IJkdijk project ("Calibration levee" in Dutch) was initiated to test the viability of modern sensor technology for the assessment of dike strength. A full scale trial embankment was built and brought to failure. The results of this test and other studies will lead to a Flood Early Warning System (FEWS) for dike stability.

Two of those different sensor techniques are now used for a pilot of a groyne alongside the Yellow river. The two used sensors for this pilot are:

(1) Tri – axially digital multi – parameter sensors containing silicon MEMS elements (Micro – Electrical – Mechanical Systems).

(2) Optical glass fibre system in geotextile.

An integrated flood management system based on the ReadyMIND platfrom (Fig. 1), will be employed to collect and analyze data on dike stability and deformations in the dike and on the risk of flooding. These will provide a basis for taking measures to prevent subsidence or dike failure as quickly as possible. This system is also able to carry out a flood prognosis and can assist decision makers in timely evacuation and will provide the manager with information about the condition and stability of the dike. The integrated system in this paper, was designed to the special structure of the groyne in relation to the hydraulic changes of the Yellow River throughout the year, in order to monitor the safety of the groyne and the underlying main dike.

2 Site conditions and sensors

2.1 Pilot location

The location of the pilot project is along the Yellow River at the Dayulan works in Wuzhi County of the Henan Province. In the outer corner of the river lies groyne #24. Most of the time the water level at this location is low (approximately 8 m minus the crown of the groyne) , but from mid june the water level of the Yellow River can come up 4 m minus the height of the crown of the groyne. The river level at this point of the Yellow River will rise because of the opening of the reservoir stream upwards and the beginning of the rainy season.

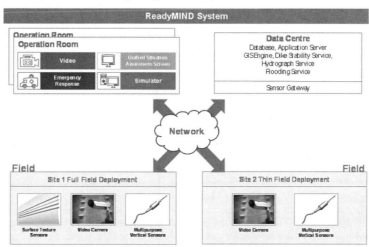

Fig. 1   ReadyMIND system architecture

In Fig. 2 the top view depicted of the pilot location including the location of the different sensors.

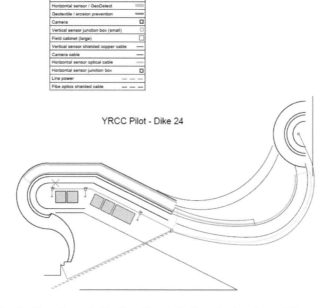

Fig. 2   Top view of pilot location including the location of the sensors

2.2 Failure mechanism and dike stability

At first, an analysis has been carried out of all the failure mechanisms which are relevant for this groyne, as dike safety is the key to prevent flooding. The result of the analysis is that there are two main threads to this groyne; erosion at the toe of the groyne and dike instability.

The slope stability analysis has been carried out to determine the location of the proposed sensors in the dike and to determine the safety factor. The Bishop – method is used to determine the slope stability, which is one of the standard modules used by the computer program D – Geo Stability. The input for the analysis are; geometry of the groyne, soil layers (including soil parameters) and the water conditions.

The critical factor of safety for any combination of geometry, geotechnical characteristics and ground water conditions is the lowest of those calculated by the program.

For the D – Geo stability calculations the following soil parameters per layer are needed; the bulk unit weight $[kN/m^3]$, cohesion $[kN/m^2]$ and the internal friction angle $[°]$.

Just before installing the MEMS, a cone penetration test has been conducted. After interpretation of the CPTs the soil construction can be derived. The soil construction of the groyne consists mainly of clayey silt and the lower layers consists of very silty fine sand layers.

Overall, it should be clear that the parameters selected for these analyses have been chosen for the current calculation are mainly on the basis of experience and the CPT. It is therefore recommended that target factors of safety of between 1.3 and 1.4 are used for the stability assessments, because of the use of material factors. Because of the uncertainties in this case, the best way to determine the degree of safety of the dike is to determine the relative safety. When Groyne #24 will fail, the main dike will not breach as a result. Based upon given information, there are no visual anomalies, during the low water season.

It can be concluded that the stability factor of 1.10 (calculated result with low river level) is a safe situation. The safety is 100%. When the slope stability (stability factor) will decrease, the relative factor will decrease as well. A decrease of 15% will result in an unsafe situation, so when the safety factor is 0.94 or lower measurements are necessary to stabilize the dike.

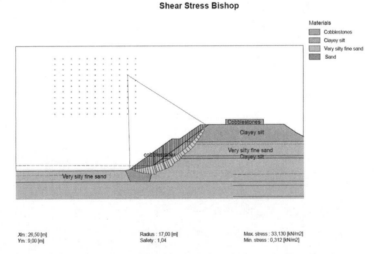

Shear Stress Bishop

Materials
- Cobblestones
- Clayey silt
- Very silty fine sand
- Sand

Xm : 26,50 [m]
Ym : 9,00 [m]

Radius : 17,00 [m]
Safety : 1,04

Max. stress : 33,130 [kN/m2]
Min. stress : 0,312 [kN/m2]

Fig. 3 Result of calculation with the Dike stability model

In Fig. 3 the result of the critical circle, after the fall from high river level to low river level, is depicted. To monitor, among other things, the water pore pressure inside the groyne MEMS are used. In 2.3 these will be discussed.

2.3 Micro Electro Mechanical Systems(MEMS)

The emerging area of MEMS has its roots in IC processing. After decades of research and development, the state of the art in MEMS processes is capable of integrating micro electronics and sensors on a single chip. The combination of micro electronics and mechanical components makes MEMS more powerful and versatile than the conventional sensors.

In this pilot MEMS sensors of the type GeoBeads ® are applied. Each instrument contains MEMS elements for sensing pore water pressure, inclination and temperature which are digitized on – board with an integrated microprocessor. This set – up enables connecting the instruments together on a serial network bus to cover wide monitoring areas.

The objective of installing the MEMS sensors (type GeoBeads) in the groyne is to acquire data that will:

(1) Be gathered frequently and transmitted in near real – time.

(2) Deliver input for geotechnical stability models.

(3) Enable relevant warning and alarming schemes.

The sensors are installed at three different cross – sections. Each cross – section consists of 6 sensor locations. These are in line with the positions needed by the dike stability model to provide the needed input on water infiltration and

Fig. 4   MEMS (type: GeoBeads)

pressure build – up in the inner body of the dike. In Fig. 5 the locations are depicted.

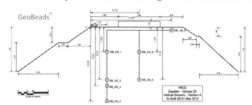

Fig. 5   Locations MEMS in cross section A Groyne #24

Besides the sensors in the groyne, also a sensor is placed in the Yellow River to measure the river level.

The MEMS are installed by a CPT truck, See Fig. 6.

Fig. 6   Method of installation, vertical sensors on top of groyne #24

2.4 Dike stability analysis

The output data from the MEMS is, after some small alternations send to the Dike Analysis Module (DAM), a platform that automatically determines the strength of a levee or the failure probability based on a given hydraulic load. It involves a semi 3 – dimensional determination of the levee strength. This means that cross sections are schematized from a 3 – dimensional terrain model and complemented with point observations of soil structure. From these cross sections, the stability can be determined.

A configuration of relevant modules can be developed for different applications. The modules are calculation models describing a specific failure mode. The applied modules are dependent on the availability of data and purpose of the analysis. The high degree of automation in DAM makes rapid analyses of levee systems on large geographical scale feasible. For this project , dike stability calculations will be performed every 10 min.

2.5 Optical glass fibre system in geotextile

The optical glass fibre system sensors (type: TenCate GeoDetect Ⓡ) combines a geotextile enabled sensor, which is a textile composite (Fig. 7) embedding optical cables.

Fig. 7　Fibre optics woven in the textile

The sensors used for this pilot consists out of a strip embedding 2×2 optical cables, 2×1 for temperature measurement (light and dark blue), and 2×1 for strain measurement (yellow and orange). One optical cable for strain and one optical cable for temperature measurement are used for this project. The two other cables are redundant for safety. A picture of the fibre optic woven in geotextile is given in Fig. 7.

The combined measurement of strain and temperature in a single strip enhance the detection of critical events along the groyne: strain changes occur when there is soil settlement or slipping, temperature changes may detect water flow through the dike if there are temperature gradients. The resolution of the sensor is 1 measurement every 1 m.

The purpose of the fibre optic surface sensors is mainly the detection and the localization of soil and stones movement along the length of the groyne, due to the erosion of the Yellow River.

Fig. 8　Method of installation, fibre optic and, CPT truck for installing the MEMS on the crest of the groyne

Other groynes next to the pilot location have shown external erosion resulting in displacement of the stones rip – rap layer and soil scour erosion.

The most critical area is the submerged part during high water level, when the water velocities and turbulences are the highest, and when nothing is visible due to the high water turbidity.

It can be observed two slopes with an intermediate platform at the middle of the height. This intermediate platform corresponds to the highest water level.

Two sensor levels are installed:

(1) One corresponding to the highest water level. Along the exposed side of groyne #24 to the water current, just below the intermediate platform to monitor the settlement of stones on the slope. In Fig. 8 the installation of the fibre optics is shown.

(2) One continued along the eroded part of the main dike on its junction.

Between groyne #23 and #24. The slope of this part is rebuilt (slope 1v:2h) and covered with a layer 50 cm thick of stones for revetment. Half of the section is protected with a geotextile for soil filtration as a comparison, corresponding to standard design rules.

3 Real time monitoring, Emergency Response Planning and Crisis Management

3.1 General

The overall management system used, ReadyMIND, has the ability to link the sensor output to various models.

The purpose of the system is:

(1) To help the operator to interpret the raw sensor data and showing it in a simple and easily understood way.

(2) To link the sensor output to dike stability analysis models, hydrograph – and flooding models.

(3) To simulate possible events based on historic, predicted and specially authored data (which can also be fed into the various models used).

The system leverages on recent advancements made in Human Interface Technology and is GIS based. The system is therefore capable of showing complex information in an easily understood way by the simple toggling of the appropriate combination of layers for display.

An alarm event may be triggered by the system, e. g. on detection of an unstable dike. The operator display will show the information related to the event, and provide the tools to manage the event as efficiently as possible.

The purpose of the flooding simulation in the ReadyMIND system is to enhance the quality of the decision made by officials. Simulation of flooding of an area can lead to changes in procedures:

(1) Prediction and simulation of the development of a current flooding event enables emergency response teams to act in a pro – active and co – ordinated manner.

(2) Simulation of a potential future event enables agencies to test preparedness and optimise procedures and resource deployment.

(3) Determine high ground which is safe to evacuate people to.

(4) Determine the effect of countermeasures to the flooding.

The ReadyMIND system may be extended to include operational procedures, such that clear guidance is provided to the operator, and a record of the actual response may be captured by the activation of automated fields. This enables an organization to review and improve procedures over time, and both reduces the level of training required and makes the training process more effective.

3.2 Output of sensors

3.2.1 MEMS

Based on currently available time series of the MEMS the measurements shows that the infiltration depth is very high. The groyne reacts very directly to the height of the water level in the river, See Fig. 9. This is caused by the very sandy layers (cone resistance > 4 MPa) layers of the lower layers of the groyne and the sandy subsoil. See a representative CPT (Cone Penetration Test) for the groyne at Fig. 10.

The CPT has been carried just before placing the MEMS at cross section B. For the dike stability model this means, that the expected most critical circle who will occur after the fall from high to low level will probably not occur.

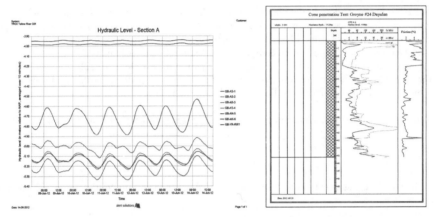

Fig. 9　Hydraulic head measurements, cross section A

3.2.2　Optical glass fibre

The first results of the optical glass fibre shows at the beginning of the project, an increase of the deformation of the intermediate platform. The cause of the initial deformation is mainly caused by the replacing of the stones directly after the installation of the sensors. In October 2012, when then pilot is finished, the result will be known and will be published.

At this moment, we expect that there will be a certain amount of deformation at the intermediate platform, caused by erosion at the toe of the groyne.

4　Conclusions

The system provides multiple value propositions, based upon the input of the two sensors in the groyne.

Dike Maintenance:

Enhanced knowledge of the internal structural stability of the dike, enables avoidance of unnecessary dike strengthening activities. In the case that dike strengthening is required the system enables specifically targeted dike strengthening activities and thus lowers potential costs.

Flood Early Warning:

Testing has shown, that dike instability may be identified and failure predicted several days in advance. This enables emergency maintenance, which may prevent dike failure altogether. Prediction of the impact of flooding allows prioritization of emergency preparations and an extended planning period to co – ordinate with Emergency services.

Crisis management:

Extended warning of flood events provides greater opportunity to plan response, evacuate population and protect key assets. The ability to predict a timeline of event development, linked to realistic scenario planning enables emergency response planning and effective crisis management, based on realistic response times. This avoids the problems of reactive crisis management based on out of date and confused information.

References

Zwanenburg C, et al. Failure of Trial Embankment on Peat in Booneschans[R], The Netherlands, Geotechnique, 2012.

Bennet V, et al. Real Time Monitoring System and Advanced Characterization Techniques for Civil Infrastructure Health Monitoring[J]. Advances in Civil Engineering, 2011.

Johari H. Development of MEMS Sensors for Measurements of Pressure, Relative Humidity and Temperature[D]. WPI, 2003.

Peters E T, van der Vliet P P. GeoBeads, Multi – parameter Sensor Network for Soil Stability Monitoring[C]// 8th International Symposium on Field Measurements in GeoMechanics, Berlin, Germany,2011.

Knoeff J G, Vastenburg E W. Automated Levee Flood Risk Management[C]// 5th International Conference on Flood Management (ICFM5), Tokyo – Japan,2011.

The Research for 3D Design Method of Hydraulic and Hydropower Engineering Based on CATIA

Li Bin, *Zhuge Meijun* and *Li Han*

Yellow River Engineering Consulting Co. Ltd. , Zhengzhou, 450003, China

Abstract: The 3D visualization design technology has been maturely applied in such industries as the aviation and machine-building, while rather lag in technology of the industry of hydraulic and hydropower engineering design, still in its initial stage. This article introduces the basic acknowledge of CATIA 3D design platform researched by the French Company Dassault Systems and its powerful aided design function, and expounds the advantages of 3D collaborative design in aspects of improving the design efficiency and product quality of the Company, shortening the design phase as lowering the cost, prolonging the life circle of hydroelectric design industry, and perfecting the image of hydroelectric design company. This article concludes the key technologies as geological modeling, template design, frame design and 3D collaborative design undertaken by applying the tool as CATIA, fundamentally confirms the work pattern for hydraulic and hydropower engineering 3D design application, to construct its basis for the practical application of CATIA platform in hydraulic and hydropower engineering. This article finally gives the initial results via applying this technology into the hydraulic and hydropower engineering design work, combining with the engineering case of Guxian Project at the Middle Reach of Yellow River, reveals the results as the 3D reservoir-region geological model and building model, automatically updating after changing the program set by the CATIA platform, and the 2D layout plan generated by 3D model, and makes clear the work pattern for hydraulic and hydropower engineering 3D design and the results' usability.

Key words: hydraulic and hydropower engineering, 3D collaborative design, CATIA, template technology, frame design

1 Introduction

The CAD technique has always been playing an important role in the engineering design area. It has been widely applied and acquired great success for its short design term, qualified design quality and less design cost, bringing the emergence of a series commercially software, such as AutoCAD, etc.

It's been quite mature for 3D visualization design platform being applied in such industries as aviation, automobile and machine-building, accompanied by the swift development of computer hardware and software technology and the gradual improvement of the application of information technology.

Though it has been greatly improved for the engineering technology level and the computer application and information technology level of the hydraulic and hydropower engineering design industry in China at present, it's still at the traditional design mode and 2D design phase, rather backward in the 3D collaborative design and simulation technique areas, being a quite big gap compared with the international advanced level. The nationwide survey and design units all have been presently explored the advanced design techniques and methods to overcome the defects of 2D plan design. It's still in the initial stage for applying the 3D design method in the hydraulic and hydropower industry.

The 3D design research application work of Yellow River Engineering Consulting Co. Ltd. has started from 2004, and it brought the CATIA 3D design platform under the support by the Ministry of Water Conservancy '948' Project: 'The Introduction, Research and Extension for Hydraulic and Hydropower Engineering 3D Design Methods' in 2009, while taking this as the turning point to entirely develop the 3D design and research work, and finally master the platform application methods and accumulate the application experiences through studying, digestion and absorption, which fundamentally determined the operating mode for the hydraulic and hydropower engineering 3D design application, being the basis of the practical application for CATIA platform using in hydraulic and hydropower engineering. This article introduces the acquired experiences from introducing the international 3D design software and techniques by the authors, and its initial achievements acquired by the practical application during the hydraulic and hydropower design work.

2　The brief introduction of CATIA

CATIA is the abbreviation of 'Computer Aided Tri-Dimensional Interface Application', a kind of the CAD/CAE/CAM integration software of the French Dassault Systems. CATIA has been emerged at the historical moment during the 1970 s, under the need by the world-famous aerospace enterprise, Dassault Aviation.

CATIA is a kind of 3D design platform, which can indirectly produce the digital mode in the design phase. Though coming from the aerospace industry, its powerful functions have been acquired the approval of other industries, and it has been widely used in such aspects as aerospace industry, automobile industry, shipbuilding, machine-building, electronics and consumer-goods-industry. Its integrated solution plan covers all the product design and manufacturing areas, while its special DMU electronic prototyping module function and hybrid modeling technique can further promote the competitiveness and productivity of enterprises to improve.

CATIA owns a special generation tool for assembly sketches, which can support the drawing of assembly sketches under constraints, and swiftly promote the idea design. It can support the modeling methods, like parametric modeling and Boolean operation, etc. , and the bidirectional data association between drawing and numerical control machining. CATIA's configuration design and style design can provide the detailing with integrated tools. This software is of the powerful surface modeling, with its peculiar integrated-development-environment, while CATIA can also proceed the finite-element analysis. For special, though the general 3D modeling software observes the detailing in 3D space, CATIA can provide the way to operate the 4D space observation, namely, this software can enter into the inside to observe the detailing by simulating the observer's view, and proceed with the assembling by reality simulation.

CATIA can provide the convenient solving plan, catering to the needs of companies with all scales in all the industry areas. Its famous clients include a host of well-known enterprises such as the Boeing, CHRYSLER, BMW and BENz. Its client group takes the decisive place in the worldwide manufacture industry. The Boeing Company employed CATIA to complete the electronic assembly of the whole Boeing 777, creating a miracle in the Industry, and further confirming the leading place of CATIA in the CAD/CAE/CAM industry. While its ruling place continuing to enhance in the automobile and aerospace fields, CATIA has been also accessing into other such industries as motorbike, train manufacturing, general machinery, household electrical appliance and hydraulic and hydropower, etc. .

Since Edition V5R19, CATIA has begun to add the terrain modeling function, as using for the original modeling architecture, design and planning for civil engineering, and foundation engineering, which can apply the CATIA 3D cooperative design platform to the overall process of the hydropower engineering project, covering different phases of pre-feasibility study, feasibility study, preliminary design and technology application, while achieving the design management for 3D digital power plant. The uniform 3D design data model can promise the timely updating for information among various specialties of the hydropower engineering, and its associated design can greatly reduce the modification work load. It's effective to well assemble the engineering design flow and experience knowledge into the model file by using the parameter and design rule of knowledge engineering module and document template, etc. , which can rapidly realize the comparison for plans of dam type and site, engineering budget, digital simulation of construction and maintaining operation, finally promoting to shorten the engineering design cycle.

3　The advantages of 3D design

3.1　The 3D collaborative design can improve the design efficiency and product quality of the Company, while shortening the design phase as lowering the cost

It not only can enhance the design efficiency and overall design ability of the company, but also can improve the design quality of product, as well as can shorten the design phase and lower the cost. For example, after applying the 3D collaborative design the foreign auto industry can lower about 20% design cost while shorten 30% ~ 50% design phase, through the design reuse of enterprise knowledge and the concurrent design. At present, the most primal problem restricting each hydropower design company has still been the shorter normal design phase and lower-efficient design efficiency. If adopting the 3D collaborative design technique, it will certainly get the better technical economic benefit, making the survey and design level and management level of hydraulic and hydropower engineering step further.

3.2　The 3D collaborative design can promote expanding the service scope of hydropower designing company and prolonging the life circle of industry

The Engineering-Procurement-Construction mode, EPC, has been presently developing rapidly in foreign countries, and has become a kind of main construction management mode, under which the projects applied in some countries has been up to 40% ~ 50%. Under this EPC mode, the investor can procure the upmost investment returns as confirming the risk, while for making the reasonable intersection and overall optimization for various phases of design, procurement, construction and commissioning, further integrating them as one organic integrity, the general contractor can shorten the design phase and lower the cost, thus acquiring enough profits. For covering the whole life circle of the power-station construction, the 3D collaborative design can thoroughly exert the leading role of design company, overcome the malpractice brought by its mutual disconnection among the traditional design, procurement, construction and commissioning, and finally achieve goals of reduction of erection time, quality guarantee and saving construction costs through proceeding the reasonable cross connection. And it can simultaneously expand the service scope of designing company to aspects of procurement, construction and commissioning, as well as the operation, maintaining and late services of power station, thus prolonging the life cycle of hydropower design industry

3.3　The application for 3D collaborative design technique can improve the image of hydropower design enterprise and promote its transition to international engineering company

The international well-known engineering company generally occupies the 3D collaborative design ability, for which not only represent the design ability and level of one company, but also its management level. Therefore, it has become an important indicator to judge one company whether being the international engineering company or not as whether mastering the 3D collaborative design technique or not. If the hydropower design company can master the 3D collaborative design technique, it not only can improve the management level and image of enterprise, but also enhance its core-competitiveness, and further accelerate the step of its transition to engineering company from a design one.

3.4　The application for 3D collaborative design technique can make for the virtual training for operating personnel before finishing construction.

The results of virtual simulation can also make for the virtual training for operating personnel before finishing construction, thus realizing aiming that the staff put to work once the construction done. It also can make for the virtual training for equipment repair and maintenance personnel, thus improving the availability of equipment and repair-and-maintenance-equipment, while reducing the cost for operation and maintenance.

4　The K\key techniques of hydraulic and hydropower engineering 3D design

There are several following key techniques using CATAI as tool, considering from such aspects as enhancing the design efficiency, optimizing the design flow, and improving the sharing and reusability of design information:

4.1　CATIA key technique for geological modeling

The CATIA software platform has a quite powerful 3D modeling ability, but it could be a new challenge to apply it to the hydraulic and hydropower engineering geological industry. There are several kinds of datum needed by CATIA geological modeling as: the geological plane with properties, the borehole log, the adit and shaft display chart and the geological section map, etc.. The geological plane mainly includes the contour marked with elevation, the stratum surveying line, contours of bedrock, the profile position, the drilling position and the hydraulic building location, etc.. The engineering geological profile mainly includes the stratigraphic boundary, the tectonic line and the Hydrogeological boundary, etc.. It should preprocess the plane to the format that can import CATIA, of which the redundant map layers ought to be closed or cleared. The data format of histogram and profile should also be processed to the corresponding geological package ones developed according to CATIA platform.

It should import the point cloud data generated by topographic contour to CATIA, thus producing the triangle mesh surface and finally completing the emergence of land surface. During the above process, it ought to proceed with the bug-seeking between sheets and the following bug-correcting, thus forming a completed 3D land surface, owing to the requirements required by precision of initial land form data and mesh.

It should model the convex body according to required model scope by utilizing the terrain model, which ought to be then carried out the Boolean operation with the land surface after meshing, thus forming the mesh geologic body without layering.

It ought to transform the general format of geological mapping material and profile of geological plane and the drilling material to the required one, which is then imported into the geologic body and can thus form the strata interface by the multi section surface or mesh morphing. After meshing, the interface can proceeded with the boolean operation in proper sequence, together with the initially-formed mesh geologic body, according to the sequence of "first the superstratum and geological structure and then the land surface", finally completing the integrated geological model.

4.2　The CATIA template design technique

It's efficient to integrate the design knowledge into the templates by utilizing the feature, parts and assembly template created by the CATIA product knowledge template. And after instantiating, the integrating can be revised by correct input and geometric reference to adapt various kinds of need and rapidly complete the repeating design work. Each template actually all contains one design thinking and process recorded by CATIA, which contains the experiences and results accumulated by one design company for years, thus storing its design experiences in CATIA system by all kinds of means. The template can also integrate several knowledge reuse, design standard and check by former experiences, etc. Fig. 1 displays the design process for an arch dam completed by using relevant templates.

The template created by CATIA has the abilities of parameterization and transformation by rules, which can adapt to various kinds of design cases. The template not only contains 3D model, but also contains the relevant 2D charts, which can easily create the 2D chart required by intersection graph, thus solving the reuse problem for similar design, improving the design speed, reducing the repeating furthest, sharing the best design method and raising the work efficiency.

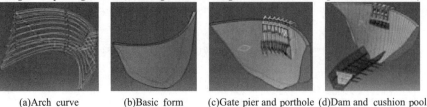

| (a)Arch curve | (b)Basic form | (c)Gate pier and porthole | (d)Dam and cushion pool |

Fig. 1   The template for Arch Dam design process

4. 3   CATIA framework design technique

The CATIA framework design is the main element wire frame design, indicating that after fully understanding and resolving the design product, it employs the control element of main wire frame and construct the efficient parameter information transfer frame and the design method of superincumbent flow by carrying out efficient overall control to the whole product mix and forming the trunk-like product design structure.

The main idea for framework design technique is to begin with the overall plan and follow by detailing design. Before carrying out the project, it's necessary to construct a qualified overall framework, which contains the key location of the whole constructions, arranges the basic standard, defines relevant dimensions among each construction, preparing well for multi-specialty data transfer of design flow. During the initial stage of overall design for power station, it's decided to consider the station design located from the upmost level of whole station system, which can consider and express the relationships among relevant positions of the whole station system and each subsystem, while deliver the design data transfer from the upstream to downstream. It's advisable to apply this technique to continue the subsequent product design work, making it more pertinent and normative.

The CATIA framework design is superior in overall control, clear logical relationship, distinct structure and convenient design modification. The framework contains the key location of overall project and basic standard arrangement, while defines the relevant important dimensions among each constructions, which also make preparations on models for professional cooperation.

4. 4   The collaborative design technique

The collaborative design is a kind of method aiming at the same project, which can make different design staff complete each design project-part at the same time while in various places, through the network sharing or collaborative design software. It can raise the design efficiency and reduce its cost furthest, which is the design objective praised highly by the design industry.

The traditional off-line design method needs lots of face-to-face communications, of which the design flow is complicated, and the efficiency and quality cannot be obtained. The collaborative design makes all the designers work under the same circumstance by adopting the on-line collaborative design method, achieving the design data synchro and unique, the design flow clear and simple, while accomplishing the relevance among different specialties or design product-parts. Under the relevant design circumstances, changing of the top specialty's product can be soon known by the down one, or after altering the 3D model, the 2D design drawing can make the amending notice and finish it automatically, ensuring its design efficiency and quality.

5　The application in Guxian Project by 3D design

5.1　The introduction for Guxian Hydraulic Multi-purpose Project

Guxian hydraulic multi-purpose project locates in the reach of North Main Stream of Yellow River between Qikou to Yumenkou, about 10.1 km above the Hukou Waterfall, of which at the left bank is Ji County, Shanxi Province, and at the right is Yichuan County, Shanxi Province. This Project controls 65% of the total area of Yellow River Reach, 80% of the Yellow River's quantity of water, and 66% of its sand sediment. It mainly functions as flood control and sediment reduction, also giving conditions to power generation, water supply and irrigation, etc. , which is a multi-purpose-used large key project. Its constructions are mainly composed by the panel stack stone dam, flood discharging tunnel, flushing tunnel, diversion tunnel, water diversion system and power plant, etc. .

5.2　The 3D design results of Guxian Hydraulic Multi-purpose Project

The following figures respectively show the part results of 3D design created by CATIA platform for Guxian hydraulic multi-purpose project, displaying the geological model for reservoir region, construction group model, and the 2D plane layout generated by the 3D model, which basically realize the automatic update after plan changing, and greatly improve the efficiency and the application of results(see Fig. 2, Fig. 3, Fig. 4).

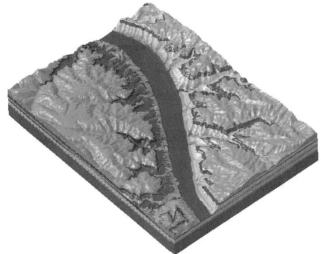

Fig. 2　The 3D Geological Model for Guxian Engineering Reservoir Region

6　Conclusions

By studying, researching and innovating about CATIA 3D collaborative design platform, the key techniques of hydraulic and hydropower engineering 3D design process has been mastered, the 3D design work pattern has been determined and the application of 3D visual design has been successfully achieved. Those techniques can rapidly carry out the plan comparison, present the calculation results of each quantities, solve different problems of each specialty design existed in 2D design, thus reduce the design cost and raise the design quality and work efficiency.

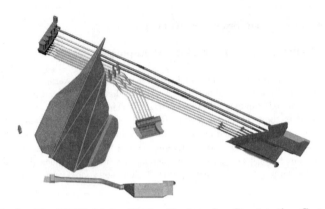

Fig. 3 The 3D Model for Guxian Engineering Construction Group

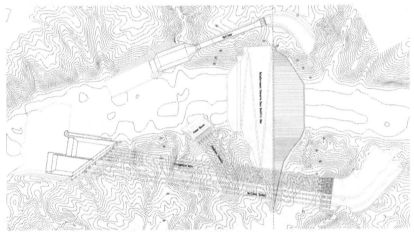

Fig. 4 The General Layout Drawing for Guxian Project generated by the 3D model

References

Li Bin, Zong Zhijian, Zheng Huichun. The Introduction and Research for Hydraulic And Hydropower Engineering 3D Design Method[J]. Yellow River, 2011.
Dassault Systemes. CATIA V5 Expert Mechanical Designer. 2005.

Basin Economic Allocation Model (BEAM1)
—An Economic Model for Water Use in Transboundary
River Basins Customized for the Aral Sea Basin

Palle Lindgaard-Jørgensen[2], Niels Riegels, DHI
Jesper Karup Pedersen, Mikkel Kromann, COWI
Vadim Sokolov, GWP CACENA[3], Anatoly Sorokin, SIC-ICWC[4]

Abstract: The water resources of the Aral Sea basin are under increasing pressure, particularly from the conflict over whether hydropower or irrigation water use should take priority. The purpose of the Aral Sea Basin Economic Allocation Model (BEAM) model is to explore the impact of changes to water allocation and investments in water management infrastructure on the overall welfare of the Aral Sea Basin.

BEAM is developed as a decision support system to facilitate putting "value on water use" along with the sustainable use of water resources in support of development. The model estimates welfare changes associated with changes to how water is allocated between the five countries in the basin (Kazakhstan, Kyrgyz Republic, Tajikistan, Turkmenistan and Uzbekistan; water use in Afghanistan is assumed to be fixed). Water is allocated according to economic optimization criteria; n other words, the BEAM model allocates water across time and space so that the economic welfare associated with water use is maximized. The model is programmed in GAMS (General Algebraic Modelling System) -software developed by a group of economists at the World Bank.

The model addresses the Aral Sea Basin as a whole – that is, the rivers Syr Darya, Amu Darya, Kashkadarya, and Zarafshan, as well as the Aral Sea. The model representation includes water resources, including 14 river sections, 6 terminal lakes, 28 reservoirs and 19 catchment runoff nodes, as well as land resources (i. e. , irrigated croplands). The model covers 5 sectors: agriculture (crops: wheat, cotton, alfalfa, rice, fruit, vegetables and others), hydropower, nature, households and industry. The focus of the model is on welfare impacts associated with changes to water use in the agriculture and hydropower sectors.

The model aims at addressing the following issues of relevance for economic management of water resources:

(1) Efficiency (estimating how investments in irrigation efficiency affect economic welfare).

(2) Effectiveness (estimating how changes in how water is allocated affect welfare).

(3) Equity (who will gain from changes in allocation of water from one sector to another and who will lose?).

Stakeholders in the region have been involved in the development of the model, and about 10 national experts, including staff from the International Fund for Saving the Aral Sea (IFAS), have been trained in using the model. The model will be publicly accessible through a web-based user interface that allows users to investigate scenarios and perform sensitivity analyses.

Preliminary results suggest that:

---

① On behalf of the International Fund for Saving the Aral Sea (IFAS) and USAID a project team consisting of experts from DHI, COWI and Global Water Partnership CACENA has developed an economic model for water use in the Aral Sea Basin, named the Aral Sea BEAM (or simply BEAM). BEAM stands for Basin Economic Allocation Model.

② Presenter of the paper.

③ Central Asia and Caucasus.

④ Scientific Information Centre Interstate Commission for Water Cooperation.

（1）At the margin, hydropower water use increases basin-wide welfare more than irrigation water use.

（2）Under normal or average hydrological conditions, water scarcity is not a significant problem in the basin.

（3）Under dry hydrological conditions, water scarcity is significant. Under these conditions, preliminary results suggest that cotton irrigation is less effective than other uses, particularly in Turkmenistan.

（4）Investments in irrigation efficiency can have a significant impact on the effectiveness of water use for irrigation, thereby increasing the welfare of irrigation regions during dry periods.

Key words：water allocation; ecnomic model; DSS; And sea basin

1 Background

More than 40% of the global population lives in basins shared by two or more countries. In addition some 280 river basins and numerous underground aquifers are shared by two or more countries. Transboundary water cooperation (TWC) has traditionally been based on agreements on sharing of water in ways perceived as equitable by the riparians. A new paradigm based on benefit-sharing is gradually being considered and introduced as the practical aspects of this concept are developed and reality checks are made. Lessons on benefit-sharing in transboundary basins are beginning to appear

Considering that more than 50% of the earth's land area is within shared river basins together with the prospect of a quarter of the world's population being exposed to extreme water scarcity within the next 25 years it is imperative to develop tools which can support decision making for management of water resources in shared basins.

One of the key components of applying the benefit sharing concept in practice is the identification of the potential benefits of cooperation. It is also important to consider carefully the question identifying winners and losers if changes are made to the way that water is managed in the basin. Providing this information in a transparent way through a combination of hydrological and economic modelling tools may help in negotiations and help persuade upstream and downstream riparians to cooperate.

2 The Aral Sea Basin its water resources and management arrangements

The main rivers feeding into the Aral Sea—the Syr Darya and the Amu Darya—have catchment areas of around 500. 000 km^2 and 800. 000 km^2, respectively (Fig. 1). The rivers are fed mainly by snow and glacier melt in the high mountain regions of Kyrgyzstan, Tajikistan and Afghanistan. This leads to a seasonally variable runoff regime with peaks in spring and summer. Inter-annual variability is also considerable; in the Syr Darya Basin, the annual runoff volume observed during the wettest year in the historical record is about four times the volume observed during the driest year. A highly developed system of dams, reservoirs and hydropower facilities regulates water flows. The operation of these facilities, many of which have impacts beyond national borders, requires sound coordination of the involved national agencies and a transboundary regulatory framework. The population In the Aral Sea Basin is 41. 8 million (2000), with agriculture being the main sector for livelihood. Water withdrawals for irrigation have led to the dessication of the Aral Sea—considered one of the world's foremost environmental disasters.

Regional water-security disputes are among the most highly contentious issues in Central Asia, impeding regional cooperation in a host of. The economic costs of sub-optimal use of shared water resources on the countries of Central Asia are high, and transcend the immediate cost costs imposed on agriculture and energy sectors as water-security tensions overshadow and impede efforts to forge closer cooperation within the region.

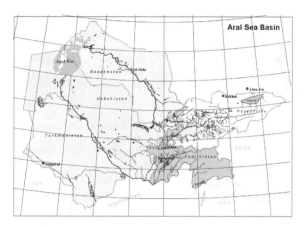

Fig. 1   Map of Aral Sea Basin

Water resources management in the Aral Sea basin remains one of the key challenges among the Central Asian countries of Kazakhstan, Kyrgyzstan, Tajikistan, Turkmenistan and Uzbekistan. Waters of Syr Darya and Amu Darya— the main rivers feeding into the Aral Sea—are critically important to the livelihoods of millions of people who rely on these rivers' resources for irrigation and hydroelectric generation. During the Soviet central planning, these rivers' resources were managed centrally under an approach that sought to maximize their economic value. Large dams and hydroelectric facilities that were constructed on both rivers in Kyrgyzstan and Tajikistan (the upstream countries) were operated primarily to supply water-storage and release services to support agriculture in downstream Uzbekistan, Turkmenistan and Kazakhstan. The arrangement involved both water and energy with the delivery of energy resources in exchange for water-storage and release services was designed to extract maximum value from the water resources.

With the dissolution of the Soviet Union, incentives to cooperate were reduced and basin-wide water and benefit sharing arrangements broke down as national governments began to pursue non-cooperative water management strategies. The result has been that upstream countries have increased winter hydropower production, resulting in inadequate supplies for water for irrigation in downstream countries during the summer irrigation season along with occasional winter flooding.

The Executive Committee of the International fund for the Aral Sea is working (EC IFAS) is working towards a comprehensive regional water resources management agreement and has supported efforts to estimate the costs and benefits of co-operative water management. These efforts have led to the development of the BEAM. BEAM is a decision support system developed to help facilitate a long-term agreement on the integrated management of water resources and the environment in the Aral Sea Basin.

3   Beam—focus on economic value of the water

BEAM belongs to class of simulation models called river basin planning models, or sometimes water resources planning models. These models are used for optimization of reservoir operations, analysis and optimization of water allocation institutions, and long-term water supply planning. The focus of these models is on representing the essential hydrological, infrastructure, and institutional components of a water resources system.

BEAM is developed as a decision support system to facilitate putting "value on water use" as well as the sustainable use of water resources in support of development. The model estimates welfare changes associated with changes to how water is allocated between the five countries in the basin: Kazakhstan, Kyrgyzstan, Uzbekistan, Tajikistan and Turkmenistan. The model assumes that water use in Afghanistan is fixed.

In the BEAM representation, water is allocated according to economic optimization criteria:

(1) The BEAM model allocates water across time and space to different uses so that the basin-wide economic value of water use is maximized .

(2) The purpose of the BEAM model is to explore whether it may be possible to change existing water allocation patterns in ways that enhance overall welfare in the Aral Sea basin

(3) The BEAM model also facilitates the estimation of the economic impact of changes to water allocation patterns on different groups within the basin, including the riparian states as well as different sectors such as irrigation and hydropower.

(4) The model further allows the user to estimate the economic impact of changes to physical infrastructure such as new reservoirs and irrigation efficiency improvements.

The model focuses on the following principles of economic management of water resources:

(1) Efficiency (assessing how efficiency within irrigation influences the economic output).

(2) Effectiveness (assessing whether re-allocation of water from one sector, such as hydropower, to another, such as agriculture, increases welfare).

(3) Equity (who will gain from changes in allocation of water from one sector to another and who will lose?).

4　The geographic scope and sectors modeled

The geographic scope of the Aral Sea BEAM consists of the rivers Syr Darya, Amu Darya, Kashkadarya, and Zarafshan, along with the Aral Sea. Together they constitute the Aral Sea basin. BEAM covers both the water resources of the basin including 14 river sections, 6 lakes, 28 reservoirs and 19 water source nodes as well as land resources including 17 planning zones (app. $60.000 \sim 80.000 \ hm^2$ each) with water demand .

The BEAM simulation runs on a monthly time step for one year. The simulation period corresponds to the hydrological year used for water planning in the Aral Sea basin. This hydrological year begins on 1 October and ends on 30 September. Representative years have been chosen for the baseline (a normal average year) and a dry year.

The model covers the following sectors:

(1) Irrigation of crops including wheat, cotton, alfalfa, rice, fruit, vegetables and others. The value added from the use of water in irrigation is assumed to equal the difference between gross production value and the sum of input costs.

(2) Hydropower production, which is simulated as a function of both flow and reservoir surface elevation (head). The value added from hydropower production is assumed to equal the cost of importing equivalent supplies. The unit cost of importing energy varies seasonally.

(3) Household water use. Household water use is assumed to be fixed and value added is not estimated.

(4) Industry water use. Industry water use is assumed to be fixed and value added is not estimated.

(5) Environmental water use. Water uses characterized as environmental water uses in the model consist of flows reaching the Aral Sea. These water uses are fixed and the value added from these uses is not estimated.

5　General Algebraic Modelling system (GAMS) as a part of the beam

BEAM is programmed in GAMS—software developed by a group of economists at the World Bank. Stakeholders throughout the region have been involved in the development of the model, and about 10 national experts, including IFAS staff, have been trained in using the model. Among others, they have learned how to do sensitivity analyses on the basis of the model, make scenarios and report on these in a standard format.

6　Main list of input/output data

The input/output data entered by the user or visible to the user include the following.

6.1 Input data

(1) Monthly estimates of basin inflows for both normal and dry years.
(2) Base year cropping patterns.
(3) Crop prices.
(4) Crop input costs (labour, capital, other).
(5) Parameters of CET transformation frontier (constraining agricultural land use change).
(6) Monthly electricity prices.
(7) Ecosystem/nature water needs.
(8) Household water needs.
(9) Industry water needs.
(10) New reservoir capacity plans.
(11) New hydropower facilities.
(12) Irrigation efficiency investments.

6.2 Output data

(1) Changes in welfare/economic value by planning zones and countries.
(2) Crop pattern change.
(3) Water use change.
(4) Reservoir discharge patterns.
(5) Water balances for all 84 nodes.
(6) Agricultural economics for 17 zones.
(7) Reservoirs' storage/discharge/income.

7 Core decision variables and scenario planning

The core decision variables are crop areas and reservoir releases.

The model input user interface can be used to modify input data and develop scenarios. A scenario is set of model assumptions defined by the user for the purpose of investigating how these assumptions might affect water use and welfare in the basin. It is possible to modify the following types of assumptions in BEAM:

(1) Crop sales prices.
(2) Crop input factor prices.
(3) A parameter affecting the extent to which cropping patterns may change from the baseline (the "flexibility" parameter).
(4) The group of crops for which cropping patterns may change from the baseline (the "flexible" crops).
(5) Allocation of water to the Aral Sea and other terminal lakes in the basin.
(6) Electricity prices.
(7) Hydrological conditions.
(8) Demographic conditions (which affect domestic and industrial water demands).
(9) Level of investment in irrigation efficiency improvements.
(10) New reservoirs and hydropower facilities.
(11) The extent to which carryover storage may be exercised during dry years.

Any combination of the types of assumptions listed above may be used to define a scenario.

8 Examples of results from aral sea beam modelling

Fig. 2 compares basin-wide value added assuming optimization at the basin scale to a baseline scenario in which water use is constrained to observed levels. The scenarios are based on hydrological and economic data observed in 2009, which was considered an average year in terms of water availabity. The figure suggests that it may be possible to improve overall welfare in the basin through co-operative management. However, the figure also suggests that water scarcity is not a major problem in the basin when water availability is consistent with the historical average (i. e. , water scarcity does not appear to place significant limits on production for both irrigated agriculture and hydropower).

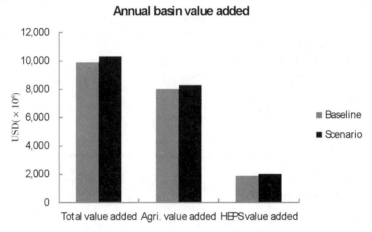

**Fig. 2 Comparison of basin-wide welfare under baseline and optimized
scenarios assuming average hydrological conditions**

Fig. 3 compares the same baseline scenario presented in Fig. 2 to an optimized scenario in which hydrological conditions are set equal to conditions observed in 2001, which was considered a dry year in the basin. The figure suggests that water scarcity limits production in the agriculture and hydropower sectors during dry years and that it may be possible to improve basin-wide welfare through co-operation during dry years.

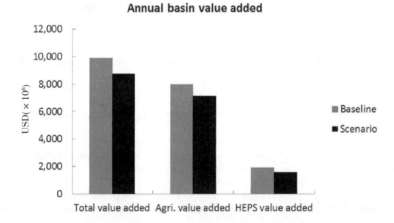

**Fig. 3 Comparison of basin-wide welfare under baseline and optimized
scenarios assuming average hydrological conditions
in the baseline and dry conditions in the optimized scenario**

Fig. 4 presents welfare changes in the agriculture sector associated with the two scenarios presented in Fig. 3. The figure suggests that, if water scarcity exists, it is economically efficient from a basin-wide perspective to continue irrigation to reduce irrigation significantly in Turkmenistan. The figure highlights the fact that welfare gains at the basin-scale are not evenly distributed at the national scale. In this case, Turkmenistan would be a loser if water were allocated to maximize welfare at the basin scale. This suggests that efforts to co-operate at the basin scale will have to be accompanied by mechanisms under which countries that gain from co-operation can compensate countries that lose.

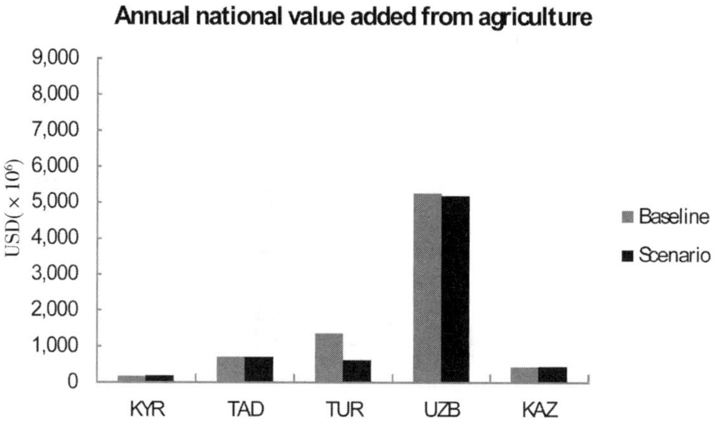

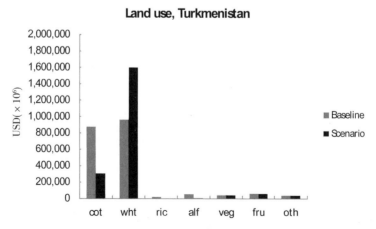

Fig. 4 **Comparison of welfare changes under baseline and optimized
scenarios assuming average hydrological conditions in the
baseline and dry conditions in the optimized scenario**

Fig. 5 presents changes in land use by crop type in Turkmenistan associated with the two scenarios presented in Fig. 3. The figure indicates that most of the reduction in value added in Turkmenistan results from the conversion of cotton production to less water-intensive wheat production.

Fig. 5 **Comparison of land use in Turkmenistan under baseline and
optimized scenarios assuming average hydrological conditions
in the baseline and dry conditions in the optimized scenario**

Fig. 6 compares the same baseline scenario presented in Fig. 2 to an optimized scenario in which hydrological conditions are set equal to dry conditions and all countries have made investments in irrigation efficiency. The figure suggests that investments in irrigation efficiency could increase the economic efficiency of irrigation in Turkmenistan under dry conditions. Contributions to investment in water-saving irrigation technologies in Turkmenistan could perhaps be a way for the winners from co-operation to compensate the losers.

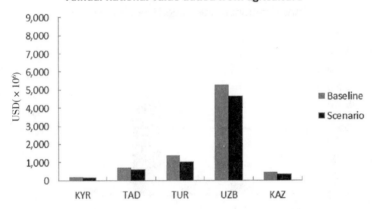

Annual national value added from agriculture

Fig. 6   Comparison of welfare changes under baseline and optimized
         scenarios assuming average hydrological conditions in the baseline and
         dry conditions and irrigation investment in the optimized scenario

9   Conclusions

BEAM will be a decision support tool to policy makers in the region when negotiating on water allocation, considering major investment decisions in the water, food and energy sectors, and when exploring consequences of climate change for the economic development in the basin.

(1) It provides input to the decision making process of setting sustainable development goals for water, food and energy —not least, because it addresses water, food and energy issues in an integrated and holistic manner. The BEAM acknowledges the strong inter-dependence between long term sustainable water resources management, food security and sustainable energy supply. It facilitates the establishment of targets to increase total food supply-chain efficiency, to increase water efficiency in agriculture and to increase water use efficiency in energy production.

(2) It provides a tool to integrate water in climate adaptation and, especially, mitigation policy at all levels. It highlights the cross-cutting nature of water and the need to ensure that water management is fully integrated into climate adaptation and mitigation measures. It eases the integration of water resource management in national climate adaptation and mitigation plans and measures.

(3) It advances a more effective cooperation on transboundary waters. Increasing population, economic growth and climate change nay intensify tensions over transboundary water resources around the globe, not only in the Aral Sea Basin. Multi-country dialogue and cooperation is imperative to address issues adequately.

(4) It promotes good water governance constituting a prerequisite for sustainable development. In particular because it will be accessible to the public, thereby promoting stakeholder participation in water resource management.

References

Phillips D, Daoudy M, McCaffrey S, et al. Trans-boundary Water Cooperation as a Tool for Conflict Prevention and Broader Benefit Sharing, Global Development Studies No. 4 [R]. Stockholm: Ministry of Foreign Affairs Sweden, 2006.

Qaddumi, Halla . Practical approaches to transboundary water benefit sharing [R]. Overseas Development Institute, London, 2008.

Rogers P, Bhatia R, Huber A. Water as a Social and Economic Good: How to Put the Principle into Practice TAC Background Papers No. 2 [M]. Stockholm: Global Water Partnership, 1998.

Sadoff C W, Grey D. Beyond the River: The Benefits of Cooperation on International Rivers [J]. Water Policy, 2002, 4: 389-403.

Victor Dukhovny, Vadim Sokolov. Lessons learned on cooperation building ato manage water conflicts in the Aral Sea [EB/OL]. http://unesdoc. unesco. org/images/0013/001332/133291e. pdf. http://www. ec-ifas. org/aral_basin/aral_sea/

Ferris M C, Munson T S. PATH 4. 6 in GAMS - The Solver Manual, GAMS Development Corporation [EB/OL], Washington. http://gams. com/dd/docs/solvers/path. pdf.

Study on the Online Real – time Irrigation Schedule for the Climate Change and Its Management Software

Ma Jianqin and *Li Ming*

North China University of Water Conservancy and Hydroelectric Power,
Zhengzhou , 450045 , China

Abstract: Aimed at the shortcomings of existing irrigation schedule, such as lack of full use of precipitation, especially lack of exercisable implementation, based on the theory of water balance, the crop's on – line simulation model is proposed, then a real – time irrigation schedule is built in the paper. In the irrigation schedule the actual precipitation and soil moisture content are fully taken into accounting with the on – line monitoring weather data and soil moisture data. Then Java Language is used to develop the software for real – time irrigation management. The winter wheat is chosen for study, and some parameters are verified by the experiment. The method and program could realize real – time remote management and help use random precipitation effectively, and it is suitable for crops in North China.

Key words: irrigation schedule, precipitation, online real – time, remote, soil moisture

The agricultural water shortage has posed a great threat for food security in China. How to deal with the effective utilization of precipitation to relieve water shortage in agriculture, and effectively improve crop yields is one of the most difficult problems which urgently needed to be solved. Real – time irrigation forecast based on the "real – time" data is the foundation for the dynamic water plan. Based on the experimental real – time monitoring data, using of water balance principle and the simulation model of crop root growth, then the precipitation and soil moisture are fully considered, the online real – time irrigation model for crop is proposed in the paper, and the corresponding online management software is also developed. All this can help to solve the problem that the real – time precipitation is not fully considered in irrigation schedule and effectively alleviate the severe water shortage in agriculture, and are suitable for use in the arid and semi – arid area in North China.

1 Introduction

Studies on irrigation system dated from 1970s. Gear et al. used neutron probe to monitor soil moisture, and proposed an irrigation forecast model by predicting the soil moisture dynamics of field; Maidment et al. proposed irrigation water requirements prediction model by using soil, crop type and years of meteorological data, in the model the daily soil water balance is simulated to forecast irrigation water demand projections for per unit area, and observations of historical water consumption data is simulated to get simple irrigation decisions; In 1985, Smith et al. proposed one method to determine the appropriate irrigation schedule which could simulate the irrigation schedule changes among years, by combining the simple crop growth model and soil water balance model; In 2002, George et al. developed a decision support system of reference crop evapotranspiration forecasting, which could select the suitable calculation method based on the input data by the means to improve prediction accuracy. Recently spatial information technology and computer simulation techniques have been used in irrigation, based on soil moisture monitoring, internet technology and 3S (RS, GIS, GPS) technology to complete the information collection, exchange and transmission , several water – saving irrigation forecast models are proposed, and irrigation water decision support systems are established to simulate crop yields and process of crop water needs.

Irrigation research began in 1980s in China. According to the soil moisture change characteristics over time, Kang established soil water prediction model with the general additive model of time series in 1990. In 1994, Li put forward field water balance model under different conditions, the crop water demand was calculated by historical meteorological data, and the estimate of crop growth trends and short – term weather forecast were used for the real – time irrigation forecasting; In 1997, on the basis of the previous research, Li proposed a daily crop water requirement calculation according to the change laws of annual historical reference crop water demand, and a general model of real – time irrigation were also posed based on the principle of water balance. In 2002, Mao et al. improved conventional forecasting methods and proposed a new model which used weather type, crop coverage of green leaves, and soil available moisture to get real – time crop water requirement which provided a more accurate basis. Based on the soil water balance model, Yang et al. added in soil moisture index to execute the soil moisture status of field forecasting, and combined with the measured meteorological data to simulate and analysis the actual evapotranspiration and soil moisture dynamics of wheat in Hexi oasis irrgation distric, and this could improve the forecast accuracy of crop irrigation time; in 2009, Cai et al. made continuous observation on crops canopy temperature, used canopy temperature indicator to determine the crop drought degree, and made real – time decision – making and control for irrigation, which achieved "real time" and "moderate" irrigation.

Existing research shows that there is few application example for foreign models and methods in China due to different crop types, national conditions and different levels of agricultural management. Most domestic irrigation models make modification to the initial value of soil moisture after rain, randomly considering the impact of rainfall forecasts before irrigation, especially lack of remote real – time control system. Therefore, establishing crops' online real – time irrigation schedule to efficiently use of precipitation for easing the shortage of agricultural water resources has become a serious problem. Based on real – time weather and soil moisture monitoring data, using field water balance principle and crop root growth simulation technique, a real – time irrigation schedule considering rainfall into is proposed for winter wheat by experiment and software debugging in the paper.

2 Crops' real – time irrigation model

2.1 Crops' real – time water requirement prediction model

2.1.1 Crops' real – time water requirement prediction

Crop water demand prediction can be gotten:

$$ET_i = ET_{0i} \cdot K_c \cdot K_w \tag{1}$$

where, ET_i is the crop water demand of the i th day, mm/d; ET_{0i} is the reference crop evapotranspiration of the i th day, mm/d, can be calculated by the modified Penman equation in which the real – time meteorological data is measured from Automatic Weather Stations (AWS); K_c is the crop coefficient, in the paper it is calculated as Eq. (2) and Eq. (3):

$$K_c = 7.346(i/I)^2 - 1.606(i/I) + 0.097,2 \quad (i/I \leqslant 0.58) \tag{2}$$
$$K_c = -3.463\ln(i/I) - 0.190,9 \quad (i/I \geqslant 0.58) \tag{3}$$

where, i is the cumulative days of growth period, d; I is the total days of growth period, d. K_w is the soil moisture correction factor, it is 1 for dry crops under full irrigation, for deficit irrigation, it is calculated as follows:

$$K_w = \ln(1 + \theta_i)/\ln101 \quad (\theta_{min} < \theta_i < \theta_{cl}) \tag{4}$$
$$K_w = \alpha \cdot \exp[(\theta_i - \theta_{min})]/\theta_{min} \quad (\theta_i < \theta_{min}) \tag{5}$$

where, θ_i is the soil water content of the ith day; θ_{cl} is the soil water maximum index of deficit irrigation, expressed as percentage of field water – holding rate θ_{max}; θ_{max} is the soil moisture minimum criteria, also expressed as percentage of θ_{max}, θ_{max} is always 60% for dry crops; α is the empirical coefficient, always be put as 0.89.

2.1.2　Effective precipitation calculation

Available precipitation is gotten as Eq. (6):

$$P_{0i} = \alpha P_i \tag{6}$$

where, P_i is the actual precipitation of the i th day, mm; P_{0i} is the effective precipitation of the i th day, mm; α is the effective precipitation utilization factor, the specific value is gotten as reference (Guo YY, 1997).

2.1.3　Soil moisture prediction

Soil moisture prediction is established as Eq. (7):

$$\theta_i = \frac{H_{i-1}}{H_i} \theta_{i-1} - (ET_{i-1} - P_{0i-1} - M_{i-1} - W_{Ti})/1,000nH_i \tag{7}$$

where, θ_{i-1} is the soil water content of the $(i-1)$ th day, expressed as percentage of soil volume; M_{i-1} is the irrigation amount of the $(i-1)$ th day, mm; H_i is the depth for roots absorbing water of the i th day, m, can be simulated every day according to the crop root growth, the root simulation of winter wheat is performed as Eq. (8); n is the soil porosity(%); W_{Ti} is the addition water due to the adds of moisten soil layer from the $(i-1)$ th day to the i th day, calculated as Eq. (9):

$$H(i) = \frac{a}{(1 + b \cdot e^{-ci})} \tag{8}$$

where, a, b and c are parameters, for winter wheat, $a = 34.588,36$, $b = 127.144,1$, $c = 0.400,458$;

$$W_{Ti} = 1,000(H_i - H_{i-1}) \cdot n \cdot \theta \tag{9}$$

where, θ is the average moisture content in the depth of$(H_i - H_{i-1})$.

2.2　Crop real – time irrigation model

When the predicted soil moisture θ_i is less than θ_{min}, then the i th day will be the projected irrigation date. Modify the calculated soil moisture with those measured each day and simulate the depth of roots using roots growth model, and forecast the irrigation date. If there is precipitation on the predicted irrigation date, then give no irrigation or postpone it. Then after irrigation, take the measured soil moisture as the initial value of the next stage, continue to predict the next irrigation date and amount.

(1) Irrigation amount can be obtained as equation (10) when the runoff is sufficient:

$$M_i = 1,000 \cdot n \cdot H_i(1 - \theta_i) \cdot \theta_{max} \tag{10}$$

(2) Irrigation amount can be performed as equation (11) when the runoff is insufficient. When the runoff is insufficient or water resource is in shortage:

$$M_i = 1,000 \cdot n \cdot H_i(\theta_{c1} - \theta_i) \cdot \theta_{max} \tag{11}$$

3　Irrigation management software

The Web – based software is development of B/S (Browse/Server) mode system. Java is the language, MyEclipse is the integrated development platform, the Struts, Spring and Hibernate framework are adopted for three – tier architecture, MYSQL is taken as the database, and Ajax technology and JFreeChart components are used to finish asynchronous data request and report display. The process of crop irrigation forecasting programming is shown in Fig. 1.

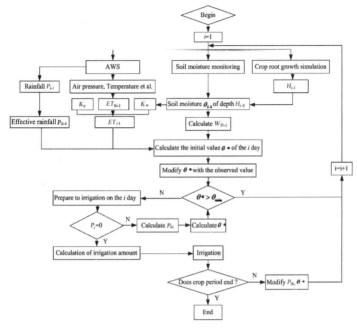

Fig. 1   Flow chat of real – time crop irrigation schedule programming

4   The application and experiment scheme

Winter wheat is chosen as testing crop, using the real – time irrigation schedule to conduct the irrigation and verify the crop index in the experimental district of North China Institute of Water Conservancy and Hydroelectric Power. The daily root length of winter wheat is calculated according to Eq. (8), the soil moisture of different depths during the growth of winter wheat can be obtained via those 20 non – interfering sensors which buried with spacing 10 cm in the $0 \sim 2$ m soil layer. Using the models in the paper and software for real – time online irrigation schedule of winter wheat to realize the function of real – time automatic transfer of the weather data, soil moisture data and real – time forecast of water demand. This model can take full advantage of precipitation for winter wheat during its growth.

5   Conclusions

Aimed at the poor real – time and inadequate use of precipitation and other issues in current irrigation schedule, based on the principles of the field water balance and crop root growth simulation theory, with the soil moisture monitoring equipment and the data of automatic weather stations, a web – based crop online real – time irrigation schedule model is proposed in this paper, and java language is used to develop the corresponding management software. The field test of winter wheat is used to check and apply the model and the software. Studies show that the online irrigation technology not only realizes the remote real – time monitoring on crop growth, but also makes fully use of precipitation resources and alleviates the shortage of agriculture water resources effectively, it can be applied in the northern region and has practical significance on agricultural water – saving and effective utilization of water resources.

Acknowledgements
The authors wish to express their thanks to the 948 project funding of Ministry of Water

Resources of P. R. China (code: 201047), The National Natural Science Funds (41071025) , Henan natural science research project of Education Department (2009A170004) , and Henan technology funds (092102310197).

References

Gear , Roy D, Campbell, et al. Irrigation Scheduling with Neutron Probe [J]. Journal of the Irrigation and Drainage Division, 1977,103(3):291 – 298.

Maidment, David R, Hutchinson, et al. Modeling Water Demands of Irrigation Projects [J]. Journal of Irrigation and Drainage Engineering, 1983,109(4):405 – 418.

Smith R C G, Steiner J L, Meyer W S, et al. Influence of Season to Season Variability in Weather on Irrigation Scheduling of Wheat: A Simulation Study [J]. Irrigation Science, 1985,6(4): 241 – 251.

George B A, Reddy B R S, Raghuwanshi N S, et al. Decision Support System for Estimating Reference Evapotranspiration [J]. Journal of lrrigation and Drainage Engineering, 2002,128 (1):1 – 10.

Inman – Bamber N G, et al. A Web – base System For Scheduling Irrigation In Sugarcane [J]. Prc. Int. Soc. Sugar cane Tech. , 2007,26:459 – 464.

Kang S Z. Stochastic Modeling of Dynamic Process of Soil Moisture [J]. Acta Pedologica Sinica, 1990,27(1):17 – 24.

Li Y H. Mothod and Application of Real – time Irrigation Forecast [J]. Journal Of Hydraulic Engineering, 1994(2):46 – 51.

Li Y H, Cui Y L. A Study on Real – time irrigation forecasting in the Zhanghe irrigation schome [J]. Advances In Water Science, 1997,8(1):72 – 74.

Mao Z, Li Y H, Li H C. Real – time Forecast of Irrigation [J]. Engineering Science, 2002,4(5): 25 – 31.

Yang J, Huang P, Wei B. Research on Irrigation Forecast Model of Wheat in the Hexi Oasis Irrigation District [J]. Journal of Gansu Agricultural University, 2007(4):118 – 122.

Cai J B, Liu Y, Li X. Online Data Collection System of Crop Infrared Canopy Temperature for Irrigation Control [J]. China Rural Water and Hydropower, 2010(2):64 – 66.

Liu B, He X L, Pu H S. Framework Design for Real – time Irrigation Dispatching System [J]. China Water Resources, 2007(15):51 – 52.

Gu S X, Li Y H. Real – time Forecasting of Crop Evapotranspiration of Huoquan Irrigation District [J]. Journal of Wuhan University of Hydraulic and Electrical Engineering, 1998,31(1): 37 – 41.

Guo Y Y. Irrigation and Drainage Engineering [M]. Beijing: China WaterPower Press, 1997.

Wang M L, He D J. Visualized Simulation of Wheat Roots Based on L – system [J]. Journal of Agricultural Mechanization Research, 2008(3):36 – 37.

Nutrient Management in the Lake Tai Basin, China —Application of Source Water Quantity and Quality Model to Dongshan Peninsula

D. K. Waters[1], *Jiao F*[2], *Chen M D*[3], *Yang J D*[3], *Pape S*[4], *Zhang H J*[5], *Li, J*[4] and *Nash D*[6]

1. eWater CRC, University of Canberra, Australia
2. Nanjing Institute of Geography and Limnology Science, Nanjing, 210000, China
3. Academy of Environmental Science, Suzhou Environment Institute, Suzhou, 215000, China
4. Earth Systems, Melbourne, Australia and Shanghai, 200000, China
5. Taihu Basin Authority, Shanghai, 200000, China
6. Department of Primary Industries, Victoria, Australia

Abstract: Lake Tai is the third largest freshwater lake in China, bordering Jiangsu and Zhejiang provinces, providing water to 3.0×10^7 residents. A severe algal bloom in 2007 led to the development of the Lake Tai Master Plan, launched by the National Development and Reform Commission (NDRC), to improve nutrient management in the basin. Under a joint Australian China Environmental Development Project, the Australian eWater Source Integrated Modelling System (IMS) was applied to model water quantity and quality for a pilot area on the Dongshan Peninsula in the Lake Tai Basin. Source is a powerful modelling platform for environmental management which can integrate many physical processes and human impacts, successfully applied in over 70 basins across Australia.

The aims of the project were to: apply the model to a small pilot area and investigate strategies to improve nutrient (total nitrogen (TN) and total phosphorus (TP)) management in the Lake Tai Basin. Source was applied to the 77 km² Dongshan Peninsula. The model consisted of 50 subcatchments with seven major landuse categories including aquaculture (43% of the area), upland and lowland fruit trees (31%), low and high density urban (12%) and vegetables (6%). The daily timestep model was run for the period 2001 to 2010. The flexible nature of the Source software enabled a number of novel modelling approaches to be applied such as simulation of the seasonal variability in runoff and nutrient export from the 33 km² of aquaculture ponds on the Peninsula.

The model was parameterised and validated drawing on local knowledge, expert opinion from the Chinese and Australian technical working group and literature from both countries. Three scenarios were investigated. The implementation of a proposed wetland to filter runoff from the aquaculture ponds could reduce TN and TP exports to the lake by 13% and 16% respectively. Other scenarios included the closure of the sewage treatment plant, and the adoption of improved fertilizer management practices in the vegetable and tree cropping areas. The project has provided valuable insights into the dominant nutrient sources exported to Lake Tai, potential nutrient reduction strategies and priorities for future research and data collection.

Key words: water quality, catchment modelling, source

1 Introduction

Lake Tai is the third largest freshwater lake in China, bordering Jiangsu and Zhejiang Provinces, providing water to 3.0×10^7 residents and covering an area of 2,338 km². Rapid industrial and agricultural development in the last 20 years have caused water eutrophication in Lake Tai watershed to become one of most serious environmental issues in Southeast China (Mao et al. , 2008). A severe algal bloom in the Lake in 2007 led to the development of the Lake Tai Master Plan (NDRC 2008), launched by the National Development and Reform Commission (NDRC), to improve nutrient management in the basin.

Through the Australia China Environment Development Partnership (ACEDP), the Aus Lake Tai Cluster (AUS) led by Earth Systems and the NDRC's International Cooperation Centre (ICC) has implemented the Lake Tai Water Pollution Treatment Project in Suzhou, Jiangsu, Huzhou and Zhejiang Provinces. The project's objective is to strengthen management and policy approaches in the Lake Tai Basin by increasing awareness of Integrated River Basin Management (IRBM) and science – based management approaches, and identify priority areas for further assistance.

Non – point source nutrient pollution was identified as a major but still poorly understood source of pollution for the Lake. Municipal officials raised concerns that while they have mobilized significant resources, understanding of the impact and effectiveness of their investments is lacking. They indicated that approaches that enable better informed and evidence based government policy and decision making are needed.

Modelling is one of a number of strategies used to assist in identifying and prioritizing management practices to reduce both point source (PS) and non – point source (NPS) pollution to the Lake. The Australian eWater Source Integrated Modelling System, Source (Delgado et al. , 2001), is a flexible modeling platform for dealing with the complex water quantity and quality management rules and integrates many physical processes and human impacts into the model. Source has been successfully applied in over 70 catchments across Australia (eWater CRC 2011).

The aim of this project was to construct a source model for a pilot area in the Lake Tai Basin, the Dongshan Peninsula, to demonstrate how this particular model could be used to investigate strategies to improve nutrient (total nitrogen (TN) and total phosphorus (TP)) management. This paper outlines the model development and presents key findings from the modelling related to a number of the planned strategies for reducing nutrient export on the peninsula.

It should be noted that the main focus of this modelling exercise, was to demonstrate how this model could be applied in China to explore alternative nutrient management strategies. Model results presented in this paper should only be taken as indicative estimates of pollutant loads for the Dongshan Peninsula. Further work will be required to refine and improve model outputs as further local data becomes available.

2　Source modelling framework

The source catchment water quantity and quality modelling framework is not a single model but a collection of models whereby the user chooses runoff, pollutant generation, filter and in stream model components that best describe the catchment hydrology and pollutant generation processes occurring in their catchment. Pollutant generation models are applied to a "functional unit", typically a specific landuse or hydrological response unit within a subcatchment. All flows and pollutant loads for each functional unit in a subcatchment are aggregated at a subcatchment outlet node. Runoff and pollutant loads are routed through a series of nodes and links (Fig. 2) representing the stream network to the catchment outlet (Delgado et al. , 2011).

In addition to the standard range of models available within Source, the user can also add their own mathematical equations or "expression" at any node or link within the model to calculate and record any range of variables. The Dongshan model was developed incorporating physical processes at a level of complexity appropriate to the current knowledge and data available for the area.

3　Donghsan Peninsula

The Dongshan Peninsula is located in the south eastern corner of Lake Tai (N 31°05', E 120° 24') (Fig. 1). The catchment area of interest covers 77 km^2. It is located in a subtropical monsoon zone. The peninsula consists of a low mountain range with steep slopes, surrounded by lowlands mostly occupied by aquaculture ponds. Average temperatures range from 4 ℃ in January to 28 ℃ in July with the average annual rainfall of 1,113 mm and average lake evaporation of

984 mm. Approximately 60% of the annual rainfall occurs over the spring and summer periods.

Fig. 1   Location of Dongshan Peninsula indicated by dashed circle

The soils of the upland ridges of the peninsula are shallow red loess, used for growing pine forests only and are protected from further development to minimize erosion. The intermediate and lower slopes, used for growing fruit and tea trees are a clay loam (30% clay fraction), have slopes between 15° ~25°, pH 5 ~6, and a soil depth of 0.40 ~1.0 m. The flat areas close to the lake shore are predominantly used for aquaculture, vegetables and paddy fields, are clay loam with some fine sands and sandy loam soils present.

The dominant landuse is aquaculture (predominantly used for crab farming) covering approximately 43% of the area. Upland and lowland fruit trees occupy 31% of the area, high and low density urban (12%), vegetables (6%) with the remainder industrial and forested land. Dongshan Peninsula has a population of 53,000, of which around 44,000 people are in rural or low density urban areas. Runoff to Lake Tai is from both rainfall and controlled drainage from the aquaculture ponds. Runoff enters the lake via a combination of streams and constructed canals. Gates are located at each exit point and many of the canals are interconnected. When lake levels increase to 3.45 m or greater, all gates are closed to prevent flooding of the low lying areas and runoff is then pumped to the lake via pump stations located at four of the 13 exit points to the lake.

4   Model development process

The eWater CRC Best Practice modelling guidelines (Black et al., 2011) were used to guide the project development process. Five of the key steps in the project development process are described in the following section.

4.1   Problem definition

It was important at the project inception to clearly describe and reach agreement on the problem to be addressed. This was achieved through iterative consultation with a range of stakeholders involved including policy staff, scientists, and other technical experts.

4.2   Technical working group

A project technical working group was established comprised of Australian experts from eWater

Cooperative Research Centre (eWater) and Earth Systems Australia & China; Chinese technical experts and scientists from Suzhou Municipality, the Taihu Basin Authority (TBA) and the Nanjing Institute of Geography and Limnology (NIGLAS). A series of workshops and training activities were conducted both in China and Australia to build capacity in the development and use of the Source modelling framework. The working group took an active role in identification of priority landuses, catchment drainage configuration, collation of local literature, validation data and scenario development.

4.3 Conceptual model development

Once all relevant information was collated, site visits and discussion with local experts completed, a conceptual model of the hydrology and nutrient generation and transport processes was constructed. This was achieved through a half day workshop to develop maps and flow diagrams on a whiteboard. There were numerous iterations from the initial design and agreement was reached on an appropriate model with a level of complexity appropriate to the data availability and project time constraints.

4.4 Scenario development

A key objective of the project was to demonstrate how the model could be applied to assess different management options to reduce nutrient export to the lake. The scenarios selected were based on Lake Tai Basin policies or projects proposed in regional plans. Numerous scenarios were considered by the working group with three selected for detailed investigation based on their practicality and potential for implementation. The three scenarios proposed were as follows.

4.4.1 Improved point source management

There were two components to this scenario, formulated from recommendations in the Lake Tai Master Plan (NDRC 2008). Firstly, decommissioning of the Sewage Treatment Plant (STP) for the Dongshan Peninsulas with sewage from Dongshan township transferred to Wuzhong District STP. Secondly, in recent years 80% of low density urban areas were upgraded with secondary treatment facilities as part of the Master Plan. In the scenario, we assume the remaining 20% of household septic systems were also upgraded.

4.4.2 Improved diffuse source management

There were two components to this scenario based on recommendations in the Lake Tai Master Plan (NDRC 2008). Firstly, the plan recommends that improved farming practices such as precision application of fertilizer, modifying the timing of fertilizer application and crop rotations could reduce nitrogen and phosphorous fertilizer inputs by 20% ~ 30% & 10% ~ 20% respectively. Currently, only a small portion of the agricultural areas (<10% of fruit tree crops for example) are implementing these practices. For this scenario the assumption was made that the nutrient generation rates were positively correlated with the rate of fertilizer application (Wang et al. , 2008) in the Lake Tai Basin. A 20% reduction in TN and 10% reduction in TP generation rates for the two farming practices occupying the greatest area on the Peninsula (lowland vegetables and upland fruit trees) were therefore applied. Secondly, with 43% of the area of the Peninsula used for crab farming it was important that this industry be included in the scenario. Local technical experts suggested that the majority of existing crab ponds were using traditional farming practices used in the district for many years. Based on local knowledge, the technical working group estimated that improved crab farming operations could reduce pond nutrient concentrations exported to the canals by as much as 25% ~ 30%. For this scenario we assumed all ponds were upgraded to reduce TN and TP exports by 25%.

4.4.3 Construction of large wetland for aquaculture pond filtering

Scenario 3 was a remediation project proposed in the Suzhou Environmental Protection Plan.

The project was in a very early stage and thus difficult to obtain detailed information and data. The proposed wetland has joint objectives for environmental remediation, landscape transfiguration and recreational use. The location of the proposed wetland is in the southern corner of the Dongshan Peninsula in an area not currently farmed (Fig. 2). The objective of this work was to provide a wetland to filter aquaculture pond water prior to discharge to the Lake. The construction is proposed to occur in three stages. Stage 1 – over an area of 3 000 mu (2 km^2), Stage 2 – 10,000 mu (6.7 km^2) and Stage 3 – 35,000 mu (23 km^2). The model was modified to divert up to 50% of the drained aquaculture pond water to the Stage 2 wetland prior to release to the Lake over the month of December. With limited details available on the design specifications for the wetland, a number of assumptions were made as to the wetland design specifications and nutrient removal rates based on Australian and Chinese literature ((Melbourne Water (2005), Lu et al., (2009), Fei et al., (2011))

4.5 Model construction

Specific details of the development of the Dongshan Peninsula Source model is detailed in Waters et al., (2012), the following is a brief outline of the model configuration. The model contained 45 subcatchments, with subcatchment boundaries based largely on topography for the upland areas, canals, streams and earthen banks on the lowland aquaculture areas. The node link network was designed to enable reporting of modelled runoff and loads at five nodes or zones of the Peninsula (Fig. 2). The model contained nine landuse categories and one STP.

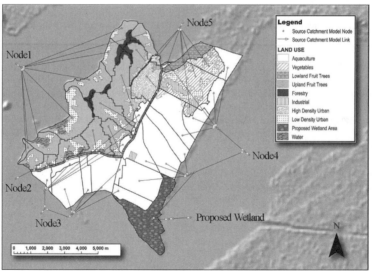

Fig. 2   Source Landuse and subcatchment configuration with the node link network overlayed

The model was run at a daily time step for a 10 year period from 2001 ~ 2010. Annual rainfall over the model run period ranged from 782 ~ 1,280 mm with an annual average of 1,068 mm close to the long term mean. Rainfall and evaporation data were supplied by the Suzhou Environment Institute, Academy of Environmental Science; and the Suzhou Environmental Protection Bureau. Source uses conceptual rainfall runoff models to generate daily runoff estimates. The inputs required are daily rainfall and potential evaporation and typically rely on measured runoff data for model calibration. SIMHYD rainfall runoff model (Chiew et al., 2002) was chosen due to its extensive use across Australia particularly in wet and temperate climates. Due to a lack of measured daily runoff data for the Peninsula, manual calibration was undertaken to ensure the proportion of runoff generated for each landuse corresponded with literature values for the region. Australian literature was also sourced particularly for urban areas where comparisons could be made as an additional source of data to validate the model results (Refer Waters et al., 2012 for further details). Through consultation with the technical working group it was agreed that Total Nitrogen (TN) and Total Phosphorus (TP) would be modelled. It was acknowledged that differentiation of the different forms of nitrogen, in particular Ammonia nitrogen ($NH_3 - N$), would be useful when considering management strategies outlined in the Master Plan and may be explored in future projects. A mean concentration was assigned to each landuse based on local literature. The input data was predominantly from farm or field scale experiments.

Aquaculture ponds are a major source of both runoff and nutrients entering Lake Tai from the Dongshan Peninsula. It was important therefore that the pond hydrology and nutrient generation processes were appropriately represented in the model. The 33 km^2 of aquaculture ponds were aggregated into four large storages with the total volume of the four storages approximating the volume of all individual ponds. Specific expressions were written to simulate filling and draining patterns of the ponds throughout the year.

An additional feature requested to be incorporated into the model was to quantify runoff to Lake Tai from either natural rainfall runoff or pumped runoff as described in section 3. Equations or expressions were written into the model to record the runoff volume and pollutant load entering the Lake as natural rainfall runoff (Lake level below 3.45 m) or via the pump stations (Lake level $\geq$ 3.45 m when gates are a closed). A timeseries file of lake levels was provided for determining pumping periods to the lake.

5 Model results and discussion

The base model provides a good representation of the hydrology and nutrient generation processes for the Dongshan Peninsula at a level of complexity appropriate to the level of data available and modelling objectives. The output from the base model provides useful information on the relative contribution of runoff, Total Nitrogen and Total Phosphorus to Lake Tai from the major landuses represented in the model. The following section summarises the model validation results, modelled runoff and nutrient load estimates and the scenario findings.

5.1 Model validation

There were two sources of data provided for model validation. The first data set was provided from in stream water quality sampling collected at two monthly intervals at two locations in the main stream of the Peninsula collected from 2008 ~ 2011 by the Suzhou Environment Institute. The mean TN concentration for the four years was 2.15 mg/L (0.24 ~ 19.0 mg/L). The Source average annual modelled estimate of TN concentrations entering the lake was double the observed concentration at 4.40 mg/L. For TP the mean measured concentration was 0.18 mg/L (0.02 ~ 2.70 mg/L) with source at 0.35 mg/L.

The second data source was from the Lake Tai Master Plan (NDRC 2008). The plan listed average annual measured TN and TP concentrations from rivers entering Lake Tai between 1998 ~ 2006 in the Changzhou, Wuxi, Suzhou and Huzhou Municipality's. Measured TN concentrations ranged from 2.57 ~ 5.75 mg/L. Source average annual TN concentration was within the range reported at 4.40 mg/L. Measured TP concentrations ranged from 0.114 ~ 0.27 mg/L with Source estimate of 0.35 mg/L slightly higher than the range reported.

The model results were encouraging given the limited local data available to parameterise and validate the model. The over estimation of modelled nutrient concentrations may be due to not accounting for in - stream nutrient losses or nutrient transformation processes. Further data collection to quantify nutrient generation rates for aquaculture, fruit trees and vegetables in conjunction with in stream nutrient sampling during summer runoff periods will greatly improve confidence in model results.

5.2 Base model runoff and load estimates

Aquaculture contributed around two – thirds (66%) of the average annual runoff with the second largest contribution from upland fruit trees (around 15%). Two – thirds of all runoff from the ponds to the Lake occurs in December each year when they are drained. Despite the aquaculture ponds contributing the majority of the runoff, the ponds and upland fruit trees are contributing similar proportion of TN load 37% and 34% respectively (Fig. 3). Whilst fruit trees occupy a smaller area, the combination of above average runoff volumes from the steep slopes and high TN generation rates resulted in similar loads being generated from both landuses. For TP, the model suggests that the majority of the load contribution is from the aquaculture ponds due to their high runoff volume in comparison to the low runoff volumes and low TP generation rates from cropping and urban areas.

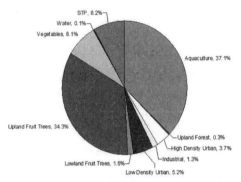

Fig. 3 Average annual modelled TN load for each landuse as a proportion of the total load entering Lake Tai from Dongshan Peninsula

5.3 Management scenario results

An important objective of the project was to demonstrate how the Source model could be applied to explore a range of management scenarios to reduce nutrient export to the Lake Tai Basin. The modelling results for the three scenarios are outlined below and in Fig. 4.

5.3.1 Improved point source managemen

The model suggests that upgrading the remaining 20% of low density urban areas from traditional septic systems to domestic waste units in conjunction with the connection of the Dongshan STP to the main sewage network could achieve a 9% reduction in average annual TN and 13% reduction in TP load to the Lake. The majority of this load (90%) is attributed to the removal of the STP.

5.3.2 Improved diffuse source management

Reduced fertilizer inputs to vegetable and upland fruit growing areas resulted in a 9% and 2% reduction in average annual TN and TP loads respectively directly related to the reduction in generation rates. Adoption of improved practices for aquaculture pond management resulted in a 9% and 23% reduction in average annual TN and TP loads respectively. Therefore the implementation of improved practices in vegetable farming, upland fruit tree areas and aquaculture pond management combined, could result in an approximate 18% reduction in TN export and 25% reduction in TP export to the Lake annually.

5.3.3 Wetland filtration

The model suggests that construction of the wetland to filter 50% of aquaculture pond runoff could reduce total exports of TN by approximately 13% and TP by 16%.

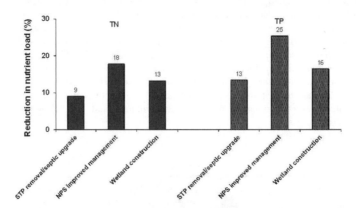

Fig. 4 Modelled TN and TP estimates of load reductions to Lake Tai as a result of implementing three different improved practices: point source management, non – point source management and wetland construction

The scenario results, demonstrate the potential application of Source to explore a range of management strategies. Whilst the results are only indicative, the assumptions used in the model were based on findings from local research and the technical working group. Results suggest that there is potential for significant improvements in water quality to help achieve the short and long term water quality objectives outlined in the Lake Tai Master Plan.

6 Conclusions

A Source model was developed for the Dongshan Peninsula with the flexibility of the model demonstrated through features such as the inclusion of the aquaculture ponds and canal pumping regimes.

The results suggest that the modelled TN and TP concentrations showed reasonable agreement with local data collected from streams within the Peninsula and a number of other rivers entering Lake Tai. The modelling suggests that the Aquaculture ponds and upland fruit trees each contribute between 30 ~ 40% of the total nitrogen load to the lake. Three scenarios were modelled with a number of potential management options identified which require further investigation.

The project has demonstrated the potential of models to inform policy and improve our understanding of the impacts and effectiveness of investment in nutrient management strategies for Lake Tai Basin. The success of this project was only possible through the collaborative efforts of the Australia China Environment Development Partnership (ACEDP), the Aus Lake Tai Cluster (AUS) led by Earth Systems, the NDRC's International Cooperation Centre (ICC), the technical working group members and the eWater CRC.

References

Black D, Wallbrink P, Jordan P, et al. eWater Cooperative Research Centre (CRC). Guidelines for Water Management Modelling: Towards Best – practice Modelling Applications.
Chiew FHS, Peel M C, Western A W. Application and Testing of the Simple Rainfall – runoff Model SIMHYD[M]// Mathematical Models of Small Watershed Hydrology and Applications (Editors: VP Singh and DK Frevert). Water Resources Publication, Littleton, Colorado, 335

−367.

Delgado P, Kelley P, Murray N, et al. Source User Guide[M]. eWater Cooperative Research Centre, Canberra, Australia.

Fei Zhong, Wei Liang, Tao Yu, et al. Removal Efficiency and Balance of Nitrogen in a Recirculating Aquaculture System Integrated with Constructed Wetlands [J]. Journal of Environmental Science and Health. Part A 46:789 – 794.

Lu S, Zhang P, Jin X, et al. Nitrogen Removal from Agricultural Runoff by a Full – scale Constructed Wetland in China[J]. Hydrobiologia, 621:115 – 126.

National Reform and Development Commission (NDRC) (2008), Master Plan for Comprehensive Management of Water Environment in the Taihu Lake Basin, Approved by the State Council, April 2008, Beijing. (Translated by the Lake Tai Water Pollution Treatment Project 2009)

Mao J, Chen Q, Chen Y, et al. Three – dimensional Eutrophication Model and Application to Taihu Lake, China [J]. Journal of Environmental Science, 2008, 20 (3): 278 – 284.

Melbourne Water (2005). Constructed Wetland Systems Design Guidelines for Developers. Version 3. November

WANG P, XU Ailan. Nitrogen Losses with Surface Runoff from Farm Lands in Polder Area Around Taihu Basin[J]. Journal of Agro – Environment Science 2008, 27(4): 1335 – 1339.

Waters D K, Jiao F, Chen M D, et al. (eWater Cooperative Research Centre) (2012). Application of Source to Investigate Nutrient Management Options for the Dongshan Peninsula, China.

Banked-up Water Analysis of Bridge Construction Based on Two Dimensional Finite Element Numerical Simulation

Zhang Huaxing[1], *Wang Lianhua[2]*, *Li Dongfeng[3] and Zhang Hongwu[4]*

1. Yellow River Conservancy Commission, Zhengzhou, 450003, China
2. Beijing Wuzi University, Beijing 101149, China
3. Zhejiang University of Water Resources and Electric Power, Hangzhou 310018, China
4. State Key Laboratory of Hydro-science and Engineering Department of Hydraulic Engineering, Tsinghua University, Beijing 100084, China

Abstract: Bridge building in the river will raise the river water levels and velocity around the bridge piers and abutment wall, and these will take great effect on river flood control. How to evaluate the water levels increment and effect to river flood is essential for river channel management. Based on the river channel two dimensional finite element numerical model, these effects are analyzed. The main indicators include velocity variation, water stage variation, around bridge piers and abutment wall. By the above data, some results can be made. The maximum water level increment of banked-up water level in front of bridge is 0. 21 m height, the data of banked-up water level by empirical formula is 0. 25 m. The difference value of maximum banked-up water level is 0. 04 m, this explains the correspondence and the reasonableness of the two methods.

Key words: banked-up water, numerical simulation, two dimensional finite element model, bridge piers

1 Introduction

Building bridge in the river is essential to the local development. Building bridges in the river causes the river water stage to increase in front of bridge and velocity increases around the bridge piers and abutment wall, these will take great effect on river level and flood control management (Li Dongfeng, et al. , 2004, 2006; Zou Bing, et al. , 2007; Lu hao and Gao Dongguang, 1996; J. N. Bradley and Zheng Huaqian, 1980). How to evaluate the flood effect is very important for the river management. The evaluation method includes field data analysis, physical model and numerical simulation method, the numerical model is common ones, so the numerical method is used in the study.

Based on the two dimensional finite element river channel numerical model (Li Dongfeng, et al. , 1999, 2004, 2009, 2010, 2010), the effect of building bridge in the river on the flood is analyzed. The main indicators include velocity variation, water stage variation, water depth variation and river channel bed deposition and erosion around bridge piers and abutment wall. By the above data, some results can be made. The maximum water stage increment of the banked up water level in front of bridge and the flood effect range are discussed. The change of flow direction is shown in a few graphs.

2 Numerical simulation basic theory methodology

2.1 The governing equations and deterministic conditions (Li Dongfeng, 1999)

Using depth-averaged planar 2-D shallow water equations as the governing equations for computation. The equations of continuity:

$$\frac{\partial Z}{\partial t} + \frac{\partial HU}{\partial x} + \frac{\partial HV}{\partial y} = 0$$

The equations of motion:

$$\frac{\partial U}{\partial t} + U\frac{\partial U}{\partial x} + V\frac{\partial U}{\partial y} + g\frac{\partial Z}{\partial x} + \frac{gn^2 U \sqrt{U^2 + V^2}}{H^{4/3}} - fV - \varepsilon\left(\frac{\partial^2 U}{\partial x^2} + \frac{\partial^2 U}{\partial y^2}\right) = 0$$

$$\frac{\partial V}{\partial t} + U\frac{\partial V}{\partial x} + V\frac{\partial V}{\partial y} + g\frac{\partial Z}{\partial y} + \frac{gn^2 V \sqrt{U^2 + V^2}}{H^{4/3}} + fU - \varepsilon\left(\frac{\partial^2 V}{\partial x^2} + \frac{\partial^2 V}{\partial y^2}\right) = 0$$

where, u, v is x, y direction components of depth averaged velocity; z, h is water level (or tidal level) and depth; g is acceleration due to gravity; ε is turbulent viscosity coefficient; C is Checy's coefficient.

C is calculated by Checy's formulation:

$$C = \frac{1}{n}R^{1/6}$$

where, n is Manning roughness coefficient, f is Coriolis force coefficient, $f = 2\omega\sin\varphi$, ω is rotation angular velocity of earth, φ is the latitude of computed reach.

2.2 The deterministic conditions

The deterministic conditions involve boundary conditions and initial conditions.

2.2.1 Boundary conditions

Boundary conditions include opening boundary and closing boundary. The former opening boundary is inlet and outlet water boundary, which is governed by inlet flow quantity process and outlet water levels process for model. The latter closing boundary is land boundary and the normal velocity is treated as zero for model.

2.2.2 Initial conditions

The initial water level and the initial velocity are given by measured tidal level or given value zero, initial conditions does not affect the precision of computed result. At initial time, tidal level, velocity and other varibles are given.

2.3 The finite element solution of the equations and verification of model

2.3.1 Numerical procedure

According to the above formulation, finite element analysis is carried out first, then composing the each element and the overall finite element equations can be obtained.

2.3.2 Establation of finite element model

The finite element method is employed to solve the equations. In this model, quadrangle grids are selected to disperse calculation zones. Applying Galerkin's weighted residual method is used in every element where weighted function is the function of interpolation, the weak formulation of the above equations are, then their finite element equations can be given.

2.3.3 Solution of finite element model

The predictor-corrector method is used to estimate the iteration equations.

2.3.4 Verification of the Model

The comparison of calculated data and filed data show that all these numerical simulation results, the water level and velocity process, are well agreement with the field ones. The detailed verification of the model is shown in References (Li Dongfeng, 2009).

3 The analysis of calculated results

3.1 Calculating conditions

According to the plan and design, one of the design and construction schemes is of three piers bridge. Owing to the flood control standard, the design flood frequency is 5%, the corresponding flow discharge at the entrence is 2,154 m³/s. The extrence control condition is the extrence water level.

Numerical simulation calculation boundary and the bridge sketch map are demonstrated below. Fig. 1 and Fig. 2 are calculation grids, one is no bridge, the other is given gridge, grids around bridge are shown in Fig. 3 and Fig. 4. In order to analyze the flow, many typical mesh point are selected along the river path and near the piers.

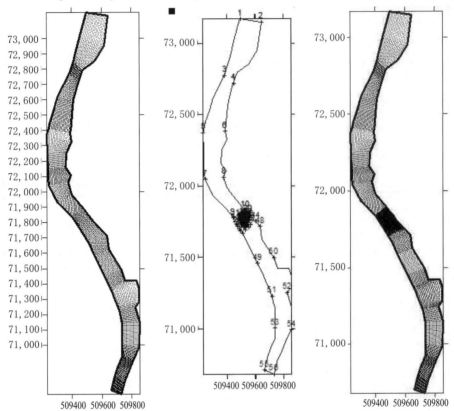

Fig. 1　No bridge grids　　**Fig. 2　Bridge location and initial grids**　　**Fig. 3　Bridge grids**

3. 2　Calculation results and analysis

Calculation results and analysis are expanded on and go into particulars below.

3. 2. 1　Velocity variation analysis
Fig. 4 and Fig. 5 are the velocity vector and isoclines distribution are also disclosed by these graphics.

These graphics show that When setting up the bridge, water flow is turn back by bridge, the flow have much new flow characteristics around piers.

3. 2. 2　Banked-up water analysis
Because of the more resistant of lengthening piers and widen bridge, the locality water level magnitude varies a great deal, The banked-up water alone river channel and map around piers graphics are shown in Fig. 6 and Fig. 7. Fig. 6 are water level variation 3D isoclines map alone river channel, Fig. 7 water level variation isoclines map around piers, these the increase value is 0. 21 m.

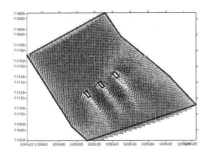 

Fig. 4   Velocity vector graphics of bridge   Fig. 5   Velocity isoclines distribution of bridge

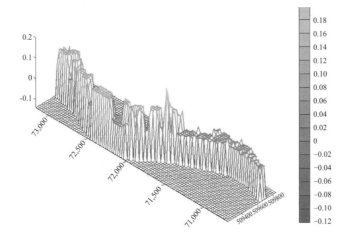

Fig. 6   *W* level variation 3D isoclines map alone river channel

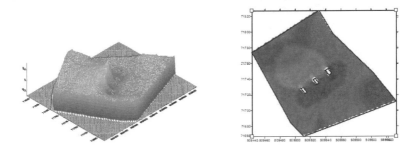

Fig. 7   *W* level variation isoclines map around piers

4   Empirical formula of banked-up water level

In order to test the reliability verification of numerical simulation, the empirical formula of banked-up water level is applied to the banked-up water level evaluation, instructions is detailed in References.

The increment of maximum banked-up water level Δz is calculated. The increment of maximum banked-up water level is 0. 25 m, the difference value of maximum banked-up water level

Δz is 0. 04 m, this explains the correspondence and the reasonableness of the two methods.

$$\Delta z = \frac{K}{2g}(v_M^2 - v_{0M}^2) \quad K = K_y K_N$$

$$K_y = \frac{0.5}{\dfrac{v_M}{\sqrt{gH_1}} - 0.1} \quad (H_1 = 1 \text{ m})$$

$$K_N = \frac{2}{\sqrt{\dfrac{v_M}{v_{0M}} - 1}}$$

$$v_M = \frac{v'_M}{1 + 0.5d_{50}^{-0.25}\left(\dfrac{v'_M}{v_c} - 1\right)}$$

5   Conclusions

The above analysis indicators the maximum water level increment of back-up water level in front of bridge is 0. 21 m height by two dimensional finite element numerical model, the data of banked-up water level by empirical formula is 0. 25 m, The difference value of maximum banked-up water level is 0. 04 m, this explains the correspondence and the reasonableness of the two methods. Some river channel flow hydrodynamics characteristics of flow over piers, water level isoclines distribution are disclosed by these graphics.

Acknowledgements

This research was supported by the National Natural Science Foundation of China (No. 51039003) and the 2011 projects the Water Resources bureau of Zhejiang Province (RC11092011), the 2009 Scientific Research Fund of Zhejiang Provincial Education Department (No. Z200909405), the Zhejiang Provincial Education Science Plan Office Project [(2009), No. SCG220] and the "325" Talent Training Program of Water Resources Department of Zhejiang Province. All authors contributed equally to this work.

References

Li Dongfeng, Zhang Hongwu, Zhong Deyu, et al. Numerical Simulation and Analysis on Tidal Current and SediMent Silting Process in Yellow River Estuary[J]. Journal of Hydraulic Engineering, 2004(11) .

Li Dongfeng, Zou Bing, Zhang Hongwu, et al. Two Dimensional Analysis of Tide Influence Upon Estuary and Riverway[J]. Journal of Zhejiang Water Conservancy and Hydropower College, 2006(1) .

Zou Bing, Li Dongfeng, Zhang Hongwu, et al. Effect Analysis of Building Bridges on River Hydraulic Dynamics Circumstances by 2D Numerical Method[J]. Journal of Hangzhou Dianzi University, 2007(3).

Lu Hao, Gao Dongguang. Bridge River Hydraulics [M]. Beijing: China Communications Press, 1996.

J. N. Bradley, Zheng Huaqian. Bridge and River Hydraulic[M]. Beijing: China Communications Press, 1980.

Li Dongfeng, Zhang Hongwu, Zhang Junhua, et al. Finite Element Method of Yellow River and Sediment Movement[J]. Journal of Sediment Research, 1999(4).

Li Dongfeng, Zhang Hongwu, Zhong Deyu, et al. 2-D Mathematical Model for Flow and Sediment Transport in the Estuary of the Yellow River[J]. Journal of Hydraulic Engineering, 2004 (6):1 -7.

Li Dongfeng, Zhang Jiyuan, Chen Bin. Flood Influence Evaluation of Chuanbuton Bridge in the Haoxi River[R]. Report of ZJWCHC, 2009.

Li Dongfeng, Zhang Hongwu. Numerical Simulation Analysis of Bridge Construction Flood Flow Effect Based on two Dimensional Finite Element Model [C]//Proceedings of 2010 International Conference on Modern Hydraulic Engineering, 2010, Xi ' an, China, 295 - 298.

Li Dongfeng, Zhang Hongwu. Dammed water level comparison of widening bridge piers based on 2D FEM numerical simulation and empirical formula[C]//Proceedings of 2010 International Conference on Modern Hydraulic Engineering, 2010, Xi' an, China, 255 - 258.

Numerical Technique and Tool
for Modelling Area-Source Pollution

Liren Yu and *Jun Yu*

ESDV, Researcher of ASSER-CESUSC, Brizil
Lenovo, PMP, Brizil

Abstract: This paper reports a quasi 3-D numerical simulation, in which the area-source discharge, produced from a slop along the south riverbank of a meandering river reach, has been well modelled. In the computation, three depth-averaged two-equation turbulence models (i. e. $\bar{k} - \tilde{\varepsilon}$, $\bar{k} - \tilde{w}$ and $\bar{k} - \tilde{\omega}$) were used to close the quasi 3-D hydrodynamic fundamental governing equations, discretized on a non – orthogonal, two-levels' body-fitted grids with spatially collocated variable arrangement. The last turbulence model however, was recently established by the authors. The advanced multi-grid iterative method has been used for accelerating iterative convergence. The aim of the computation is to enhance the modelling ability of a professional numerical tool (namely Q3drm1.0), developed recently by the authors, which can be used to simulate and analyze various problems in engineering, related to flow, mixing and contaminant transport.

Key words: numenical techniqua; area – source pollution; quasi3 – D

1   Introduction

Almost all flows in natural waters are turbulence. Dealing with the problems of turbulence related tightly to flow and pollutant transport is challenging both for scientists and engineers, because of their damaging effect on our limited resource and fragile environment. It is important to develop adequate mathematical models, turbulence closure models, numerical methods and algorithms as well as corresponding analytical tools for timely simulating and predicting transport behaviours in natural and artificial waters.

It is well known that various treaded or non-treated contaminants are usually discharged into waters in three typical situations: side-discharge, point-source discharge and area-source discharge, respectively. Most of reported depth-averaged numerical computations can easily simulate side-discharge through traditional boundary condition treatment; some of them also can well deal with point-source discharge and point-sink as inner boundary condition by using Rodi's vertical linear source concept and source-sink method . However, the simulation of area-source pollution from a slope along waterside, also often encountered in practice, is rarely reported in literature.

The area-source discharge in this computation is realized by arranging a series of vertical linear source at the centre points of CV near the south riverbank. For each vertical linear source on given grid level, the index, position, hydraulic parameters (flow-rate, temperature and concentration) and jet directions (i. e. , the angle of submerged discharge direction to horizontal plane—α and the angle of horizontal discharge direction to the x-axis—β) should be assigned in advance. These values would be determined suitably according to the site situation.

Current numerical testing for modelling area-source pollution is partially based on a previous flow modelling in a complex natural river, in which the calculated depth-averaged velocity profile by using depth-averaged $\bar{k} - \tilde{\varepsilon}$ turbulence model closure has compared with site-data, reported by Yu and Salvador.

2   Hydrodynamic fundamental governing equations

The complete, non-simplified fundamental governing equations of quasi three-dimensional computation, in terms of coordinate-free vector forms derived by using vertical Leibniz integration for a CV (an arbitrary quadrilateral with center point P) , considering the variation of the bottom topography and water surface and neglecting minor terms in the depth-averaging procedure, can be written as follows:

$$\frac{\partial}{\partial t}\int_{\Omega}\rho h\bar{\phi}\mathrm{d}\Omega + \int_{S}\rho h\bar{\phi}\vec{v}\cdot\vec{n}\mathrm{d}S = \int_{S}\Gamma h\mathrm{grad}\bar{\phi}\cdot\vec{n}\mathrm{d}S + \int_{\Omega}\bar{q}_{\phi}\mathrm{d}\Omega \tag{1}$$

where, Ω is the CV's volume; S is the face; $\bar{v}$ is the depth-averaged velocity vector, the superscript " – " indicates that the value is strictly depth-averaged; $\bar{\phi}$ is any depth-averaged conserved intensive property (for mass conservation, $\bar{\phi}=1$; for momentum conservation, $\bar{\phi}$ is the components of $\vec{v}$ in different directions; for conservation of a scalar, $\bar{\phi}$ is the conserved property per unit mass) ; Γ is the diffusivity for the quantity $\bar{\phi}$; $\bar{q}_{\phi}$ denotes the source or sink of $\bar{\phi}$; and h and ρ are local water depth at P and density, respectively.

For the momentum conservation of Eq. (1) , $\Gamma = \tilde{\mu}_{\mathrm{eff}}$ (depth-averaged effective viscosity) ; for temperature or concentration transport, $\Gamma = \tilde{\Gamma}_{\phi,t}$ (temperature or concentration diffusivity) , where the superscript " ~ " indicates the quantity characterizing depth-averaged turbulence. The source (sink) term $\bar{q}_{\phi}$ for momentum conservation may include surface wind shear stresses, bottom shear stresses, pressure terms and additional point sources (or point sinks).

The continuity and momentum equations as well as the transport equation of the scalar have been reported in details by Yu and Yu .

3 Depth-averaged turbulence closure models

The depth-averaged effective viscosity $\tilde{\mu}_{\mathrm{eff}}$ and diffusivity $\tilde{\Gamma}_{\phi,t}$ appeared in Eq. (1) , are dependent on the molecular dynamic viscosity $\tilde{\mu}_{t}$ and depth-averaged eddy viscosity $\tilde{\mu}$: $\tilde{\mu}_{\mathrm{eff}}=\mu+\tilde{\mu}_{t}$ and $\tilde{\Gamma}_{\phi,t}=\tilde{\mu}_{t}/\sigma_{\phi,t}$, where $\sigma_{\phi,t}$ is the turbulence Prandtl number for temperature diffusion or Schmidt number for concentration diffusion, and $\tilde{\mu}_{t}$ is a scalar property and normally determined by two extra transported parameters.

The used first two-equation turbulence model for depth-averaged calculation ($\tilde{k}-\tilde{\varepsilon}$ model) was suggested by McGuirk, and Rodi as early as in 1977:

$$\frac{\partial(\rho h\tilde{k})}{\partial t} + div(\rho hk\vec{v}) = div(h(\mu+\frac{\tilde{\mu}_{t}}{\sigma_{k}})\mathrm{grad}\tilde{k}) + hP_{k} - \rho h\tilde{\varepsilon} + \rho hP_{kv} + \bar{S}_{k} \tag{2}$$

$$\frac{\partial(\rho h\tilde{\varepsilon})}{\partial t} + div(\rho h\tilde{\varepsilon}\vec{v}) = div(h(\mu+\frac{\tilde{\mu}_{t}}{\sigma_{\varepsilon}})\mathrm{grad}\tilde{\varepsilon}) + C_{1}hP_{k}\frac{\tilde{\varepsilon}}{\tilde{k}} - C_{2}\rho h\frac{\tilde{\varepsilon}^{2}}{\tilde{k}} + \rho hP_{ev} + \bar{S}_{\varepsilon} \tag{3}$$

where, $\bar{S}_{k}$ and $\bar{S}_{\varepsilon}$ are the source-sink terms, P_{k} is the production of turbulent kinetic energy due to interactions of turbulent stresses with horizontal mean velocity gradients, $\tilde{\mu}_{t}$ can be expressed as:

$$\tilde{\mu}_{t} = \rho C_{\mu}\tilde{k}^{2}/\tilde{\varepsilon} \tag{4}$$

where, $\tilde{k}$ and $\tilde{\varepsilon}$ stand for depth-averaged turbulent kinetic energy parameter and dissipation rate parameter of $\tilde{k}$. The values of empirical constants C_{μ} , σ_{k} , σ_{ε} , C_{1} and C_{2} in Eq. (2) ~ Eq. (4) are the same as the 'standard' $k-\varepsilon$ model, i. e. equal to 0. 09, 1. 0, 1. 3, 1. 44 and 1. 92, respectively. The additional source terms P_{kv} and P_{ev} in Eq. (2) and Eq. (3) are mainly produced by the vertical velocity gradients near the bottom, and can be expressed as follows:

$$P_{kv} = C_{k}u_{*}^{3}/h, \quad P_{ev} = C_{\varepsilon}u_{*}^{4}/h^{2} \tag{5}$$

where, the local friction velocity u_{*} is equal to $\sqrt{C_{f}(\bar{u}^{2}+\bar{v}^{2})}$, the empirical constants C_{k} and C_{ε} for open channel flow and rivers are:

$$C_{k} = 1/\sqrt{C_{f}}, \quad C_{\varepsilon} = C_{2}C_{\mu}^{1/2}/(C_{f}^{3/4}\times e^{*1/2}) \tag{6}$$

where, C_{f} represents an empirical friction factor and e^{*} is the dimensionless diffusivity of the empirical formula for undisturbed channel/river flows $\tilde{\mu}_{t} = e^{*}U_{*}h$ with U_{*} being the global friction velocity.

In 1989, the first author of the present paper and his colleague developed a depth-averaged second-order closure model , $\tilde{k}-\tilde{w}$, which originated from the revised $k-w$ model developed by Ilegbusi and Spalding in 1982 and has been adopted as the second turbulence closure model in the paper. The turbulence parameter equations ($i.\ e.$, the $\bar{k}$ – Eq. and $\bar{k}$-eq.) should be:

$$\frac{\partial(\rho h \bar{k})}{\partial t} + \mathrm{div}(\rho h \bar{k} \vec{v}) = \mathrm{div}(h(\mu + \frac{\mu_t}{\sigma_k})\,\mathrm{grad}\bar{k}) + h P_k + \rho h P_{kv} - C_\mu \rho h \bar{k} \bar{w}^{1/2} + \bar{S}_k \tag{7}$$

$$\frac{\partial(\rho h \bar{w})}{\partial t} + \mathrm{div}(\rho h \bar{w} \vec{v}) = \mathrm{div}(h(\mu + \frac{\mu_t}{\sigma_\varepsilon})\,\mathrm{grad}\bar{w}) + C_{1w}\tilde{\mu}h(\,\mathrm{grad}\Omega)^2 - C_{2w}\rho h \bar{w}^{3/2} f +$$

$$C_{3w} h \frac{\bar{w}_t}{\bar{k}} P_k + \bar{S}_w \tag{8}$$

where $\bar{S}_k$ and $\bar{S}_w$ are the source-sink terms; function $f = 1 + C'_{2w}(\,\partial L/\partial x_i\,)$ and L is the characteristic distance of turbulence; Ω stands for mean movement vorticity. In $\bar{k} - \bar{w}$ model, the turbulent viscosity is defined as:

$$\tilde{\mu}_+ = \rho \bar{k}^2/\tilde{w}^{1/2} \tag{9}$$

where, $\tilde{w}$ is depth-averaged time-mean-square vorticity fluctuation parameter of turbulence. The turbulence parameter equations ($\bar{k}$-Eq. and $\tilde{w}$-eq.) should be solved in this model as well. The values of empirical constants C_μ, σ_k, σ_w, C_{1w}, C_{2w}, C_{2w} and C_{3w} are the same as those of 'standard' $k - w$ model, $i.\ e.$, equal $0.\ 09$, $1.\ 0$, $1.\ 0$, $3.\ 5$, $0.\ 17$, $17.\ 47$ and $1.\ 12$, respectively. The corresponding additional source terms P_{kv} and P_{wv}, also mainly due to the vertical velocity gradients near the bottom, and can be expressed as:

$$P_{kv} = C_k u_*^3/h, \quad P_{wv} = C_w u_*^3/h^3 \tag{10}$$

The empirical constants C_w for open channel flow and rivers can be written as:

$$C_W = C_{2w}/(C_\mu^{3/2} \times C_f^{3/4} \times e^{*\,3/2}) \tag{11}$$

Recently, the authors have established a new depth-averaged model, $\bar{k} - \bar{\omega}$, based on the most common standard $k - \omega$ model (in which ω is the special dissipation rate), originally introduced by Saffman but popularized by Wilcox in 1998 . In the depth – averaged $\bar{k} - \bar{\omega}$ turbulence model, the turbulent viscosity is expressed by:

$$\tilde{\mu}_t = \rho \bar{k}/\bar{\omega} \tag{12}$$

where, $\bar{\omega}$ is the special dissipation rate parameter of turbulence kinetic energy in the depth-averaged sense. As the used third two-equation closure turbulence model, $\bar{k}$ and $\bar{\omega}$ are also determined by solving two turbulence parameter transport equations, $i.\ e.$, the $\bar{k}$ -Eq. and $\bar{\omega}$-eq. :

$$\frac{\partial(\rho h \bar{k})}{\partial t} + \mathrm{div}(\rho h \bar{k} \vec{v}) = \mathrm{div}(h(\mu + \frac{\tilde{\mu}_t}{\sigma_k^*})\,\mathrm{grad}\bar{k}) + h P_k - \rho \beta^* h \bar{k} \bar{\omega} + \rho h P_{kv} + \bar{S}_k \tag{13}$$

$$\frac{\partial(\rho h \bar{\omega})}{\partial t} + \mathrm{div}(\rho h \bar{\omega} \vec{v}) = \mathrm{div}(h(\mu + \frac{\tilde{\mu}_t}{\sigma_\omega^*})\,\mathrm{grad}\bar{\omega}) + \alpha h \frac{\bar{\omega}}{\bar{k}} P_k - \rho h \beta \bar{\omega}^2 + \rho h P_{kv} + \bar{S}_\omega \tag{14}$$

where, $\bar{S}_k$ and $\bar{S}_\omega$ are the source-sink terms. The values of empirical constants α, β, β^*, σ_k^*, and σ_ω^* in Eq. (13) through Eq. (14) are the same as in the 'standard' $k - \omega$ model: $5/9$, $0.\ 075$, $0.\ 9$, 2, and 2, respectively. According to the dimensional analysis, the additional source terms P_{kv} in the k – Eq. (13) and P_{wv} in the ω – Eq. (14) are:

$$P_{kv} = C_k u_*^3/h, \quad P_{\omega v} = C_\omega u_*^2/h^2 \tag{15}$$

while the empirical constant C_ω for open channel flow and rivers can be expressed as:

$$C_\omega = \beta/(C_\mu \times e^* \times C_f^{1/2}) \tag{16}$$

4   Expressions of source/sink for quasi 3-D modelling

Q3drm1. 0 software adopts the Pantankar's method of source – term linearization to the discretization process of the fundamental governing equations. Because the source-sink term $\bar{S}$ depends on $\bar{\phi}$, therefore, one can express the dependence in a linear form given by:

$$\bar{S} = S_C + S_P \bar{\phi}_P \tag{17}$$

where, S_C represents the constant part of the linearized source-sink term, $S_P(S_P < =0)$ stands for the coefficient of $\bar{\phi}_P$, and P denotes central nodal point of CV under consideration.

In the programming of *flow-solvers*, Q3drm1. 0 software uses H_P to express the local water-depth at P point and Vol to represent the area of corresponding CV, thus, the additional source-sink term for solving each discretized fundamental governing equation on non-orthogonal grid with collocated variable arrangement can then be written as follows:

in $\bar{u}$ – eq.
$$S_C = \frac{\rho Q_H \bar{u}_P}{H_P Vol} \quad S_P = -\frac{\rho Q_P}{H_P Vol} \tag{18}$$

in $\bar{v}$ – eq.
$$S_C = \frac{\rho Q_H \bar{v}_P}{H_P Vol} \quad S_P = -\frac{\rho Q_P}{H_P Vol} \tag{19}$$

in pressure – correction equation
$$S_C = \frac{\rho Q_P}{H_P Vol} \tag{20}$$

in concentration transport equation – 1
$$S_C = -\frac{\rho Q_P \bar{C}_{1P}}{H_P Vol} \quad S_P = -\frac{\rho Q_P}{H_P Vol} \tag{21}$$

in $\bar{k}$ – eq.
$$S_C = \frac{\rho Q_P \bar{k}_P}{H_P Vol} \quad S_P = -\frac{\rho Q_P}{H_P Vol} \tag{22}$$

in $\tilde{\varepsilon}$ – eq.
$$S_C = \frac{\rho Q_P \tilde{\varepsilon}_P}{H_P Vol} \quad S_P = -\frac{Q_P}{H_P Vol} \tag{23}$$

in $\tilde{w}$ – eq.
$$S_C = \frac{\rho Q_P \tilde{w}_P}{H_P Vol} \quad S_P = -\frac{\rho Q_P}{H_P Vol} \tag{24}$$

in $\tilde{\omega}$ – eq.
$$S_C = \frac{\rho Q_P \tilde{\omega}_P}{H_P Vol} \quad S_P = -\frac{\rho Q_P}{H_P Vol} \tag{25}$$

where, $\bar{\mu}_P$, $\bar{v}_P$, $\bar{C}_{1P}$, $\bar{k}_P$, $\tilde{\varepsilon}_P$, $\bar{w}_P$ and $\bar{\omega}_P$ represent velocity components in x – directions and y – directions, concentration, and turbulence parameters at the nodal point P.

5 Grid generation

A reach of a typical meandering river was selected, which curved riverbank contours can be obtained from the site map. With the help of the developed Graphical User Interface of Q3drml1.0 software, it is possible to determine the scale of map, to collect conveniently geometrical data, including the positions of two curved riverbanks in the computational domain, and finally to generate one file with the suffix '.GIN'. The generated file contains all of messages, which illustrate necessary control variables and characteristic parameters, including those on four exterior boundaries (north-east inlet section, south-west outlet section, south and north riverbanks), and can be read in by a *grid-generator* (written by Fortran Language) to generate the expectant coarse and fine grids (two levels' grids were used). In fact, the developed *Interface* divided the considered river reach into 58 sub-reaches with 59 short cross-river lines. In addition, at the nodal points along these lines, the water-depthes need to be assigned, in order to form bottom topography dataset.

Fig. 1 presents the generated non-orthogonal boundary-fitted coarse grid, with a grid solution of 147 nodal points in i-direction and 22 nodal points in j-direction, respectively. The resolution of corresponding fine grid, displayed by Fig. 2, is 292×42. In the generated meshes, the nodal points in transversal grid lines are uniform. The total length of calculated river reach is 5.159 km. The *grid-generator* can generate an unformatted file with the suffix '.grd', in which all of geometric data needed in computation should be storied and can be read in by the flow-solver. Fig. 3 represents the bottom topography of computational domain.

6 Solutions of flow and area-source discharge

The behaviours of flows and contaminant transport were simulated by using the developed *flow-solver* (also written by Fortran Language), in which the SIMPLE (*Semi-Implicit Method for Pressure-Linked Equation*) algorithm for FVA (*Finite Volume Approach*), Guass' divergence theorem, ILU (*Incomplete Lower-Upper*) decomposition, PWIM (*Pressure Weighting Interpolation Method*), SIP (*Strongly Implicit Procedure*), under relaxation and multi-grid iterative method have been used. The discretized fundamental governing equations were solved at each grid level in the following sequence:

two momentum equations ($\overline{u}$ – eq. and $\overline{v}$ – eq.) , one pressure – correction equation (p' – eq.) , one concentration transport equation ($\overline{C}_1$ – eq.) , and two turbulence parameter transport equations ($\overline{k}$ – eq. and $\widetilde{\varepsilon}$ – eq. ; or $\overline{k}$ – eq. and $\widetilde{w}$ – eq. ; or $\overline{k}$ – eq. and $\widetilde{\omega}$ – eq.) , respectively.

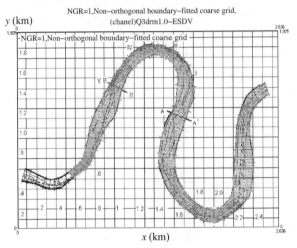

Fig. 1　Coarse grid

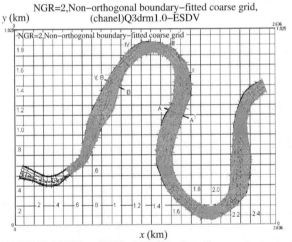

Fig. 2　Fine grid

The calculated main stream flow – rate is 184 m³/s, while the width, area and mean water-depth of the inlet section are 118.1 m, 326.6 m² and 2.76 m. The empirical friction factor (C_f) equals 0.002,685. On the outlet section, the variables satisfy constant gradient condition. The area-source discharge can be realized by arranging 81 vertical linear sources at the centre points of CV near the south riverbank ($i.e.$ $j = 2$) , from $i = 65$ to $i = 145$ with the flow-rate (= 0.05 m³/s) , difference of concentration (= 100 ppm) , angle $\alpha = 90°$ and angle $\beta = 0°$ for each source on the coarse grid. Three depth-averaged two-equation closure turbulence models, $i.e.$ $\overline{k} - \widetilde{\varepsilon}, \overline{k} - \widetilde{\omega}$ and $\overline{k} - \widetilde{\omega}$ models, are adopted to close the quasi three-dimensional hydrodynamic model. The corresponding turbulence parameters of three turbulence models at the inlet sections can be calculated by empirical formulae, i. e., $\overline{k}_0$, $\widetilde{\varepsilon}_0$, $\widetilde{w}_0$, $\widetilde{\omega}_0$ are 0.008,8 m²/s², 0.000,15 m²/s³,

0. 100,5 s^{-2} and 0. 190,7 s^{-1}, respectively. The wall function approximation has been used for determining the values of velocity components and turbulence parameters at the nodal points in the vicinity of riverbanks. In the calculation, the water-depths at all nodal points change in the range from 0. 2 m to 7. 3 m.

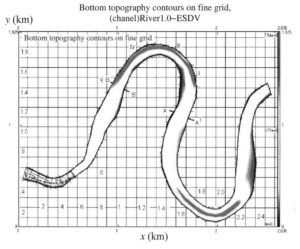

Fig. 3　Bottom topography

The used values of the under-relaxation factors for velocity components, pressure, concentration and two turbulence parameters in the multi-grid iterative method are 0. 6, 0. 6, 0. 1, 0. 7, 0. 7 and 0. 7, respectively. The maximum allowed numbers of inner iteration for solving velocity components, pressure, concentration and two turbulence parameters are 1, 1, 20, 1, 1 and 1. The convergence criterions for inner iteration are 0. 1, 0. 1, 0. 01, 0. 1, 0. 01 and 0. 01, respectively. The Stone's solver parameter is equal to 0. 92. The normalize residuals for solving velocity, pressure, concentration and turbulence parameter fields are all less than pre-determined convergence criterion (1. e-4).

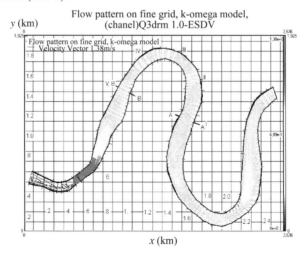

Fig. 4　Flow pattern ($\bar{k} - \tilde{\omega}$).

The simulation obtained various distributions of flow, pressure, concentration and turbulence parameter, which are useful to analyze interested problems in engineering. A part of results, simulated by using $\bar{k} \sim \tilde{\varepsilon}$, $\bar{k} \sim \tilde{w}$ and $\bar{k} \sim \tilde{\omega}$ models on fine grid, are presented from Fig. 4 to Fig. 18. Fig. 4 demonstrates the color flow pattern, in which the flow direction is from The East to The West. Fig. 5 and Fig. 6 display the streamlines and color-filled pressure field. Fig. 7 ~ Fig. 9 illustrate the calculated depth-averaged concentration distributions with 100 ppm difference between the main stream and confluent area-source pollutant, discharged from the inlet to A - A section. It is clear that the pollution plume well forms and gradually develops along the left riverbank. Fig. 10 ~ Fig. 12 represent the distributions of depth-averaged eddy viscosity $\tilde{\mu}_t$. Fig. 13 displays the comparison between the field data and depth-averaged velocity profiles on cross-section V, also calculated by using these three turbulence models, but without considering the flow-rate, caused by area-source discharge. Fig. 14 ~ Fig. 16 illustrate turbulent parameters: $\bar{k}$, $\tilde{\varepsilon}$, $\tilde{w}$ and $\tilde{\omega}$ as well as eddy viscosity $\tilde{\mu}_t$ on cross-section V. Fig. 17 and Fig. 18 represent the concentration profiles, calculated by using three different turbulence models, on the cross-section V and outlet section, while the maximum concentration value at the left riverbank decreases from 8 ppm to 4.5 ppm and the minimum value at the right riverbank increases from 0.5 ppm to 0.8 ppm, respectively.

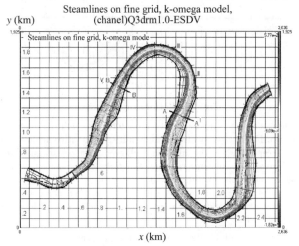

Steamlines on fine grid, k-omega model,
(chanel)Q3drm1.0-ESDV

Fig. 5　Streamlines $(\bar{k} - \tilde{\omega})$

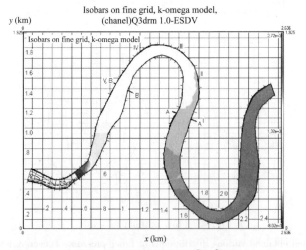

Isobars on fine grid, k-omega model,
(chanel)Q3drm 1.0-ESDV

Fig. 6　Pressure field $(\bar{k} \sim \tilde{\omega})$

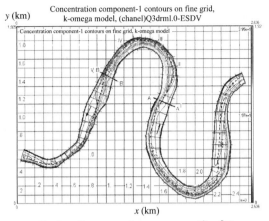

Fig. 7　Concentration contours ($\tilde{k} - \tilde{\omega}$)

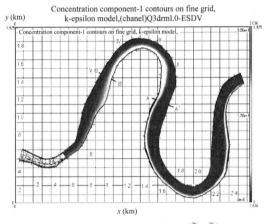

Fig. 8　Concentration field ($\tilde{k} - \tilde{\varepsilon}$)

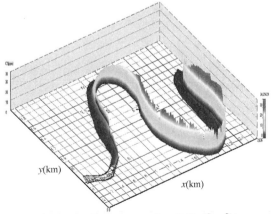

Fig. 9　3 – D concentration field ($\tilde{k} - \tilde{w}$)

446

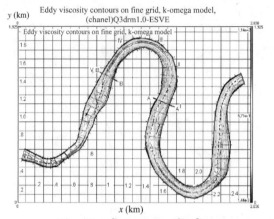

Fig. 10　$\widetilde{\mu}_t$ contours ($\bar{k} - \widetilde{\omega}$)

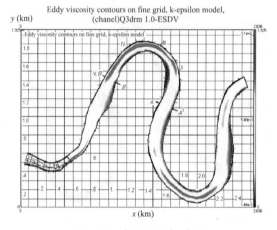

Fig. 11　$\widetilde{\mu}_t$ field ($\bar{k} - \widetilde{\varepsilon}$)

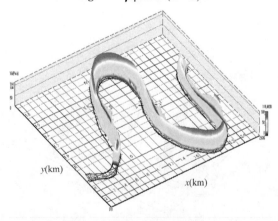

Fig. 12　3 – D $\widetilde{\mu}_t$ field ($\bar{k} - \widetilde{w}$)

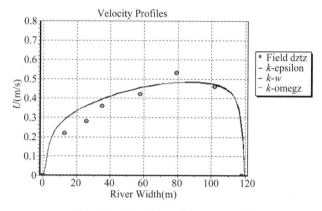

Fig. 13   Velocity profiles on sec. V

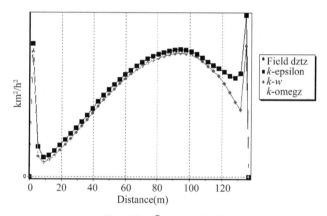

Fig. 14 $\bar{k}$ on section V

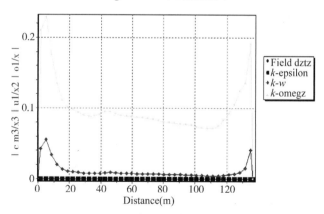

Fig. 15 $\tilde{\varepsilon}, \tilde{w}$, and $\tilde{\omega}$ on section V

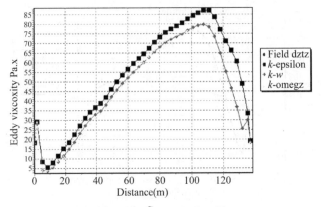

Fig. 16 $\tilde{\mu}_t$ **on section V**

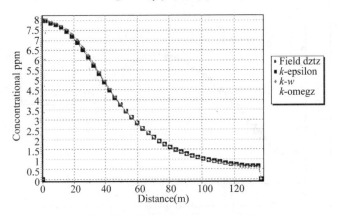

Fig. 17 C1 profiles on section V

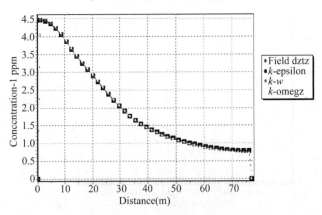

Fig. 18 C1 profiles on outlet

7 Discussions

With the help of a series of vertical linear source, being arranged at the centre points of CV near the south riverbank where the area-source discharge happens, current computation successfully simulates the contaminant confluence from the slop along the riverbank. In this example, the area-source discharge parameters at each nodal point are simply assumed to be equal each other. However, they are variable in case. Three used depth-averaged two-equation turbulence models can provide reasonable velocity profiles and concentration plumes (see Fig. 7 ~ Fig. 9, Fig. 13, Fig. 17 and Fig. 18) in the sufficiently mixing and strongly turbulent river, though the used turbulence closure models are different. All of these technologies, mentioned above, have been integrated into an available professional software, namely Q3drm1.0, for refinedly modelling quasi 3-D flow and temperature/contaminant transport using three selectable depth-averaged two-equation closure models (see www.esdv.webs.com).

The number of grid levels in the current simulation equals two, i. e. the computations were firstly performed on the generated coarse grid, and secondly operated on the fine grid. The simulation on two levels' grids can satisfy the demands of the simulation. By setting the number of grid level = 3, for example, the computation, not only on the coarse and fine grids, but also on the finest grid can then be realized. The selection of the number of grid levels depends on the solved problem and computational requirement.

The solved depth-averaged concentration variable in the current computation is a concentration difference between the confluent water, caused by area-source discharge, and the clean water of main stream. Actually, other contaminant index, such as COD, BOD and so on, also can be taken as the solved variable. The developed Q3drm1.0 software possesses the ability to simultaneously solve two concentration components in one calculation (i. e. to solve $\bar{C}_1$ – eq. and $\bar{C}_2$ – eq. together), caused by three discharge situations: side-discharge, point-source/sink discharge and area-source discharge, respectively.

The distributions of turbulent parameters: $\tilde{\varepsilon}$, $\tilde{w}$ and $\tilde{\omega}$, shown in Fig. 15, are different each other and vary quite sharply. The order of magnitude of $\tilde{\varepsilon}$, however, is smaller than the order of magnitude of $\tilde{w}$, and much smaller than the order of magnitude of $\tilde{\omega}$. It is notable that these three turbulence parameters all appear in the denominators of Eqs. (4), Eq. (9) and Eq. (12). They were used to calculate turbulent eddy viscosity. For numerical simulation, the occurrence of numerical error is unavoidable, especially in the place near irregular boundary. It is clear that a small numerical error, caused by solving $\tilde{\varepsilon}$ – eq. for example, will bring on larger error for calculating eddy viscosity than the same error, caused by solving other two equations ($\tilde{w}$ – eq. and $\tilde{\omega}$ – eq.). Without doubt, the elevation of the order of magnitude of the second turbulence parameter, such as $\tilde{\omega}$, reflecting the advance of two-equation closure models, provides a possibility to elevate computational precision. The insufficiency of classical depth-averaged $\bar{k} - \bar{\varepsilon}$ turbulence model for modelling in irregular domain may be avoided by adopting other turbulence models appeared recently, such as depth-averaged $\bar{k} - \tilde{\omega}$ model.

References

Yu L, Yu J Quasi 3-D Refined Flow and Contaminant Transport Software and Its Application in River Water Mixing in the Amazon River [C]// Proceedings of the 4th International Yellow River Forum (IYRF) on Ecological Civilization and River Ethics. Zhengzhou: Yellow River Conservamcy Press, 2009: 37 – 43.

Yu L, Giorgetti M F. Hydrodynamic Analysis of Flow Patterns and Estimation of Retention Time for a Polluted Reservoir [C]. Journal of Mechanical Engineering Science, Proceedings Part , London, UK, 2000(214): 873 – 880.

Yu L, Salvador N N B. Flow Modelling for Natural River [C]// Proceedings of the 2nd International Yellow River Forum (IYRF) on Keeping Healthy Life of the River. Zhengzhou: Yellow River Conservancy Press, 2005.

Yu L, Yu J. Numerical Research on Flow and Thermal Transport by Using Three Turbulence Depth – Averaged Models in a Cooling Pool of Electrical Power Station [J]. Journal of Water Science and Engineering, 2 (3), 1 – 12.

McGuirk J J, Rodi W, A Depth – averaged Mathematical Model for Side Discharges into Open Channel Flow, SFB 80/T/88, Universität Karlsruhe.

Yu L, Zhang S. A New Depth – Averaged Two – Equation ($\tilde{k} - \tilde{w}$) Turbulent Closure Model [J]. Journal of Hydrodynamics China Ocean Press, Ser. B, 1989(3): 47 – 54.

Wilcox D C, Turbulence Modeling for CFD, DCW Industries, Inc., La Canada, California.

Patankar S V. Numerical Heat Transfer and Fluid Flow[M]. Hemisphere Pub, 1980.

Розовскйй, И. Л., "Гидрологии и Гидротехники", Движение Воды На Повопоме Омкрымого Русла. Издательство Академии Наук Украин, 1959.

The Application Research of Multiple Linear Regression Analysis in the Yellow River Diversion of the Lower Yellow River

Li Zhenquan[1] , *Wang Xiwen*[1] , *Xu Jinjin*[1] and *Zhu Zhu*[2]

1. Water Supply Bureau, YRCC, Zhengzhou,450003, China
2. The University of New South Wales, Sydney,2052, Australia

Abstract: The Yellow River diversion is a complicated synthesis system, and its sustainable development is an important part in keeping the river healthy. To effectively promote the Yellow River diversion development, we need to find the influential factors out, to define the main factors by evaluating their importance, and then to present the corresponding proposals. This paper first proposes the multiple linear regression mathematic models for the Yellow River diversion, based on the multiple linear regression analysis. Then after some significant tests, the paper also indicates some main influential factors of the regression analysis, including the measured stream flow of river section, the local utilization of water resource, the local rainfall amount and the irrigation area. In the end, this paper gives a result that the influence degree of measured stream flow is similar with that of local rainfall amount in the Yellow River diversion.

Key words: the Yellow River diversion, multiple linear regression analysis, F – Test, regression coefficients

The variables of regression analysis are often classified as independent variables and dependent variables. When the dependent variable is non – time continuous one, multiple linear regression analysis is an important tool of the study of dependency relations between variables. The basic content of this analysis includes: to establish the multiple linear regression equation between dependent variables and independent ones that are based on their measured values; to inspect and analyze synthesis and single linear influential significance between each independent variables and dependent variable; to establish an optimizing multiple linear regression equation by only selecting the independent variables which have a significant linear effect on dependent variables; to evaluate relative importance of the influence of each independent variables on dependent variable, then to determine the deviation degree of the optimizing multiple linear regression equation and so on.

1   Establishment of multiple linear regression equation for yellow river diversion of the lower Yellow River

Step 1: Establish the multiple linear regression mathematic model.
Given dependent variable y and independent variables x_1 , x_2 , $\cdots$,x_m in n groups(see Tab. 1):

Tab. 1   Variable in multiple linear regression equation

Serial	Variable				
	y	x_1	x_2	$\cdots$	x_m
1	y_1	x_{11}	x_{21}	$\cdots$	x_{m1}
2	y_2	x_{12}	x_{22}	$\cdots$	x_{m2}
$\vdots$	$\vdots$	$\vdots$	$\vdots$	$\cdots$	$\vdots$
n	y_n	x_{1n}	x_{2n}	$\cdots$	x_{mn}

Given existing linear relation between dependent variable y and independent variables x_1 ,

$x_2, \cdots, x_m$, the mathematic model is:

$$y_j = \beta_0 + \beta_1 x_{1j} + \beta_2 x_{2j} + \cdots + \beta_m x_{mj} + \varepsilon_j \quad (j = 1, 2, \cdots, n) \tag{1}$$

where, x_1, x_2, $\cdots$, x_m are observable general variables; y is observable random variables, and changes with x_1, x_2, $\cdots$, x_m as well as test error influence; ε_j is a random variable which is relative independent and is subject to $N(0, \sigma^2)$.

We can estimate β_0, β_1, β_2, $\cdots, \beta_m$ and variance σ^2 according to practical values.

Step 2: Establish the multiple linear regression equation.

Given m − linear regression relation between x and y:

$$\hat{y} = b_0 + b_1 x_1 + b_2 x_2 + \cdots + b_m x_m$$

where, b_0, b_1, b_2, $\cdots$, b_m are the least − square values of β_0, β_1, β_2, $\cdots, \beta_m$.

The graph for m − linear regression equation is $(m + 1)$ − dimension space in a plane, called regression plane; b_0 is regression constant; b_1 ($i = 1, 2, \cdots, m$) is partial regression coefficient of independent variable x_i of dependent variable y.

Step 3: Determine the deviation degree of multiple linear regression equation.

According to the established multiple linear regression equation, sum of square of deviations $\sum (y - \hat{y})^2$ shows deviation degree between measured points and regression plane, so sum of square of deviations is also called sum of deviation regression square. In m − linear regression analysis, the degree of freedom of sum of deviation regression square is $(n - m - 1)$. In that case, mean square of deviation regression is $\sum (y - \hat{y})^2 / (n - m - 1)$, which is estimated value of σ^2 in model (1). The square root of mean square of deviation regression is called standard deviation of regression $S_{y.12\ldots m}$.

Therefore:

$$S_{y.12\ldots m} = S_e = \sqrt{\sum (y - \hat{y})^2 / (n - m - 1)} \tag{2}$$

Standard deviation of regression $S_{y.12\ldots m}$ presents deviation degree between measured points and regression plane, that is, deviation degree of regression equation.

2 Significant test of multiple linear regression for the Yellow River diversion of the lower Yellow River

Step 1: Ssignificantly test of multiple linear regression relation.

After the establishment of multiple linear regression equation, we need to significantly test this relation. So we use F − Test.

The same as one − dimensional linear regression, in multiple linear regression analysis, total sum of dependent variable y square SS_y can be divided into two parts, regression sum of square SS_R and deviation regression sum of square SS_r that is:

$$SS_y = SS_R + SS_r \tag{3}$$

The df_y, total degree of freedom of dependent variable y, also can be divided into two parts, regression degree of freedom df_R and deviation regression degree of freedom df_r, that is:

$$df_y = df_R + df_r \tag{4}$$

Eq. (3) and Eq. (4) are called partition or split type of the sum of square of multiple linear regression and its degree of freedom.

Significant test of multiple linear regression relation is the test, to evaluate whether partial regression coefficients, β_i ($i = 1, 2, \cdots, m$) of each independent variables are equal to zero at the same time or not.

The null and alternative hypotheses of significant test are:

$H_0: \beta_1 = \beta_2 = \cdots = \beta_m = 0$, $H_A: \beta_1$, β_2, $\cdots, \beta_m \neq 0$ at the same time

With H_0 assumption,

$$F = \frac{MS_R}{MS_r}, \quad (df_1 = df_R, df_2 = df_r) \tag{5}$$

After F – Test, significant test of multiple linear regressions can be finished.

Step 2: Significantly test of partial regression coefficients.

When multiple linear regression relation is significant or extremely significant after significant test, we also need to significantly test each partial regression coefficient to evaluate whether linear influence of each independent variable of dependent variable is significant or not. After that, we need to reject those variables which are not significant from the regression equation to re – establish a simpler multiple linear regression equation. We use T – Test as a tool in significant test.

Step 3: Reject some independent variables and re – establish multiple linear regression equation.

If each partial regression coefficient presents significant after significant test, single influence of each independent variable on dependent variable is significant. If one or more partial regression coefficients show non – significant, these corresponding independent variables are not important or their influence on dependent variable are non – significant in regression equation. In that case, we need to re – establish multiple linear regression equation and re – significantly test each new partial regression coefficient by rejecting one of those independent variables. And then, we need to repeat this process till both multiple linear regression equation and each partial regression coefficient are significant. After these processes, the multiple linear regression equation is optimizing.

3   The application of multiple linear regression analysis in the Yellow River diversion

Main factors, such as water condition, project condition, water utilization ways and water supply management have some effect on project scale of the Yellow River diversion of the lower Yellow River. Water condition mainly refers to inflow condition of river channel, regional rainfall, local water utilization and water quality situation. Project condition mainly means the change of river bed scouring and depositing, culvert diversion capacity, water transport and distribution in irrigation area. Water utilization ways mainly refers to planting structure of irrigation area, saving water measurement, and peasant's saving water consciousness. Water supply management mainly means management system in canal head and irrigation area, water charge and its price. We will make both quantitative and qualitative analysis of these factors in chapters as following.

The trend of changes in project scale of the Yellow River diversion of the lower Yellow River. From inter – annual view, the scale increases before 1990s, and decreases after that time. Meanwhile, the scale of Shandong Province has the most effect on the whole one. From annual view, the scale reaches to the peak at two periods, spring irrigation and summer irrigation. What is more, from 1970s to 1990s, the period in spring irrigation during which the highest point is attained is advanced, and that in summer irrigation is delayed. But the annual water supply roughly presents homogenization. From classification purpose view, after 1990s, the water usage for agriculture descends and that for industry ascends.

The trend of changes in project scale of the Yellow River diversion of irrigation districts in the lower Yellow River. Overall, except the decrease in irrigation area of a few irrigation districts of the Henan Province, the whole irrigation area steadily increases from 33 million mus (1 mu = 666.67 m^2) to 40 million mus. And the trend of changes in irrigation scale of the Shandong Province is roughly identical with that of the lower Yellow River. Meanwhile, the total irrigation area of the Henan Province rapidly goes up, but mainly in insufficient water usage region of the lower irrigation districts.

The trend of changes in water use for agriculture per mu of the Yellow River irrigation districts of the lower Yellow River. With the decrease in total water supply amount in the Yellow River diversion of the lower Yellow River while the increase in corresponding irrigation scale, water use for agriculture per mu goes down from 222 m^3 per mu, the average figure in 1990s, to 148 m^3 per mu in the early period of the 21th century. After 1990s, irrigation frequency drops and we only use irrigation at the crucial moment.

3.1 Regression analysis

To figure out the influence degree of each factor on changes in water supply scale of the Yellow River diversion of the lower Yellow River, we use multiple linear regression analysis method to evaluate the factors after making both quantitative and qualitative analysis.

Considering characteristics of multiple regression analysis and main influential factors, we choose the measured stream flow of river section, the local utilization of water resource, the local rainfall amount and the irrigation area as the analysis objects.

Tab. 2 shows regression analysis results between the measured stream flow of Huayuankou section, the local utilization of water resource, the local rainfall amount as well as the irrigation area and water supply scale of the Yellow River diversion of the lower Yellow River. And the result that correlation coefficient $R = 0.90$ means a significant correlation between four factors and the water supply amount.

Tab. 2 Regression equation parameters of water supply of irrigation districts of the lower Yellow River

Factor	Coefficient	Standard error	R	F	T's statistical value
Intercept	54.643	43.004			1.271
Rainfall amount(P)	-0.062	0.027			-2.286
Stream flow(R)	0.181	0.054	0.9	8.43	3.357
Irrigation area(S)	-0.02	0.007			-2.631
Local water utilization(W local)	1.756	0.805			2.181

From Tab. 2, regression equation between four factors and corresponding water supply scale:
$$W_{\text{supply}} = -0.062P + 0.181R - 0.020S + 1.756W_{\text{local}} + 54.643$$

(1) F – Test: given sample capacity $n = 13$, free variable $p = 4$ and at 95% confidence level ($\alpha = 0.05$), the critical value of F_{α} is:
$$F_{\alpha}(p, n - p - 1) = F_{0.05}(4,8) = 3.84$$

According to $F_{\alpha} = 3.84 < 8.43$, there is a significant correlation between four factors and the water supply.

(2) T – Test: at $\alpha = 0.05$, the critical value of t_{α} is:
$$t_{\alpha}(n - p - 1) = t_{0.05}(8) = 2.306$$

Compared to T's statistical values in Tab. 2, the local water utilization has no important effect on the water supply amount. According to the criterion of multiple regressions, that is, the evaluation process by rejecting one of those independent variables again and again, we know that the measured stream flow of river section and local rainfall amount are the main impact factors. And Tab. 3 shows the corresponding regression equation parameters below.

Tab. 3 Corresponding regression equation parameters between the water supply scale and measured stream flow as well as local rainfall amount

Factor	Coefficient	Standard error	R	F	$F_{0.05}$	T's statistical value	$T_{0.05/2}$
Intercept	105.254,3	14.713,04				7.154	
Rainfall amount(P)	-0.105	0.026,746	0.8	8.74	4.1	-3.934	2.23
Stream flow(R)	0.195	0.059,337				3.289	

From Tab. 3, the regression equation is:
$$W_{\text{supply}} = -0.105P + 0.119,5R + 105.254,3$$

Compare F with $F_{0.05}$ and T's statistical value with $T_{0.05/2}$ in Tab. 3, and then we know that the regression equation is a significant correlation, and there is a significant linear relation between water supply scale of the Yellow River diversion and two parameters, the measured stream flow of river section and local rainfall amount.

3.2 Significance evaluation of the influential factors

In correlation analysis, the correlation coefficient refers to the degree of linear relation between dependent variable and independent variables while the coefficient means the degree of linear relation between two variables. But the relation between two variables in correlation is often under the effect of other variables, so the coefficient can not reflect the level of independent variables' influence on dependent variable, that is, it can not reflect the correlation between them. But the correlation coefficient can reflect this correlation, regardless of other factors. Therefore, we can do significance evaluation of the influential factors by calculation and then comparison in size of the correlation coefficients of each variable.

The calculation formula of correlation coefficient is:

$$r_{y1 \cdot 2} = \frac{r_{y1} - r_{y2} r_{12}}{\sqrt{(1 - r_{y2}^2)(1 - r_{12}^2)}} \tag{6}$$

$$r_{y2 \cdot 1} = \frac{r_{y2} - r_{y1} r_{12}}{\sqrt{(1 - r_{y1}^2)(1 - r_{12}^2)}} \tag{7}$$

where, $r_{y1 \cdot 2}$ is the correlation coefficient between stream flow and water supply amount; $r_{y2 \cdot 1}$ is the correlation coefficient between rainfall amount and water supply amount; r_{y1} is the coefficient between stream flow and water supply amount; r_{y2} is the coefficient between rainfall amount and water supply amount; r_{12} is the coefficient between rainfall amount and stream flow.

After calculation, the correlation coefficient between local rainfall amount as well as measured stream flow and water supply amount is -0.78 and 0.72. Consequently, the influence degree of measured stream flow is similar with that of local rainfall amount in the Yellow River diversion.

4 Conclusions

Changes in scales of water supply and irrigation in the Yellow River diversion is a complicated system, and is influenced by many factors. When the water supply amount of the Yellow River diversion is a non – time continuous variable (independent variables can be discrete or continuous), relation analysis between many influential factors is very difficult. In that case, this paper investigates the multiple linear regression analysis method's performance in the analysis and evaluation of influential factors on the project scale, which shows that the multiple linear regression analysis in the evaluation of the Yellow River diversion can give stable results. This analysis method can also be applied in other system evaluations.

References

Wang Huiwen, Meng Jie. Predictive Modeling on Multivariate Linear Regression[J]. Journal of Beijing University of Aeronautics and Astronautics, 2007, 33(4).

XieYu. Regression Analysis [M]. Beijing: Social Sciences Academic Press, 2010.

Chen Ping, Li Zhaohe. The Modern Theory of Statistics[M]. Guangzhou: Sun Yat – sen University Press, 2004.

The Measurement of River Surface Discharge Using Continuous – wave Microwave Radar Techniques

Chien-jung Liu, Yu-ding Cai, Zhi-wei Huang, Jin-huang Huang,
Ying-song Hsu, Guang-zhi Zhang and Bing-Chang Cao

Research Center for Science and Technology across the Taiwan Strait, Taichung, 40763, China

Abstract: In Taiwan, rainfall is concentrated and unevenly distributed, and the river discharge volume is altered drastically by the strength and concentration of rainfall during the typhoon periods. Traditional means of manually observing high – volume discharge are time – consuming and dangerous. A non – contact continuous – wave microwave radar apparatus would be more suitable for measuring the velocity of river surface currents with a high discharge and further estimating the discharge volume. In this study, we applied a method using a linear prediction coefficient (LPC) to replace traditional signal – averaging by fast Fourier transform (FFT) analysis. The objective of this study is to determine a more accurate surface velocity measure by analyzing the backscattered signals produced by the actual river surface currents and filtering the interferences produced by the environment.

Key words: surface velocity, microwave radar, LPC, FFT

1　Introduction

Taiwan is located in a region of Southeast Asia where typhoons frequently occur. Every year, several typhoons pass over Taiwan bringing an abundance of rain. However, climate change has caused the flooding from typhoons to become increasingly frequent over the years. Because of these occurrences, disaster prevention and river regulation planning are necessary, which enhances the importance of measuring and understanding river discharge. Discharge measurements primarily focus on measuring river velocity and river cross – sections. Regarding river velocity, the non – contact continuous – wave microwave radar velocimeter (CWRV) has been used as the primary form of measurement and development in recent years. River velocity is determined by collecting wave divergence signals from river surface currents reflecting transmitted coherent microwave signals. The velocity and direction of river surface currents are influenced by environmental interferences (such as wind and rain droplets) during onsite measurements, producing results that occasionally differ from the actual surface current discharge generated by the river flow. The influences cause the measurements to fluctuate. An average measurement result of 100 s using current measuring equipment is recommended; then, by applying averaging methods, fluctuation can be averaged out. However, using this approach, not only the river surface current discharge but also environmental interferences that affect river discharge are calculated. Thus, we aimed to develop a corresponding autoregressive (AR) filter coefficient through autocorrelation analysis. This coefficient, which is known as the linear prediction coefficient (LPC), was then used to filter out noise from the original signals. The main objective of this study was to determine the actual surface current discharge of water flow conditions by testing various settings and conditions and further applying the LPC to filter out interferences.

2　Literature review

Crombie (1955) suggested the presence of Bragg resonance in the microwave radar echo signals of water surfaces in the Doppler spectrum (as shown in Fig. 1). Wright (1966), Wright and Keller (1971), Bass et al. (1968), Barrick (1972), and others verified the presence of Bragg resonance in the microwave radar echo signals of water surfaces in the Doppler spectrum, and applied radar signals to sea surface current measurements. Plant and Keller (1990) developed an analysis model that reduces the wind interferences to accurately obtain Bragg resonant frequencies.

Measurements of river discharge using a microwave radar mainly consists of two types, namely, the pulsed Doppler radar (Costa et al. , 2000; Melcher et al. , 2002) and the continuous – wave radar (CW) (Kimura et al. , 2005; Costa et al. , 2006; Fukami et al. , 2008).

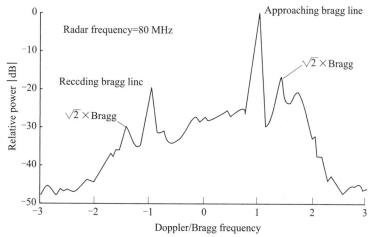

Fig. 1 Doppler spectrum of radar wave bragg scattering from surface currents (data source: geernaert and plant, 1990)

When an incident wavelength is $2n$ times the surface wavelength, Bragg scattering occurs through the interaction of radar signals and the surface currents, generating Bragg resonance in the echo signals of the Doppler spectrum. Surface velocity can be calculated using the Doppler peak frequency (f_D), which is obtained by applying the river current movement to Eq. (1). The f_D required to calculate surface velocity was traditionally obtained by directly analyzing the radar echo signals with FFT and determining the resonance peak of the power spectrum density (PSD).

$$v = \frac{C}{2f_0 \cdot \cos\theta} f_D \qquad \cos\theta = \frac{\cos\theta_1 + \cos\theta_2}{2} \qquad (1)$$

where, v is the surface velocity of the river, C is the spread velocity of electromagnetic waves in the air (3×10^8 m/s), f_0 is the frequency of the radar antenna (10.525×10^9 Hz), f_D is the Doppler peak frequency produced by the surface current movement, is the angle of depression between the transmitted waves and the water surface, and is the azimuth angle of the transmitted waves and the flow direction. As shown in Eq. (1), the spread velocity of electromagnetic waves in the air, the antenna frequency, the angle of the transmitted waves and the water surface, and the angle of the transmitted waves and the flow direction are all fixed conditions when measuring surface current velocity. Additionally, the Doppler peak frequency (f_D), which is produced by surface current movement, is the main parameter of velocity divergence. If surface currents are exposed to external interference, the Doppler peak frequency (f_D) (calculated using radar echo signals) would cause the surface velocity (calculated using Eq. (1)) to fluctuate unstably and inaccurately. Therefore, noise should be analyzed and filtered out.

3 Methods

3.1 Signal analysis theory

As the echo signals of surface velocity radars are narrowband and time – varying, the external interferences of water currents cannot be directly identified and filtered out by the primitive PSD. To analyze the actual surface velocity in echo signals and the intermittent occurrence of external interferences, we proposed an autocorrelation analysis method to establish a corresponding

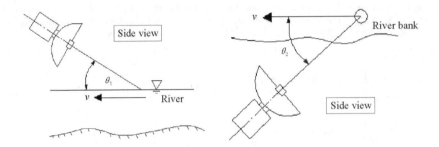

Fig. 2   Schematic diagrams of depression and azimuth angles

autoregressive filter coefficient in this study. By applying this coefficient, we identified and filtered out interferences from the primitive echo signals. The analysis procedures conducted in this study are as follows:

We first considered the radar echo primitive signals to be composed of a coherent microwave and noise. After analysis of the primitive radar signals, we obtained the CW structure, as shown in Eq. (2).

$$S(n) = \sum_{k=1}^{p} a_k y(n - k) + GS_{\text{noise}}(n) \tag{2}$$

where, $S(n)$ is the primitive short radar echo signals analyzed; $S_{\text{noise}}(n)$ is the noise within the primitive signals; n is the total signal echo points; a_k is the filter coefficients obtained after determining the CW structure in the primitive signal, and p is the number of each obtained filter coefficient.

If a_k is a linear filter and noise is not considered, then Eq. (2) can be changed to Eq. (3).

$$\tilde{S}(n) = \sum_{k=1}^{p} a_k y(n - k) \tag{3}$$

We applied the theory of fast Fourier transform to segment a duration of time into m number of $S(m)$ short – term signals, as shown in Eq. (4) below.

$$S(m) = S(m + n)w(m), \quad 0 \leqslant m \leqslant L - 1 \tag{4}$$

where, L is a given length of time; $w(m) = 0.54 - 0.46\cos(\frac{2\pi n}{L - 1})$ is a window function using the Hanning window.

If $\tilde{S}$ is close to S, the object function $J(n)$ is obtained. When $\frac{\partial J(n)}{\partial a_k} = 0$, Eq. (6) is obtained, as shown below.

$$J(n) = [S(n) - \tilde{S}(n)]^2 = \left[S(n) - \sum_{k=1}^{p} a_k y(n - k)\right]^2 \tag{5}$$

$$\sum_{m=0}^{L-1+p} S[m - i]S[m] = \sum_{k=1}^{p} a_k \sum_{m=6}^{L-1+p} y[m - i]y[m - k], i = 1, 2, \cdots, p \tag{6}$$

Between the intervals of $m = 0 \sim L - 1 + p$ in Eq. (6), an autocorrelated finite difference sequence is defined, as shown below.

$$\psi(i,k) = \sum_{m=0}^{L-1+p} y(m - i)y(m - k) \quad \begin{cases} 1 \leqslant i \leqslant p \\ 0 \leqslant k \leqslant p \end{cases} \tag{7}$$

When $i = 0$ and $p = -(i - k)$ are defined, the autocorrelated finite difference is changed to Eq. (8), which minimizes the target function F of Eq. (9), further obtaining Eq. (10). Subsequently, Eq. (10) is spread wide to obtain Eq. (11), which is solved by applying the Levinson – Durbin algorithm to define coefficient a_k.

$$\phi(k) = \sum_{m=0}^{L-1+k} y(m) y(m+k) \tag{8}$$

$$\min F - \sum_{m=0}^{L-1+p} \left[\psi(i,0) - \sum_{k=1}^{p} u_k \psi(i,k) \right]^2 - \sum_{m=0}^{L-1+p} \left[y(m) - \sum_{m=0}^{L-1+p} y(m-i) y(m-k) \right]^? \tag{9}$$

$$\phi(i) = \sum_{k=1}^{p} a_k \phi(i-k) \tag{10}$$

$$\begin{bmatrix} \phi(0) & \phi(1) & \phi(2) & \cdots & \phi(p-1) \\ \phi(1) & \phi(0) & \phi(1) & \cdots & \phi(p-2) \\ \phi(2) & \phi(1) & \phi(0) & \cdots & \phi(p-3) \\ \vdots & \vdots & \vdots & & \vdots \\ \phi(p-1) & \phi(p-2) & \phi(p-3) & \cdots & \phi(0) \end{bmatrix} \begin{bmatrix} a_1 \\ a_2 \\ a_3 \\ \vdots \\ a_p \end{bmatrix} = \begin{bmatrix} \phi(1) \\ \phi(2) \\ \phi(3) \\ \vdots \\ \phi(p) \end{bmatrix} \tag{11}$$

By applying Eq. (7) to Eq. (11) determine the autocorrelation learning of the signal structure, we found that the predicted filter was linear. Therefore, the method of analysis proposed in this study is known as the linear prediction coefficient (LPC) method. After analyzing the signal structure learning of the actual primitive radar echo signals, we obtained the order of coefficient a_k in the filter. Additionally, we determined the characteristic function $S_H(a_k,k)$ by conducting PSD analysis on signal $y(n)$. By applying PSD analysis to $S_H(a_k,k)$, we can obtain the corresponding frequency f_D of the energy peak value A_{fD}, as shown in Eq. (12). Finally, after obtaining f_D, we can then calculate the surface velocity v by applying Eq. (1) and determining noise S_{noise} through Equation (13). The calculation for the signal analysis is shown in Fig. 3.

$$A_{fD} = \max PSD \left[S_H(a_k,k) \right] \tag{12}$$

$$GS_{noise}(n) = S(n) - S_H(a_k,k) \tag{13}$$

where, A_{fD} is the energy peak value obtained after applying PSD analysis on $S_H(a_k,k)$, f_D is the frequency corresponding to the energy peak value; $S_H(a_k,k)$ is the signal data after applying the LPC filter to obtain the filter coefficient a_k; $S(n)$ is the primitive radar echo signal data; and S_{noise} is the interference in radar echoes.

Fig. 3　Radar signals and surface velocity calculation procedures

3.2　Analysis plan

To analyze the degree to which environmental interferences affect the measurement results, we separately measured the surface velocity of the outdoor Shimen Canal, Taoyuan, and an indoor trough without environmental interference, and further analyzed the obtained radar echoes. The indoor trough used in this study was an adjustable – speed vehicle (0.5 m/s to 6 m/s), which we considered an adjustable – velocity water surface opposed to the static water within, as shown in Fig. 4. To ensure a constant vehicle speed, the acceleration and deceleration phases of the vehicle movement were disregarded, and only the middle 65 m were employed as the considered surface velocity.

4　Analysis and discussion

The subject of this analysis was the outdoor field measurement at Shimen Canal and an indoor experimental trough. The blue signal data in Fig. 5 and Fig. 6 were the spectra analysis results for the Shimen Canal obtained by applying FFT to radar echo signals. We determined from the spectra analysis results that fluctuation was significant when applying FFT methods. We analyzed primitive velocity signals learning by separately applying various autoregressive process orders (Order 9 and 19) proposed by the LPC method in this study, which is denoted as red signal data in Fig. 5 and

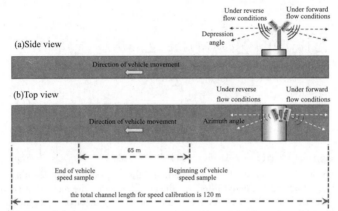

Fig. 4　Schematic diagram of a trough with adjustable direction, velocity, and depression angle

Fig. 6. Through the analysis of characteristic function learning, we found that the spectra signals were smoothed. Additionally, we accurately predicted the position of the Doppler frequency f_D, which is caused by surface current movement. Using these analysis results, we can determine the spectrum coefficients that exist in the flow of rivers. Additionally, when intermittent changes occur in the radar echoes because of external interferences, we can apply the characteristic function to identify and filter out these interferences.

Fig. 7 and Fig. 8 show the surface velocity measurement results for the indoor trough under movement speeds of 0.2 m/s and 2.5 m/s. As shown in Fig. 7, fluctuation occurs in the surface velocity measurements when FFT analysis is directly applied. By contrast, when we applied the LPC method proposed in this study, we obtained the actual corresponding radar echo signals for the vehicle speed. Additionally, the intermittent external influences were automatically filtered out, allowing us to obtain a convergence analysis result similar to the actual speed of the vehicle.

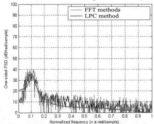

Fig. 5　Spectrum analysis using autoregressive order 9

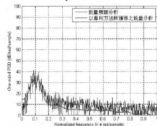

Fig. 6　Spectrum Analysis using Autoregressive Order 19

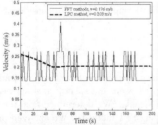

Fig. 7　Results of measurementa analysis with a movement speed of 0.2 m/s

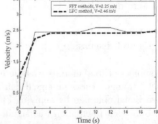

Fig. 8　Results of measurement analysis with a movement speed of 2.5 m/s

5 Conclusions

Using the LPC analysis method proposed in this study, we can analyze signals of the primitive data of surface velocity and filter out intermittent interferences. By applying this method, we can effectively filter out the occurrence of interference signals when using a surface velocimeter to conduct onsite measurements. We also obtained actual surface velocity signals to accurately determine the surface velocity, which could be used to accurately evaluate the river flow discharge.

Acknowledgements

We thank the Water Resources Agency, Ministry of Economics, and the Automatic Monitoring Technology Development Plan for River Flow program for funding this study. Additionally, we appreciate the support of the Hydrology Division, the many teaching assistants at the Research Center for Science and Technology Across the Taiwan Strait, Feng Chia University, and Pymble Tech. for their assistance in conducting the experiments.

References

Bass F G, Fuks I M, Kalmykov A I, et al. Very High Radiowave Scattering by a Disturbed Sea Surface, II, Scattering From an Actual Sea Surface[J]. IEEE Trans. Antennas Propag., 1968, AP – 16:560 – 568.

Barrick D E. First – order Theory and Analysis of MF/HF/VHF Scatter From the Sea[J]. IEEE Trans. Antennas Propag., 1972, AP – 20:2 – 10.

Costa J E, Spicer K R, Cheng R T, et al. Measuring Stream Discharge by Non – contact Methods: a Proof – of – concept Experiment[J]. Geophy. Res. Let., 2000,7(4):553 – 556.

Costa J E, Cheng R T, Haeni F P, et al. Use of Radars to Monitor Stream Discharge by Noncontact methods[J]. Water Resources Research, 2006.

Crombie D D. Doppler Spectrum of Sea Echo at 13.56 Mc/s[J]. Nature, 1955(175):681 – 682.

Fukami K, Yamaguchi T, Iamura H, et al. Curretn Status of River Discharge Observation Using Non – contact Current Neter for Operational Use in Japan[R]. World Environmental and Water Resources Congress 2008 Alupua'a, 2008:1 – 10.

Kimura Y, Satoshi I, Hiroaki M. River Flow Information System with RF Current Sensors[C]// 21st International Conference on Interactive Information Processing (IIPS) for Meteorology, Oceanography, and Hydrology, Session 10.8. The 85th AMS Annual Meeting(San Diego, CA).

Melcher N B, Costa J E, Haeni F P, et al. River Discharge Measurements by Using Helicopter – mounted Radar[J]. Geophysical Research Letters, 2009,29(22).

Yamaguchi T, Niizato K. Flood Discharge Observation Using Radio Current Meter[J]. Doboku Gakkai Rombun – Hokokushu / Proc. of JSCE, 1994(497):41 – 50.

Wright J W. Backscattering from Capillary Waves with Application to Sea Clutter[J]. IEEE Trans. Antennas Propag., 1966, AP – 14:749 – 754.

Wright J W, Keller W C. Doppler Spectra in Microwave Scattering from Wind Waves[J]. Phys. Fluid, 1971(14):466 – 473.

Chu Mushou. Non – contact Measurements and Accuracy Analyses for High Stages Open Channel Discharge[J]. PhD Thesis, National Cheng Kung University, 2010.

Predicting the Typical Year of Precipitation Based on the Fuzzy Pattern Recognition and R/S Analysis Model

Huang Xin and *Qiu Lin*

Institute of Environmental and Municipal Engineering, North China University of
Water Resources and Electric Power, Zhengzhou, 450011, China

Abstract: The hydrological system is randomness and fuzziness. Integrating many methods to reform the prediction model will improve the accuracy of the typical year. Uncertain with the R/S analysis whose characteristics is long – term correlation and statistical self – similarity, the fuzzy pattern recognition and R/S analysis model is the time series predicting model by coupling fuzzy pattern recognition. It avoids the arbitrariness and subjectivity by artificially dividing the known data, and mines the more abundant information in the time series than the traditional predicting method of certainty and randomness, and can be applied in the detection of the long – term memory effect and the non – gauss process with peak and thick tail. The result shows that the model is convenient and current, with the high prediction precision and popularization using value.
Key words: fuzzy pattern recognition, R/S analysis model, coupling, typical year, predict

1 Introduction

The abnormality of precipitation usually causes incredible loss in agricultural production, and also directly affects people's life and work. Since 2010, the drought happened in many northern provinces and five southwest provinces in China, which puts forward higher requirements for related departments to master the situation of future precipitation. The hydrological system is a complex, high nonlinear, uncertain system, and because of its randomness and fuzziness, the traditional linear predicting method must constrain the research of precipitation prediction. The self – similarity of some characteristic value of hydrological system showed in statistical significance determines that using the single method impossibly improves the prediction reliability. Therefore, by organically combining the various methods, namely coupling, can improve the prediction model and undoubtedly greatly increase the prediction precision. Hence, the paper explores to predict the typical year by coupling the analysis and its method of fuzzy pattern recognition and R/S analysis, and illustrates in detail the recognition criterion of minimum sum of squares weighted distance, the mathematics and physical meaning of its relevant iterative formula. The fuzzy pattern recognition and R/S analysis model is the time series predicting model by coupling fuzzy set theory with the R/S analysis whose characteristics is long – term correlation and statistical self – similarity. The result shows that the prediction model for typical year is indeed feasible, with the high prediction precision and popularization using value.

2 The model of fuzzy pattern recognition

Let n samples compose set, and each sample has m evaluation index. Then the index matrix is $X_{m \times n} = (x_{ij})$, where, $i = 1, 2, \cdots, m$; $j = 1, 2, \cdots, n$; x_{ij} is measured value of sample j' index i. m index are evaluated by c grade standard, then the index standard matrix is $Y_{m \times c} = (y_{ih})$, where, $h = 1, 2, \cdots, c$, y_{ih} is standard value of standard h' index i.

Let matrix $X_{m \times n}$ and $Y_{m \times c}$ change into corresponding $R_{m \times n} = (r_{ij})$ which is measured index relative membership degree matrix and $S_{m \times c} = (s_{ih})$ which is index standard relative membership degree matrix. The sample j belongs to standard h by relative membership degree u_{hj}, then the sample relative membership degree matrix is $U_{c \times n} = (u_{hj})$, which meets the constraint requirement $\sum_{h=1}^{c} u_{hj} - 1 = 0$, $\forall j$, $\sum_{j=1}^{n} u_{hj} > 0$, $\forall h$. Index relative membership degree matrix $R_{m \times n}$ not only describes the over standard, but also express the weight. Here the column normalized matrix $W_{m \times n} = (w_{ij})$ is defined as over standard weight matrix, which meets $\sum_{i=1}^{m} w_{ij} = 1$, $\forall j$.

Next solve u_{hj}. The difference of sample j and standard h can be expressed by general weighted length, namely:

$$D(r_j, s_k) = u_{kj}^{\ p} \sqrt{\sum_{i=1}^{m} (w_{ij} | r_{ij} - s_{ih} |)^p} \tag{1}$$

With the objective function of minimizing sum of squares general weighted length seeks optimal relative membership degree of sample j to standard h, namely:

$$\min \{ F(u_{hj}) = u_{hj}^2 [\sum_{i=1}^{m} (w_{ij} | r_{ij} - s_{ih} |)^p]^{\frac{2}{p}} \} \tag{2}$$

where, distance parameter $p = 1$ is hamming distance, $p = 2$ is Euclidean distance.

So

$$u_{hj} = \cfrac{1}{\sum_{k=1}^{c} \left[\cfrac{\sum_{i=1}^{m} (w_{ij} | r_{ij} - s_{ih} |)^p}{\sum_{i=1}^{m} (w_{ij} | r_{ij} - s_{ik} |)^p} \right]^{\frac{2}{p}}} \tag{3}$$

The value range of k is dynamic, which is determined by compared sample j' index m in the matrix $R_{m \times n}$ with matrix $S_{m \times c}$, when using Eq. (3) to evaluated. Let the relative membership degree value of sample j' index m is $r_j = (r_{1j}, r_{2j}, \cdots, r_{mj})^T$. Let compare the relative membership degree value r_{ij} of $i(i = 1, 2, \cdots, m)$ with each grade standard's relative membership degree value $s_i = (s_{i1}, s_{i2}, \cdots, s_{ic})$ of index i in the matrix $S_{m \times c}$ one by one. If the minimum grade that index i falls into the range of s_i is a_j, and the maximum grade is b_j, the value range of k will be a_j and b_j, then $1 \leqslant a_j < b_j \leqslant c$. The value range of h in the Eq. (4) is the same with k, obviously when $h < a_j$ or $h > b_j$, $u_{hj} = 0$, and $\sum_{h=a_j}^{b_j} u_{hj} = 1$.

Conclusion demonstrated that the complete form of evaluation model of fuzzy pattern recognition can be expressed as:

$$\begin{cases} u_{hj} = \cfrac{1}{\sum_{k=a_j}^{b_j} \left[\cfrac{\sum_{i=1}^{m} (w_{ij} | r_{ij} - s_{ih} |)^p}{\sum_{i=1}^{m} (w_{ij} | r_{ij} - s_{ik} |)^p} \right]^{\frac{2}{p}}}, \quad h \in [a_j, b_j] \\ u_{hj} = 0, \ h < a_j \ 或 \ h > b_j \\ u_{hj} = 1, \ r_{ij} = s_{ik} \end{cases} \tag{4}$$

where, $j = 1, 2, \cdots, n$; $1 \leqslant a_j < b_j \leqslant c$; $h = a_j, a_j + 1, \cdots, b_j$.

When $p = 1, 2$, the relative membership degree matrix of sample set corresponding grade h is $_1U_{c \times n} = (_1 u_{hj})$ and $_2U_{c \times n} = (_2 u_{hj})$ respectively. Here takes the average relative membership degree matrix $\overline{U}_{c \times n} = (_1 u_{hj} + _2 u_{hj})/2 = (\overline{u}_{hj})$ of sample set corresponding grade h for the two distance parameters to clear the influence of different value p for evaluation.

From the average relative membership degree matrix we can get the grade characteristic value vector of n samples belong to each grade as:

$$H = (1, 2, \cdots, c) \cdot (\overline{u}_{hj}) = (H_1, H_2, \cdots, H_n) \tag{5}$$

According to the grade characteristic value of samples, we can conclude a comprehensive evaluation for samples belong to grade.

3 R/S analysis model

The atmospheric precipitation was manifested as discontinuous points distribution along the time axis, namely, a structured Cantor set. Here is using Fractional Brownian Motion model to predict the typical year of precipitation. Its correlation coefficient is:

$$\gamma(t) = 2^{2H-1} - 1 \tag{6}$$

where, H is Hurst index.

Mandelbrot extends index H to the range of 0 to 1, and gets

$$R(i)/S(i) = (ai)^H \tag{7}$$

$$\bar{x} = \frac{1}{m}\sum_{i=1}^{m} x(i) \quad m = 1,2,\cdots,n \tag{8}$$

$$X(i,m) = |x(i) - \bar{x}| \quad i = 1,2,\cdots,m \tag{9}$$

$$R(i) = \max X(i,m) - \min X(i,m) \quad (1 \leqslant i \leqslant m) \tag{10}$$

$$S(i) = \sqrt{\frac{1}{m}\sum_{i=1}^{m}[x(i) - \bar{x}]^2} \tag{11}$$

where, a is a constant; $X(i,m)$ is accumulated deviation; $R(i)$ is a range; $S(i)$ is standard deviation.

From Eq. (7) to Eq. (11), we can get the typical years.

4 Application example

Now taking an example of ten days rainfall data of certain irrigation area from the Yellow River during the years of 1960 ~ 2009 to analyze and predict the typical year.

Here the typical year is divided into 5 standard types, that is, wet year, partial wet year, middle year, partial dry year, and dry year. According to the basic theory of fuzzy pattern recognition, the sample set is composed of 50 samples (from 1960 to 2009), and each sample has 36 evaluation indexes (ten days). x_{ij} is actual measured precipitation of each ten days. According to the 5 standard types, the 36 evaluation indexes per year are evaluated by 5 standards. y_{ih} is the critical value of each standard. In the view of the above, we can get $R_{36 \times 50} = (r_{ij})$ and $S_{36 \times 5} = (s_{ih})$. Eq. (4) and Eq. (5)'s solution will determine the annual type of 1960 ~ 2009.

According to the result of fuzzy pattern recognition, the wet years' series are used as the initial value of R/S analysis prediction typical year model.

From 1960, there are 13 wet years in the irrigation area. According to its year, annual rainfall and corresponding $R(i)/S(i)$, we can draw the relation graph of $\ln i \sim \ln[R(i)/S(i)]$. The graph is nearly straight line, so the sequence of rainfall meets the condition of Fractional Brownian Motion model.

Firstly, here we consider 1960 as the zero point of calculation, and obtain the time series noted as $(x_i) = (1,3,4,7,9,11,13,30,33,34,38,45)$. Where, 2009 which is corresponding to the 49 time series is kept as verification of the result. We put x_i into Eq. (7) to Eq. (11), and get the time series $x_{13} = 49$ of next wet year, and feed back the zero point of calculation, that is, 1960 + 49 = 2009. From the result of fuzzy pattern recognition, the prediction value coincides with the actual year. Then, the year of 2009 used as new information, and added into the original time series $\{x_i\}$. We can get the time series $x_{14} = 52$ of next wet year, and feed back the zero point of calculation, that is, 1960 + 52 = 2012. Consequently, the next type year is determined.

If the time series of partial wet year, middle year, partial dry year and dry year are used, we can predict the other type year. The followed is that the type year can be predicted year by year.

5 Conclusions

With the rapid and steady development of the economical society, the accurate prediction for typical year is very important in both theory and practice to instruct agricultural irrigation, improve availability of water resources, and bring the economic, social and environmental benefit of the irrigation area.

The model by coupling fuzzy pattern recognition with the R/S analysis avoids arbitrariness and subjectivity by artificially dividing the known data. The fuzzy pattern recognition based on ten days precipitation makes the year type more reasonable and accurate. The model mines the more abundant information in the time series than the traditional predicting method of certainty and randomness. It can be applied in the detection of the long – term memory effect and the non – gauss process with peak and thick tail. When the time scale is too small and the observed samples are greatly related, the model has estimated error. The result shows that the prediction model for typical year is indeed feasible, with the high prediction precision and popularization using value.

References

Hu Ping, Kang Ling. Application of A Kind of Nonlinear Combination Forecast Method in Run off Prediction[J]. China Rural Water and Hydropower, 2004(3): 38 – 40.

Xia Jun. The Quasi – analysis and Application of Model Identifiability for Hydrological System[J]. Journal of Hydraulic Engineering, 1988, 4: 1 – 8.

Chen Shouyu. The Fuzzy Sets Theory and Practice for Engineering Hydrology and Water Resources System[M]. Dalian: Dalian University of Technology Press, 1998.

Qiu Lin, Huang Xin, Li Hongliang. Application of Precipitation Forecast Based on Fuzzy R/S Analysis Model in Agriculture[J]. China Rural Water and Hydropower, 2006(10): 20 – 23.

Cai Fuxin, Qiu Lin, Xie Xinmin. Real Time Optimal Dispatch of Water Resources for Irrigation Area[J]. Journal of Hydraulic Engineering, 2007, 38(6): 710 – 716.

Huang Xin. The theory, Model and Application of Real – time Optimal Operation of Water Resources in Irrigation Area[D]. Zhengzhou: North China University of Water Resources and Electric Power, 2007.

Research on the Image Deployment Scheme for Remote Sensing and Monitoring of the Water and Soil Conservation of the Loess Plateau

Zhao Bangyuan[1], *Ma Ning*[1], *Zhai Ran*[2], *Ma Hongbin*[1], *Li Zhihua*[1] and *Wang Qinxiang*[1]

1. Upper and Middle Yellow River Bureau, YRCC, Xi'an, 710021, China
2. School of Earth Sciences and Engineering, Hohai University, Nanjing, 211100, China

Abstract: This paper identifies the scope of remote sensing and monitoring of the water and soil conservation of the loess plateau, including terrace, vegetation, artificial grassland, check dams and dammed land, ground material composition, types of soil erosion, etc; analyzes the requirements of the remote sensing and monitoring of water and soil conservation for the spatial resolution, spectral characteristics and time phase of remote sensing images; evaluates the technical feasibility of the remote sensing and monitoring of water and soil conservation by means of the multi – spectral remote sensing images not less than 30 m spatial resolution; and proposes that the multi – spectral remote sensing images of not less than 2 m, 2 ~ 5 m or 5 ~ 10 m between June and September shall be selected respectively for remote sensing and monitoring of the water and soil conservation of the Loess Plateau on 1:10,000, 1:25,000 or 1:50,000 scale through comparison of the indexes (spectral characteristics, spatial resolution, shortest acquiring period, provision of DEM, accuracy, investment, etc) of typical remote sensing images at not less than 10 m spatial resolution, which can assure both monitoring accuracy and economic rationality. Further investigations are warranted on determination of remote sensing images by comparing image availability, cost performance, monitoring accuracy and so forth in actual monitoring work. In addition, the remote sensing and monitoring of water and soil conservation shall be carried out by taking integrated methods.

Key words: water and soil conservation, monitoring, remote sensing, image, deployment

1 Introduction

The loess plateau region of the China's Yellow River basin is one of the regions subject to the most serious soil erosion and exposed to the most fragile ecological environment in the world. Serious soil erosion worsens ecological environment, restricts sustainable social and economic development, silts up lower riverbed of the Yellow River, threatens its flood control security, and restrains the utilization of its limited water resources. Water and soil conservation is an effective way to cure soil erosion and an important means to protect ecological security and uphold the sustainable economic and social development of the region exposed to serious soil erosion. Timely and accurate access to regional soil and water conservation information provides an important foundation for analysis and evaluation of the effectiveness of soil conservation measures and for scientific decision of water loss and soil erosion control. Remote sensing is an important means and the only way to quickly acquire a large range of high precision information of regional water and soil conservation. The information extracted from the remote sensing images of different spatial resolutions, different spectral characteristics and different time phases is different in type and accuracy. Different monitoring targets have different requirements for image resolution. Therefore, a study on what type of images shall be selected for remote sensing and monitoring of the water and soil conservation of a specific object is essential to achieving both suitable precision and reasonable investment.

2　Requirements of water and soil conservation remote sensing and monitoring for remote sensing images

2.1　Monitoring targets for water and soil conservation

2.1.1　Taxonomy of monitoring targets

（1）Soil erosion factors.

Monitoring shall cover land use types, vegetation, terrain, distribution and area of ground substances, and so forth.

In the loess plateau region, land use types mainly include arable land (irrigable land, dry land, etc), garden plot (fruit ranch, etc), woodland (shrubbery land, forest land, etc), grassland (natural pasture, artificial pasture, etc), residential area, water area as well as the lands used for industrial and mining, transport and water conservancy facilities, etc. Dry land includes dry flat land, terrace land, sloping land and gully – dammed land. Shrubbery land and forest land includes natural and artificial types. Other woodland includes open forest land, immature forest land, nursery, etc. Other grassland includes natural meadow, artificial meadow, wild grass ground, etc. Other land includes bare land, etc.

The monitoring targets for vegetation include species (trees, shrubs, grass, and mingled forest), type (Chinese pine, sea backthern, etc.), coverage, shade density, etc.

Topographical factors include inclination, exposure, etc, which can be extracted from topographic data such as DEM.

The monitoring targets for ground substances include types of soil, earth rock, sand and rock.

（2）Soil erosion.

The type, intensity, distribution and hazard of soil erosion are concerned. Erosion type is generally classified into water erosion, wind erosion, freeze – thaw erosion and man – made erosion. Erosion intensity is generally obtained through calculation of erosion factors on an erosion model or through actual measurement. Distribution refers to the area and location of erosion factors and intensity. Hazard means the impact of soil erosion on ecological environment, social economy, river channel sedimentation, water resource utilization, etc.

（3）Prevention and control of soil erosion.

The type, distribution, quantity and quality of soil and water conservation measures are concerned. The measures for the soil and water conservation of Loess Plateau involve five kinds: sloping land regulation (including terrace land and soil conservation based cultivation), wasteland management (including growing trees and grass and enclosing), gully control (including gully head protection, check dam, warp land dam, small reservior, and slope collapse control), windy desert area regulation (checkerboard protection, shelter belt for checking winds and fixing drifting sand, forest network for farmland protection, growing trees and grass, building farmland by diverting river water and transporting sand, etc.), and small water storage, drainage and diversion works (including intercepting drain, impounding basin, drainage ditch, water kiln, pond, dyke, irrigation works with torrential flood).

2.1.2　Selection of monitoring targets

According to monitoring targets taxonomy, and evaluation methodology for investment and benefits of soil and water conservation measures, in order to give top priority to what is the most important and quickly acquire a large range of regional water and soil conservation information, the following main monitoring targets for soil and water conservation was determined through classification of monitoring targets and pick – out of the objects unsuitable for remote sensing: land use types (irrigable land, dry flat land, terrace land, sloping land, gully – dammed land, fruit ranch, shrubbery land, forest land, open forest land, immature forest land, nursery, natural meadow, artificial meadow, wild grass ground, bare land, residential area, water area as well as the lands used for industrial and mining, transport and water conservancy facilities), vegetation (species, type, coverage, shade density, etc.), ground substances, soil erosion types and so forth. Monitoring shall focus on terrace land, vegetation species, vegetation coverage, artificial meadow, check dams and dammed land, ground substances, soil erosion types, water area, residential area and industrial and mining land.

2.2　Requirements of different monitoring targets for remote sensing images

2.2.1　Spatial resolution requirement

Monitoring accuracy generally refers to the accuracy of national basic scale topographic map series. Distinguishing and withdrawal of ground objects have a higher requirement for the spatial resolution of remote sensing images. Suppose that at least three pixels are required to recognize a ground object, then the minimum requirements for spatial resolution shall be 1/3 of the minimum dimension of the object. For instance, for a terrace land with minimum width of 5 m, the spatial resolution of its remote sensing images should generally be not less than 1.5 m.

2.2.2　Spectral Characteristic Requirement

Distinguishing and extraction of vegetation for species, type, coverage, shade density, soil and/or water have a higher requirement for the spectral characteristic of remote sensing images. Within the spectrum segment range of remote sensing images are mainly visible light and near infrared light. They are generally divided into five bands, i. e. blue, green, red, near infrared, and microwave. Blue band is mainly used for distinguishing between evergreen vegetation and deciduous vegetation and between vegetation and soil. Green band is mainly used for identifying vegetation form. Red band is mainly used for identifying such information as vegetation form, coverage, soil, topographical features, etc. Near infrared band is mainly used for identifying vegetation form, coverage and soil. Microwave is mainly used for extracting such information as soil humidity and ground temperature. The spectrum segment range of remote sensing images shall be selected in light of monitoring targets.

2.2.3　Time phase requirement

For the remote sensing and monitoring of loess plateau water and soil conservation, the multispectral remote sensing images of the season (e. g. June to September) when ground attachments such as crops and vegetations present a dramatic variation characteristics and are easily identified shall be selected in light of monitoring targets.

3　Research on image deployment scheme for remote sensing and monitoring of water and soil conservation

The remote sensing and monitoring of water and soil conservation cover a lot of targets. We select the following major targets for analysis: terrace land, vegetation, artificial meadow, check dams and dammed land, ground substances, soil erosion types, water area, residential area and industrial and mining land.

3.1　Technical feasibility analysis

The technical feasibility for developing the remote sensing and monitoring of water and soil conservation by means of the multi – spectral remote sensing images not less than 30 m spatial resolution was analyzed based on specific sample survey and in combination with practical experience on the remote sensing and monitoring of Huangfuchuan and Gushanchuan watersheds in the rich and coarse sediment area of the middle Yellow River, as shown in Tab. 1. In actual work, the remote sensing images that are technically feasible and with qualified accuracy, multi – spectrum and suitable time phase shall be selected for remote sensing and monitoring in light of the size of monitoring objects.

Tab. 1 Technical feasibility analysis for remote sensing and monitoring of water and soil conservation by means of multi – spectral remote sensing images

Monitoring targets	Spatial resolution of remote sensing images				
	< 2 m	2 ~ 5 m	5 ~ 10 m	10 ~ 20 m	20 ~ 30 m
Terrace land	More than 5 m in width	More than 5 ~ 15 m in width	More than 15 ~ 30 m in width	?	×
Artificial meadow				?	×
Check dams and dammed land	√	√	√	?	?
Water area	√	√	√	√	?
Residential area and industrial/ mining land	√	√	√	?	?
Vegetation type	√	√	√	?	?
Vegetation coverage	√	√	√	√	√
Ground substances	√	√	√	√	√
Soil erosion type	√	√	√	√	√

Note: The symbols, √, ? and ×, respectively represent "Feasible", "Not necessarily feasible" and "Not feasible".

3.2 Analysis of typical remote sensing image indices

According to the results of technical feasibility analysis, the typical remote sensing images that are made available at the present time are selected for comparison. Considering the images less than 10 m spatial resolution cannot clearly identify terrace land or artificial meadow, or provide an insufficient monitoring accuracy, plus with significant price discrepancy, it is often used in actual monitoring work. No more analysis is done here. Tab. 2 presents the comparative analysis of typical image indices.

Tab. 2 Comparison of typical remote sensing image indices

Spatial resolution	Image type	Spectral characteristics	Spatial resolution of images	Fastest cycle time for image acquirement	DEM and accuracy	Investment
<2 m	Quickbird fusion image	Red, green, blue, near infrared and panchromatic	0.6 m	About 1 week	None	Filing, 195 Yuan/km²; programming, 227 Yuan/km²; post – treatment extra
	IKONOS fusion image	Red, green, blue, near infrared, panchromatic	1 m	About 3 days	1:10,000 scale	Stereo – pair: 510 Yuan/km²; post – treatment extra. Not Stereo – pair: filing, 135 Yuan/km²; programming, 198 Yuan/km²; post – treatment extra
	GeoEye – 1 image	Red, green, blue, near infrared, panchromatic	Panchromatic 0.41 m, multispectral 1.65 m	About 2 ~ 3 days	1: 10,000 scale	Stereo – pair: 640 Yuan/km²; post – treatment extra. Not Stereo – pair: filing, 168 Yuan/km²; programming, 220 Yuan/km²; post – treatment extra

Conbinued Tab. 2

Spatial resolution	Image type	Spectral characteristics	Spatial resolution of images	Fastest cycle time for image acquirement	DEM and accuracy	Investment
	DMC, UCX/UCD, ADS40/80, LIDARdigital aerial image	Red, green, blue, near infrared, panchromatic	Generally higher than 0.6 m	Depend on local weather	Not lower than 1:10,000 scale	Digital aerial photogrammetry, about 150 Yuan/km²; post - treatment, about 250 Yuan/km².
	Quickbird multispectral image	Red, green, blue, near infrared, panchromatic	2.4 m	About 1 week	None	Filing, 195 Yuan/km²; programming, 227 Yuan/km²; post - treatment extra
	IRS - P5 & RapidEye fusion image	Red, green, blue, near infrared, panchromatic, red edge	2.5 m	5 days	About 1:25,000 scale	IRS - P5 single, 20 Yuan /km²; programming, 3,000 Yuan/view. IRS - P5 stereo - pair, 50 Yuan/km²; programming, 6,000 Yuan/pair; RapidEye and post - treatment extra
2 ~ 5 m	IKONOS multispectral image	Red, green, blue, near infrared	4 m	About 3 days	About 1:10,000 scale	Stereo - pair: 510 Yuan/km²; post - treatment extra. Not Stereo - pair: filing, 135 Yuan/km²; programming, 198 Yuan/km²; post - treatment extra
	RapidEye image	Red, green, blue, near infrared, red edge	5 m	Every day	None	Filing, about 10 Yuan/km²; programming, about 12 Yuan/km²; post - treatment extra
5 ~ 10 m	SPOT fusion image SPOT multispectral image	Red, green, blue, near infrared, shortwave infrared	5 m 10 m	About 5 days	About 1:25,000 scale	Filing, about 10 Yuan/km²; programming, about 12 Yuan/km²; post - treatment extra Filing, about 5 Yuan/km²; programming, about 6 Yuan/km²; post - treatment extra

Note: All DEMs are defined by a nominal accuracy except digital aerial photogrammetry, which may not be in consistency with the spatial resolution of images.

3.3 Image deployment scheme for remote sensing and monitoring

The typical remote sensing images listed in Tab. 2 generally are all available with red, green, blue, near infrared, and panchromatic bands with similar spectral characteristics. The fastest cycle time for image acquirement is generally within a week. Except Quickbird and RapidEye which do not provide DEM service, others can provide DEM service with nominal accuracy of 1:10,000 ~ 1:25,000 scales. Image prices normally increase exponentially with the improvement of spatial resolution. According to Tab. 1, the remote sensing images with suitable spatial resolution and time phase shall be selected depending on the minimum dimension of monitoring objects (giving priority to terrace land and artificial meadow) so that both accuracy and cost – effectiveness are satisfactory. Tab. 3 shows the image deployment schemes for remote sensing and monitoring of water and soil conservation at 1: 10,000, 1: 25,000 and 1: 50,000 scales.

Tab. 3 Image deployment schemes for remote sensing and monitoring of water and soil conservation

Monitoring accuracy	Monitoring purpose	Spatial resolution and optional images	Monitoring targets	Remark
1: 10,000 scale	Planning, design, regulation, acceptance, research and soil erosion survey (including man – made soil erosion distribution) for water and soil conservation of minor watersheds or larger areas	Not lower than 2 m, including 0.6 m Quickbird fusion image, 1 m IKONOS fusion image, 0.41 m/1.65 m GeoEye – 1 fusion image and multispectral image, and DMC/UCX/UCD/ADS40/80//LIDAR digital aerial image 2 ~ 5 m, including	Land use type, vegetation, water and soil conversion measures, ground substances, erosion type, etc.	The topographic factors such as slope and exposure shall be monitored with stereo – pair images. The minimum dimension of monitoring object shall be relied on to determine the spatial resolution of remote sensing images
1: 25,000 scale	Planning, regulation, acceptance, research and soil erosion survey (including man – made soil erosion distribution) for water and soil conservation of county regions, minor watersheds or larger areas	2.4 m Quickbird multispectral image, 2.5 m IRS – P5 and RapidEye fusion image, 4 m IKONOS multispectral image		
1: 50,000 scale	Planning, research and soil erosion survey for water and soil conservation of county regions or larger areas	5 ~ 10 m, including RapidEye image, 5 m SPOT fusion image, 10 m SPOT multispectral image		

Note: For the remote sensing images (e. g. 4 m IKONOS, 5 m RapidEye) whose spatial resolution is at the bound of monitoring accuracy, their monitoring accuracy is related to the terrain of the monitor area, and not fully agrees with those listed above, and shall be experimentally determined in actual work.

4 Conclusions and Discussions

The multi – spectral remote sensing images of not less than 2 m, 2 ~ 5 m or 5 ~ 10 m spatial resolution between June and September shall be selected respectively for the remote sensing and monitoring of loess plateau water and soil conservation on 1: 10,000, 1: 25,000 or 1: 50,000 scale, which can assure both monitoring accuracy and economic rationality. However, for the remote sensing and monitoring of water and soil conservation in the loess plateau region which differ significantly in topographical features, further investigations shall be made to determine one or several types of remote sensing images to be used by comparing image availability, cost performance, monitoring accuracy and so forth in light of monitoring targets (e. g. whether topographic factors and/or terrace land size need to be monitored or not). On the other hand, multi – spectral remote sensing has its limitations. It may not be able to distinguish all monitoring objects no matter how high spatial resolution and spectral resolution are. So radar or other remote sensing shall be applied in this case. For the monitoring objects which are undersized, or aren't major monitoring targets, or cannot or shouldn't be found by using remote sensing, they shall be monitored by a comprehensive way such as specific sample survey or data collection (especially statistical data collection) instead, provided monitoring accuracy can be assured, so as to make up for the deficiency of remote sensing, give play to the strong points of different monitoring methods, guarantee accuracy, and save investment.

References

Ministry of Water Resources of P. R. China. SL 277—2002. Technical Code of Practice on Water and Soil Conservation [S]. Beijing :China Water Power Press, 2002.

Ministry of Water Resources of P. R. China. SL 449—2009. Compilation Rules on Preliminary Design Report of Soil and Water Conservation Project [S]. Beijing : China Water Power Press, 2009.

Ministry of Water Resources of P. R. China. SL 190—2007 . Soil Erosion Classification and Categorization Standards[S]. Beijing:China Water Power Press, 2008.

State Administration of Quality Supervision, Inspection and Quarantine of P. R. China, China National Standardization Management Committee. GB/T 16453. 1 ~ 5—2008. Technical Specification for Comprehensive Control of Soil and Water Conservation [S]. Beijing: Standards Press of China, 2009.

State Administration of Quality Supervision, Inspection and Quarantine of P. R. China, China National Standardization Management Committee. GB/T 15774—2008. Efficiency Calculation Method for Soil and Water Conservation[S]. Beijing :Standards Press of China, 2009.

Application of Drought Disaster Loss Assessment Based on Improved Grain Disaster Loss Rate Model

Li Yan[1], *Liang Zhongmin*[1], *Zhao Weimin*[2] and *Liu Xiaowei*[2]

1. College of Hydrology and Water Resources, Hohai University, Nanjing, 210098, China
2. Hydrology Bureau of Yellow River Conservancy Commission, Zhengzhou, 450004, China

Abstract: In this article, the factors affecting grain yield are analysed, which mainly consider the planting area and yield per unit area. An improved grain disaster loss rate model was bult, and the evaluation of grain loss based on the improved model was studied. For the typical area Xi'an, using 1992 ~ 2008 grain growth statistical material, the grain drought loss rate of eight counties (districts) under the typical area were evaluated. The results show that grain drought loss rate is at an average of almost 3.9% for nearly 17 years. Among them, drought loss rate in Chang'an is the highest, being 5.89%, Huxian's is the lowest, being 2.76%.

Key words: planting area, yield per unit area, disaster loss rate, drought loss evaluation

1 Introduction

In recent years, natural disaster occurred frequently, which has brought inestimable loss to social economic development. From a global perspective, drought is considered to be one of the world's most serious types of natural disaster with the most extensive sphere of influence and the largest economic losses. China is one of the countries most severely affected by drought disasters. Frequent drought causes serious harm to China's industrial and agricultural production. The impact is increasing in energy demand, environmental protection and economic construction. With the socio – economic development and population growth, drought and water shortage are worsening, and drought level is gradually increasing. In the context of global warming, water shortage has become one of the most urgent problems which are needed to improve and solve with the national territory. Therefore, drought research and quantitative assessment of the grain losses have important practical significance for a comprehensive understanding of the drought disaster and its impact on agricultural production.

At present, the study of the impact of drought on agricultural production is relatively high. Some researchers have put forward variety of models and methods to compute the grain disaster yield loss. According to the statistics materials such as the affected area, inundated area and no any output area, Liu Yingqiu et al calculate the drought grain output reduction using integrated reduction percentage method. Considering the impact of planting area and yield per unit area changes on grain yield, Zheng Jingyun builds assessment model of natural disaster grain losses for quantitative calculation and qualitative evaluation of grain disaster loss rate. Because the details in the agricultural production are not easy to obtain, there are some limitations of using statistical methods. Nowadays, the methods of comparing with the previous year or perennial to estimate grain disaster yield loss are adopted. However, the trend growth ability of the grain production, which will result in lack of accuracy of the evaluation results are often ignored. The article undertakes from the factors that affect grain yield, based on the view of grain yield analysis, put forward an improved grain disaster loss rate model, and calculate the grain disaster loss rate caused by the drought to agricultural production, which provide theoretical support for drought disaster mitigation department decision – making.

2 Improved grain disaster loss rate model

Grain yield is influenced by grain varieties, technical level of agricultural production, soil and climate and other factors. There are highly complex linear and nonlinear relationship between these factors and grain yield. This study mainly considers the influence of grain unit yield fluctuations and planting area changes on grain yield.

2.1 Trend unit yield calculation model

Grain unit yield is affected by social factors, material inputs, scientific technological level and natural factors. Domestic and foreign researchers have undertaken extensive research on factors affecting grain unit yield fluctuations. The factors are mainly divided into three categories: agricultural technological measures, weather conditions and random noise. When we build grain yield forecasting model, we commonly adopt statistical methods to extract meteorological yield, and consider that the trend yield is the yield of grains under normal climate conditions. Typically, after eliding random noise which has small impact, the grain yield is mainly comprised by the trend yield and weather yield. Namely, the grain unit yield model is:

$$y_t = \hat{y}_t + y_w \tag{1}$$

where, y_t is the actual grain unit yield; $\hat{y}_t$ is the agricultural technological trend yield; y_w is the weather yield.

Among, $\hat{y}_t$ reflects the social factors, material inputs, scientific technological level; y_w reflects the impact of natural factors on grain yield.

Depend on the different weather conditions, weather yield can be divided into two kinds: under favorable and unfavorable conditions, that means $y_w = y_a - y_d$, substitute into Eq. (1) have:

$$y_t = \hat{y}_t + y_w = \hat{y}_t + y_a - y_d \tag{2}$$

where, y_a stands for the grain yield under favorable weather condition; y_d stands for the grain yield under unfavorable weather condition, namely grain yield loss causing by natural disaster.

Because the sum of $\hat{y}_t$ and y_a stands for grain trend unit yield under favorable weather conditions, we can establish the following trend unit yield calculation model:

$$\hat{y}_p = \hat{y}_t + y_a \tag{3}$$

2.1.1 Grain agricultural technological trend yield estimation

Because agricultural technological trend yield is closely related to promotion of improved varieties, fertilizer use, the increasing popularity of film and resilience enhancement, we can determine the key factors through the local agricultural technological factors analysis, get agricultural technological factors series matrix. Then combined with the trend yield series to establish a multiple regression model, different years' agricultural technological trend yield can be calculated. Due to data limitations, this study simulate the agricultural technological trend unit yield mainly by polynomial approximation method, which is calculated as:

$$\hat{y}_t = \sum_{i=0}^{n} a_i \cdot t^i \tag{4}$$

where, t is time; a_i is undetermined coefficient; n is polynomial order, the value can be determined by statistical reliability and simulation results.

In general, $n = 2$ or 3 can meet the requirement in a small range area (such as county, region).

2.1.2 Grain yield under then favorable weather condition estimation

Grain yield y_a under favorable weather conditions can be determined by the grain yield in the non – disaster year. Identify non – disaster year is the year which actual yield y is large than agricultural technological trend unit yield, so y_a is equal to y_w which is in non – disaster year. In determining the yield y_a under the favorable weather condition, the chosen non – disaster years may be more than one year. Gong Deji et al regard the actual grain yield in non – disaster year as expect yield in optimal weather condition, fit the upper envelope of grain yield trend by using actual yield curve in non – disaster year as calendar year grain yield under the favorable weather condition. But because there is less non – disaster year, estimated optimal yield trend may cause large errors. The article uses the weather condition yield mean in multiple non – disaster years as calculated value of y_a.

2.2 Improved grain disaster loss rate model

As Zheng Jingyun(1994) described, grain yield consists of four parts: the yield has reached in the previous year, growth ability of grain yield based on the previous year, grain yield variation caused by planting area change, and grain disaster yield loss caused by natural disaster. We can establish the following grain yield loss model:

$$Y_d = Y_{t-1} + \hat{Y}_p + Y_s - Y_t \tag{5}$$

where, Y_d means grain disaster yield loss; Y_{t-1} means the previous year's grain yield; $\hat{Y}_p$ means trend yield based on the previous year; Y_s means grain yield variation caused by planting area change; Y_t means current actual yield.

As Zheng Jingyun(1994) described, author takes $\hat{Y}_p = (\hat{y}_p - y_{t-1}) \cdot S_t$ and $Y_s = y_t \cdot (S_t - S_{t-1})$. Bring them into Eq. (5), have:

$$Y_d = Y_{t-1} + (\hat{y}_{t-1}) \cdot S_t + y_t \cdot (S_t - S_{t-1}) - Y_t \tag{6}$$

where, $\hat{y}_p$ is grain trend yield under the favorable weather condition; y_t is the actual yield in the t th year; y_{t-1} is the actual yield in the $t-1$ th year; S_t is the actual planting area in the t th year; S_{t-1} is the actual planting area in the $t-1$ th year.

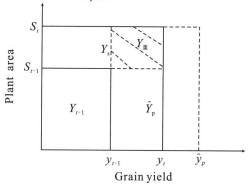

Fig. 1   The decompose diagram of grain yield

Mapping shows that there is overlap between $\hat{Y}_p$ and Y_s, in which, $Y_\text{重} = (y_t - y_{t-1}) \cdot (S_t - S_{t-1})$. As shown in Fig. 1, Eq. (6) regards the overlap as grain disaster yield loss. Which will lead to the calculation result is larger. Deducting the overlap, we can establish the improved grain disaster yield loss model:

$$Y_d = Y_{t-1} + (\hat{y}_p - y_{t-1}) \cdot S_t + y_t \cdot (S_t - S_{t-1}) - Y_t - (y_t - y_{t-1}) \cdot (S_t - S_{t-1}) \tag{7}$$

where, Y_d represents the absolute quantity of grain disaster yield loss, so these values can not be compared in different agricultural production level. In order to reflect the influence of natural disaster grain yield under the same natural conditions and different productivity levels, we use grain disaster loss rate (the ratio of disaster yield loss and total yield) Y_d / Y_t to measure. Then establish the improved grain disaster loss rate model:

$$L_d = \frac{Y_{t-1} + (\hat{y}_p - y_{t-1} \cdot S_t + y_t \cdot (S_t - S_{t-1}) - Y_t - (y_t - y_{t-1}) \cdot (S_t - S_{t-1})}{Y_t} \tag{8}$$

3 Drought loss assessment based on improved grain disaster Loss rate model

In order to accurately evaluate the influence of the drought on the grain yield in Xi'an area, now select Xi'an as the typical study area, combined with the above – mentioned improved grain disaster loss rate model, using the actual grain statistics from 1992 to 2008 in the eight counties (districts) which are under the study area jurisdiction, and do examples of applied research on the assessment of the grain drought loss rate.

3.1 Overview of the study area

Xi'an is located in the middle of the Yellow River Basin(Guanzhong basin), which belongs to a mid – latitude warm and humid zone semi – arid sub – humid continental monsoon climate zone, and has cold and warm wet and dry with four distinct seasons. Winter is cold, less windy, foggy, and has less rain and snow. Spring is warm, dry, windy and climate variability. Summer is hot and rainy, has late summer highlight and the multi – thunderstorm winds. Autumn is cool and the temperature downhill, autumn leaching is obvious. Annual precipitation is 522. 4 ~ 719. 5 mm, increases from north to south, July and September are the two obvious precipitation peak month. During the year, there are main meteorological disasters such as droughts, storms, floods, dust, and low – temperature damage et al. Due to its location within the mainland, this area can't be regulated from the marine role in temperature and rainfall; the south being effected by barrier function of the Qinling Mountains, rainfall is not rich and uneven distribution during the year, resulting in drought disaster occurred more frequently in Xi'an area.

3.2 Grain drought loss assessment in Xi'an area

According to eight counties (districts) grain statistics of Xi'an area from 1992 to 2008, combined with the polynomial approximation method, determine the grain calendar year agricultural technology trend yield. Then by comparing with the actual yield and the findings of disaster situations in the Xi'an area over the years, select the non – disaster year in the study area, and determine grain yields in the favorable weather condition, then determine the grain trend yield. The field research results show that drought grain loss in Xi'an region account for about 50% of the loss of natural disasters. Combined with improved grain disaster loss rate model, calculate and obtain the eight counties (districts) grain drought loss rate in the study area, the calculation results shown in Tab. 1.

Tab. 1 Eight counties (districts) grain drought loss rate over the years in Xi'an region

Year	Xi'an	Yanliang	Chang'an	Lintong	Zhouzhi	Huxian	Lantian	Gaoling
1993	0.031,8	0.030,8	0.063,2	0.043,1	0.045,1	0.030,8	0.052,5	0.034,0
1994	0.036,2	0.035,4	0.081,5	0.056,4	0.058,5	0.035,7	0.079,3	0.030,1
1995	0.031,6	0.030,1	0.078,2	0.042,6	0.048,8	0.029,3	0.068,5	0.029,0
1996	0.034,1	0.032,1	0.067,2	0.043,3	0.045,1	0.032,1	0.057,8	0.034,0
1997	0.028,9	0.031,3	0.063,0	0.038,9	0.044,9	0.028,6	0.050,3	0.029,5
1998	0.028,5	0.028,2	0.055,3	0.039,1	0.040,0	0.027,0	0.044,9	0.029,3
1999	0.029,2	0.029,1	0.060,8	0.039,1	0.042,3	0.028,4	0.049,3	0.029,1
2000	0.029,8	0.030,0	0.054,9	0.040,6	0.043,9	0.028,2	0.048,0	0.029,7

Continued Tab. 1

Year	Xi'an	Yanliang	Chang'an	Lintong	Zhouzhi	Huxian	Lantian	Gaoling
2001	0.030,4	0.030,2	0.060,2	0.039,4	0.042,4	0.024,7	0.052,3	0.029,9
2002	0.029,6	0.031,7	0.055,6	0.040,1	0.042,6	0.025,1	0.051,3	0.030,5
2003	0.030,5	0.030,2	0.059,3	0.044,6	0.044,8	0.028,3	0.048,4	0.031,2
2004	0.029,7	0.026,8	0.052,1	0.038,1	0.040,2	0.025,8	0.041,8	0.030,2
2005	0.030,3	0.026,4	0.048,4	0.040,5	0.036,7	0.024,5	0.039,1	0.030,3
2006	0.029,6	0.025,8	0.045,7	0.037,1	0.036,8	0.023,4	0.036,5	0.029,8
2007	0.034,6	0.029,1	0.051,8	0.042,3	0.041,6	0.026,7	0.040,6	0.034,8
2008	0.028,2	0.024,5	0.045,4	0.036,7	0.037,1	0.023,5	0.035,1	0.029,5
Average	0.030,8	0.029,5	0.058,9	0.041,4	0.043,2	0.027,6	0.049,7	0.030,7

From the date in Tab. 1, we can find that Xi'an average drought grain loss rate is 3.9%, among them, Chang'an drought loss rate is the highest, up to 5.89%, Huxian is the lowest (2.76%).

4 Conclusions

The occurrence of natural disasters had a serious impact on socio – economic, which restrict the steady development of society pace. As one of the worst natural disasters, drought disasters has taken indelible damage to China's agricultural production, caused some areas to agricultural grains. These phenomena have caused widespread concern of the community and all sectors.

(1) From analyzing factors that influence grain yields, combined with the actual grain planting area and yield per unit area data, the article construct an improved grain disaster loss model, and take Xi'an Shanxi province as a typical example of applied research, then evaluate the eight counties (districts) grain disaster damage respectively.

(2) The results show that nearly 17 a the grain drought loss rate of Xi'an area is about 3.9% on average, among them, Chang'an drought loss rate is the highest, up to 5.89%, Huxian is the lowest 2.76%.

References

Sun Guangzhong, Wang Ansheng, Zhang Wangyuan. The Natural Disaster in China [M]. Academic Books and Periodicals Publishing Company, 1990.

Zheng Yuanchang. Overview of the Global Natural Disaster [J]. Disaster Reduction in China, 2000, 10 (1): 14 – 19.

Liu Yingqiu, Song Jiangjun, Zhang Qingjie. Study on Impact of Drought Disasters to Our Social and Economic [M]. Beijing: China Water and Power Press, 2005.

Zheng Jingyun. Assessment Model of Natural Disaster Grain Yield Loss and Estimation of China's Grain Disaster Yield Loss [J]. Agricultural Meteorology, 1994, 15(6): 7 – 10.

Goodwin BK. Nonparametric Estimation of Grain Yield Distributions: Implications for Rating Group – risk Grain Insurance Contracts [J]. AP Ker – American Journal of Agricultural Economics, 1998. JSTOR.

Li Maosong, Li Zhangcheng, Wang Daolong, et al. Impact of Natural Disasters Change on Grain Yield in China in the Past 50 Years [J]. Journal of Natural Disasters, 2005, 14(2): 55 – 60.

Han Yongxiang, Yin Dong. Study on a New Method of Grain Yield Forecast [J]. Agricultural Research in the Arid Areas, 2002, 20(3): 124 – 127.

Li Yonghua, Gao Yanghua, Liao Liangbing, et al. Changes of Maize Meteorological Yields and an

Analysis of Main Climatic Affecting Factors in Chongqing [J]. Journal of Southwest University (Natural Science Edition), 2007, 29(3):104 – 109.

Gong Deji, Chen Suhua. A Loss Estimation Method for Agrometeorological Disasters and Its Application in Yield Prediction [J]. Quarterly Journal of Applied Meteorology, 1999, 10 (1):66 – 71.

Wang Yuan, Fang Xiuqi, Xu Tan. A Method for Calculating the Climatic Yield of Grain under Climate Change [J]. Journal of Natural Resources, 2004, 19(4):531 – 536.

Liu Jing, Wang Liangxi, Ma Liwen, et al. A Loss Estimation Method of Monitoring and Estimating the Yield Loss of Wheat by Drought in Dry Farming Areas in Northwest of China [J]. Scientia Agriclutura Sinica, 2004, 37(2):201 – 207.

Hydrological Simulation and Comparative Analysis Using HIMS and SWAT in the Upstream of the Heihe River Basin

Li Ying, *Wu Mengying* and *Wang Zhonggen*[1]

Key Laboratory of Water Cycle and Related Land Surface Processes,
Institute of Geographic Sciences and Natural Resources Research,
CAS, Beijing, 100101, China

Abstract: In this article, the China – developed Hydro-Information Modeling System (HIMS), and the overseas developed Soil and Water Assessment Tool (SWAT) are applied. The study area is focused on the arid inland, which is located in the catchment above Yingluo Valley Hydrologic Station in Heihe River Basin. The daily and monthly runoff of the study area is simulated from 1990 to 2008 in using the aforesaid two models. The results indicated that the two models can be applied in the arid inland river basins. In general, the simulation result of HIMS system is better than the SWAT model both in the daily and monthly runoff. Moreover, the E_{ns} of validation and calibration in monthly runoff simulation using HIMS is up to 0.96, which is a comparatively outstanding result. Furthermore, the HIMS requires less data sources than the SWAT does. As a result, HIMS could be applied better in arid inland area.

Key words: HIMS, SWAT, hydrological simulation, Heihe Basin, arid inland river

The Heihe, with a drainage area of approximately 143,000 km², is the second largest inland river in northwest China, which flows through Qinghai, Gansu and Inner Mongolia provinces. The upstream of the Heihe Basin, accounting for more than 80% of the entire river basin water, lies above the Yingluo Valley hydrological station. In recent years, the increase of population, water consumption and unreasonable use of water resources leads the ecological degradation in the middle and lower streams. The downstream of Dongjuyan Sea is even drying up continuously; grassland is gradually desertificated (Wang L C, Zhang X Y, 2010). Consequently, the research of hydrological modeling and analysis of the Heihe Basin upstream becomes an important task for the development and utilization of water resources and ecological environmental protection. At present, many scholars have built different hydrologic models for the upstream of the Heihe Basin; many differences among these models appear in terms of data requirements, simulation efficiency and suitability. Carrying out the suitable comparative analysis of different models is an important tool to develop the hydrologic model for the Heihe Basin. In this paper, the HIMS (Hydro-Information Modeling, System) and the SWAT (Soil and Water Assessment Tool) were applied for simulation and comparative analysis in daily and monthly runoff process in upstream of the Heihe Basin. in order to compare the suitability, advantages and disadvantages of using HIMS and SWAT models in the arid inland river basins.

1 Models introduction

1.1 Brief description of the SWAT model

In 1994, Dr. Jeff Amonld of the Agricultural Research Center of the Agriculture Department of the U. S. A. (USDA) developed Soil and Water Assessment Tool (SWAT). SWAT is physics based model, which can simulate in a continuous time series. Hydrological simulation process of SWAT is divided into land surface part of the water cycle (part of runoff yield and slope concentration) and the water part of the water cycle (part of river concentration). The former

① Fundation item: National basic research program of China(973) (No:2009CB421305)

controls the main input channel of sands, nutrients and chemicals, the later determines the movements of water, sands and other substances from the river to the basin outlet (Jin J L, Wang F R, 2010). The whole water circulation system follows the laws of water balance. Flow hydrological processes simulation of SWAT model relates to precipitations, runoff, soil water, groundwater, evapotranspiration and flow concentration, and so on. It involves nearly a thousand of various process equations for calculations. Since the development of the model, it has been applying widely in places and countries of North America, Europe and Asia. and has been improving constantly during applications. The SWAT model is constantly being amended and supplemented to the latest version of SWAT 2009.

1.2 Brief description of the HIMS model

HIMS, a water cycle multiple element generalized simulation system, is jointly developed by LIU Changming and WANG Zhonggen, and follows of Institute of Geographic Sciences and Natural Resources, which mainly focuses on simulations of different scales of , water sources, water environment, water ecology, water disaster and water management. HIMS integrated various hydrological simulation methods in runoff producing and conflux. Runoff producing process is considered to be a multiple hydrological cycle process, such as precipitation, snowmelt, interception, infiltration, evaporation and so on. HydroLib, the HIMS model base, includes 9 basin water cycle processes, integrated more than 110 models, involving over 600 parameters (Liu Changming, et al. ,2000). In comparison of other models, the important feature of HIMS is that the model can customize hydrological model according to the research purpose. The customized function provides various model selections which can be specialized in different space-time scale water cycle process simulation according to different natural circumstances, human environments and different runoffs producing and conflux way (Liu Changming, et al. ,2010). At present, HIMS has been achieving good experimental results in parts of basins in China home and abroad, such as Yellow River in China, Australia and California (Liu Changming, et al. ,2010).

2 Materials and methods

2.1 Study area

The Heihe Basin is the second largest inland river basin in Northwest China, located in the central of Gansu Corridor, geographically located roughly around 96 ° 42′ E ~ 102 ° 00′ E, 37 ° 41′ N ~ 42 ° N. The Heihe originates in the southern of Qilian Mountains, which is divided into two strings (the West and the East). The east string is the main stream, upstream is divided into east and west forks. East fork is Erbo River, also known as Eightbao River, originates from Jinyang Ridge, Erbo Beach East, and the length of it from the east to the west flow is about 80 km. West fork is Yeniugou, originates from the Tieligan Mountain, and the length of it from the west to the east is about 190 km. The east and the west fork meets at Huangzang temple, and the north of it is named Ganzhou river. The river enters the corridor plain from Yingluo Valley station. The upper reaches of the Heihe Basin, and the entire river length is about 303 km (Fig. 1). The upstream is a good ecological environment, where has plenty of precipitations, which is from glacier and snow melting. It is the main runoff yield of the Heihe Basin, and produces more than 80% of water quantity of the total watershed. The midstream and downstream have little rainfall and large evaporation. The uneven spatial distributions of water resources have caused the different ecological evolutions of the Heihe Basin. The Heihe River basin can also be divided into the mountain forest belt, plain oasis belt and gobi desert zone from the south to the north. The upper stream of the Heihe Basin is cold and damp, the altitude is from 1,700 m to 5,564 m, annual average temperature is from − 5 ℃ to 4 ℃, annual precipitation is about 300 ~ 600 mm. Annual precipitation of local alpine paragraph is up to 600 ~ 700 mm, and the long year average precipitation is about 350 mm. The precipitation is decreasing from the east to the west, the height of snow line increases gradually from the east to the west.

DEM
- 4,977~5,011
- 4,277~4,643
- 3,910~4,276
- 3,543~3,909
- 3,176~3,542
- 2,809~3,175
- 2,442~2,808
- 2,075~2,441
- 1,707~2,074
- No Data

50   0   50 mi

Fig. 1　DEM map of upstream of Heihe Basin

2.2　Data source

As shown in Tab. 1, the data of HIMS simulation includes the minimum and maximum temperatures, rainfall data of meteorological stations within the basin from 1990 to 2008, the yearly runoff volume of Yingluo Valley hydrological station from 1990 to 2008. SWAT simulation includes Digital Elevation Map (DEM), land use map of 2000 year, soil map (1：100,000) and the corresponding physical and chemical characters, digital river network, meteorological data of meteorological stations in the basin from 1990 to 2008, yearly runoff volume of Yingluo Valley hydrological stations from1990 to 2008. The HIMS is more convenient for application in the areas of lack of information than the SWAT as it requires less data sources.

Tab. 1　Data of SWAT and HIMS model

Information type	SWAT	HIMS	Data source
Fundamental Geographic Information	DEM, Land use map(2000 year) Soil map(1：1,000,000), Digital river net map		Environmental sciences data center of Chinese Academy of sciences.
Meteorological Data	Maximum and minimum air temperature, wind speed, precipitation, relative humidity	Maximum and minimum air temperature, precipitation	China meteorological data sharing service system
Hydrological data	Daily and monthly runoff of Yingluo Valley hydrological station (1990 ~ 2008)	Daily and monthly runoff of Yingluo Valley hydrological station (1990 ~ 2008)	Water year book

Digital elevation map and the digital river network diagram as shown in Fig. 1, the highest elevation within the study area is 5,011 m, the minimum is 1,673 m. Land use map as shown in Fig. 2, the types of land use are agriculture, grass, forest, water, urban land, and unused land, the proportion of which accounts for the entire watershed are 0.32%, 51%, 21.19%, 2.66%,

0.1% and 24.74% respectively. The mainly land use type are forest, grassland and unused land. The soil map of the study area is shown in Fig. 3, the type of soil map are glacier and snow (BCXB), grass carpet soil (CZT), frost soil(HDT), frigid calcic soil (of HGT), gray brown soil (HHT), the calcareous meadow soil (SHXCDT), rocky soil (SZT) and bog soil (ZZT), the proportion of which accounts for the entire watershed are 0.99%, 69.73%, 8.71%, 5.00%, 3.94%, 0.90%, 5.71% and 5.07% respectively. The main soil type is grass felt soil. The soil type and land use type of SWAT model are developed mainly according to the actual situation of the U.S.A. The soil database and land use database are required to be modified according to the reality. In this article, the particle size conversion of soil properties database and the soil property parameters are convertible with SPAW model.

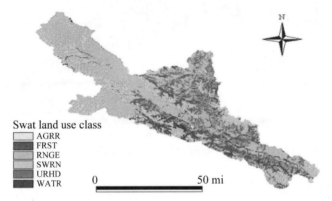

Fig. 2   Land use map of study areas in Heihe Basin

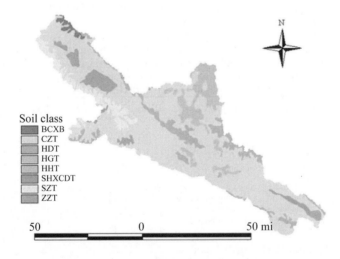

Fig. 3   Soil map of upstream of Heihe Basin

2.3   Evaluation method

In this article, the objective function of parameters calibration included 3 kinds of method:

Nash-Sutcliffe efficiency coefficient (E_{ns}) 、correlation coefficient (r) and volumetric error(V_e), the specific methods are as follows.

(1) Nash-Sutcliffe efficiency coefficient :

$$E_{ns} = 1 - \frac{\sum (Q_{obs,i} - Q_{sim,i})^2}{\sum (Q_{obs,i} - \overline{Q_{obs}})^2} \qquad (1)$$

where, Q_{sim} is simulation value; Q_{obs} is observation value; and n is the total number of observations.

The greater of E_{ns}, indicated that the better simulation results. When $E_{ns} = 1$, the best simulation result.

(2) Correlation coefficient:

$$r = \frac{\sum (Q_{obs,i} - \overline{Q_{obs}})(Q_{sim,i} - \overline{Q_{sim}})}{\sqrt{\sum (Q_{obs,i} - \overline{Q_{obs}})^2 (Q_{sim,i} - \overline{Q_{sim}})^2}} \qquad (2)$$

where, Q_{sim} is simulation value; Q_{obs} is observation value; and n is the total number of observations.

The value of Q_{sim} is between 0 and 1. The value was bigger; the correlation between simulation and observation was higher, and vice versa.

(3) Volume error:

$$V_s = \frac{\overline{Q_{sim}} - \overline{Q_{obs}}}{\overline{Q_{obs}}} \times 100\% \qquad (3)$$

where, Q_{sim} is simulation value; Q_{obs} is observation value.

When V_e is less than 10% , runoff of model is balanced basically.

3　Results

3.1　SWAT parameters calibration and sensitivity analysis

The sensitivity analysis is carried out with the method of Latin the hypercube one-factor-at-a-time (LH-OAT), which is included in SWAT 2005. Eight most sensitive parameters were selected and grouped in descending order as SCS curve number (CN2), saturated hydraulic conductivity (SOL_K), available water capacity (SOL_AWC), snow pack temperature lag factor (TIMP), soil evaporation compensation factor (ESCO), maximum canopy storage (canmx), threshold water depth in the shallow aquifer for flow (GWQMN), SOL_ALB. The parameter sensitivity results indicate that the main factors affect the runoff of the Heihe Basin are precipitation, soil water, groundwater and snow melt.

The validation is carried out with the aforesaid eight sensitive parameters, with the goal of optimizing the objective functions which were Nash-Sutcliffe efficiency coefficient, correlation coefficient and volume error.

3.2　Daily runoff simulation results

The HIMS and SWAT were calibrated from 1990 to 2000, and validated from 2001 to 2008. The daily runoff simulation results of the two models were shown in Tab. 2. For the calibration period (1990 ~ 2000), E_{ns}, V_e and r of HIMS were 0.81, -9.1 and 0.82 separately, while those of SWAT were 0.61, -0.02 and 0.82 separately. For the validation period (2001 ~ 2008), E_{ns}, V_e, and r of HIMS were 0.73, -8.0 and 0.76 separately, while those of SWAT model were 0.65, -0.02 and 0.85 superlatively. In general, when R^2 is above 0.5 and E_{ns} is above 0.6, the precision meets the simulation requirements. Simulation results indicate that both of HIMS and SWAT meet daily runoff hydrologic simulation requirements in arid area. However, according to Nash-Sutcliffe Coefficient and r, the day runoff simulation results of HIMS are significantly better than that of SWAT.

Tab. 2   The daily simulation results of HIMS and SWAT

Objective function	Calibration(1990 ~ 2000)			Validation(2001 ~ 2008)		
	E_{ns}	$V_e(\%)$	r	E_{ns}	$V_e(\%)$	r
HIMS	0.81	– 9.1	0.82	0.73	– 8.0	0.76
SWAT	0.61	– 0.02	0.82	0.65	– 0.02	0.85

Fig. 4 and Fig. 5 show the calibration and validation results of daily runoff simulation of the upstream of the Heihe Basin in applying the methods of HIMS and SWAT. According to the results, the daily runoff simulation result of HIMS is better than SWAT in arid inland areas. The runoff simulation trends of the two models were both ideal. It was consistent with the actual situation of the basin. However, the SWAT has worse simulation effect than the HIMS in terms of base flow and the maximum flood peak. The reason of lower base flow may be related to poor snowmelt simulation. And the reason of the maximum flood peak simulation error may be related to rainfall – runoff simulation.

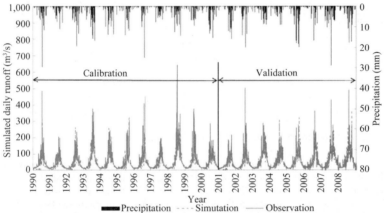

Fig. 4   Daily runoff simulations of upstream of Heihe Basin by HIMS during calibration (1990 ~ 2000) and validation (2001 ~ 2008)

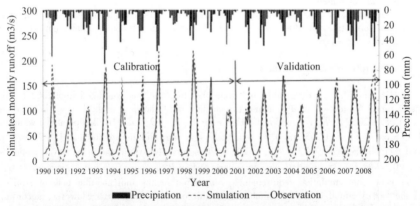

Fig. 5   Daily runoff simulations of upstream of Heihe Basin by SWAT during calibration (1990 ~ 2000) and validation (2001 ~ 2008)

3.3 Monthly runoff simulation results

Similar to daily runoff simulation, the HIMS and SWAT were calibrated from 1990 to 2000 monthly, and validated from 2001 to 2008 monthly. The calibration and validation results of the two models were shown in Tab. 3. For calibration period (1990 ~ 2000), E_{ns}, V_e and r of HIMS were 0.97, 9.8 and 0.97 respectively, and those of SWAT were 0.84, -0.09 and 0.95. For validation period (2001 ~ 2008), E_{ns}, V_e and r of HIMS were 0.96, 11.1 and 0.96 respectively, and those of SWAT were 0.81, -0.07 and 0.95. From the simulation results, both models met the monthly runoff hydrologic simulation of the basin. According to the Nash-Sutcliffe efficient and r, monthly runoff efficiency of HIMS and SWAT was improved greatly compared with the daily runoff simulation efficiency. And the E_{ns} of HIMS in calibration period and validation period were up to 0.97 and 0.96, with very high precision.

Tab. 3 Monthly simulation results of HIMS and SWAT

Objective function	Calibration(1990 ~ 2000)			Validation(2001 ~ 2008)		
	E_{ns}	$V_e(\%)$	r	E_{ns}	$V_e(\%)$	r
HIMS	0.97	9.8	0.97	0.96	11.1	0.96
SWAT	0.84	-0.09	0.95	0.81	-0.07	0.95

Fig. 6 and Fig. 7 show the calibration and validation results of monthly runoff simulation of the Heihe with HIMS and SWAT. The runoff simulation curve matched basically with the actual situation. The time of flood appeared and its trend was consistent. The runoff change trend of observation was also consistent with that of simulation basically. As to monthly runoff simulation, the SWAT simulation got the lower simulation efficiency than that of HIMS. The main reason was mainly due to lower base flow simulation, which may be caused by poor snowmelt simulation.

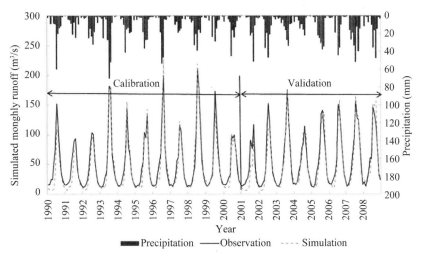

Fig. 6 Monthly runoff simulations of upstream of Heihe Basin by HIMS during calibration (1990 ~ 2000) and validation (2001 ~ 2008)

486

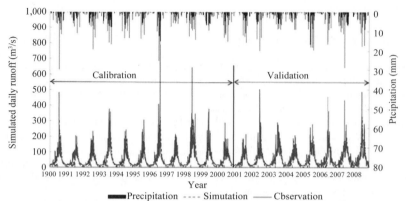

Fig. 7   Monthly runoff simulations of upstream of Heihe Basin by SWAT during calibration (1990 ~ 2000) and validation (2001 ~ 2008)

4   Conclusions

In this paper, the HIMS, which is independently researched and developed in China and the SWAT, which is widely applied internationally are applied for daily and monthly runoff simulation in the upper reaches of the Heihe Basin. Calibration and verification results of the two model show that both HIMS and SWAT meet river runoff simulation in arid inland areas. While simulation efficiency of HIMS is significantly higher than SWAT, which mainly because of the poor base flow simulation and the poor maximum flood peak simulation.

SWAT require many parameters, which is mainly based on topography, land use and soil type. Theoretically, it can be reflect the actual situation. However, in the actual application: ①the land use type database and soil type database of SWAT model were more attributes for United States, and were different from domestic division of standards. There are some errors in the data conversion process, which affects the accuracy of the model; ②SWAT model requires more data; the input uncertainty of each data will affect the simulation accuracy. Lack of regional applications, the simulation accuracy in the data will be reduced. Based on the above reasons, the SWAT model cannot accurately establish the quantitative relationship between the parameters and vegetation or soil properties. There are still some difficulties for parameters determination.

In terms of the model structure, HIMS have simple structure, fewer parameters, lower requirements of input data, wide range of applications, and convenience for application. SWAT requires many parameters and the complex relationship between the various modules lead a difficult situation for application. Therefore, HIMS is fully considered to be widely used in the domestic hydrological characteristics, which has a strong applicability in the northwest arid region. To improve the application of SWAT model in China's northwest arid inland, the snowmelt module need to be improved, the model database need to improved, establish the unified soil classification system, vegetation database and replace procedures (Tian Y J, et al. , 2012).

References

Jin Jingliang, Wang Feier. Research Progress on SWAT Model and its Application and Improvement[J]. Journal of Northeast Forestry University, 2010, 38(12): 111 –112.

Liu Changming, Wang Zhonggen, Zheng Hongxing, et al. , HIMS System and its Development and Application [J]. Science in China (Series E). 2008,38(3): 350 – 360.

Liu Changming, Zheng Hongxing, Yang Shengtian, et al. Multi – scale Integrated Simulation of Hydrological Processes Using HIMS with Verified Case Studies [J]. Journal of Beijing Normal University (Natural Science). 2010,46(3):268 – 273.

Tian Yanjie, Wang Zhirong, Zhang Xiaoxiao. Development and Ddvances in Application of SWAT Model [J]. Journal of Anhui Agriculture Science. 2012,40(6) :3480 – 3483,3486.

Wang Lucang, Zhang Xiaoyu. Effect of the Recent Climate Change on Water Resource in Heihe River Basin [J]. Journal of Arid Land Resources and Environment. 2010, 24(4): 60 – 65.

Method of Water Resources Assessment in Information Limited Areas Based on the SWAT Model

Liang Lili[1] , *Hu Yufeng*[1] , *Ye Yuntao*[1, 2] and *Zhang Yunhui*[1]

1. China institute of water resources and hydropower research, Beijing, 100038, China
2. State Key Laboratory of Hydrology – Water Resources and Hydraulic Engineering, Nanjing Hydraulic Research Institute, Nanjing, 210029, China

Abstract: Climate change and human activities remarkably affect the water resources in watersheds, which challenge the traditional method of water resources assessment. In connection with the problems of "restore" distortion of water quantity, the separate assessment of water quality and quantity, as well as groundwater and surface runoff, and other issues of traditional water resources assessment methods under changing environment in the information limited areas, this paper adopts an assessment method of water resources based on the distributed hydrological model of SWAT, and then explores its assessment principle, key steps, the applicability in the information limited areas. Taking a northwest inland watershed – mountainous area of Manasi River basin as an example, it compares the differences of the results between the traditional water resources assessment method and adopted method based on SWAT model. The results show that the adopted method has more significant advantages on the accuracy of assessment results, the integrity of assessment objects and elements of water cycle, as well as the flexibility of the assessment period of water resources assessment in the information limited area. It is expected to achieve the dynamic, efficient and rapid water resources assessment using a distributed hydrological model based on physical mechanism, and its assessment principle is hopeful of providing a new idea and support to the improvement of water resources assessment methods.

Key words: SWAT model, water resources assessment, information limited area, Manasi River basin

1 Introduction

Water resources assessment is a series of comprehensive analysis and assessment processes of water quantity, water quality, the temporal and spatial distribution of water resources and its conditions for development and utilization in a watershed or a region, so it is the basic work to the water resources planning, development, utilization, protection and management, and the assessment results are important foundations to the water – related activities and decision – making. The main content of the water resources assessment includes assessment of water quantity and water quality, assessment of water resources development and utilization, and the comprehensive assessment of water resources. Hydrological data integration and water quantity statistics carried out by the United States and the former Soviet Union in the mid – 20th century can be seen as the prototype of the water resources assessment. With the problems of water shortage and hydro – ecological environment deterioration, the technologies and methods of water resource assessment receive more attention and are rapidly being developed at home and abroad. Early water resources assessment focused on the use of statistical methods for the water exploitation and utilization assessment and supply – demand analysis. Since the 1980s, the method has got a progressive development based on the water budget: considering the "artificial section" of water used by human activities on the basis of the measured data of the monitoring sites, to assessment the water quality and water quantity separately, as well as the groundwater and surface runoff, and respectively calculating precipitation, runoff, evaporation and other hydrological elements. But in this method, it largely relies on the personal experience of the assessment workers, and it has heavy assessment workload, long assessment period and difficult test work, especially hard to use in the information limited regions. With the increasing impact on water resources of the dual factors of climate change and human activities in a watershed or a region, it poses a new challenge to the traditional water resources assessment techniques in the water resources management. Firstly, in favor of the propounding of the concept of "atmospheric water – surface water – soil water – groundwater" conversion and the requirements of unified allocation of water resources, it not only should assess the surface water and groundwater, but also should strengthen the assessment of atmospheric water and soil water. Secondly, under the requirements of ecological and environmental protection, it needs a unified assessment of the water quality and water quantity. Thirdly, in the constant demand for fine assessment regions and shortened assessment periods, the traditional lumped and long – period water resources assessment can't satisfy the needs of the modern water resources management. Especially in the regions or basins where the data are scarce, the traditional method of manually sketching the precipitation or runoff contour lines to assessment the water resources now can't satisfy the needs of the modern water resources management, so it urgently needs to establish a set of dynamic water resources assessment methods that have strong theoretical basis and practicability, high calculation speed, flexible and easy to operate, to reflect the natural and human activities effects.

The information limited areas of China mostly locate in the remote mountainous areas of northwest and southwest China. In these areas, it is much difficult to place hydrological stations, thus a large number of measurement and monitoring data are lack, then the work of water resources assessment is done by drawing and using precipitation contours, but the precipitation contours are relatively rough, causing much deviation to the results of the water resources assessment and also no reliable basis for water quality assessment; meanwhile, the groundwater and surface water are evaluated respectively. In recent decades, with the application and rapid development of Global Positioning System (GPS), Remote Sensing (RS) and Geographic Information System (GIS) including the Digital Elevation Model (DEM) in water resources management, as well as advances in radar and satellite cloud image technology, it is easier to access to the spatial information on meteorological and hydrological data, characteristics of underlying surface and the effect of human activities within watershed scale, and it provides a strong support to quantitatively describe the distribution of water resources using the spatial data. All these advanced technologies and applications above make water resources assessment using distributed hydrological models in information limited areas become possible. This paper firstly analyzes the principle, key steps and the applicability of water resources assessment based on a distributed hydrological model of SWAT in regions lack of information, and then taking the mountainous area of Manasi River in inland river basin of northwest China as an example, it analyses and compares the assessment results between the traditional method and the adopted method based on the SWAT model when assessing water resources quantity, in order to explore the practical water resources assessment techniques in the information limited areas.

2 Principle and techniques of water resources assessment in information limited areas based on the SWAT Model

The distributed hydrological model based on physical mechanisms can be use to more detailed hydrologic analysis and prediction combined with the digital watershed by aerial surveying and new technologies of remote sensing and GIS, et al. , and to be able to analyze the hydrological conditions of water resources under different surface conditions. When applying in water resources assessment, the main functions of distributed hydrological model are the composition analysis of water resources and water quality assessment at the level of both sub – basins and Hydrological Response Units (HRUs) according to the different underlying surface conditions, making the precipitation as statistical caliber and the water budget as the basis, and considering the water consume of natural vegetation. So it can meet the real – time and detail management needs of water resources. The SWAT model is a distributed hydrological model physically based, which is developed by Agricultural Research Service (ARS) of USDA going through more than 40 years, to predict the impact of land management practices on water, sediment and agricultural chemical yields in large complex watersheds with varying soils, land use and management conditions over long periods of time. This model has been widely applied to runoff simulation, water quality assessment and water resources management in areas of cold zones of USA and Canada, and also upper and middle reaches of watersheds of the Yangtze River, the Yellow River and Heihe River in China. Now a growing number of hydrological researchers are in favor of this model because of its open source code, modular structure, refined and improving functions, a relatively less parameters and flexible operation mode, etc.

2.1　Principle of water resources assessment

Deterministic mathematical models can be divided into two categories. The models belonging to the first category are established on the basis of quality and energy conservation laws, having strict physical and mathematical formulas; The second category models are nevertheless established based on the mass and momentum conservation equations, but are conceptual hydrological models that not apply strict equations, only combining the water cycle processes conceptually. SWAT model belongs to the first category of semi – distributed hydrological models which have more clear physical and mathematical meanings.

In information limited areas where it has few hydrological stations or none, under the premise of full use of spatial data and information provided by GIS and RS, the SWAT model can automatically add the missing meteorological data in accordance with the principle of random statistic simulation and allocate the available data to each HRU using spatial distribution methods coming with the model or improved, only according to the data of a small number of hydro – meteorological monitoring sites in the watershed or outside. The work of water balance calculation, runoff simulation, hydrological forecasting, and water quality assessment in the river or on the sloping surfaces of the watershed are done subsequently. In areas where monitoring data of groundwater is lack, with the filtering technique to split the base flow and calibration, the SWAT model also provides a new method for the study of groundwater resources.

The Hydrological Response Unit is the smallest computing unit in the SWAT model. It delineates the sub – basin and HRUs according the topography, soil, vegetation and slope of the watershed, and each sub – basin may contain a number of HRUs. Firstly, precipitation, vegetation interception, evapotranspiration, surface runoff, soil water, groundwater and other elements of water balance are calculated at the HRU level, and then through the runoff confluence and sediment routing phases on the slope and in the river, it can get the flows of the outlets and other components of water resources at the sub – basin level, and finally by the concentration of channels of sub – basins, the components of water resources and water balance calculation are done at the watershed level. Water balance calculation at the HRU level in each sub – basin uses the lumped mode while the distributed mode at the sub – basins level. That is, it has the hydraulic connection among sub – basins. Water resources in a HRU consist of four parts: precipitation (including snowfall and snowmelt), soil water in the soil layer of 0 ~ 2 m, runoff of shallow aquifers (2 ~ 20 m) and runoff of deep aquifer (> 20 m). A soil profile can be divided into several layers, and the movement of soil water includes infiltration, evaporation, vegetation interception, lateral flow and recharge percolating to the groundwater from the bottom of soil profile. The groundwater recharge in the SWAT model is the sum of percolation of the soil profile, transport losses along the river and the seepage of reservoirs and ponds or wetlands. Specific calculation principle can refer to the literature (Neitsch et al, 2011).

2.2　Key steps for water resources assessment based on the SWAT model

It has four key steps for water quantity and water quality assessment using the SWAT model, including preparation of model data, the main processes before model running, model calibration and validation, as well as calculation of water quantity and quality assessment.

On the step of the model data preparation, the key points are that: ① the rectification, mosaic, cutting, projection and transformation of the vector maps of DEM, vegetation, soil, stream network and other maps; ②the databases preparation of user – defined vegetation, user, soil and so on, such as inquiry and modification of physiological and ecological parameters of vegetation, input and modification of the parameters of physical and chemical properties of soil profiles, and preparation of link tables between databases and their properties; ③ the production of hydro – meteorological databases and data input forms, such as the vectorization of the hydro – meteorological stations and sheets of their situation properties, the input format conversion of daily rainfall, daily maximum and minimum temperature, daily wind speed and other meteorological data, and the missing data automatic interpolation by the model; ④ other processes such as generalization of the reservoir, pond or wetland, the treatment of hydraulic connection, water consume or taking away by social economy and human living, irrigation systems and management measures within the watershed, and the production of vector map of large – scale stream network and so on.

On the step of model running, the key steps are involved with: watershed delineation and stream network extraction based on DEM data; HRU definition based on vectorgraph of vegetation, soil and slope; input of meteorological and hydrological data and its link tables; the spatial distribution of the existing hydro – meteorological data; then setting of the basic parameters within watershed, sub – basin and HRU, such as the moisture condition II curve number (CN2), plant uptake compensation factor(EPCO), soil evaporation compensation factor (ESCO), delay time for aquifer recharge (Gw_Delay) and other basic parameters, and also the period of simulation, the parameters of rainfall distribution and printout frequency, calculation methods of evaporation and transpiration and water routing; finally writing data and setup SWAT run.

On the step of model calibration and validation, with the principles of upstream to downstream, the tributaries to the main stream and water quality simulation after water quantity, much time is spent modifying the relevant characters of the database of soil, vegetation, user – defined weather stations, slope, underlying surface, sediment, nutrients and pollutants, etc, according to the characteristics of the HRUs and sub – basins, and also the monitored runoff or water quality data of controlling stations (generally the stations near the mountain mouth in the upper reach of a river for the information limited areas) , then re – running the model, calibrating and validating until getting the satisfactory results.

On the step of water quantity and quality assessment, the main work is to find precipitation, actual evapotranspiration, infiltration, lateral recharge, surface runoff, soil water, groundwater and so on at the sub – basin level and the watershed level according to the outputs of the final simulation, and to do the necessary water balance calculation on the basis of the relationships among the items above, thus to get the water quantity assessment results of the watershed over a long period of time or during one year, so as to the water quality assessment of sediment, pollutants and so on.

2. 3　Applicability analysis of the SWAT model in water resources assessment

The SWAT model is developed as a distributed hydrological model for a large complex watershed with a long period. Seen from the model structure, it uses the traditional lumped hydrological model to calculate the effective precipitation in a single sub – basin, and then through confluence and flow routing along different sub – basins, it finally obtains the runoff of outlets in cross section. From the modeling techniques, the SWAT model uses the advanced modular structure: each component element of the water cycle corresponds to a sub – module to facilitate the expansion and modification of the model. On the operating mode, the SWAT model uses a unique controlling mode of command codes to control the flow routing in sub – basins and the river; and the source codes compiled with FORTRAN language are open to researchers for free, easily to modify parts of the model and to add the new modules. With the interface development, it combines with the GIS interface, so the pre – processes are visual, convenient and easy to operate. As for the model theory, the model takes full account of the spatial variability of factors of soil, land use, human activities and the underlying surface, etc. , making the simulation more rational; meanwhile, the simulation mechanisms of runoff, soil water, shallow groundwater and deep groundwater adopt the theoretical or semi – empirical algorithms, with which the model can be applied within different climatic and underlying surface conditions.

In addition, Romanowicz pointed that the SWAT model also had the following advantages: ①requiring less input parameters relatively, and the basic data that needed is easier to collect, such as meteorological and topographical data, soil, vegetation and management measurements information, part of the meteorological data can be automatically generated according to the existing data, especially suiting for the areas lack of information or data; ②an integrated hydrological model that can be able to quantitatively simulate and analyze components of water resources; and can also be used for quantitative impact assessment with changed distributed parameters of climate, vegetation cover, land management conditions, etc; ③the model comes with a database, users can fully take advantage of this existing database or modify it, only taking a little time and energy can realize the simulation and prediction of a large complex watershed, having high computational efficiency and applicability; ④ needing less calibration, it can implement the simulation and prediction of a watershed over long periods of time .

It can be drawn that from the above, the SWAT model has a clear physical mechanism, using the "3S"(GPS, DIS, RS) technologies to obtain the spatial information, it can make an accurate watershed delineate in the information limited areas; and the model can add the missing data according to existing data and needs less inputs, which makes up the disadvantages of inadequate data; also the model has high simulation accuracy applied to runoff simulation with mountainous areas, and thus the water quality simulation accuracy is also higher, which shows great advantages in the simulation and assessment of non – point source pollution; the runtime of a simulated watershed depending on the number of hydrological response units, compared with the traditional methods of water resources assessment, it takes less time but has high accuracy and good timeliness. However, it also has some shortcomings, such as slightly simple treatment of deep groundwater, but this deficiency can be made up and it can give full play to its advantages of water resources assessment in the mountainous areas with limited information, where have less human activities and no underground infundibulum or excessive extraction areas of groundwater.

3 Application of the water resources assessment method based on the SWAT model in the mountainous area of Manasi River Basin

3.1 Overview of mountainous watershed of Manasi River basin

Manasi River, the longest mountainous inland river in southern margin of Junggar Basin, rises in the mountains of Kala Wu Cheng (Mongolian), Yi Lian Habi Er Ga(Mongolian) and Bi Yi Da Ke(Mongolian) in the middle of the North Tianshan Mountains, springs from the glaciers above the altitude of 5,000 m, and finally disembogues into the Manasi Lake. Manasi River basin locates in the hinterland of Eurasia, the middle of the Tianshan Mountains, and the southern margin of Junggar Basin, and east of it is the TaXi River basin, while west is Ba Yin Gou River basin, separated from Hejing County by the Yi Lian Habi Er Ga Mountain. It has a catchment area of about 19,600 km², of which mountainous area is about 5,156 km²(above the hydrological station of Hong Shanzui). Annual temperature is about 4.7 ~ 7.9 ℃, and annual potential evaporation is about 1,500 ~ 2,100 mm, while 527 ~ 565 kJ/cm² of solar radiation, ≥10 ℃ accumulated temperature of 3,400 ~ 3,600 ℃ and frost – free period of 170 ~ 190 d. The basin has the high mountain area above 3,500 ~ 5,000 m, most of which is covered by snow and modern glaciers; it is rich of precipitation at the elevation of 1,800 ~ 3,500 m, and annual precipitation may be up to 600 mm; and melt water of snow and glacier and rainfall constitute the recharge sources of the runoff and groundwater in mountainous watershed. Mountainous vegetation can be divided into types of the alpine cushion plants, alpine sparse plants, cool temperature zone and temperature zone conifer forest, subalpine deciduous broad – leaved shrub, alpine meadow of Kobresia humilis and forbs, alpine grassland of grasses and sedges, halophytic meadow of grasses and forbs and so on. Mountainous soil types can be divided into four subcategories of frigid frozen soil, felty soil, dark felty soil and chernozems. Groundwater types are mainly structural fissure water stored in the weathered rocks, fissures, diaclases, ruptures and intersection parts of the cracks, and the water quality is better, having low salinity, generally less than 1 g/L.

3.2　Data sources and processing

DEM data of Manasi River Basin is cut out from the radar topology data of SRTM23 (a resolution of 3 arcsec, 90 m × 90 m), surveying and mapping by space shuttle of the National Geospatial – Intelligence Agency (NGA) and National Aeronautics and Space Administration (NASA) of the United States, conducted in the year of 2000. The soil map and vegetation map, resolution of 1:1,000,000, are intercepted from the data set provided by Environmental and Ecological Science Data Center for West China (http://westdc. westgis. ac. cn), National Natural Science Foundation of China. The data sets of daily precipitation and daily average temperature of four hydrological stations of QingSHuiHe, KenSiWaTe, HongSHanZui and MeiYao, as well as the measured stream network, are provided by Bureau of Water Resources of Xinjiang Uygur Autonomous Region. The meteorological data of daily precipitation, daily maximum and minimum temperature, relative humidity, wind speed and solar radiation of three weather stations of ShiHeZi, KeLaMaYi and WuSu are provided by web site of China Meteorological Administration (http://cdc. cma. gov. cn).

DEM data is used after geometrical and radiometric correction, and the soil and vegetation images and relevant data stored as vector, text and other desired types are processed and prepared according to the model, while the meteorological and hydrological data are processed and input into the model under the format. The physiological and ecological parameters of plants can be get from relevant literatures, and physical and chemical parameters of soils can access to the books of*Soil geography of Xinjiang* and*Soil record of Manasi* and so on. Then inputting the various types of prepared data into SWAT model and setting the model run, after model calibration and validation, it can carry out water resources assessment using the series of outputs.

3.3　Results comparison between the two assessment methods

The SWAT model can take out the simulation results of yearly, monthly and daily water quantity and quality data not only at the level of hydrological response units, but also at the level of sub – basins. Taking the water quantity assessment of mountainous areas in Manasi River Basin as an example, this article compares the assessment results of traditional assessment method and that based on the SWAT model. The hydrological station of Kensiwate is the controlling site of the mountainous area in Manasi River basin, and the study area is divided into 55 sub – basins and 1,141 hydrological response units (the simulation results of sub – basins and stream network can be seen in Fig. 1). The whole study area only has three hydrological stations of Kensiwate, Qingshuihe and Meiyao, and the monitoring data of QingShuiHe began from 1980, while Meiyao has moved for several times, so only Kensiwate has a series of observation data from 1954 to1974 and from 1976 to 2009. It uses the basically coherent time series between the method based on the SWAT model and the traditional water resources assessment method adopted by the*General Planning Report of Manasi River Basin (1997)*. The time series of the former is from 1958 to 1989, while the latter is from 1956 to 1989. The period of model calibration is from 1970 to 1985, and the Nash coefficient

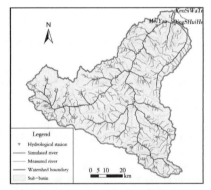

Fig. 1　Simulation results of sub – basins and stream network of mountainous area in Manasi River Basin

is 0. 89 and annual average relative error (Re) is -1.63; while the period of model validation is from 1958 to 1989, and the Nash coefficient is 0. 64 and annual average relative error (Re) is -0.88. The assessment results of the two methods can be seen in the following Tab. 1.

Tab. 1　Results comparison between the two assessment methods

Comparison items	Control area(km^2)	Annual runoff ($\times 10^8 \ m^3$)	Groundwater ($\times 10^8 \ m^3$)	Soil water ($\times 10^8 \ m^3$)	Melt water ($\times 10^8 \ m^3$)	Actual evapotranspiration ($\times 10^8 \ m^3$)	Precipitation ($\times 10^8 \ m^3$)
Traditional method	4,637	11. 7	0. 37	—	—	15. 58	28. 10
Method adopted	5,118. 3	11. 83	0. 01	1. 05	3. 98	14. 41	28. 20
Percentage of difference(%)	10. 38	1. 08	-97.36	—	—	-7.49	0. 36

Note: In the traditional method, groundwater is the natural recharge of mountainous area, including lateral recharge and infiltration of precipitation but deducted the repeated amount, which are taken from the *General Planning Report of Manasi River Basin (1997)*. "—" indicates no relevant data. In the method adopted, groundwater refers to the deep groundwater.

It can be seen from Tab. 1, there is a slight deviation of the control area of KenSiWaTe station between simulation and measurement in the main stream of Manasi River, which relates to the simulation location of the station, the simulation results should be more accurate because of the GPS data. There is less deviation between simulation and measured data of surface runoff, precipitation and evaporation, due to the model calibration using measured surface runoff of Kensiwate station. Because the traditional assessment method can't separate groundwater and soil water, the two methods have no comparability on the two elements. Moreover, soil water and snow melt are not easy to calculate by the traditional assessment method, while it is easier using the adopted method based on distributed hydrological model. In contrast, the adopted method has many advantages: according to the results of a single simulation, it can get the assessment results of surface runoff, groundwater, soil water, precipitation, evaporation, and other elements of water cycle, and takes less time but high accuracy; in addition, the assessment period can be selected flexibly.

4　Conclusions

With the rapid development of fine and strict management of China's water resources, in the information limited areas of China's southwest and northwest regions, the existing conventional water resources assessment methods have been unable to meet the needs of water resources planning and management in the watersheds under the changing environment. As to the mountainous areas where have fewer meteorological and hydrological stations, fully using the "3S" technologies and spatial information obtained by these technologies, this paper presents a water resources assessment method based on distributed hydrological model of SWAT, and then discusses its assessment principle, key steps and the applicability. Taking the water quantity assessment in the upstream watershed of Manasi River as an example, it analyzes and evaluates the proposed method on the full elements and dynamic assessment of water resources in the areas lack of information, illustrating that the method adopted has significant superiority in the selection of simulation periods, and can flexibly select the assessment elements, so the water quantity and water quality assessment of surface water and groundwater can achieve simultaneously. But the application and improvement of this assessment method are restricted by the development of distributed hydrological model, such as the improvement of accuracy and integrity of the data, the coupled simulation of surface and ground, model uncertainty problems, etc. With the development of technologies and methods of the distributed hydrological model, it is expected to achieve dynamic, efficient and rapid water resources assessment using a distributed hydrological model having physical mechanism, and it also will be the development trend of water resources assessment method in the future.

Acknowledgements

The project is financially supported by "Open Foundation of State Key Laboratory of Hydrology – Water Resources and Hydraulic Engineering" (NO: 2011491911), "National Key Technology R&D Program of China" (NO: 2008BAB42B09), and "National Natural Science Foundation of China" (NO: 51079160).

References

Wang Hao, Wang Jianhua, Qin Dayong, et al. Theory and Methodology of Water Resources Assessment Based on Dualistic Water Cycle Model [J]. Journal of Hydraulic Engineering, 2006, 37(12): 1496 – 1502.

Jia Yangwen, Wang Hao, Qiu Yaqin, et al. Generalized Water Resources Assessment Based on Watershed Hydrological Cycle Model I. Assessment Approach [J]. Journal of Hydraulic Engineering, 2006, 37(9): 1051 – 1055.

Wang Hao, Wang Jianhua, Jia Yangwen, et al. A Study on the Method of Water Resources Assessment in River Basin under the Present Environment[J]. Journal of China Hydrology, 2006, 26(3): 18 – 22.

Wang Hao, Yang Guiyu, Jia Yangwen, et al. Connotation and Assessment Index of Soil Water Resources [J]. Journal of Hydraulic Engineering, 2006, 34(4): 389 – 394.

Liang Lili, Wang Dangxian, Wang Fang. Discussion on SWAT Model and Its Application[J]. Journal of China Institute of Water Resources and Hydropower Research, 2007, 5(2): 125 – 131.

Li Jianfeng. Quantity Assessment of Water Resources and Its Application in a Watershed Based on GIS[D]. Zhengzhou: Document of Bachelor of Engineering of Zhengzhou University, 2005.

Wang Zhongjing, Yang Fen, Zhao Jianshi, et al. New Approach of Water Resources Assessment Based on Distributed Hydrological Model[J]. Journal of Hydraulic Engineering, 2008, 39 (12): 1279 – 1285.

Yang Guilian, Hao Fanghua, Liu Changming, et al. The Study on Base Flow Estimation and Assessment in SWAT Luohe Basin as An Example [J]. Progress in Geography, 2003, 22 (5): 463 – 471.

Neitsch S L, Arnold J G, Williams J R. Soil and Water Assessment Tool Theoretical Documentation Version 2009[R]. Texas Water Resources Institute, 2011.

Wang Zhonggen, Liu Changming, Huang Youbo. The Theory of SWAT Model and its Application in Heihe Basin[J]. Progress in Geography, 2003, 22(1): 79 – 86.

Romanowicz A A, Vanclooster M, Rounsevell M, et al. Sensitivity of the SWAT Model to the Soil and Landuse Data Parametrisation: A Case Study in the Thyle Catchment, Belgium [J]. Ecological Modelling, 2005, 87(1): 27 – 39.

Arnold J G, Williams J R, Srinivasan R, et al. Large Area Hydrologic Modeling and Assessment Part I: Model Development[J]. Journal of the American Water Resources Association, 1998, 34(1): 73 – 89.